Lecture Notes in Computer Science 8222

Commenced Publication in 1973
Founding and Former Series Editors:
Gerhard Goos, Juris Hartmanis, and Jan van Leeuwen

FoLLI Publications on Logic, Language and Information
Subline of Lectures Notes in Computer Science

Claudia Casadio   Bob Coecke
Michael Moortgat   Philip Scott (Eds.)

Categories and Types in Logic, Language, and Physics

Essays Dedicated to Jim Lambek
on the Occasion of His 90th Birthday

 Springer

Volume Editors

Claudia Casadio
Gabriele D'Annunzio University
Department of Philosophy, Education and Quantitative Economics
Chieti, Italy
E-mail: casadio@unich.it

Bob Coecke
Oxford University, Department of Computer Science
Oxford, UK
E-mail: bob.coecke@cs.ox.ac.uk

Michael Moortgat
Utrecht University, Utrecht Institute of Linguistics OTS
Utrecht, The Netherlands
E-mail: m.j.moortgat@uu.nl

Philip Scott
University of Ottawa, Department of Mathematics
Ottawa, ON, Canada
E-mail: scpsg@uottawa.ca

ISSN 0302-9743   e-ISSN 1611-3349
ISBN 978-3-642-54788-1   e-ISBN 978-3-642-54789-8
DOI 10.1007/978-3-642-54789-8
Springer Heidelberg New York Dordrecht London

Library of Congress Control Number: 2014933796

LNCS Sublibrary: SL 1 – Theoretical Computer Science and General Issues

Typesetting: Camera-ready by author, data conversion by Scientific Publishing Services, Chennai, India

Printed on acid-free paper

Springer is part of Springer Science+Business Media (www.springer.com)

Preface

We start with a little quiz. A maths book: *Lectures on Rings and Modules*, 1966. A linguistics book: *From Word to Sentence*, 2008. And in the future a forthcoming physics book with an as-yet undisclosed title. What do these books have in common? The same question now for the following papers. A pioneering linguistics paper: "The Mathematics of Sentence Structure," 1958. A paper paving the way for computational types: "Deductive Systems and Categories," 1968. A paper exposing connections between natural language and physics: "Compact Monoidal Categories from Linguistics to Physics," 2011 (written in 2006).

What they have in common is their author: Jim Lambek. What they also have in common is their respective great impact on the relevant research area. But what they definitely do not have in common is a research area! And what they also do not have in common, is their date of publication! What we have here are works in a wide variety of scientific disciplines, spanning an area of some 60 years, all with a sustained very high level of quality and innovation, extreme clarity, and a unique humility and elegance in style.

This volume brings together a series of papers by leading researchers that span the full spectrum of these diverse research areas. It is not the first celebratory volume: limiting ourselves to the past decade, there were two journal special issues, the *Studia Logica* issue *The Lambek Calculus in Logic and Linguistics* on the occasion of Jim's 80th birthday (Vol 71(3), 2002, Wojciech Buszkowski and Michael Moortgat, editors) and the complete 2010 volume of *Linguistic Analysis* (Vol 36(1–4), Johan van Benthem and Michael Moortgat, editors). In addition, CSLI Publications in 2004 issued a collection, *Language and Grammar*, edited by Claudia Casadio, Philip Scott, and Robert Seely.

These publications all have in-depth introductions discussing the significance of Lambek's work in its broader context. Rather than repeating this type of information, we have opted here for some personal recollections, recalling the different ways we have come to know Jim.

CLAUDIA CASADIO — My acquaintance with Jim Lambek goes back to 1985, during the conference Categorial Grammars and Natural Language Structures in Tucson, Arizona. At the time I was working on the developments of categorial grammars and Lambek was one of the leading authors in the field, as were Ajdukiewicz and Bar-Hillel before him. Coming to know him directly, more than 20 years after his 1958 paper on the Syntactic Calculus, was a wonderful experience, offering the opportunity of discussing the many dimensions of language from the formal, and from the empirical and historical, point of view. From that time on, I have always enjoyed the natural capacity of Jim Lambek to understand linguistic problems and to find interesting answers to a variety of challenging questions. In the following years our friendship and scientific cooperation have constantly grown in connection with a number of conferences and events in Europe and particularly in Italy, during the meetings on logic and grammar known as the "Rome Workshops."

There is another event I would like to point out: my first visit to Montreal, in 1997, where I presented a paper on linguistic applications of non-commutative linear logic at the McGill category theory seminar. Lambek traces back his intuitions about Pregroups to this event, although one can easily recognize that he was thinking of a theoretical reworking of his Syntactic Calculus for many years, particularly in connection with the developments of linear logic. Afterwards, numerous papers were written by Jim Lambek, in cooperation with several colleagues, on the calculus of pregroups and their applications to a variety of human languages; I had the pleasure of co-authoring three of them, and I greatly appreciated his intelligent application of mathematical methods to linguistic problems and his widespread historical and literary competence.

BOB COECKE — I first met Jim when I invited him in 1999 for a conference in Brussels on quantum logic. At the time, people in the area were becoming interested in quantales as well as linear logic, and both of these could be traced back (at least in part) to Lambek's 1958 pioneering linguistics paper that we mentioned here. The conference consisted of two parts, one of which took place in a somewhat unusual anarchistic setting: in a swimming pool in a squatted building under a hand-made balloon and with a hand-crafted blackboard. A bunch of artists were also attending the event. The balloon was supposed to keep the heat in, but this plan did not work that well, so there was a cosy chill about. To the organizers' great surprise, there were complaints by several participants, but not by Jim, the oldest participant, who loved that bit of anarchy!

The next year in a somewhat surprising turn of events I became a postdoc at McGill in Lambek's group, and lodged for a while in Lambek's house. One day there was an extreme snow storm, of the kind that makes one hesitate even about leaving one's house. We had planned to have lunch together, but it seemed impossible that Jim, who had to come all the way from Westmount to McGill, would be able to make it. He arrived, and did not really get what the problem was.

A few years later, returning to McGill from a category theory seminar talk, I spoke about the then freshly baked topic of categorical quantum mechanics. Jim immediately realized and mentioned during the talk that the structures underpinning quantum theory and natural language are in fact essentially the same, which he wrote down in the paper mentioned before, and which inspired the more recent quantum linguistics, which combines Lambek's pregroups with the Hilbert space categorical semantics.

MICHAEL MOORTGAT — Although I had met him before at one of the Amsterdam Colloquium conferences, I really got to know Jim better at the legendary 1985 Tucson conference that Claudia mentioned already. This conference, which was a turning point regarding the appreciation of Lambek's logical view on categorial grammar, was organized by Dick Oehrle, together with Emmon Bach and Deirdre Wheeler, in a quite pleasant resort, the Arizona Inn. After talks, one could find Jim there relaxing on a sunbed under the arcades near the wonderful swimming pool (filled with water this time, unlike Bob's Brussels specimen), discussing the finer points of the distinction between *categorical* and *categorial* grammars — a perfect modern incarnation of a Greek philosopher.

At that time, I was trying to make up my mind about a thesis subject, oscillating between Gazdar's generalized phrase structure grammars and Lambek-style categorial grammar. After the Tucson conference, the choice was clear. When Jim reviewed the book edition of the thesis (in *Journal of Symbolic Logic*, Vol 57(3)), with characteristic modesty he states that soon after his 1958 paper, he "developed a skeptical attitude towards the practicality of [this paper's] program," although he admits I had *almost* (emphasis mine) succeeded in reconverting him from his apostasy.

A few years later, again in Tucson, Dick Oehrle and I were coteaching a course on the Lambek Calculus at the 1989 LSA Linguistic Institute. For me, this was the first experience with teaching at an international summer school of this type. One day, while I was presenting a session on the cut-elimination algorithm, I spotted, high up in the back row of the auditorium, the author of the Syntactic Calculus himself! Most of the time, Jim gave the impression of being asleep (as I learned later, he had just returned from a conference in Moscow, and was fighting jet lag). But then all of a sudden, he would open his eyes, and ask a piercing question, which certainly helped to keep the lecturer alert.

PHILIP SCOTT — I first met Jim when I was a graduate student at the University of Waterloo in the early 1970s. Jim gave some beautiful lectures there on categorical proof theory, and I was immediately struck both by the elegance of his work and by the "Montreal school" of categorical logic that was developing then. After I gave my first talk at a conference in 1976, I was awestruck when Jim kindly invited me to become part of his extended troupe of "Lambek postdocs" (which, by the way, became an important career step for many generations of students in categorical logic). For a small-town boy like me, the city of Montreal seemed like a far-off world of mathematical as well as cultural sophistication. Jim introduced me to both. Jim and I worked intently in those days on categorical proof theory and the higher-order internal logic of toposes, much of it becoming the material that appeared in our book *Introduction to Higher-Order Categorical Logic*. The ideas and viewpoints of this project have deeply influenced me throughout my career.

But Jim has continued to work unabated on an extraordinarily large range of projects, from mathematics to linguistics to physics. He has the uncanny ability to discover original and deep insights in all these fields. Yet his approach is not the way of a mere technician, nor of abstraction for its own sake. Rather, he searches for something else: simplicity and elegance. And this is probably the most difficult albeit enticing goal that Jim challenges us with.

While this Festschrift was being assembled, the Centre de Recherches Mathématiques in Montreal hosted a workshop From Categories to Logic, Linguistics and Physics: A Tribute for the 90th Birthday of Joachim Lambek. The workshop, which was held on September 21, 2013, was organized by Prakash Panangaden

(McGill University), Philip Scott (Université d'Ottawa), and Robert A.G. Seely (John Abbott College) and featured the following talks.

- Michael Makkai (McGill University), Multicategories.
- Bob Coecke and Aleks Kissinger (University of Oxford), The Truth Unveiled by Quantum Robots: Lambek Does Not Like Lambic!
- Peter Selinger (Dalhousie University), Control Categories and Duality.
- Michael Moortgat (Utrecht University), Multidimensional Dyck Languages.
- Claudia Casadio (Università degli Studi "G. d'Annunzio", Chieti) and Mehrnoosh Sadrzadeh (Queen Mary University of London), Cyclic Properties in Linear Logic vs. Pregroups – Theoretical Insights and Linguistic Analysis.
- Wojciech Buszkowski (Adam Mickiewicz University, Poznan), Full Lambek Calculus in Logic and Linguistics.
- Philip Scott (Université d'Ottawa), From Gödel to Lambek: Studies in the Foundations of Mathematics.

The workshop, like the papers in this volume, amply demonstrated how Jim Lambek has been a profoundly inspirational mathematician for more than 60 years, with groundbreaking contributions to algebra, category theory, linguistics, theoretical physics, logic, and proof theory.

We hope that Jim will enjoy reading the papers collected here as much as we enjoyed putting this volume together.

December 2013

Claudia Casadio
Bob Coecke
Michael Moortgat
Philip Scott

Table of Contents

Semantic Unification: A Sheaf Theoretic Approach to Natural
Language 1
  Samson Abramsky and Mehrnoosh Sadrzadeh

On Residuation 14
  V. Michele Abrusci

Type Similarity for the Lambek-Grishin Calculus Revisited 28
  Arno Bastenhof

NP-Completeness of Grammars Based Upon Products of Free
Pregroups.. 51
  Denis Béchet

Distributional Semantics: A Montagovian View 63
  Raffaella Bernardi

A Logical Basis for Quantum Evolution and Entanglement 90
  Richard F. Blute, Alessio Guglielmi, Ivan T. Ivanov,
  Prakash Panangaden, and Lutz Straßburger

Learning Lambek Grammars from Proof Frames 108
  Roberto Bonato and Christian Retoré

Multi-Sorted Residuation 136
  Wojciech Buszkowski

Italian Clitic Patterns in Pregroup Grammar: State of the Art 156
  Claudia Casadio and Aleksandra Kiślak-Malinowska

On Associative Lambek Calculus Extended with Basic Proper
Axioms 172
  Annie Foret

Classical Structures Based on Unitaries 188
  Peter Hines

Initial Algebras of Terms with Binding and Algebraic Structure 211
  Bart Jacobs and Alexandra Silva

Abstract Tensor Systems as Monoidal Categories.................... 235
  Aleks Kissinger

On Canonical Embeddings of Residuated Groupoids 253
  Mirosława Kołowska-Gawiejnowicz

L-Completeness of the Lambek Calculus with the Reversal Operation
Allowing Empty Antecedents 268
  Stepan Kuznetsov

A Note on Multidimensional Dyck Languages....................... 279
  Michael Moortgat

Extended Lambek Calculi and First-Order Linear Logic 297
  Richard Moot

A Categorial Type Logic 331
  Glyn Morrill

Chasing Diagrams in Cryptography 353
  Dusko Pavlovic

The Monotone Lambek Calculus Is NP-Complete 368
  Mati Pentus

A Mathematical Analysis of Masaccio's *Trinity* 381
  Gonzalo E. Reyes

Conjoinability in 1-Discontinuous Lambek Calculus.................. 393
  Alexey Sorokin

The Hidden Structural Rules of the Discontinuous Lambek Calculus 402
  Oriol Valentín

Author Index 421

Semantic Unification

A Sheaf Theoretic Approach to Natural Language

Samson Abramsky[1] and Mehrnoosh Sadrzadeh[2]

[1] Department of Computer Science, University of Oxford
[2] School of Electronic Engineering and Computer Science,
Queen Mary University of London
samson.abramsky@cs.ox.ac.uk, mehrnoosh.sadrzadeh@qmul.ac.uk

Abstract. Language is contextual and sheaf theory provides a high level mathematical framework to model contextuality. We show how sheaf theory can model the contextual nature of natural language and how gluing can be used to provide a global semantics for a discourse by putting together the local logical semantics of each sentence within the discourse. We introduce a presheaf structure corresponding to a basic form of Discourse Representation Structures. Within this setting, we formulate a notion of *semantic unification* — gluing meanings of parts of a discourse into a coherent whole — as a form of sheaf-theoretic gluing. We illustrate this idea with a number of examples where it can used to represent resolutions of anaphoric references. We also discuss multivalued gluing, described using a distributions functor, which can be used to represent situations where multiple gluings are possible, and where we may need to rank them using quantitative measures.

Dedicated to Jim Lambek on the occasion of his 90th birthday.

1 Introduction

Contextual models of language originate from the work of Harris [12], who argued that grammatical roles of words can be learnt from their linguistic contexts and went on to test his theory on learning of morphemes. Later, contextual models were also applied to learn meanings of words, based on the frequency of their occurrence in document copora; these gave rise to the distributional models of meaning [8]. Very recently, it was shown how one can combine the contextual models of meaning with formal models of grammars, and in particular pregroup grammars [15], to obtain a compositional distributional semantics for natural language [6].

One can study the contextual nature of language from yet another perspective: the inter-relationships between the meanings of the properties expressed by a discourse. This allows for the local information expressed by individual properties to be glued to each other and to form a global semantics for the whole discourse.

C. Casadio et al. (Eds.): Lambek Festschrift, LNCS 8222, pp. 1–13, 2014.

A very representative example is anaphora, where two language units that may occur in different, possibly far apart, sentences, refer to one another and the meaning of the whole discourse cannot be determined without resolving what is referring to what. Such phenomena occur in plenty in everyday discourse, for example there are four anaphoric pronouns in the following extract from a BBC news article on 16th of May 2013:

> One of Andoura's earliest memories is making soap with his grandmother. She was from a family of traditional Aleppo soap-makers and handed down a closely-guarded recipe [···] to him. Made from mixing oil from laurel trees [···], it uses no chemicals or other additives.

Anaphoric phenomena are also to blame for the complications behind the infamous Donkey sentences 'If a farmer owns a donkey, he beats it.' [9], where the usual Montague-style language to logic translations fail [18] . The first widely accepted framework that provided a formal solution to these challenges was Discourse Representation Theory (DRT) [14]. DRT was later turned compositional in the setting of Dynamic Predicate Logic (DPL) [11] and extended to polarities to gain more expressive power, using actions of modules on monoids [19]. However, the problem with these solutions is the standard criticism made to Montague-style semantics: they treat meanings of words as vacuous relations over an indexical sets of variables.

The motivation behind this paper is two-fold. Firstly, the first author has been working on sheaf theory to reason about contextual phenomena as sheaves provide a natural way of gluing the information of local sections to obtain a consistent global view of the whole situation. Originally introduced in algebraic topology, recently they have been used to model the contextual phenomena in other fields such as in quantum physics [3,5] and in database theory [2]. Based on these and aware of the contextual nature of natural language, the first author conjectured a possible application of sheaves to natural language. Independently, during a research visit to McGill in summer of 2009, the second author was encouraged by Jim Lambek to look at DRT and DPL as alternatives to Montague semantics and was in particular pointed to the capacities of these dynamic structures in providing a formal model of anaphoric reference in natural language. In this paper, we bring these two ideas together and show how a sheaf theoretic interpretation of DRT allows us to unify semantics of individual discourses via gluing and provide semantics for the whole discourse. We first use the sheaf theoretic interpretation of the existing machinery of DRT and apply the setting to resolve *constraint-based* anaphora. We then show how the composition of the sheaf functor with a probability distribution functor can be used to resolve the so called *preferential* anaphora. In such cases, more than one possible resolution is possible and frequencies of occurrences of discourse units from document corpora and the principle of maximal entropy will help choose the most common solution.

2   Sheaves

We recall some preliminary definitions. A category $\mathcal{C}$ has objects and morphisms. We use A, B, C to denote the objects and f, g to denote the morphisms. Examples of morphisms are $f \colon A \to B$ and $g \colon B \to C$. Each object A has an identity morphism, denoted by $Id_A \colon A \to A$. The morphisms are closed under composition: given $f \colon A \to B$ and $g \colon B \to C$, there is a morphism $g \circ f \colon A \to C$. Composition is associative, with identity morphisms as units.

A covariant functor F from a category $\mathcal{C}$ to a category $\mathcal{D}$ is a map $F \colon \mathcal{C} \to \mathcal{D}$, which assigns to each object A of $\mathcal{C}$ an object $F(A)$ of $\mathcal{D}$ and to each morphism $f \colon A \to B$ of $\mathcal{C}$, a morphism $F(f) \colon F(A) \to F(B)$ of $\mathcal{D}$. Moreover, it preserves the identities and the compositions of $\mathcal{C}$. That is, we have $F(Id_A) = Id_{F(A)}$ and $F(g \circ f) = F(g) \circ F(f)$. A contravariant functor reverses the order of morphisms, that is, for $F \colon \mathcal{C} \to \mathcal{D}$ a contravariant functor and $f \colon A \to B$ in $\mathcal{C}$, we have $F(f) \colon F(B) \to F(A)$ in $\mathcal{D}$.

Two examples of a category are the category **Set** of sets and functions and the category **Pos** of posets and monotone maps.

A presheaf is a contravariant functor from a small category $\mathcal{C}$ to the category of sets and functions, which means that it is a functor on the *opposite* (or dual) category of $\mathcal{C}$:

$$F \colon \mathcal{C}^{op} \to \mathbf{Set}$$

This functor assigns a set $F(A)$ to each object A of $\mathcal{C}$. To each morphism $f \colon A \to B$ of $\mathcal{C}$, it assigns a function $F(f) \colon F(B) \to F(A)$, usually referred to as a *restriction map*. For each $b \in F(B)$, these are denoted as follows:

$$F(f)(b) = b \mid_f .$$

Since F is a functor, it follows that the restriction of an identity is an identity, that is for $a \in A$ we have:

$$F(Id_A)(a) = a \mid_{Id_A} = a.$$

Moreover, the restriction of a composition $F(g \circ f) \colon F(C) \to F(A)$ is the composition of the restrictions $F(f) \circ F(g)$ for $f \colon A \to B$ and $g \colon B \to C$. That is for $c \in C$ we have:

$$F(g \circ f)(c) = c \mid_{g \circ f} = (c \mid_g) \mid_f .$$

The original setting for sheaf theory was topology, where the domain category $\mathcal{C}$ is the poset of open subsets of a topological space X under set inclusion. In this case, the arrows of $\mathcal{C}$ are just the inclusion maps $i \colon U \hookrightarrow V$; and restriction along such a map can rewritten unambiguously by specifying the domain of i; thus for $U \subseteq V$ and $s \in F(V)$, we write $s|_U$.

The elements of $F(U)$ — 'the presheaf at stage U' — are called *sections*. In the topological case, a presheaf is a sheaf iff it satisfies the following condition:

Suppose we are given a family of open subsets $U_i \subseteq U$ such that $\bigcup_i U_i = U$, i.e. the family $\{U_i\}$ covers U. Suppose moreover that we are given a family of sections $\{s_i \in F(U_i)\}$ that are compatible, that is for all i, j the two sections s_i and s_j agree on the intersection of two subsets U_i and U_j, so that we have:

$$s_i \mid_{U_i \cap U_j} = s_j \mid_{U_i \cap U_j}.$$

Then there exists a unique section $s \in F(U)$ satisfying the following *gluing* condition:

$$s \mid_{U_i} = s_i \quad \text{for all } i.$$

Thus in a sheaf, we can always unify or glue compatible local information together in a unique way to obtain a global section.

3 Discourse Representation Theory and Anaphora

We shall assume a background first-order language $\mathcal{L}$ of relation symbols. There are no constants or function symbols in $\mathcal{L}$.

In Discourse Representation Theory (DRT), every discourse K is represented by a Discourse Representation Structure (DRS). Such a structure is a pair of a set U_K of discourse referents and a set Cond_K of DRS conditions:

$$(U_K, \mathsf{Cond}_K).$$

Here we take U_K to be simply a finite subset of Var, the set of first-order variables. For the purpose of this paper, we can restrict this set to the set of referents.

A *basic DRS* is one in which the condition Cond_K is a set of first-order literals, *i.e.* atomic sentences or their negations, over the set of variables U_K and the relation symbols in $\mathcal{L}$.

The full class of DRS[1] is defined by mutual recursion over DRS and DRS conditions:

- If X is a finite set of variables and C is a finite set of DRS conditions, (X, C) is a DRS.
- A literal is a DRS condition.
- If K and K' are DRS, then $\neg K$, $K \Rightarrow K'$ and $K \vee K'$ are DRS conditions.
- If K and K' are DRS and x is a variable, $K(\forall x)K'$ is a DRS condition.

Our discussion in the present paper will refer only to basic DRS. However, we believe that our approach extends to the general class of DRS. Moreover, our semantic unification construction to some extent obviates the need for the extended forms of DRS conditions.

---

[1] Note that we write DRS for the plural 'Discourse representation Structures', rather than the clumsier 'DRSs'.

The structure corresponding to a discourse followed by another is obtained by a merge and a unification of the structures of each discourse. The merge of two DRS K and K' is defined as their disjoint union, defined below:

$$K \oplus K' := (U_K \uplus U_{K'}, Cond_K \uplus Cond_{K'})$$

A merge is followed by a unification (also called matching or presupposition resolution), where certain referents are equated with each other. A unification is performed according to a set of accessibility constraints, formalising various different ways linguistics deal with endophora resolution. These include constraints such as as c-commanding, gender agreement, syntactic and semantic consistency [17].

An example where anaphora is fully resolved is 'John owns a donkey. He beats it.'. The merge of the DRS of each discourse of this example is:

$$\Big(\{x, y\}, \{John(x), Donkey(y), Own(x, y)\}\Big) \quad \oplus \quad \Big(\{\underline{v}, \underline{w}\}; \{Beat(\underline{v}, \underline{w})\}\Big)$$

$$= \Big(\{x, y, \underline{v}, \underline{w}\}, \{John(x), Donkey(y), Own(x, y), Beat(\underline{v}, \underline{w})\}\Big)$$

Here, $\underline{v}$ can access x and has agreement with it, hence we unify them by equating $\underline{v} = x$. Also $\underline{w}$ can access y and has agreement with it, hence we unify them as well by equating $\underline{w} = y$. As a result we obtain the following DRS:

$$\Big(\{x, y\}, \{John(x), Donkey(y), Own(x, y), Beat(x, y)\}\Big)$$

An example where anaphora is partially resolved is 'John does not own a donkey. He beats it.', the DRS of which is as follows:

$$(\{x\}, \{John(x), \neg (\{y\}, \{Donkey(y), Own(x, y)\})\}) \oplus (\{\underline{v}, \underline{w}\}, \{Beat(\underline{v}, \underline{w})\})$$

Here $\underline{v}$ can be equated with x, but $\underline{w}$ cannot be equated with y, since y is in a nested DRS and cannot be accessed by $\underline{w}$. Hence, anaphora is not fully resolved.

The unification step enables the DRT to model and resolve contextual language phenomena by going from local to global conditions: it will make certain properties which held about a subset of referents, hold about the whole set of referents. This is exactly the local to global passage modelled by gluing in sheaves.

4   From Sheaf Theory to Anaphora

4.1   A Presheaf for Basic DRS

We begin by defining a presheaf $\mathcal{F}$ which represents basic DRS.

We define the category $\mathcal{C}$ to have as objects pairs (L, X) where

- $L \subseteq \mathcal{L}$ is a finite vocabulary of relation symbols.
- $X \subseteq \mathsf{Var}$ is a finite set of variables.

A morphism $\iota, f : (L, X) \longrightarrow (L', X')$ comprises

- An inclusion map $\iota : L \hookrightarrow L'$
- A function $f : X \longrightarrow X'$.

Note that we can see such functions f as performing several rôles:

- They can witness the inclusion of one set of variables in another.
- They can describe relabellings of variables (this will become of use when quantifiers are introduced).
- They can indicate where variables are being identified or merged; this happens when $f(x) = z = f(y)$.

We shall generally omit the inclusion map, simply writing morphisms in $\mathcal{C}$ as $f : (L, X) \longrightarrow (L', X')$, where it is understood that $L \subseteq L'$.

The functor $\mathcal{F} : \mathcal{C}^{\mathrm{op}} \longrightarrow \mathbf{Set}$ is defined as follows:

- For each object (L, X) of $\mathcal{C}$, $\mathcal{F}(L, X)$ will be the set of deductive closures of consistent finite sets of literals over X with respect to the vocabulary L.
- For each morphism $f : (L, X) \to (L', Y)$, the restriction operation $\mathcal{F}(f) : \mathcal{F}(L', Y) \to \mathcal{F}(L, X)$ is defined as follows. For $s \in \mathcal{F}(Y)$ and L-literal $\pm A(\boldsymbol{x})$ over X:

$$\mathcal{F}(f)(s) \vdash \pm A(\boldsymbol{x}) \iff s \vdash \pm A(f(\boldsymbol{x})).$$

The functoriality of $\mathcal{F}$ is easily verified. Note that deductive closures of finite sets of literals are finite up to logical equivalence. Asking for deductive closure is mathematically convenient, but could be finessed if necessary.

The idea is that a basic DRS (X, s) with relation symbols in L will correspond to $s \in \mathcal{F}(L, X)$ in the presheaf — in fact, to an object of the *total category* associated to the presheaf [16].

4.2   Gluing in $\mathcal{F}$

Strictly speaking, to develop sheaf notions in $\mathcal{F}$, we should make use of a Grothendieck topology on $\mathcal{C}$ [16]. In the present, rather short and preliminary account, we shall work with concrete definitions which will be adequate to our purposes here.

We shall consider *jointly surjective* families of maps $\{f_i : (L_i, X_i) \longrightarrow (L, X)\}_{i \in I}$, *i.e.* such that $\bigcup_i \mathrm{Im} f_i = X$; and also $L = \bigcup_i L_i$.

We can think of such families as specifying *coverings* of X, allowing for relabellings and identifications.

We are given a family of elements (sections) $s_i \in \mathcal{F}(L_i, X_i)$, $i \in I$. Each section s_i is giving information local to (L_i, X_i). A *gluing* for this family, with respect to the cover $\{f_i\}$, is an element $s \in \mathcal{F}(L, X)$ — a section which is *global* to the whole of (L, X) — such that $\mathcal{F}(f_i)(s) = s_i$ for all $i \in I$.

We shall interpret this construction as a form of *semantic unification*. We are making models of the meanings of parts of a discourse, represented by the family $\{s_i\}$, and then we glue them together to obtain a representation of the meaning of the whole discourse. The gluing condition provides a general and mathematically robust way of specifying the adequacy of such a representation, with respect to the local pieces of information, and the identifications prescribed by the covering.

We have the following result for our presheaf $\mathcal{F}$.

Proposition 1. *Suppose we are given a cover* $\{f_i : (L_i, X_i) \longrightarrow (L, X)\}$. *If a gluing* $s \in \mathcal{F}(X)$ *exists for a family* $\{s_i \in \mathcal{F}(L_i, X_i)\}_{i \in I}$ *with respect to this cover, it is unique.*

Proof. We define s as the deductive closure of

$$\{\pm A(f_i(\boldsymbol{x})) \mid \pm A(\boldsymbol{x}) \in s_i, i \in I\}.$$

If s is consistent and restricts to s_i along f_i for each i, it is the unique gluing.

Discussion and Example. Note that, if the sets L_i are *pairwise disjoint*, the condition on restrictions will hold automatically if s as constructed in the above proof is consistent. To see how the gluing condition may otherwise fail, consider the following example. We have $L_1 = \{R, S\} = L_2 = L$, $X_1 = \{x, u\}$, $X_2 = \{y, v\}$, and $X = \{z, w\}$. There is a cover $f_i : (L_i, X_i) \longrightarrow (L, X)$, $i = 1, 2$, where $f_1 : x \mapsto z, u \mapsto w$, $f_2 : y \mapsto z, v \mapsto w$. Then the sections $s_1 = \{R(x), S(u)\}$, $s_2 = \{S(y), R(v)\}$ do not have a gluing. The section s constructed as in the proof of Proposition 1 will e.g. restrict along f_1 to $\{R(x), S(x), R(u), S(u)\} \neq s_1$.

4.3   Linguistic Applications

We shall now discuss a number of examples in which semantic unification expressed as gluing of sections can be used to represent resolutions of anaphoric references.

In these examples, the rôle of *merging* of discourse referents in DRT terms is represented by the specification of suitable cover; while the gluing represents merging at the semantic level, with the gluing condition expressing the semantic correctness of the merge.

Note that by Proposition 1, the 'intelligence' of the semantic unification operation is in the choice of cover; if the gluing exists relative to the specified cover, it is unique. Moreover, the vocabularies in the covers we shall consider will always be disjoint, so the only obstruction to existence is the consistency requirement.

Examples

1. Consider firstly the discourse 'John sleeps. He snores.' We have the local sections

$$s_1 = \{John(x), sleeps(x)\} \in \mathcal{F}(\{John, sleeps\}, \{x\}),$$
$$s_2 = \{snores(y)\} \in \mathcal{F}(\{snores\}, \{y\}).$$

To represent the merging of these discourse referents, we have the cover

$$f_1 : \{x\} \longrightarrow \{z\} \longleftarrow \{y\}.$$

A gluing of s_1 and s_2 with respect to this cover is given by

$$s = \{John(z), sleeps(z), snores(z)\}.$$

2. In intersentential anaphora both the anaphor and antecedent occur in one sentence. An example is 'John beats his donkey'. We can express the information conveyed in this sentence in three local sections:

$$s_1 = \{John(x)\}, \quad s_2 = \{donkey(y)\}, \quad s_3 = \{owns(u,v), beats(u,v)\}$$

over $X_1 = \{x\}$, $X_2 = \{y\}$ and $X_3 = \{u,v\}$ respectively.
We consider the cover $f_i : X_i \longrightarrow \{a,b\}$, $i = 1,2,3$, given by

$$f_1 : x \mapsto a, \quad f_2 : y \mapsto b, \quad f_3 : u \mapsto a, v \mapsto b.$$

The unique gluing $s \in \mathcal{F}(\{John, donkey, owns, beats\}, \{a,b\})$ with respect to this cover is

$$s = \{John(a), donkey(b), owns(a,b), beats(a,b)\}.$$

3. We illustrate the use of negative information, as expressed with negative literals, with the following example: 'John owns a donkey. It is grey.' The resolution method for this example is agreement; we have to make it clear that 'it' is a pronoun that does not refer to men. This is done using a negative literal. Ignoring for the moment the ownership predicate (which would have been dealt with in the same way as in the previous example), the local sections are as follows:

$$s_1 = \{John(x), Man(x)\}, \quad s_2 = \{donkey(y), \neg Man(y)\}, \quad s_3 = \{grey(z)\}\}.$$

Note that a cover which merged x and y would not have a gluing, since the consistency condition would be violated. However, using the cover

$$f_1 : x \mapsto a, \quad f_2 : y \mapsto b, \quad f_3 : z \mapsto b,$$

we do have a gluing:

$$s = \{John(a), Man(a), donkey(b), \neg Man(b), grey(b)\}.$$

4. The following example illustrates the situation where we may have several plausible choices for covers with respect to which to perform gluing. Consider 'John put the cup on the plate. He broke it'. We can represent this by the following local sections

$$s_1 = \{John(x), Cup(y), Plate(z), PutOn(x,y,z)\}, \quad s_2 = \{Broke(u,v)\}.$$

We can consider the cover given by the identity map on $\{x,y,z\}$, and $u \mapsto x, v \mapsto y$; or alternatively, by $u \mapsto x, v \mapsto z$.
In the next section, we shall consider how such multiple possibilities can be ranked using quantitative information within our framework.

5   Probabilistic Anaphora

Examples where anaphora cannot be resolved by a constraint-based method are plentiful, for instance in 'John has a brother. He is happy', or 'John put a cd in the computer and copied it', or 'John gave a donkey to Jim. James also gave him a dog', and so on. In such cases, although we are not sure which unit the anaphor refers to, we have some preferences. For instance in the first example, it is more likely that 'he' is referring to 'John'. If instead we had 'John has a brother. He is nice.', it would be more likely that 'he' would be referring to 'brother'. These considerations can be taken into account in a probabilistic setting.

To model degrees of likelihood of gluings, we compose our sheaf functor with a distribution functor as follows:

$$\mathcal{C}^{\mathrm{op}} \xrightarrow{\mathcal{F}} \mathbf{Set} \xrightarrow{D_R} \mathbf{Set}$$

The distribution functor is parameterized by a commutative semiring, that is a structure $(R, +, 0, \cdot, 1)$, where $(R, +, 0)$ and $(R, \cdot, 1)$ are commutative monoids, and we have the following distributivity property, for $x, y, z \in R$:

$$x \cdot (y + z) = (x \cdot y) + (x \cdot z).$$

Examples of semirings include the real numbers $\mathbb{R}$, positive real numbers $\mathbb{R}^+$, and the booleans $\mathbf{2}$. In the case of the reals and positive reals, $+$ and $\cdot$ are addition and multiplication. In the case of booleans, $+$ is disjunction and $\cdot$ is conjunction.

Given a set S, we define $D_R(S)$ to be the set of functions $d : S \to R$ of finite support, such that

$$\sum_{x \in S} d(x) = 1.$$

For the distribution functor over the booleans, $D(S)$ is the set of finite subsets of S, hence D becomes the finite powerset functor. To model probabilities, we work with the distribution functor over $\mathbb{R}^+$. In this case, $D_R(S)$ is the set of finite-support probability measures over S.

The functorial action of D_R is defined as follows. If $f : X \to Y$ is a function, then for $d \in D_R(X)$:

$$D_R(f)(y) = \sum_{f(x)=y} d(x).$$

This is the direct image in the boolean case, and the image measure in the probabilistic case.

5.1   Multivalued Gluing

If we now consider a family of probabilistic sections $\{d_i \in D_R\mathcal{F}(L_i, X_i)\}$, we can interpret the probability assigned by d_i to each $s \in \mathcal{F}(L_i, X_i)$ as saying how likely this condition is as the correct representation of the meaning of the part of the discourse the local section is representing.

When we consider this probabilistic case, there may be several possible gluings $d \in D_R\mathcal{F}(L, X)$ of a given family with respect to a cover $\{f_i : X_i \longrightarrow X\}$. We can use the principle of maximal entropy [13], that is maximizing over $-\sum_{s\in\mathcal{F}(L,X)} d(s) \log d(s)$, to find out which of these sections is most probable. We can also use maximum entropy considerations to compare the likelihood of gluings arising from different coverings.

In the present paper, we shall study a more restricted situation, which captures a class of linguistically relevant examples. We assume that, as before, we have a family of deterministic sections $\{s_i \in \mathcal{F}(L_i, X_i)\}$, representing our preferred candidates to model the meanings of parts of a discourse. We now have a number of possible choices of cover, representing different possibilities for resolving anaphoric references. Each of these choices c will give rise to a different deterministic gluing $s_c \in \mathcal{F}(L, X)$. We furthermore assume that we have a distribution $d \in D_R\mathcal{F}(L, X)$. This distribution may for example have been obtained by statistical analysis of corpus data.

We can then use this distribution to rank the candidate gluings according to their degree of likelihood. We shall consider an example to illustrate this procedure.

Example

As an example consider the discourse:

> John gave the bananas to the monkeys. They were ripe. They were cheeky.

The meanings of the three sentences are represented by the following local sections:

$$s_1 = \{John(x), Banana(y), Monkey(z), Gave(x, y, z)\},$$
$$s_2 = \{Ripe(u)\},$$
$$s_3 = \{Cheeky(v)\}.$$

There are four candidate coverings, represented by the following maps, which extend the identity on $\{x, y, z\}$ in the following ways:

$$c_1 : u \mapsto y, v \mapsto y \quad c_2 : u \mapsto y, v \mapsto z \quad c_3 : u \mapsto z, v \mapsto y \quad c_4 : u \mapsto z, v \mapsto z.$$

These maps induce four candidate global sections, $t_1, \ldots, t_4$. For example:

$$t_1 = \{John(x), Banana(y), Monkey(z), Gave(x, y, z), Ripe(y), Cheeky(y)\}.$$

We obtain probability distributions for the coverings using the statistical method of [7]. This method induces a grammatical relationship between the possible antecedents and the anaphors and obtains patterns for their possible instantiations by substituting the antecedents and anaphors into their assigned roles. It then counts how many times the lemmatised versions of the patterns obtained from these substitutions have occurred in a corpus. Each of these patterns correspond to a possible merging of referents. The events we wish to assign probabilities to are certain combinations of mergings of referents. The probability of each such event will be the ratio of the sum of occurrences of its mergings to the total number of mergings in all events. Remarkably, these events correspond to the coverings of the sheaf model.

In our example, the sentences that contain the anaphors are predicative. Hence, the induced relationship corresponding to their anaphor-antecedent pairs will be that of "adjective-noun". This yields the following four patterns, each corresponding to a merging map, which is presented underneath it:

$$\text{'ripe bananas', 'ripe monkeys', 'cheeky bananas', 'cheeky monkeys'}$$
$$u \mapsto y \qquad\qquad u \mapsto z \qquad\qquad v \mapsto y \qquad\qquad v \mapsto z$$

We query the *British News corpus* to obtain frequencies of the occurrences of the above patterns. This corpus is a collection of news stories from 2004 from each of the four major British newspapers: Guardian/Observer, Independent, Telegraph and Times. It contains 200 million words. The corresponding frequencies for these patterns are presented below:

'ripe banana'	14
'ripe monkey'	0
'cheeky banana'	0
'cheeky monkey'	10

The events are certain pairwise combinations of the above, namely exactly the pairs whose mappings form a covering. These coverings and their probabilities are as follows:

Event	Covering	Probability
'ripe banana' , 'cheeky banana'	$c_1 : u \mapsto y, v \mapsto y$	$14/48$
'ripe banana' , 'cheeky monkey'	$c_2 : u \mapsto y, v \mapsto z$	$(14+10)/\ 48$
'ripe monkey' , 'cheeky banana'	$c_3 : u \mapsto z, v \mapsto y$	0
'ripe monkey' , 'cheeky monkey'	$c_4 : u \mapsto z, v \mapsto z$	$10/48$

These probabilities result in a probability distribution $d \in D_R \mathcal{F}(L, X)$ for the gluings. The distribution for the case of our example is as follows:

i	t_i	$d(t_i)$
1	$\{John(x), Banana(y), Monkey(z), Gave(x, y, z), Ripe(y), Cheeky(y)\}$	0.29
2	$\{John(x), Banana(y), Monkey(z), Gave(x, y, z), Ripe(y), Cheeky(z)\}$	0.5
3	$\{John(x), Banana(y), Monkey(z), Gave(x, y, z), Ripe(z), Cheeky(y)\}$	0
4	$\{John(x), Banana(y), Monkey(z), Gave(x, y, z), Ripe(z), Cheeky(z)\}$	0.205

We can now select the candidate resolution t_2 as the most likely with respect to d.

6   Conclusions and Future Work

We have shown how sheaves and gluing can be used to model the contextual nature of language, as represented by DRT and unification. We provided examples of the constraint-based anaphora resolution in this setting and showed how a move to preference-based cases is possible by composing the sheaf functor with a distribution functor, which enables one to choose between a number of possible resolutions.

There are a number of interesting directions for future work:

- We aim to extend our sheaf-theoretic treatment of DRT to its logical operations. The model-theoretic semantics of DRS has an intuitionistic flavour, and we aim to develop a sheaf-theoretic form of this semantics.
- The complexity of anaphora resolution has been a concern for linguistics; in our setting we can approach this matter by characterizing the complexity of finding a gluing. The recent work in [4] seems relevant here.
- We would like to experiment with different statistical ways of learning the distributions of DRS conditions on large scale corpora and real linguistic tasks, in the style of [10], and how this can be fed back into the sheaf-theoretic approach, in order to combine the strengths of structural and statistical methods in natural language semantics.

References

1. Aone, C., Bennet, S.W.: Applying machine learning to anaphora resolution. In: Wermter, S., Riloff, E., Scheler, G. (eds.) IJCAI-WS 1995. LNCS, vol. 1040, pp. 302–314. Springer, Heidelberg (1996)
2. Abramsky, S.: Relational databases and Bells theorem. In: Tannen, V. (ed.) Festschrift for Peter Buneman (2013) (to appear); Available as CoRR, abs/1208.6416
3. Abramsky, S., Brandenburger, A.: The sheaf-theoretic structure of non-locality and contextuality. New Journal of Physics 13(11), 113036 (2011)
4. Abramsky, S., Gottlob, G., Kolaitis, P.: Robust Constraint Satisfaction and Local Hidden Variables in Quantum Mechanics. To Appear in Proceedings of IJCAI 2013 (2013)
5. Abramsky, S., Hardy, L.: Logical Bell Inequalities. Physical Review A 85, 062114 (2012)
6. Coecke, B., Sadrzadeh, M., Clark, S.: Mathematical foundations for a compositional distributional model of meaning. Linguistic Analysis 36, 345–384 (2010)
7. Dagan, I., Itai, A.: Automatic processing of large corpora for the resolution of anaphora references. In: Proceedings of the 13th International Conference on Computational Linguistics (COLING 1990), Finland, vol. 3, pp. 330–332 (1990)
8. Firth, J.R.: A synopsis of linguistic theory 1930-1955. Studies in Linguistic Analysis, Special volume of the Philological Society. Blackwell, Oxford (1957)
9. Geach, P.T.: Reference and Generality, An examination of some medieval and modern theories, vol. 88. Cornell University Press (1962)
10. Grefenstette, E., Sadrzadeh, M.: Experimental Support for a Categorical Compositional Distributional Model of Meaning. In: Proceedings of the Conference on Empirical Methods in Natural Language Processing, EMNLP 2011 (2011)
11. Groenendijk, J., Stokhof, M.: Dynamic Predicate Logic. Linguistics and Philisophy 14, 39–100 (1991)

12. Harris, Z.S.: Mathematical structures of language, Interscience Tracts in Pure and Applied Mathematics, vol. 21. University of Michigan (1968)
13. Jaynes, E.T.: Information theory and statistical mechanics. Physical Review 106(4), 620 (1957)
14. Kamp, H., van Genabith, J., Reyle, U.: Discourse Representation Theory. In: Handbook of Philosophical Logic, vol. 15, pp. 125–394 (2011)
15. Lambek, J.: Type Grammars as Pregroups. Grammars 4, 21–39 (2001)
16. Lane, S.M., Moerdijk, I.: Sheaves in geometry and logic: A first introduction to topos theory. Springer (1992)
17. Mitkov, R.: Anaphora Resolution. Longman (2002)
18. Dowty, D.R., Wall, R.E., Peters, S.: Introduction to Montague Semantics. D. Reidel Publishing Company, Dodrecht (1981)
19. Visser, A.: The Donkey and the Monoid: Dynamic Semantics with Control Elements. Journal of Logic, Language and Information Archive 11, 107–131 (2002)

On Residuation

V. Michele Abrusci

Department of Philosophy, University Roma Tre, Rome, Italy
abrusci@uniroma3.it

Abstract. In this paper we explore the residuation laws that are at the basis of the Lambek calculus, and more generally of categorial grammar. We intend to show how such laws are characterized in the framework of a purely non-commutative fragment of linear logic, known as Cyclic Multiplicative Linear Logic.

Keywords: Lambek calculus, categorial grammar, linear logic, proof-net, cyclicity.

Introduction

In this paper we consider the *residuation laws*, that are at the basis of categorial grammar, and particularly, of the Lambek calculus, in the framework of the *cyclic multiplicative proof-nets* (CyM-PN).

In section 1 we show several presentations of residuation laws: under the most usual presentation these rules are treated as equivalences between statements concerning operations of categorial grammar, and under another presentation as equivalences between sequents of a sequent calculus for categorial grammar.

In section 2 we deal with the concept of proof-net and cyclic multiplicative proof-net. Proof-nets are proofs represented in a geometrical way, and indeed they represent proofs in Linear Logic. Cyclic multiplicative proof-net (CyM-PNs) represents proofs in Cyclic Multiplicative Linear Logic, a purely non-commutative fragment of linear logic.

In section 3, we show how the conclusions of a CyM-PN may be described in different ways which correspond to different sequents of CyMLL. In particular, there are 15 possibile ways to read the conclusions of an arbitrary CyM-proof-net with three conclusions.

In section 4 we consider a particular point of view on CyM-PNs and on the format of the sequents which describe the conclusions of CyM-PNs, the regular intuitionistic point of view. We shall show that the sequents considered equivalent in the presentation of the residuation laws in the sequent calculus style are exactly all the possible different ways to describe the conclusions of the same CyM-proof-net with three conclusions, assuming a regular intuitionistic point of view.

C. Casadio et al. (Eds.): Lambek Festschrift, LNCS 8222, pp. 14–27, 2014.

1   Residuation Laws

Residuation laws are basic laws of categorial grammar, in particular the Lambek Calculus and its variant called Non Associative Lambek Calculus [10,11,6,7,13].

Residuation laws may be presented in a pure algebraic style (and this is the most usual presentation) and in a sequent calculus style.

In a pure algebraic style, the residuation laws involve

- a binary operation on a set M: $\cdot$ (the *residuated* operation, called *product*);
- two binary *residual* operations on the same set M: $\backslash$ (the *left residual* operation of the product) and $/$ (the *right residual* operation of the product);
- a partial ordering on the same set M: $\leq$.

In this algebraic style, the residuation laws state the following equivalences for every $a, b, c \in M$ (where M is equiped with the binary operations $\cdot$, $\backslash$, $/$, and with a binary relation $\leq$):

$$\text{(RES)} \quad a \cdot b \leq c \ \text{ iff } \ b \leq a \backslash c \ \text{ iff } \ a \leq c/b$$

An ordered algebra $(M, \leq, \cdot, /, \backslash)$ such that $(M, \leq)$ is a poset and $\cdot$, $/$, $\backslash$ are binary operations on M satisfying (RES) is called a *residuated groupoid* or, in the case the product $\cdot$ is associative, a *residuated semigroup* (see e. g. [6, p. 17], [12, pp. 670-71]) .

In a sequent calculus style, the residuation laws concern sequents of the form $E \vdash F$ where E and F are formulas of a formal language where the following binary connectives occur:

- the *residuated* connective, the *conjuntion*, denoted by $\cdot$ or by $\otimes$ in linear logic;
- the *left residual* connective, the left implication, denoted by $\backslash$ or by $\multimap$ in linear logic;
- the *right residual* connective, the right implication, denoted by $/$ or by $\circ\!\!-$ in linear logic.

In a sequent calculus style, the residuation laws state the following equivalences between *contest-free* sequents of such a formal language: for every formula A, B, C

$$\text{(RES)} \quad A \cdot B \vdash C \ \text{ iff } \ B \vdash A \backslash C \ \text{ iff } \ A \vdash C/B$$

or (using the linear logic symbols):

$$\text{(RES)} \quad A \otimes B \vdash C \ \text{ iff } \ B \vdash A \multimap C \ \text{ iff } \ A \vdash C \circ\!\!- B$$

2   Cyclic Multiplicative Proof-Nets, CyM-PN

2.1   Multiplicative Proof-Nets and CyM-PN

A multiplicative proof-net is a graph such that:

- the nodes are decorated by formulas of the fragment of Linear Logic which is called *multiplicative linear logic* (without units), i.e. the nodes are decorated by formulas constructed by starting with atoms by means of the binary connectives $\otimes$ (*multiplicative conjunction*) and $\invamp$ (*multiplicative disjunction*), where

  - for each atom X there is another atom which is the dual of X and is denoted by $X^{\perp}$, in a way such that, for every atom X, $X^{\perp\perp} = X$;
  - for each formula A the linear negation $A^{\perp}$ is defined as follows, in order to satisfy the principle $A^{\perp\perp} = A$:

    - if A is an atom, $A^{\perp}$ is the atom which is the dual of A,
    - $(B \otimes C)^{\perp} = C^{\perp} \invamp B^{\perp}$
    - $(B \invamp C)^{\perp} = C^{\perp} \otimes B^{\perp}$;

- edges are grouped by *links* and the links are:

  - the *axiom-link*, a binary link (i.e. a link with two nodes and one edge) with no premise, in which the two nodes are conclusions and each node is decorated by the linear negation of the formula decorating the other one; i.e. the conclusions of an axiom link are decorated by two formulas A, $A^{\perp}$

  - the *cut-link*, another binary link (i.e. a link with two nodes and one edge) where there is no conclusion and the two nodes are premises: each node is decorated by the linear negation of the formula decorating the other one, i.e. the premises of an axiom link are decorated by two formulas A, $A^{\perp}$

  - the $\otimes$-*link*, a ternary link (i.e. a link with three nodes and two edges), where two nodes are premises (the first premise and the second premise) and the other node is the conclusion, there is an edge between the first premise and the conclusion and another edge between the second premise and the conclusion, and the conclusion is decorated by a formula $A \otimes B$, where A is the formula decorating the first premise and B is the formula decorating the second premise

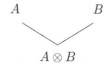

- the $\mathfrak{R}$-*link*, another ternary link (i.e. a link with three nodes and two edges), where two nodes are premises (the first premise and the second premise) and the other node is the conclusion, there is an edge between the first premise and the conclusion and another edge between the second premise and the conclusion, and the conclusion is decorated by a formula $A \mathfrak{R} B$, where A is the formula decorating the first premise and B is the formula decorating the second premise

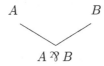

- each node is the premise of at most one link, and is the conclusion of exactly one link; the nodes which are not premises of links are called the *conclusions of the proof-net*;
- for each "switching" the graph is acyclic and connected, where a "switching" of the graph is the removal of one edge in each $\mathfrak{R}$-link of the graph.

We point out that *left* and *right* residual connectives may be defined as follows, by means of the linear negation and the multiplicative disjunction:

$$A \multimap C = A^\perp \mathfrak{R} C \quad C \mathbin{\circ\mkern-4mu-} A = C \mathfrak{R} A^\perp$$

A *cyclic multiplicative proof-net* (CyM-PN) is a multiplicative proof-net s.t.

- the graph is *planar*, i.e. the graph may be drawn on the plane with no crossing of edges,
- the conclusions are in a *cyclic order*, induced by the "trips" inside the proof-net (as defined in [1]; trips are possibile ways to visit the graph); this cyclic order of the conclusions corresponds to the order of the conclusions (from left to right, when the graph is written on the plane as a planar graph, i.e. with no crossing of edges) by adding that the "rightmost" conclusion is before the "leftmost" one.

As shown in [1], we may represent a CyM-proof-net π as a planar graph as follows:

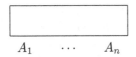

where $A_1, \ldots, A_n$ are the conclusions of π in their cyclic order (A_2 is the immediate successor of A_1, $\ldots$, A_n is the immediate successor of A_{n-1}, A_1 is the immediate successor of A_n). There are other representations of the same CyM-proof-net π as a planar graph, i.e. for each conclusion A of π, we may represent π as a planar graph in such a way that A is the first conclusion going from left to right. For example, we may represent π in such a way that the first conclusion (from the left to the right) is A_2 and the last conclusion is A_1, i.e. :

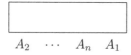

$$A_2 \quad \cdots \quad A_n \quad A_1$$

A CyM-PN is *cut-free* iff it contains no cut-link.

An important theorem (*cut-elimination theorem* or *normalization theorem* for proof-nets) states that every CyM-PN can be transformed in a cut-free CyM-PN with the same conclusions. We may therefore restrict our attention to cut-free CyM-PN.

2.2   Terminal Links in CyM-PN. Irreducible CyM-PN

A ternary link of a CyM-PN π is *terminal* iff the conclusion of the link is also a conclusion of π.

It is immediate, from the definition of CyM-PN, to prove the following propositions on terminal γ-links (see also [15]):

- if π is a CyM-PN and we remove from π a terminal γ-link with conclusion $A \,\gamma\, B$, by keeping the premises A and B which become conclusions of the graph, then we obtain a CyM-PN where, in the cyclic order of its conclusions, the conclusion $A \,\gamma\, B$ is replaced by the two conclusions A, B, with B the immediate successor of A;
- if π is a CyM-proof-net, A and B are two conclusions of π and A is the immediate predecessor of B, in the cyclic order of the conclusions of π, then by adding to π a terminal γ-link with first premise A and second premise B, we obtain a CyM-proof-net where, in the cyclic order of its conclusions, the pair of consecutive conclusions A, B is replaced by the conclusion $A \,\gamma\, B$.

Remark that this proposition does not hold for terminal $\otimes$-links, so that there is a very strong geometrical difference between $\otimes$-links and γ-links in CyM-PN.

Therefore we may remove one, more than one, or all the terminal γ-links from a CyM-PN π and we still obtain a CyM-PN ψ, and from ψ we may return back to π. Similarly we may add to a CyM-PN ψ a new terminal γ-link (where the premises are two conclusions, and the conclusion which is the first premise is the immediate predecessor of the conclusion which is the second premise), and we still obtain a CyM-PN π, and from π we may return back to ψ.

It is important to realize that the act of adding a terminal γ-link to a CyM-PN, when the second premise is the immediate successor of the first one, and the act of removing a terminal γ-link, do not fundamentally *modify* the proof-net.

On the basis of these properties, we may introduce the following definitions.

- Two CyM-PN π and ψ are *equivalent* iff each CyM-PN can be obtained from the other one by removing or by adding terminal γ-links in the way indicated above.
- A CyM-PN is *irreducible*, iff no terminal link is a γ-link.

We may then express the properties introduced above in the following way:

– *every CyM-PN is equivalent to a unique irreducible CyM-PN.*

Remark that if a CyM-PN π is equivalent to an irreducible CyM-PN ψ, then the conclusions of π differ from the conclusions of ψ as follows: some consecutive conclusions of ψ are replaced - as a conclusion of π - by the formula constructed from these conclusions by using the connective $\invamp$ and by preserving the cyclic order of these conclusions. E.g. a CyM-PN π with conclusions $A, B \invamp (C \invamp D), E \invamp F, G$, when A, B, C, D, E, F, G are formulas in which the main connective is not $\invamp$, is equivalent to the irreducible CyM-PN ψ with conclusions A, B, C, D, E, F, G.

We may limit ourself to only dealing with *irreducible* CyM-PN's, considering every CyM-PN π as a different way to read the conclusions of the unique irreducible CyM-PN ψ equivalent to π. In the above example, the addition of terminal $\invamp$-links to the irreducible CyM-PN ψ in order to get the CyM-PN π, may be considered as a way of reading the conclusions A, B, C, D, E, F, G of ψ in the form $A, B \invamp (C \invamp D), E \invamp F, G$.

2.3   Focusing on Conclusions: Outputs and Inputs

When π is a CyM-PN, we may focus on one of the conclusions of π and consider it as the *output* of π, whereas the other conclusions play the role of *inputs*; i.e. we may say that they are nodes waiting for something (waiting for some formulas) in order to get the focused conclusion of π.

Let us denote by $A^\perp$ the conclusion of a CyM-PN π, when this conclusion is considered as waiting for the formula A. Remark that each conclusion of a CyM-PN may be considered as waiting for a formula: this possibility is given by the cut-rule that establishes the communication between two formulas, one of which is the dual of the other, where each formula is waiting for its dual.

Except in the case of a CyM-PN with only one conclusion - the choice to focus on a conclusion C is arbitrary and may be revised: i.e. each conclusion may be focused! Indeed, if we focus on a conclusion C of a CyM-PN π, this conclusion may be read as the output of π and, as a consequence, all the other conclusions have to be considered as inputs of π. But the nature of a CyM-PN allows to change the focus, i.e. to change the choice of the conclusion which is considered as an output. Every conclusion of a CyM-proof-net may be considered as an output, and the choice may be changed. This possibility corresponds also to the logical nature of a proof. A proof of B from the hypothesis A is a proof with conclusions B and $A^\perp$: a proof with output B, waiting for an input A, or a proof with ouput $A^\perp$ (the negation of A), waiting for an input $B^\perp$ (the negation of B).

Moreover, when π is a CyM-PN and C is a conclusion of π, we get a CyM-PN with just the unique conclusion C in the case in which, for each other conclusion $A^\perp$, there is a corresponding CyM-proof-net with conclusion A: it is enough to apply the cut rule n times, where $n + 1$ is the number of the conclusions of π:

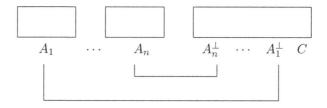

Considering this example remark that, in the planar representation of π from left to right, if the conclusion $A_i^\perp$ occurs before the conclusion $A_j^\perp$, then the proof-net with conclusion A_j occurs before the proof-net with conclusion A_i.

Given a CyM-PN π, we may also focus on more than one conclusion, in particular on more than one consecutive conclusions; in this way the focused conclusions of π are considered as *outputs* and the other conclusions of π as *inputs*.

The focus on one conclusion or on more than one consecutive conclusions of a CyM-PN does not modify the graph, but it is simply a way to consider the graph, a way to describe the graph, in terms of some inputs and some outputs (e.g. in the represntation of a CyM-PN by an intuitionistic sequent, as we shall show in section 4).

2.4 Schematic CyM-PN

The *schema* of a cut-free CyM-PN π, with conclusions occurring in the cyclic order A_1, ..., A_n, is what we get from π by removing all the decorations of the nodes and by denoting the conclusions (in their cyclic order) by the integers $1, \ldots, n$.

A *schematic CyM-PN* is the schema of a cut-free CyM-PN.

Remark that - if π is a CyM-PN - then the decoration of the nodes is induced from the decoration of the conclusions of the axiom links in the schema of π.

A schematic CyM-PN with n conclusions will be represented as

where the order of the conclusions is the cyclic order $1, \cdots, n$ (i.e. the conclusion $i+1$ is the immediate successor of the conclusion i for $i \neq n$ and the conclusion 1 is the immediate successor of the conclusion n). Every decoration of the axiom links in such a schematic cut-free CyM-PN produces a CyM-PN with conclusions decorated by formulas, i.c.

3 Sequents of CyMLL: Ways of Describing the Conclusions of CyM-PNs

3.1 Sequents of CyMLL and Conclusions of CyM-PN

A sequent of CyMLL is an expression of the form

$$\vdash \Gamma$$

where Γ is a finite sequence of formulas of the language of CyMLL.

A sequent $\vdash \Gamma$ of CyMLL is *irreducible* iff no formula in Γ is of the form $A \,\invamp\, B$.

An important theorem (the *sequentialisation theorem*, [1]) states: in the sequent calculus for CyMLL one is able to prove a sequent $\vdash \Gamma$ iff there is a CyM-PN where the conclusions are in the cyclic order induced by Γ, i.e. by taking the first formula in Γ as the immediate successor of the last formula in Γ.

The above considerations are summarized in the following statement: each sequent may be considered as the list of all the conclusions of a possible CyM-PN, by starting with one of the conclusions and by listing all the conclusions on the basis of their cyclic order, and each provable sequent is the list of all the conclusions of a real CyM-PN.

A sequent $\vdash \Gamma$ is *derivable* from a sequent $\vdash \Delta$ in CyMLL iff in the sequent calculus for CyMLL one is able to derive $\vdash \Delta$ from $\vdash \Gamma$, i.e. iff from every CyM-PN with conclusions in the cyclic order induced by Γ one gets also a CyM-PN with conclusions in the cyclic order induced by Δ.

If Γ is a finite sequence of formulas, then we denote by $\Gamma^{\perp}$ the finite sequence of the linear negations of each formula of Γ in the reverse order; i.e., if Γ is the sequence $A_1, \ldots, A_n$, then $\Gamma^{\perp}$ is the sequence $A_n^{\perp}, \ldots, A_1^{\perp}$.

When π is a CyM-PN and we focus on the conclusion C of π as an output so that all the other conclusions are expressed by the linear negations $A^{\perp}$ of formulas A belonging to a finite sequence of formulas of CyMLL, the derivable sequent corresponding to π is of the form $\vdash \Gamma^{\perp}, C$. It is usual to write this sequent also as $\Gamma \vdash C$, i.e. by putting the inputs before $\vdash$ and the output after $\vdash$.

This means that, if a CyM-PN has two conclusions, then we may focus on a conclusion B and consider the other conclusion as an input, i.e. as $A^{\perp}$, therefore writing $A \vdash B$; or we may focus on the conclusion $A^{\perp}$ and consider B as waiting for $B^{\perp}$, i.e. reading B as $B^{\perp\perp}$, therefore writing $B^{\perp} \vdash A^{\perp}$.

The above considerations are summarized in the following statement: each sequent of the form $\Gamma \vdash C$ may be considered as the reading of a possible CyM-PN modulo the *focalization* on one of the conclusions (the conclusion C on the right side of the sequent).

3.2 Equivalent Sequents

Two sequents $\vdash \Gamma$ and $\vdash \Delta$ are *equivalent* in CyMLL iff $\vdash \Gamma$ is derivable from $\vdash \Delta$, and viceversa, in the sequent calculus for CyMLL.

In other terms, two sequents $\vdash \Gamma$ and $\vdash \Delta$ are *equivalent* iff

- from every CyM-PN with conclusions in the cyclic order induced by Γ, we get a CyM-PN with conclusions in the cyclic order induced by Δ;
- from every CyM-PN with conclusions in the cyclic order induced by Δ, we get also a CyM-PN with conclusions in the cyclic order induced by Γ.

It is easy to verify that

- each sequent of the form $\vdash \Gamma, A \otimes B, \Delta$ is equivalent in CyMLL to the sequent $\vdash \Gamma, A, B, \Delta$; therefore each sequent of CyMLL is equivalent to an *irreducible* sequent;

$$\Gamma \quad A \,\otimes\, B \quad \Delta \quad \cong \quad \Gamma \quad A \quad B \quad \Delta$$

- the elements of each equivalence class of sequents, under the equivalence relation defined above, are

  - *irreducible sequents* which induce the same cyclic order,

  - all the sequents which are derivable from one of the irreducible sequents of the same class, by replacing two or more consecutive conclusions by a single conclusion which is obtained putting the connective $\otimes$ between these formulas - according to their order - under an arbitrary use of brackets.

Let us consider a CyM-PN π. We may describe the cyclic order of its conclusions by means of a sequent $\vdash \Gamma$, where Γ is a sequence of formulas which contains exactly the conclusions of π and induces the cyclic order of the conclusions of π; i.e. Γ is the sequence of the conclusions of π in a planar representation of π (from left to right). Moreover, if Δ induces the same cyclic order as Γ, then $\vdash \Gamma$ and $\vdash \Delta$ are both descriptions of the cyclic order of the conclusions of π, the difference between $\vdash \Gamma$ and $\vdash \Delta$ being only a different way to consider (to see) the conclusions of π, and no modification of π is performed when we prefer the description $\vdash \Delta$ instead of $\vdash \Gamma$.

Therefore, the cyclic order of the conclusions of a CyM-PN may be described in several ways which include all the sequents $\vdash \Gamma$ such that Γ induces the cyclic order of the conclusions of π. So, if a CyM-PN π has n conclusions, there are at least n sequents which are descriptions of the cyclic order of the conclusions of π, and all these sequents are *equivalent*.

But there are other ways to describe the conclusions of a CyM-PN π: these ways are all the other sequents which are equivalent to the sequents $\vdash \Gamma$ where Γ induces the cyclic order of the conclusions of π. They are exactly all the sequents obtained by putting the connective $\otimes$ between these conclusions - according to their order - under an arbitrary use of brackets.

Therefore, two sequents $\vdash \Gamma$ and $\vdash \Delta$ are *equivalent* in CyMLL iff $\vdash \Gamma$ and $\vdash \Delta$ are two different ways to describe the conclusions of the same possible CyM-PN. Two different ways to describe the conclusions of a possible CyM-PN by means of two different but equivalent sequents may have one of the following features:

- both the sequents describe the same cyclic order of the conclusions, but in two different ways, i.e. by starting with two different conclusions;
- two or more consecutive conclusions of one sequent are replaced in the other sequent by a single conclusion which is obtained by putting the connective ⅋ between these formulas - according to their order - under an arbitrary use of brackets.

3.3   The Case of CyM-PNs with Three Conclusions

Let us consider a schematic CyM-PN π with three conclusions denoted by $1, 2, 3$ and let us suppose that the cyclic order of the conclusions is that 2 comes after 1, 3 comes after 2, and 1 comes after 3, i.e.

Let us consider the equivalent sequents which are different ways to describe the conclusions of such a schematic CyM-PN π.

- The following equivalent and irreducible sequents are descriptions of the cyclic order of the conclusions of π:

$$\vdash 1, 2, 3 \quad \vdash 3, 1, 2 \quad \vdash 2, 3, 1$$

- On this basis, all the possible descriptions of the conclusions of π are the following equivalent sequents:

$\vdash 1, 2, 3$	$\vdash 1 ⅋ 2, 3$	$\vdash 1, 2 ⅋ 3$	$\vdash (1 ⅋ 2) ⅋ 3$	$\vdash 1 ⅋ (2 ⅋ 3)$
$\vdash 3, 1, 2$	$\vdash 3 ⅋ 1, 2$	$\vdash 3, 1 ⅋ 2$	$\vdash (3 ⅋ 1) ⅋ 2$	$\vdash 3 ⅋ (1 ⅋ 2)$
$\vdash 2, 3, 1$	$\vdash 2 ⅋ 3, 1$	$\vdash 2, 3 ⅋ 1$	$\vdash (2 ⅋ 3) ⅋ 1$	$\vdash 2 ⅋ (3 ⅋ 1)$

where:

- the sequents in the first columm are irreducible and induce the same cyclic order of the conclusions $1, 2, 3$
- in each row there are the sequents obtained from the first sequent (an irreducible sequent) by adding a ⅋ between the first two conclusions (second column), between the last two conclusions (third column), between the two conclusions of the second sequent, and between the two conclusions of the third sequent;
- for each sequent of the second column there is a sequent in the third column such that both the sequents induce the same cyclic order;

- the sequents in the fourth anf fifth colums are all the sequents which allow to express the cyclic order of the conclusions of π by means of an unique expression constructed by using twice the operation $\mathord{\invamp}$.

Thus, there are 15 different ways of describing the conclusions of the graph π represented above. Remark that the schematic CyM-PN is not really modified when we prefer one of these ways, since the introduction of terminal $\mathord{\invamp}$ links does not really modify a schematic CyM-PN π.

4   Residuation Laws as Regular Intuitionistic Descriptions of Conclusions of Intuitionistic CyM-PNs

4.1   Intuitionistic CyM-PNs

Let us call *intuitionistic* a CyM-PN π in which the focus is placed on only one conclusion of π.

The denomination *intuitionistic* is approriate, since in an intuitionistic CyM-PN there is exactly one conclusion and an arbitrary finite number of inputs, as required by the intuitionistic point of view of programs and proofs.

Therefore, in each intuitionistic CyM-PN π:

- we cannot focus on more than one conclusion;
- the change of the focus is the change to another intuitionistic CyM-PN;
- there is exactly one conclusion which is considered as output - i.e. the focused conclusion - whereas all the other conclusions are considered as inputs.

Let us label with a formula C the unique focused conclusion of an intuitionistic CyM-PN π, whereas any other conclusion of π is waiting for something and is then labeled by $A^{\perp}$ where A is a a formula (a type).

We wish to emphasize that an intuitionistic CyM-PN is simply the addition of a fixed focus on one conclusion of a CyM-PN. As a result, each intuitionistic CyM-PN is also a CyM-PN, and each CyM-PN may be considered (when we add a focus on one conclusion) as an intuitionistic CyM-PN.

4.2   Regular Intuitionistic Description of CyM-PNs

Of course, the possible descriptions of the conclusions of an intuitionistic CyM-PN π are the descriptions of π by means of equivalent sequents of CyMLL.

But the specific character of an intuitionistic CyM-PN, i.e. the focus on exactly one conclusion, imposes to write under a special format, the *intuitionistic format*, the sequents which describe the conclusions of the CyM-PN.

Let $\vdash \Gamma$ be a sequent which represents a way to describe the conclusions of a intuitionistic CyM-PN π with focused conclusion C: $\vdash \Gamma$ is of the form $\vdash \Delta^{\perp}, D, \Lambda^{\perp}$ where D is the formula C or a formula obtained from several conclusions of π including the focused conclusion C by means of the connective $\mathord{\invamp}$. The *intuitionistic format* of $\vdash \Gamma$ is the expression $\Delta, \Lambda \vdash D$.

An *intuitionistic description* of the conclusions of an intuitionistic CyM-PN is the intuitionistic format of a sequent which is a description of the conclusions of π.

An intuitionistic description of the conclusions of an intuitionistic CyM-PN is *regular* iff it is of the form $E \vdash D$ where E and D are formulas.

4.3   The Case of Intuitionistic CyM-PNs with Three Conclusions

Every intuitionistic CyM-PN π with three conclusions may be represented as

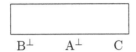

$$B^\perp \qquad A^\perp \qquad C$$

where the conclusion denoted by C is the one that is treated as the output of π and the other two conclusions are those considered as inputs of π.

Remark that we may represent π also as

$$A^\perp \qquad C \qquad B^\perp$$

or as

$$C \qquad B^\perp \qquad A^\perp$$

The 15 equivalent sequents which are the possible descriptions of the conclusions of π are the following:

$$
\begin{array}{lllll}
\vdash B^\perp, A^\perp, C & \vdash B^\perp \mathbin{\invamp} A^\perp, C & \vdash B^\perp, A^\perp \mathbin{\invamp} C & \vdash (B^\perp \mathbin{\invamp} A^\perp) \mathbin{\invamp} C & \vdash B^\perp \mathbin{\invamp} (A^\perp \mathbin{\invamp} C) \\
\vdash C, B^\perp, A^\perp & \vdash C \mathbin{\invamp} B^\perp, A^\perp & \vdash C, B^\perp \mathbin{\invamp} A^\perp & \vdash (C \mathbin{\invamp} B^\perp) \mathbin{\invamp} A^\perp & \vdash C \mathbin{\invamp} (B^\perp \mathbin{\invamp} A^\perp) \\
\vdash A^\perp, C, B^\perp & \vdash A^\perp \mathbin{\invamp} C, B^\perp & \vdash A^\perp, C \mathbin{\invamp} B^\perp & \vdash (A^\perp \mathbin{\invamp} C) \mathbin{\invamp} B^\perp & \vdash A^\perp \mathbin{\invamp} (C \mathbin{\invamp} B^\perp)
\end{array}
$$

The intuitionistic format of these sequents is as follows, representing each formula $(E^\perp \mathbin{\invamp} F^\perp)$ with its dual formula $(F \otimes E)^\perp$:

$$
\begin{array}{lllll}
A, B \vdash C & A \otimes B \vdash C & B \vdash A^\perp \mathbin{\invamp} C & \vdash (A \otimes B)^\perp \mathbin{\invamp} C & \vdash B^\perp \mathbin{\invamp} (A^\perp \mathbin{\invamp} C) \\
A, B \vdash C & A \vdash C \mathbin{\invamp} B^\perp & A \otimes B \vdash C & \vdash (C \mathbin{\invamp} B^\perp) \mathbin{\invamp} A^\perp & \vdash C \mathbin{\invamp} (A \otimes B)^\perp \\
A, B \vdash C & B \vdash A^\perp \mathbin{\invamp} C & A \vdash C \mathbin{\invamp} B^\perp & \vdash (A^\perp \mathbin{\invamp} C) \mathbin{\invamp} B^\perp & \vdash A^\perp \mathbin{\invamp} (C \mathbin{\invamp} B^\perp)
\end{array}
$$

Observe that all the sequents of the first column receive the same intuitionistic format, and that the second and the third columns contain the same sequents in the intuitionistic format. Thus, all the intuitionistic descriptions of the conclusions of the intuitionistic CyM-PN π are 10 (one in the first column, 3 in the second and third column, 3 in the fourth column and 3 in the last column).

Among these 10 intuitionistic descriptions of the intuitionistic CyM-NP π the *regular* ones are the sequents occurring in the second column or, equivalently, in the third column, i.e. there are only 3 regular intuitionistic descriptions of the CyM-NP π :

$$A \otimes B \vdash C \quad B \vdash A^\perp \,\mathbin{\rotatebox[origin=c]{180}{\&}}\, C \quad A \vdash C \,\mathbin{\rotatebox[origin=c]{180}{\&}}\, B^\perp$$

If we replace every formula $E^\perp \,\mathbin{\rotatebox[origin=c]{180}{\&}}\, F$ by $E \multimap F$ and every formula $E \,\mathbin{\rotatebox[origin=c]{180}{\&}}\, F^\perp$ by $E \mathbin{\circ\!\!-} F$, then we obtain that all the possible regular intuitionistic representations of an intuitionistic CyM-PN with three conclusions, in which the focus is on the conclusion C, are the following equivalent sequents:

$$A \otimes B \vdash C \quad B \vdash A \multimap C \quad A \vdash C \mathbin{\circ\!\!-} B$$

i.e.

$$A \bullet B \vdash C \quad B \vdash A \backslash C \quad A \vdash C / B$$

i.e. the sequents considered equivalent when the residuation laws of categorial grammar are represented in the sequent calculus style.

Therefore, we may say that residuation laws - when presented in the sequent calculus style - express the equivalence between the 3 sequents which are all the possible *regular intuitionistic descriptions of the conclusions* of the same CyMLL-PN with three conclusions:

$$A \otimes B \vdash C \quad B \vdash A \multimap C \quad A \vdash C \mathbin{\circ\!\!-} B$$

More generally, we may consider as *general residuation laws* the equivalence between the 10 sequents which are all the possible *intuitionistic descriptions of the conclusions* of the same CyMLL-PN with three conclusions:

$$A, B \vdash C$$

$A \otimes B \vdash C$	$B \vdash A \multimap C$	$A \vdash C \mathbin{\circ\!\!-} B$
$\vdash A \otimes B \multimap C$	$\vdash (A \multimap C) \mathbin{\circ\!\!-} B$	$\vdash (C \mathbin{\circ\!\!-} B) \mathbin{\circ\!\!-} A$
$\vdash C \mathbin{\circ\!\!-} A \otimes B$	$\vdash B \multimap (A \multimap C)$	$\vdash A \multimap (C \mathbin{\circ\!\!-} B)$

Conclusion

As a conclusion of our work, we would like to present the lines for further investigations as a generalization of the results obtained in this paper.

Residuation laws are the most simple examples of a large class of laws which are considered in categorial grammars, the class containing e.g. the following rules: monotonicity rules, application rules, expansion rules, transitivity rules, composition rules [10,8,9].

It would be very interesting to extend the present investigation to the full set of categorial grammar rules by adopting the same methodology presented here:

one starts by representing these rules in a sequent calculus style, and then shows that they correspond to properties or transformations of proof-nets (under a particular point of view).

In this way, we will be able to discover and represent the geometrical properties of the set of categorial grammar rules, having been facilitated in the investigation of the logical properties of these rules by the techniques and results of the theory of proof-nets (and viceversa).

References

1. Abrusci, V.M., Ruet, P.: Non-commutative Logic I: The multiplicative fragment. Ann. Pure Appl. Logic 101(1), 29–64 (1999)
2. Abrusci, M.: Classical Conservative Extensions of Lambek Calculus. Studia Logica 71, 277–314 (2002)
3. Bernardi, R., Moortgat, M.: Continuation Semantics for the Lambek-Grishin Calculus. Inf. Comput. 208(5), 397–416 (2010)
4. Buszkowski, W.: Type Logics and Pregroups. Studia Logica 87(2/3), 145–169 (2007)
5. Buszkowski, W., Farulewski, M.: Nonassociative Lambek Calculus with Additives and Context-Free Languages. In: Grumberg, O., Kaminski, M., Katz, S., Wintner, S. (eds.) Francez Festschrift. LNCS, vol. 5533, pp. 45–58. Springer, Heidelberg (2009)
6. Buszkowski, W.: Lambek Calculus and Substructural Logics. Linguistic Analysis 36(1-4), 15–48 (2010)
7. Buszkowski, W.: Interpolation and FEP for Logics of Residuated Algebras. Logic Journal of the IGPL 19(3), 437–454 (2011)
8. Casadio, C., Lambek, J.: A Tale of Four Grammars. Studia Logica 71, 315–329 (2002)
9. Casadio, C., Lambek, J. (eds.): Recent Computational Algebraic Approaches to Morphology and Syntax. Polimetrica, Milan (2008)
10. Lambek, J.: The Mathematics of Sentence Structure. The American Mathematical Monthly 65(3), 154–170 (1958)
11. Lambek, J.: From Word to Sentence. A Computational Algebraic Approach to Grammar. Polimetrica International Publisher, Monza (2008)
12. Lambek, J.: Logic and Grammar. Studia Logica 100(4), 667–681 (2012)
13. Moortgat, M.: Symmetric Categorial Grammar: Residuation and Galois Connections. Linguistic Analysis 36(1-4), 143–166 (2010)
14. Moortgat, M., Moot, R.: Proof-nets for the Lambek-Grishin Calculus. CoRR abs/1112.6384 (2011)
15. Retoré, C.: Handsome proof nets: perfect matchings and cographs. Theoretical Computer Science 294(3) (2003)

Type Similarity for the Lambek-Grishin Calculus Revisited

Arno Bastenhof

Utrecht Institute of Linguistics[*]

1 Introduction

The topic of this paper concerns a particular extension of Lambek's syntactic calculus [5] that was proposed by Grishin [4]. Roughly, the usual residuated family $(\otimes, /, \backslash)$ is extended by a *co*residuated triple $(\oplus, \oslash, \obackslash)$ mirroring its behavior in the inequality sign:

$$A \otimes B \leq C \text{ iff } A \leq C/B \text{ iff } B \leq A \backslash C$$
$$C \leq A \oplus B \text{ iff } C \oslash B \leq A \text{ iff } A \obackslash C \leq B$$

A survey of the various possible structural extensions reveals that besides same-sort associativity and/or commutativity of $\otimes$ and $\oplus$ independently, there exist as well interaction laws mixing the two vocabularies. We may categorize them along two dimensions, depending on whether they encode mixed associativity or -commutativity, and on whether they involve the tensor $\otimes$ and par $\oplus$ (type I, in Grishin's terminology) or the (co)implications (type IV):

	Type I	Type IV
Mixed associativity	$(A \oplus B) \otimes C \leq A \oplus (B \otimes C)$ $A \otimes (B \oplus C) \leq (A \otimes B) \oplus C$	$(A \backslash B) \oslash C \leq A \backslash (B \oslash C)$ $A \obackslash (B/C) \leq (A \obackslash B)/C$
Mixed commutativity	$A \otimes (B \oplus C) \leq B \oplus (A \otimes C)$ $(A \oplus B) \otimes C \leq (A \otimes C) \oplus B$	$A \obackslash (B \backslash C) \leq B \backslash (A \obackslash C)$ $(A/B) \oslash C \leq (A \oslash C)/B$

While our motivation for the classification into types I and IV may seem rather ad-hoc, one finds that the combined strength of these two groups allows for the either partial or whole collapse (depending on the presence of identity elements, or *units*) into same-sort commutativity and -associativity. Given that this result is hardly desirable from the linguistic standpoint, there is sufficient ground for making the distinction. Moortgat [8] thus proposed a number of calculi, jointly referred to by *Lambek-Grishin* (**LG**), which he considered of particular interest to linguistics. While all reject same-sort associativity and -commutativity, they adopt either one of the type I or IV groups of postulates, the results denoted **LG**$_I$ and **LG**$_{IV}$ respectively. On occasion, one speaks as well of **LG**$_\varnothing$, in reference to the minimal base logic with no structural assumptions.

---

[*] This paper was written while the author was working on his thesis. See Bastenhof, *Categorial Symmetry*, PhD Thesis, Utrecht University, 2013. [Editors' note].

C. Casadio et al. (Eds.): Lambek Festschrift, LNCS 8222, pp. 28–50, 2014.

Having explained the Lambek-Grishin calculi, we next discuss the concept of type-similarity. Besides model-theoretic investigations into derivability, people have sought to similarly characterize its symmetric-transitive closure under the absence of additives. Thus, we consider A and B *type similar*, written $\vdash A \sim B$, iff there exists a sequence of formulae $C_1 \ldots C_n$ s.t. $C_1 = A$, $C_n = B$, and either $C_i \leq C_{i+1}$ or $C_{i+1} \leq C_i$ for each $1 \leq i < n$.[1] For the traditional Lambek hierarchy, one finds their level of resource sensitivity reflected in the algebraic models for the corresponding notions of $\sim$, as summarized in the following table:

CALCULUS	MODELS	REFERENCE
NL	quasigroup	Foret [3]
L	group	Pentus [10]
LP	Abelian group	Pentus [10]

With $\mathbf{LG}_{IV}$, however, Moortgat and Pentus (henceforth M&P, [9]) found that, while same-sort associativity and -commutativity remain underivable, the latter principles do hold at the level of type similarity. More specifically, we find there exist formulas serving as common ancestors or descendants (in terms of $\leq$):

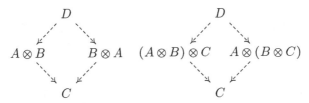

In general, we may prove (cf. [10]) that type similarity coincides with the existence of such *meets* D or *joins* C, as referred to by M&P (though not to be confused with terminology from lattice theory). With respect to linguistic applications, these findings suggest the possibility of tapping in on the flexibility of $\mathbf{L}$ and $\mathbf{LP}$ without compromising overall resource-sensitivity, simply by assigning the relevant joins or meets when specifying one's lexicon.

The current article defines a class of models w.r.t. which we prove soundness and completeness of type similarity in $\mathbf{LG}_\varnothing$ extended by type I or IV interactions, both with and without units. While M&P already provided analogous results of $\mathbf{LG}_{IV}$ inside Abelian groups, we here consider a notion of model better reflecting the presence of dual (co)residuated families of connectives, allowing for simpler proofs overall. Such results still leave open, however, the matter of deciding the existence of joins or meets. We first solve this problem for the specific case of $\mathbf{LG}_I$ together with units 0 and 1, taking a hint from M&P's Abelian group interpretation for $\mathbf{LG}_{IV}$. Decidability for type similarity in the remaining incarnations of $\mathbf{LG}$ is derived as a corollary.

We proceed as follows. First, §2 covers a general introduction to $\mathbf{LG}$ and to our formalism for describing the corresponding concept of derivation. We next illustrate type similarity in §3 with some typical instances in $\mathbf{LG}_I$ and $\mathbf{LG}_{IV}$.

---

[1] Our terminology is adapted from [9], revising that of Pentus [10], who previously spoke of *type conjoinability*.

Models for ~ in the presence of type I or IV interactions are defined in §4, along with proofs of soundness and completeness. Finally, an algorithm for generating joins inside $\mathbf{LG}_I$ in the precense of units is detailed in §5.

2 Lambek-Grishin Calculus

Lambek's non-associative syntactic calculus ([5], **NL**) combines linguistic inquiry with the mathematical rigour of proof theory. Corresponding to multiplicative, non-associative, non-commutative intuitionistic linear logic, its logical vocabulary includes a multiplicative conjunction (*tensor*) $\otimes$ with unit 1, along with direction-sensitive implications $\backslash$ and $/$. Grishin [4] first proposed adding DeMorgan-like duals (cf. remark 1 below), including multiplicative disjunction $\oplus$ (the *par*) with unit 0, as well as *subtractions* $\oslash$ and $\oslash$.

Definition 1. *Given a set of atoms $p, q, r, \ldots$, formulas are defined thus:*

$$
\begin{array}{lll}
A..E ::= p & & Atoms \\
\mid (A \otimes B) \mid (A \oplus B) & & Tensor\ vs.\ par \\
\mid (A/B) \mid (B \oslash A) & & Right\ division\ vs.\ left\ subtraction \\
\mid (B\backslash A) \mid (A \oslash B) & & Left\ division\ vs.\ right\ subtraction \\
\mid 1 \mid 0 & & Units
\end{array}
$$

The associated concept of duality $^\infty$ is defined as follows:

$$
\begin{array}{llll}
p^\infty := p & (A \otimes B)^\infty := B^\infty \oplus A^\infty & (A \oplus B)^\infty := B^\infty \otimes A^\infty \\
1^\infty := 0 & (A/B)^\infty := B^\infty \oslash A^\infty & (B \oslash A)^\infty := A^\infty / B^\infty \\
0^\infty := 1 & (B\backslash A)^\infty := A^\infty \oslash B^\infty & (A \oslash B)^\infty := B^\infty \backslash A^\infty
\end{array}
$$

Remark 1. Note that if $^\infty$ is interpreted as negation, its defining clauses for the binary connectives read as De Morgan laws. That said, while $^\infty$ is indeed involutive, it is not, like negation, fixpoint-free, seeing as $p^\infty = p$.

While derivability may be characterized algebraically using inequalities $A \le B$, we here instead use a *labelled deductive system* [6], adding an extra label f to further discriminate between different deductions.

Definition 2. *Fig.1 defines judgements $f : A \to B$, referred to by arrows.*

While we shall use arrows merely as a means of encoding derivations, one should note that, similar to λ-terms for intuitionistic logic, they may as well be considered amendable to computation (cf. [7]). The following is an easy observation:

Lemma 1. *If $f : A \to B$, then there exists $g : B^\infty \to A^\infty$.*

In practice, we use a more compact representation of derivations, compiling away monotonicity and composition. To this end, we first require the notions of positive and negative (formula) contexts.

Preorder laws

$$i_A : A \to A \qquad \frac{f : A \to B \quad g : B \to C}{(g \circ f) : A \to C}$$

Monotonicity

$$\frac{f : A \to B}{\begin{array}{c} (f \otimes C) : A \otimes C \to B \otimes C \\ (f/C) : A/C \to B/C \\ (C\backslash f) : C\backslash A \to C\backslash B \end{array}} \qquad \frac{f : A \to B}{\begin{array}{c} (f \oplus C) : A \oplus C \to B \oplus C \\ (f \oslash C) : A \oslash C \to B \oslash C \\ (C \oslash f) : C \oslash A \to C \oslash B \end{array}}$$

$$\frac{f : A \to B}{\begin{array}{c} (C \otimes f) : C \otimes A \to C \otimes B \\ (C/f) : C/B \to C/A \\ (f\backslash C) : B\backslash C \to A\backslash C \end{array}} \qquad \frac{f : A \to B}{\begin{array}{c} (C \oplus f) : C \oplus A \to C \oplus B \\ (C \oslash f) : C \oslash B \to C \oslash A \\ (f \oslash C) : B \oslash C \to A \oslash C \end{array}}$$

(Co)evaluation

$$e^{/}_{A,B} : (A/B) \otimes B \to A \qquad e^{\oslash}_{A,B} : A \to (A \oslash B) \oplus B$$
$$e^{\backslash}_{A,B} : B \otimes (B\backslash A) \to A \qquad e^{\oslash}_{A,B} : A \to B \oplus (B \oslash A)$$
$$h^{/}_{A,B} : A \to (A \otimes B)/B \qquad h^{\oslash}_{A,B} : (A \oplus B) \oslash B \to A$$
$$h^{\backslash}_{A,B} : A \to B\backslash(B \otimes A) \qquad h^{\oslash}_{A,B} : B \oslash (B \oplus A) \to A$$

Units

$$1_{A\otimes} : A \to A \otimes 1 \qquad 0_{A\oplus} : A \oplus 0 \to A$$
$$1_{\otimes A} : A \to 1 \otimes A \qquad 0_{\oplus A} : 0 \oplus A \to A$$
$$1^{*}_{A\otimes} : A \otimes 1 \to A \qquad 0^{*}_{A\oplus} : A \to A \oplus 0$$
$$1^{*}_{\otimes A} : 1 \otimes A \to A \qquad 0^{*}_{\oplus A} : A \to 0 \oplus A$$

Type I interactions   Type IV interactions

$$a^{\oplus\otimes}_{A,B,C} : (A \oplus B) \otimes C \to A \oplus (B \otimes C) \qquad \alpha^{\backslash\oslash}_{A,B,C} : (A\backslash B) \oslash C \to A\backslash(B \oslash C)$$
$$a^{\otimes\oplus}_{A,B,C} : A \otimes (B \oplus C) \to (A \otimes B) \oplus C \qquad \alpha^{\oslash/}_{A,B,C} : A \oslash (B/C) \to (A \oslash B)/C$$
$$c^{\otimes\oplus}_{A,B,C} : A \otimes (B \oplus C) \to B \oplus (A \otimes C) \qquad \gamma^{\oslash\backslash}_{A,B,C} : A \oslash (B\backslash C) \to B\backslash(A \oslash C)$$
$$c^{\oplus\otimes}_{A,B,C} : (A \oplus B) \otimes C \to (A \otimes C) \oplus B \qquad \gamma^{/\oslash}_{A,B,C} : (A/B) \oslash C \to (A \oslash C)/B$$

Fig. 1. Lambek-Grishin calculi presented using labelled deduction

Definition 3. *Define, by mutual induction, positive and negative contexts:*

$$X^{+}[], Y^{+}[] ::= [] \mid (X^{+}[] \otimes B) \mid (A \otimes Y^{+}[]) \mid (X^{+}[] \oplus B) \mid (A \oplus Y^{+}[])$$
$$\mid (X^{+}[]/B) \mid (B\backslash X^{+}[]) \mid (X^{+}[] \oslash B) \mid (B \oslash X^{+}[])$$
$$\mid (A/Y^{-}[]) \mid (Y^{-}[]\backslash A) \mid (A \oslash Y^{-}[]) \mid (Y^{-}[] \oslash A)$$

$$X^{-}[], Y^{-}[] ::= (X^{-}[] \otimes B) \mid (A \otimes Y^{-}[]) \mid (X^{-}[] \oplus B) \mid (A \oplus Y^{-}[])$$
$$\mid (X^{-}[]/B) \mid (B\backslash X^{-}[]) \mid (X^{-}[] \oslash B) \mid (B \oslash X^{-}[])$$
$$\mid (A/Y^{+}[]) \mid (Y^{+}[]\backslash A) \mid (A \oslash Y^{+}[]) \mid (Y^{+}[] \oslash A)$$

Evidently, given some $X^+[]$, $Y^-[]$ and $f : A \to B$, we have $X^+[f] : X^+[A] \to X^+[B]$ and $Y^-[f] : Y^-[B] \to Y^-[A]$. In practice, we often depict said arrows as an inference step, using the following shorthand notation:

$$\frac{X^+[\boxed{A}]}{X^+[A']}\, f \qquad \frac{Y^-[\boxed{A'}]}{Y^-[A]}\, f$$

having avoided writing $X^+[f]$ ($Y^-[f]$) by informally singling out the source of f using a box, thus unambiguously identifying the surrounding context $X^+[]$ ($Y^-[]$). Composition of arrows is compiled away by chaining inference steps. In practice, we often leave out subscripts in f, being easily inferable from context.

Our notation comes close to Brünnler and McKinley's use of *deep inference* for intuitionistic logic [1]. Contrary to their concept of derivability, however, our inference steps need not be restricted to primitive arrows. We next survey several definable arrows, proving useful in what is to follow.

Definition 4. *Lifting is defined*

$$
\begin{aligned}
l^{/}_{A,B} &:= ((e^{\backslash}_{B,A}/i_{A\backslash B}) \circ h^{/}_{A,A\backslash B}) &&: A \to B/(A\backslash B)\\
l^{\backslash}_{A,B} &:= ((i_{B/A}\backslash e^{/}_{B,A}) \circ h^{\backslash}_{A,B/A}) &&: A \to (B/A)\backslash A\\
l^{\oslash}_{A,B} &:= (h^{\oslash}_{A,A\oslash B} \circ (e^{\oslash}_{B,A} \oslash i_{A\oslash B})) &&: B \oslash (A \oslash B) \to A\\
l^{\obackslash}_{A,B} &:= (h^{\obackslash}_{A,B\obackslash A} \circ (i_{B\obackslash A} \obackslash e^{\obackslash}_{B,A})) &&: (B \obackslash A) \obackslash B \to A
\end{aligned}
$$

Using the notation introduced in our previous discussion:

$$\frac{\dfrac{A}{\boxed{(A\otimes(A\backslash B))}/(A\backslash B)}\, h^{/}}{B/(A\backslash B)}\, e^{\backslash} \qquad \frac{\dfrac{A}{(B/A)\backslash\boxed{((B/A)\otimes A)}}\, h^{\backslash}}{(B/A)\backslash B}\, e^{/}$$

$$\frac{\dfrac{(B\oslash A)\oslash\boxed{B}}{(B\oslash A)\oslash((B\oslash A)\oplus A)}\, e^{\oslash}}{A}\, h^{\oslash} \qquad \frac{\dfrac{\boxed{B}\obackslash(A\obackslash B)}{(A\oplus(A\obackslash B))\oslash(A\obackslash B)}\, e^{\obackslash}}{A}\, h^{\obackslash}$$

Definition 5. *Grishin type I and IV interactions can alternatively be rendered*

$$
\begin{aligned}
a^{\oslash}_{A,B,C} &: (A \otimes B) \oslash C \to A \otimes (B \oslash C) & \alpha^{\oslash}_{A,B,C} &: A \otimes (B \oslash C) \to (A \otimes B) \oslash C\\
a^{\obackslash}_{A,B,C} &: A \obackslash (B \otimes C) \to (A \obackslash B) \otimes C & \alpha^{\obackslash}_{A,B,C} &: (A \obackslash B) \otimes C \to A \obackslash (B \otimes C)\\
c^{\oslash}_{A,B,C} &: (A \otimes B) \oslash C \to (A \oslash C) \otimes B & \gamma^{\oslash}_{A,B,C} &: (A \oslash B) \otimes C \to (A \otimes C) \oslash B\\
c^{\obackslash}_{A,B,C} &: A \obackslash (B \otimes C) \to B \otimes (A \obackslash C) & \gamma^{\obackslash}_{A,B,C} &: A \otimes (B \obackslash C) \to B \obackslash (A \otimes C)\\
a^{/}_{A,B,C} &: A \oplus (B/C) \to (A \oplus B)/C & \alpha^{/}_{A,B,C} &: (A \oplus B)/C \to A \oplus (B/C)\\
a^{\backslash}_{A,B,C} &: (A\backslash B) \oplus C \to A\backslash(B \oplus C) & \alpha^{\backslash}_{A,B,C} &: A\backslash(B \oplus C) \to (A\backslash B) \oplus C\\
c^{/}_{A,B,C} &: (A/B) \oplus C \to (A \oplus C)/B & \gamma^{/}_{A,B,C} &: (A \oplus B)/C \to (A/C) \oplus B\\
c^{\backslash}_{A,B,C} &: A \oplus (B\backslash C) \to B\backslash(A \oplus C) & \gamma^{\backslash}_{A,B,C} &: A\backslash(B \oplus C) \to B \oplus (A\backslash C)
\end{aligned}
$$

For Type I, we have the following definitions:

$$a^{\oslash}_{A,B,C} := h^{\oslash}_{A\otimes(B\oslash C),C} \circ ((a^{\otimes\oplus}_{A,B\oslash C,C} \circ (i_A \otimes e^{\oslash}_{B,C})) \oslash i_C)$$

$$a^{\oslash}_{A,B,C} := h^{\oslash}_{(A\oslash B)\otimes C,A} \circ (i_A \oslash (a^{\oplus\otimes}_{A,A\oslash B,C} \circ (e^{\oslash}_{B,A} \otimes i_C)))$$

$$c^{\oslash}_{A,B,C} := h^{\oslash}_{(A\oslash C)\otimes B,C} \circ ((c^{\oplus\otimes}_{A\oslash C,C,B} \circ (e^{\oslash}_{A,C} \otimes i_B)) \oslash i_C)$$

$$c^{\oslash}_{A,B,C} := h^{\oslash}_{B\otimes(A\oslash C),A} \circ (i_A \oslash (c^{\otimes\oplus}_{B,A,A\oslash C} \circ (i_B \otimes e^{\oslash}_{C,A})))$$

$$a'_{A,B,C} := (((i_A \oplus e'_{B,C}) \circ a^{\oplus\otimes}_{A,B,C,C})/i_C) \circ h'_{A\oplus(B/C),C}$$

$$a\backslash_{A,B,C} := (i_A\backslash((e\backslash_{B,A} \oplus i_C) \circ a^{\otimes\otimes}_{A,A\backslash B,C})) \circ h\backslash_{(A\backslash B)\oplus C,A}$$

$$c'_{A,B,C} := (((e'_{A,B} \oplus i_C) \circ c^{\oplus\otimes}_{A/B,C,B})/i_B) \circ h'_{(A/B)\oplus C,B}$$

$$c\backslash_{A,B,C} := (i_B\backslash((i_A \oplus e\backslash_{C,B}) \circ c^{\otimes\otimes}_{B,A,B\backslash C})) \circ h\backslash_{A\oplus(B\backslash C),B}$$

While for $\boldsymbol{LG}_{IV}$, we have:

$$\alpha^{\oslash}_{A,B,C} := e\backslash_{(A\oslash B)\oslash C,A} \circ (i_A \otimes (\alpha^{\backslash\oslash}_{A,A\oslash B,C} \circ (h\backslash_{B,A} \oslash i_C)))$$

$$\alpha^{\oslash}_{A,B,C} := e\backslash_{A\oslash(B\otimes C),C} \circ ((\alpha^{\oslash/}_{A,B\otimes C,C} \circ (i_A \oslash h'_{B,C})) \otimes i_C)$$

$$\gamma^{\oslash}_{A,B,C} := e\backslash_{(A\otimes C)\oslash B,C} \circ ((\gamma^{/\oslash}_{A\otimes C,C,B} \circ (h'_{A,C} \oslash i_B)) \otimes i_C)$$

$$\gamma^{\oslash}_{A,B,C} := e\backslash_{B\oslash(A\oslash C),A} \circ (i_A \otimes (\gamma^{\oslash\backslash}_{B,A,A\oslash C} \circ (i_B \oslash h\backslash_{C,A})))$$

$$\alpha'_{A,B,C} := (i_A \oplus ((h^{\oslash}_{B,A}/i_C) \circ \alpha^{\oslash/}_{A,A\oplus B,C})) \circ e^{\oslash}_{(A\oplus B)/C,A}$$

$$\alpha\backslash_{A,B,C} := (((i_A\backslash h^{\oslash}_{B,C}) \circ \alpha^{\backslash\oslash}_{A,B\oplus C,C}) \oplus i_C) \circ e^{\oslash}_{A\backslash(B\oplus C),C}$$

$$\gamma'_{A,B,C} := (((h^{\oslash}_{A,B}/i_C) \circ \gamma^{/\oslash}_{A\oplus B,C,B}) \oplus i_B) \circ e^{\oslash}_{(A\oplus B)/C,B}$$

$$\gamma\backslash_{A,B,C} := (i_B \oplus ((i_A\backslash h^{\oslash}_{C,B}) \circ \gamma^{\oslash\backslash}_{B,A,B\oplus C})) \circ e^{\oslash}_{A\backslash(B\oplus C),B}$$

In practice, use of both type I and IV interactions may prove undesirable, given that their combined strength licenses same-sort associativity and -commutativity. To illustrate, we have the following arrow from $B\backslash(A\backslash C)$ to $A\backslash(B\backslash C)$:

$$(B \oslash 0_{\oplus(A\backslash C)}) \circ a\backslash_{B,0,A\backslash C} \circ \gamma\backslash_{A,B\backslash 0,C} \circ (A\backslash(\alpha\backslash_{B,0,C} \circ (B\backslash 0^*_{\oplus C})))$$

with similar problems arising in the absence of units as well. In addition, units themselves are also suspect, sometimes inducing overgeneration where linguistic applications are concerned. In light of these remarks, we shall restrict our discussion to the following four calculi of interest:

	GrI	GrIV	Units
LG_I	✓		
LG_{IV}		✓	
$LG_I^{0,1}$	✓		✓
$LG_{IV}^{0,1}$		✓	✓

We use the following notational convention:

$$\mathbf{T} \vdash A \to B \text{ iff } \exists f, f : A \to B \text{ in } \mathbf{T}, \text{ where } \mathbf{T} \in \{\mathbf{LG}_I, \mathbf{LG}_{IV}, \mathbf{LG}_I^{0,1}, \mathbf{LG}_{IV}^{0,1}\}$$

In the case of statements valid for arbitrary choice of $\mathbf{T}$, or when the latter is clear from context, we simply write $\vdash A \to B$.

3 Diamond Property and Examples

Definition 6. *Given* $T \in \{LG_I, LG_{IV}, LG_I^{0,1}, LG_{IV}^{0,1}\}$, *we say* A, B *are type similar in* T, *written* $T \vdash A \sim B$, *iff* $\exists C, T \vdash A \to C$ *and* $T \vdash B \to C$.

Following [10] and [9], we say that the C witnessing $T \vdash A \sim B$ is a *join* for A, B, not to be confused with the notion of joins familiar from lattice theory. Keeping with tradition, we write $\vdash A \sim B$ in case a statement is independent of the particular choice of T. We have the following equivalent definition.

Lemma 2. *Formulas* A, B *are type similar in* T *iff there exists* D *s.t.* $T \vdash D \to A$ *and* $T \vdash D \to B$.

Proof. The following table provides for each choice of T the solution for D in case the join C is known, and conversely. Note q refers to an arbitrary atom.

T	Solution for C	Solution for D
LG_I	$((B \otimes B) \oplus (B \oslash A))/D$	$C \oslash ((A/B) \otimes (B \oplus B))$
LG_{IV}	$((D/B)\backslash q) \oplus ((D/A)\backslash(q \oslash D))$	$((C/q) \oslash (A \oslash C)) \otimes (q \oslash (B \oslash C))$
$LG_I^{0,1}$	$(1 \oslash D) \oplus (A \otimes B)$	$(B \oplus A) \otimes (C\backslash 0)$
$LG_{IV}^{0,1}$	$(A \otimes B)/(1 \oslash (D\backslash 0))$	$((1 \oslash C)\backslash 0) \oslash (B \oplus A)$

Fig.2 shows the derivations for the joins, assuming $f : D \to A$ and $g : D \to B$, those concerning the solutions for D being essentially dual under $^{\infty}$.

Lem.2 is commonly referred to by the *diamond property*, in reference to the following equivalent diagrammatic representation:

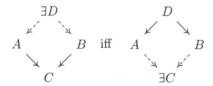

The formula D is also referred to as a *meet* for A, B. If C is known, we write $A \sqcap_C B$ for the meet constructed in Lem.2, while conversely we write $A \sqcup_D B$ for the join obtained from D. Clearly, if $\vdash A \sqcup_D B \to E$ ($\vdash E \to A \sqcap_C B$), then also $\vdash A \to E$ ($\vdash E \to A$), $\vdash B \to E$ ($\vdash E \to B$) and $\vdash D \to E$ ($\vdash E \to C$).

Remark 2. M&P provide an alternative solution for LG_{IV}, defining $A \sqcap_C B = (A/C) \otimes (C \oslash (B \oslash C))$ and $A \sqcup_D B = ((D/B)\backslash D) \otimes (D\backslash A)$. Though smaller in size compared to ours, the latter allows for easier generalization. For example, in the following event, suppose we wish to find a meet for A_1 and A_4:

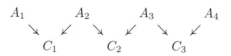

$$\cfrac{\cfrac{\cfrac{\cfrac{A}{(\boxed{A}\otimes B)/B}\ h^{/}}{((B\oplus(B\oslash A))\otimes B)/\boxed{B}}\ e^{\oslash}}{((B\oplus(B\oslash A))\otimes B)/D}\ g}{((B\otimes B)\oplus(B\oslash A))/D}\ a^{\otimes\oplus}
\qquad
\cfrac{\cfrac{\cfrac{\cfrac{B}{(B\otimes\boxed{A})/A}\ h^{/}}{(B\otimes(B\oplus(B\oslash A)))/\boxed{A}}\ e^{\oslash}}{(B\otimes(B\oplus(B\oslash A)))/D}\ f}{((B\otimes B)\oplus(B\oslash A))/D}\ a^{\otimes\oplus}$$

$$\cfrac{\cfrac{\cfrac{\cfrac{\cfrac{\cfrac{A}{(\boxed{B}/A)\backslash B}\ l^{\backslash}}{(D/A)\backslash\boxed{B}}\ g}{(D/A)\backslash((D/B)\backslash\boxed{D})}\ l^{\backslash}}{(D/A)\backslash\boxed{((D/B)\backslash(q\oplus(q\oslash D)))}}\ e^{\oslash}}{(D/A)\backslash(((D/B)\backslash q)\oplus(q\oslash D))}\ a^{\backslash}}{((D/B)\backslash q)\oplus((D/A)\backslash(q\oslash D))}\ \gamma^{\backslash}
\qquad
\cfrac{\cfrac{\cfrac{\cfrac{\cfrac{\cfrac{B}{(\boxed{A}/B)\backslash A}\ l^{\backslash}}{(D/B)\backslash\boxed{A}}\ f}{(D/B)\backslash\boxed{((D/A)\backslash D)}}\ l^{\backslash}}{(D/B)\backslash(q\oplus(q\oslash((D/A)\backslash D)))}\ e^{\oslash}}{((D/B)\backslash q)\oplus\boxed{(q\oslash((D/A)\backslash D))}}\ a^{\backslash}}{((D/B)\backslash q)\oplus((D/A)\backslash(q\oslash D))}\ \gamma^{\oslash\backslash}$$

$$\cfrac{\cfrac{\cfrac{\cfrac{B}{\boxed{1}\otimes B}\ 1}{((1\oslash D)\oplus D)\otimes B}\ e^{\oslash}}{(1\oslash D)\oplus(\boxed{D}\otimes B)}\ a^{\oplus\otimes}}{(1\oslash D)\oplus(A\otimes B)}\ g
\qquad
\cfrac{\cfrac{\cfrac{\cfrac{A}{A\otimes\boxed{1}}\ 1}{A\otimes((1\oslash D)\oplus D)}\ e^{\oslash}}{(1\oslash D)\oplus(A\otimes\boxed{D})}\ c^{\otimes\otimes}}{(1\oslash D)\oplus(A\otimes B)}\ f$$

$$\cfrac{\cfrac{\cfrac{\cfrac{\cfrac{\cfrac{\cfrac{B}{\boxed{(B\otimes(1\oslash(D\backslash 0)))}/(1\oslash(D\backslash 0))}\ h^{/}}{(\boxed{(B\otimes 1)}\oslash(D\backslash 0))/(1\oslash(D\backslash 0))}\ \alpha^{\oslash}}{(\boxed{B}\oslash(D\backslash 0))/(1\oslash(D\backslash 0))}\ 1}{((A\backslash\boxed{(A\otimes B)})\oslash(D\backslash 0))/(1\oslash(D\backslash 0))}\ h^{\backslash}}{((\boxed{A}\backslash((A\otimes B)\oplus 0))\oslash(D\backslash 0))/(1\oslash(D\backslash 0))}\ 0}{(\boxed{(D\backslash((A\otimes B)\oplus 0))}\oslash(D\backslash 0))/(1\oslash(D\backslash 0))}\ f}{(\boxed{(((A\otimes B)\oplus(D\backslash 0))\oslash(D\backslash 0))}/(1\oslash(D\backslash 0))}\ \gamma^{\backslash}\ \big/\ (A\otimes B)/(1\oslash(D\backslash 0))\ h^{\oslash}}$$

$$\cfrac{\cfrac{\cfrac{\cfrac{\cfrac{\cfrac{\cfrac{A}{(A\otimes B)/\boxed{B}}\ h^{/}}{(A\otimes B)/(1\oslash(B\oslash\boxed{1}))}\ l^{\oslash}}{(A\otimes B)/(1\oslash\boxed{(B\oslash(D\backslash(D\otimes 1)))})}\ h^{\backslash}}{(A\otimes B)/(1\oslash(D\backslash(B\oslash\boxed{(D\otimes 1)})))}\ \gamma^{\oslash\backslash}}{(A\otimes B)/(1\oslash(D\backslash(B\oslash\boxed{D})))}\ 1^{*}}{(A\otimes B)/(1\oslash(D\backslash(B\oslash\boxed{B})))}\ g}{(A\otimes B)/(1\oslash(D\backslash\boxed{(B\oslash(B\oplus 0))}))}\ 0\ \big/\ (A\otimes B)/(1\oslash(D\backslash 0))\ h^{\oslash}}$$

Fig. 2. Derivations for the joins constructed in Lem.2

Normally, we would suffice by repeated applications of the diamond property:

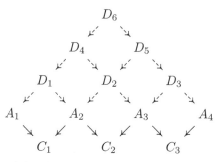

Note D_6 derives each of A_1, A_2, A_3, A_4. Restricting to A_1, A_4, we have a shorter solution obviously generalizing $A \sqcap_C B = ((C/q) \oslash (A \otimes C)) \otimes (q \oslash (B \otimes C))$:

$$((C_2/q) \oslash (A_1 \otimes C_3)) \otimes (((q/q) \oslash (A_4 \otimes C_1)) \otimes (((q/q) \oslash (A_2 \otimes C_2)) \otimes (q \oslash (A_3 \otimes C_2))))$$

Lemma 3. *[9] Already in the base logic $\mathbf{LG}_\varnothing$, type similarity satisfies*

1. *Reflexivity. $\vdash A \sim A$*
2. *Transitivity. $\vdash A \sim B$ and $\vdash B \sim C$ imply $\vdash A \sim C$*
3. *Symmetry. $\vdash A \sim B$ implies $\vdash B \sim A$*
4. *Congruence. $\vdash A_1 \sim A_2$ and $\vdash B_1 \sim B_2$ imply $\vdash A_1 \,\delta\, B_1 \sim A_2 \,\delta\, B_2$ for any $\delta \in \{\otimes, /, \backslash, \oplus, \oslash, \oslash\}$.*

We next illustrate $\sim$'s expressivity. While some examples were already considered by M&P for $\mathbf{LG}_{IV}$, we here provide alternative (often shorter) solutions. For reasons of space, we often omit the derivations witnessing our claims.

Lemma 4. *Neutrals. $\vdash C\backslash C \sim D/D$*

Proof. We have a join $(((C\backslash C) \oslash D) \oplus ((D \oslash C) \oplus C))/(C\backslash C)$ for $\mathbf{LG}_I$, as well as a meet $(C \oslash (C \oslash (C \otimes D))) \otimes (D \oslash (D \oslash (C \otimes D)))$ for $\mathbf{LG}_{IV}$.

The next few lemmas detail associativity and commutativity properties; underivable, but still valid at the level of type similarity.

Lemma 5. *Symmetry. $\vdash A\backslash B \sim B/A$*

Proof. For $\mathbf{LG}_I$ we have a join $(((A\backslash B) \oslash A) \oplus B)/(A\backslash B)$,

$$
\cfrac{
\cfrac{
\cfrac{A\backslash B}{((A\backslash B) \oslash A) \oplus \boxed{A}}\ e^\oslash
}{
\cfrac{((A\backslash B) \oslash A) \oplus (B/(A\backslash B))}{(((A\backslash B) \oslash A) \oplus B)/(A\backslash B)}\ a^/
}\ h^/
\qquad
\cfrac{
\cfrac{
\cfrac{
\cfrac{\dfrac{B/A}{((B/A) \otimes \boxed{A\backslash B})/(A\backslash B)}\ h^/}{((B/A) \otimes ((A\backslash B) \oslash A) \oplus A))/(A\backslash B)}\ e^\oslash
}{(((A\backslash B) \oslash A) \oplus \boxed{(B/A) \otimes A})/(A\backslash B)}\ c^{\otimes\oplus}
}{(((A\backslash B) \oslash A) \oplus B)/(A\backslash B)}\ e^/
}
}{}
$$

while for $\mathbf{LG}_{IV}$, we have a meet $A \oslash (B \oslash (A \otimes A))$.

$$\cfrac{\cfrac{\boxed{A}\oslash(B\oslash(A\otimes A))}{\cfrac{(A\backslash(A\otimes A))\oslash(B\oslash(A\otimes A))}{A\backslash\boxed{((A\otimes A)\oslash(B\oslash(A\otimes A)))}}\,h^{\backslash}}{A\backslash B}\,\alpha^{\backslash\oslash}}{\,l^{\oslash}}$$

$$\cfrac{\cfrac{\boxed{A}\oslash(B\oslash(A\otimes A))}{\cfrac{((A\otimes A/A)\oslash(B\oslash(A\otimes A))}{\boxed{((A\otimes A)\oslash(B\oslash(A\otimes A)))}/A}\,h^{/}}{B/A}\,\gamma^{/\oslash}}{\,l^{\oslash}}$$

Lemma 6. *Rotations.* $\vdash A\backslash(C/B)\sim(A\backslash C)/B$ *and* $\vdash A\backslash(B\backslash C)\sim B\backslash(A\backslash C)$

Proof. In $\mathbf{LG}_I$, we have $A\backslash(B\backslash((B\oslash A)\oplus((A\oslash B)\oplus C)))$ as a join for $A\backslash(B\backslash C)$ and $B\backslash(A\backslash C)$. To derive $\vdash A\backslash(C/B)\sim(A\backslash C)/B$, we proceed as follows:

    1. $\mathbf{LG}_I\vdash A\backslash(C/B)\sim(C/B)/A$ (Lem.5)
    2. $\mathbf{LG}_I\vdash(C/B)/A\sim(C/A)/B$ (shown above)
    3. $\mathbf{LG}_I\vdash(C/A)/B\sim(A\backslash C)/B$ (Lem.5 and L.3(4))
    4. $\mathbf{LG}_I\vdash A\backslash(C/B)\sim(A\backslash C)/B$ (Lem.3(2), 1,2,3)

For $\mathbf{LG}_{IV}$, we have a meet $((C\oslash(C/B))\oslash q)\otimes((C\oslash(A\backslash C))\oslash(q\backslash C))$ witnessing $\mathbf{LG}_{IV}\vdash A\backslash(C/B)\sim(A\backslash C)/B$, as well as $\mathbf{LG}_{IV}\vdash A\backslash(B\backslash C)\sim B\backslash(A\backslash C)$ with meet $((C\oslash(A\backslash C))\oslash q)\otimes((C\oslash(B\backslash C))\oslash(q\backslash C))$.

Lemma 7. *Distributivity.* $\vdash A\otimes(B/C)\sim(A\otimes B)/C$

Proof. For $\mathbf{LG}_I$, note $\vdash A\otimes(B/C)\sim A\otimes(C\backslash B)$ and $\vdash(A\otimes B)/C\sim C\backslash(A\otimes B)$ by Lem.5 and Lem.3(4). Thus, it suffices to show $\mathbf{LG}_I\vdash A\otimes(C\backslash B)\sim C\backslash(A\otimes B)$, fow which we have a join $C\backslash((A\oslash C)\oplus(C\otimes B))$.

$$\cfrac{\cfrac{\cfrac{\cfrac{\cfrac{A\otimes(C\backslash B)}{C\backslash(C\otimes(\boxed{A}\otimes(C\backslash B)))}\,h^{\backslash}}{C\backslash(C\otimes\boxed{((A\oslash C)\oplus C)\otimes(C\backslash B)})}\,e^{\oslash}}{C\backslash\boxed{(C\otimes((A\oslash C)\oplus(C\otimes(C\backslash B))))}}\,a^{\oplus\otimes}}{C\backslash((A\oslash C)\oplus(C\otimes\boxed{(C\otimes(C\backslash B))}))}\,c^{\otimes\oplus}}{C\backslash((A\oslash C)\oplus(C\otimes B))}\,e^{\backslash}$$

$$\cfrac{\cfrac{\cfrac{C\backslash(\boxed{A}\otimes B)}{C\backslash\boxed{((A\oslash C)\oplus C)\otimes B}}\,e^{\oslash}}{C\backslash((A\oslash C)\oplus(C\otimes B))}\,a^{\oplus\otimes}}{}$$

In $\mathbf{LG}_{IV}$, we have meet $A\otimes((A\oslash(B\oslash(A\otimes B)))\otimes(B\oslash(((A\otimes B)/C)\oslash(A\otimes B))))$.

Lemma 8. *Commutativity.* $\vdash A\otimes B\sim B\otimes A$

Proof. We have a join $(A\oslash B)\oplus(B\otimes B)$ for $\mathbf{LG}_I$,

$$\cfrac{\cfrac{\boxed{A}\otimes B}{((A\oslash B)\oplus B)\otimes B}\,e^{\oslash}}{(A\oslash B)\oplus(B\otimes B)}\,a^{\oplus\otimes}
\qquad
\cfrac{\cfrac{B\otimes\boxed{A}}{B\otimes((A\oslash B)\oplus B)}\,e^{\oslash}}{(A\oslash B)\oplus(B\otimes B)}\,c^{\otimes\oplus}$$

as well as a meet $(B\otimes B)\oslash(B\otimes((B\oplus A)\otimes B))$. For $\mathbf{LG}_{IV}$, we have a meet $(((A/B)\oslash((A\otimes B)\oslash(A\otimes A)))\otimes(B\oslash(B\oslash(A\backslash(A\otimes B)))))\otimes A$.

Lemma 9. *Associativity.* $\vdash(A\otimes B)\otimes C\sim A\otimes(B\otimes C)$

Proof. For $\mathbf{LG}_I$, we have meet $(A \otimes A) \oslash (A \otimes ((A \otimes (A \oplus B)) \otimes C))$, and join $(B \oslash ((A \backslash q) \oplus (Q/C))) \oplus (q \oplus q)$. In $\mathbf{LG}_{IV}$, use the diamond property after getting a meet D_1 from Lem.7 for $((A \otimes (B \otimes C))/C) \otimes C$ and $(A \otimes ((B \otimes C)/C)) \otimes C$:

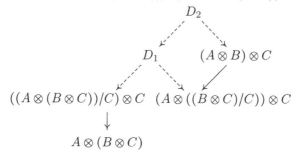

$$((A \otimes (B \otimes C))/C) \otimes C \quad (A \otimes ((B \otimes C)/C)) \otimes C$$

$$A \otimes (B \otimes C)$$

Remark 3. While the above lemmas immediately extend to $\mathbf{LG}_I^{0,1}$ and $\mathbf{LG}_{IV}^{0,1}$, the presence of units often allows for simpler joins and meets. For example, we have the following joins (J) and meets (M) in $\mathbf{LG}_I^{0,1}$ and $\mathbf{LG}_{IV}^{0,1}$:

Lemma	$\mathbf{LG}_I^{0,1}$	$\mathbf{LG}_{IV}^{0,1}$
Neutrals	J. $C \backslash ((1 \oslash D) \oplus (C \otimes D))$	J. $((1 \oslash C) \backslash 0) \backslash ((C \otimes D)/D)$
Symmetry	J. $(1 \oslash A) \oplus B$	M. $1 \oslash (B \oslash A)$
Commutativity	J. $(1 \oslash (1/A)) \oplus B$	M. $((1 \oslash A) \backslash 0) \otimes B$
Associativity	J. $(1 \oslash (1/A)) \oplus (B \otimes C)$	M. $A \otimes (((1 \oslash B) \backslash 0) \otimes C)$

4 Completeness Results

We consider models built upon algebraic structures featuring two binary operations $\times$ and $+$, related by linear distributivity. Their definition derives from the linear distributive categories of [2] by turning their arrows into equivalences.

Definition 7. *A linearly distributive algebra is a 6-tuple $\mathscr{A} = \langle A, \times, +, ^{\perp}, \top, \perp \rangle$ of type $\langle 2, 2, 1, 0, 0 \rangle$ satisfying*

1. *Associativity.* $(A \times B) \times C = A \times (B \times C)$; $(A + B) + C = A + (B + C)$
2. *Commutativity.* $A \times B = B \times A$; $A + B = B + A$
3. *Units.* $A \times \top = A$; $A + \perp = A$
4. *Inverses.* $A^{\perp} \times A = \perp$; $A^{\perp} + A = \top$
5. *Linear distributivity.* $A \times (B + C) = (A \times B) + C$

Definition 8. *A model $\mathscr{M}$ for $\sim$ is a pair $\langle \mathscr{A}, v \rangle$ extending $\mathscr{A}$ with a valuation v mapping atoms into $\mathscr{A}$, extended inductively to an interpretation $[\![\cdot]\!]$:*

$$[\![p]\!] := v(p) \quad [\![A \otimes B]\!] := [\![A]\!] \times [\![B]\!] \quad [\![A \oplus B]\!] := [\![A]\!] + [\![B]\!]$$
$$[\![1]\!] := \top \quad [\![A/B]\!] := [\![A]\!] + [\![B]\!]^{\perp} \quad [\![B \oslash A]\!] := [\![B]\!]^{\perp} \times [\![A]\!]$$
$$[\![0]\!] := \perp \quad [\![B \backslash A]\!] := [\![B]\!]^{\perp} + [\![A]\!] \quad [\![A \oslash B]\!] := [\![A]\!] \times [\![B]\!]^{\perp}$$

Note that, for arbitrary A, $[\![{}^1 A]\!] = [\![A^1]\!] = [\![{}^0 A]\!] = [\![A^0]\!] = [\![A]\!]^{\perp}$. E.g., $[\![A^1]\!] = [\![A \oslash 1]\!] = [\![A]\!]^{\perp} \times \top = [\![A]\!]^{\perp}$, and $[\![{}^0 A]\!] = [\![0/A]\!] = \perp + [\![A]\!]^{\perp} = [\![A]\!]^{\perp}$. M&P as well conducted model-theoretic investigations into type similarity for $\mathbf{LG}_{IV}$.

Their interpretation, however, takes as target the *free* Abelian group generated by the atomic formulae and an additional element $\oslash$,

$$[\![p]\!]' := p \quad [\![A \otimes B]\!]' := [\![A]\!]' \cdot [\![B]\!]' \quad [\![A \oplus B]\!]' := [\![A]\!]' \cdot \oslash^{-1} \cdot [\![B]\!]'$$
$$[\![A/B]\!]' := [\![A]\!]' \cdot [\![B]\!]'^{-1} \quad [\![B \oslash A]\!]' := [\![B]\!]'^{-1} \cdot \oslash \cdot [\![A]\!]'$$
$$[\![B \backslash A]\!]' := [\![B]\!]'^{-1} \cdot [\![A]\!]' \quad [\![A \oslash B]\!]' := [\![A]\!]' \cdot \oslash \cdot [\![B]\!]'^{-1}$$

writing 1 for unit and $^{-1}$ for inverse. While not reconcilable with Def.8 in that it does not offer a concrete instance of a linearly distributive algebra, the decidability of the word problem in free Abelian groups implies the decidability of type similarity as a corollary of completeness. The current investigation rather aims at a concept of model that better reflects the coexistence of residuated and coresiduated triples in the source language. While we can still prove type similarity complete w.r.t. the freely generated such model, as shown in Lem.14, the inference of decidability requires additional steps. Specifically, we will use Moortgat and Pentus' models in as inspiration in §5 to define, for each formula, a 'normal form', possibly involving units, w.r.t. which it is found type similar. We then decide type similarity at the level of such normal forms by providing an algorithm for generating joins, settling the word problem in the freely generated linear distributive algebra as a corollary, ensuring, in turn, the desired result.

Lemma 10. *If* $\vdash A \to B$, *then* $[\![A]\!] = [\![B]\!]$ *in every model.*

Proof. By induction on the arrow witnessing $\vdash A \to B$.

Theorem 1. *If* $A \sim B$, *then* $[\![A]\!] = [\![B]\!]$ *in every model.*

Proof. If $A \sim B$, we have a join C for A and B. By Lem.10, $\vdash A \to C$ and $\vdash B \to C$ imply $[\![A]\!] = [\![C]\!]$ and $[\![B]\!] = [\![C]\!]$, and hence $[\![A]\!] = [\![B]\!]$.

To prove completeness, we define a syntactic model wherein the interpretations of formulae are (constructively) shown to coincide with their equivalence classes under $\sim$. In defining said model, we use the following lemmas.

Lemma 11. *We have* $\vdash (A \backslash A) \oslash A \sim (A \oslash A)/A$.

Proof. Lem.5 gives meets D_1, D_2 for $\vdash (A \backslash A) \oslash A \sim A \oslash (A \backslash A)$ and $\vdash (A \oslash A)/A \sim A \backslash (A \oslash A)$. As such, we have a join C witnessing $\vdash A \oslash (A \backslash A) \sim A \backslash (A \oslash A)$, so that another use of the diamond property provides the desired meet D_3:

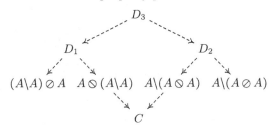

Lemma 12. $1 \oslash A \sim A \oslash 1 \sim 0/A \sim A \backslash 0$ *in* $\mathbf{LG}_I^{0,1}$ *and* $\mathbf{LG}_{IV}^{0,1}$.

Proof. That $\vdash 1 \oslash A \sim A \otimes 1$ and $\vdash 0/A \sim A\backslash 0$ are immediate consequences of Lem.5. Furthermore, $\mathbf{LG}_{IV}^{0,1} \vdash 1 \oslash A \to A\backslash 0$, as shown on the left, while for $\mathbf{LG}_I$ we apply the diamond property, as shown on the right,

$$
\begin{array}{c}
\dfrac{1 \oslash A}{A\backslash\boxed{(A \otimes (1 \oslash A))}} \ h\backslash \\[2ex]
\dfrac{}{A\backslash(\boxed{(A \otimes 1)} \oslash A)} \ \alpha^{\oslash} \\[2ex]
\dfrac{}{A\backslash(\boxed{A} \oslash A)} \ 1^* \\[2ex]
\dfrac{}{A\backslash\boxed{((0 \oplus A) \oslash A)}} \ 0^* \\[2ex]
\dfrac{}{A\backslash 0} \ h^{\oslash}
\end{array}
$$

Definition 9. *We construct a syntactic model by building a linearly distributive algebra upon the set of equivalence classes* $[A]_\sim := \{ B \mid \ \vdash A \sim B \}$ *of formulae w.r.t.* $\sim$*. The various operations of the algebra are defined as follows:*

$$
\begin{aligned}
[A]_\sim \times [B]_\sim &:= [A \otimes B]_\sim & \top &:= [A\backslash A]_\sim & [A]_\sim^\perp &:= [(A\backslash A) \oslash A]_\sim \\
[A]_\sim + [B]_\sim &:= [A \oplus B]_\sim & \bot &:= [A \oslash A]_\sim & &= [(A \oslash A)/A]_\sim
\end{aligned}
$$

For $\boldsymbol{LG}_I^{0,1}$ *and* $\boldsymbol{LG}_{IV}^{0,1}$*, the following simpler definitions suffice:*

$$
\begin{aligned}
[A]_\sim \times [B]_\sim &:= [A \otimes B]_\sim & \top &:= [1]_\sim & [A]_\sim^\perp &:= [1 \oslash A]_\sim = [A \oslash 1]_\sim \\
[A]_\sim + [B]_\sim &:= [A \oplus B]_\sim & \bot &:= [0]_\sim & &= [0/A]_\sim = [A\backslash 0]_\sim
\end{aligned}
$$

Finally, we define the valuation by $v(p) := [p]_\sim$ *for arbitrary atom* p*.*

Lemma 13. *The syntactic model is well-defined.*

Proof. We check the equations of Def.7. Definition unfolding reduces (1) to showing $\vdash (A \otimes B) \otimes C \sim A \otimes (B \otimes C)$ and $\vdash (A \oplus B) \oplus C \sim A \oplus (B \oplus C)$. Both follow from Lem.9, noting that for the latter we can take the dual of a meet (join) for $\vdash C^\infty \otimes (B^\infty \otimes A^\infty) \sim (C^\infty \otimes B^\infty) \otimes A^\infty$ under ∞. Similarly, (2) and (4) are immediate consequences of Lem.8 and Lem.11 (Lem.12 in the presence of units), while (3) is equally straightforward. This leaves (5), demanding $\vdash A \otimes (B \oplus C) \sim (A \otimes B) \oplus C$. We have $\mathbf{LG}_I \vdash A \otimes (B \oplus C) \to (A \otimes B) \oplus C$, while for $\mathbf{LG}_{IV}$ we use the diamond property:

While we could proceed to prove $[\![A]\!] = [A]_\sim$ in the syntactic model for arbitrary A, we prove a slightly more involved statement, the increase in complexity paying off when proving decidability of type similarity in Thm.4. Write $\mathscr{A}(Atom)$ for the linear distributive algebra freely generated by the atoms.

Lemma 14. *If* $[\![A]\!] = [\![B]\!]$ *in* $\mathscr{A}(Atom)$*, then also* $\vdash A \sim B$*.*

Proof. We follow the strategy pioneered by Pentus [10]. Consider the homomorphic extension h of $p \mapsto [p]_\sim$ (cf. Def.9). We prove, for arbitrary A, that $h(\llbracket A \rrbracket) = [A]_\sim$, taking $\llbracket A \rrbracket$ to be the interpretation of A in $\mathscr{A}(Atom)$. Hence, if $\llbracket A \rrbracket = \llbracket B \rrbracket$ in $\mathscr{A}(Atom)$, then also $h(\llbracket A \rrbracket) = h(\llbracket B \rrbracket)$, so that $[A]_\sim = [B]_\sim$, and thus $\vdash A \sim B$. Proceeding by induction, the cases $A = p$, $A = 1$ and $A = 0$ follow by definition, while simple definitional unfolding suffices if $A = A_1 \otimes A_2$ or $A = A_1 \oplus A_2$. The cases $A = A_1/A_2$, $A = A_2 \backslash A_1$, $A = A_1 \oslash A_2$ and $A = A_2 \oslash A_1$ are all alike, differing primarily in the number of applications of Lem.5. We demonstrate with $A = A_1/A_2$. In $\mathbf{LG}_I$ and $\mathbf{LG}_{IV}$, we have

$$h(\llbracket A_1/A_2 \rrbracket) = h(\llbracket A_1 \rrbracket) + h(\llbracket A_2 \rrbracket)^\perp = [A_1]_\sim + [A_2]_\sim^\perp = [A_1 \oplus ((A_2 \oslash A_2)/A_2)]_\sim$$

Thus, we have to show $\vdash A_1 \oplus ((A_2 \oslash A_2)/A_2) \sim A_1/A_2$:

1. $\vdash A_2 \oslash A_2 \sim A_1 \oslash A_1$ (Lem.4)
2. $\vdash A_1 \oslash A_1 \sim A_1 \oslash A_1$ (Lem.5)
3. $\vdash A_2 \oslash A_2 \sim A_1 \oslash A_1$ (Transitivity, 1, 2)
4. $\vdash A_1 \oplus ((A_2 \oslash A_2)/A_2) \sim A_1 \oplus ((A_1 \oslash A_1)/A_2)$ (Congruence, 3)
5. $\vdash A_1 \oplus ((A_1 \oslash A_1)/A_2) \sim (A_1 \oplus (A_1 \oslash A_1))/A_2$
6. $\vdash (A_1 \oplus (A_1 \oslash A_1))/A_2 \leftarrow A_1/A_2$
7. $\vdash A_1 \oplus ((A_2 \oslash A_2)/A_2) \sim A_1/A_2$ (Transitivity, 4, 5, 6)

In the presence of units, we have to show instead $\vdash A_1 \oplus (0/A_2) \sim A_1/A_2$, the desired proof being essentially a simplification of that found above.

Theorem 2. *If $\llbracket A \rrbracket = \llbracket B \rrbracket$ in every model, then $\vdash A \sim B$.*

Proof. If $\llbracket A \rrbracket = \llbracket B \rrbracket$ in every model, then in particular in $\mathscr{A}(Atom)$, and hence $\vdash A \sim B$ by Lem.14.

5   Generating Joins

We next present an algorithm for generating joins and meets in $\mathbf{LG}_I^{0,1}$, deriving decidability for the remaining incarnations of $\mathbf{LG}$ as a corollary. We proceed in two steps. First, we define for each formula A a 'normal form' $\|A\|^\circ$ w.r.t. which it is shown type similar by some join C_A. Whether or not any A and B are type similar is then decided for $\|A\|^\circ$ and $\|B\|^\circ$, an affirmative answer, witnessed by some meet D, implying the existence of a join C for A and B by the diamond property. The following figure summarizes the previous discussion.

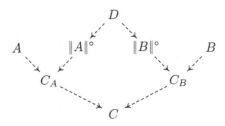

Definition 10. *We define the maps $\|\cdot\|^\circ$ and $\|\cdot\|^\bullet$ by mutual induction:*

$$
\begin{aligned}
\|p\|^\circ &:= p & \qquad \|p\|^\bullet &:= 1/p \\
\|1\|^\circ &:= 1 & \|1\|^\bullet &:= 1 \\
\|0\|^\circ &:= 0 & \|0\|^\bullet &:= 1/0 \\
\|A \otimes B\|^\circ &:= \|A\|^\circ \otimes \|B\|^\circ & \|A \otimes B\|^\bullet &:= \|A\|^\bullet \otimes \|B\|^\bullet \\
\|A/B\|^\circ &:= \|A\|^\circ \otimes \|B\|^\bullet & \|A/B\|^\bullet &:= \|A\|^\bullet \otimes \|B\|^\circ \\
\|B\backslash A\|^\circ &:= \|A/B\|^\circ & \|B\backslash A\|^\bullet &:= \|A/B\|^\bullet \\
\|A \oplus B\|^\circ &:= \|A\|^\circ \otimes ((1/0) \otimes \|B\|^\circ) & \|A \oplus B\|^\bullet &:= \|A\|^\bullet \otimes (\|B\|^\bullet \otimes 0) \\
\|A \oslash B\|^\circ &:= \|A\|^\circ \otimes (0 \otimes \|B\|^\bullet) & \|A \oslash B\|^\bullet &:= \|A\|^\bullet \otimes ((1/0) \otimes \|B\|^\circ) \\
\|B \obslash A\|^\circ &:= \|A \oslash B\|^\circ & \|B \obslash A\|^\bullet &:= \|A \oslash B\|^\bullet
\end{aligned}
$$

Compare the above definition to the Abelian group interpretation of M&P: multiplications $A \cdot B$ and inverses A^{-1} are rendered as $A \otimes B$ and $1/A$, while 0 replaces the special atom $\oplus$. We now need only solve the problem of generating joins for the formulas in the images of $\|\cdot\|^\circ$ and $\|\cdot\|^\bullet$, relying on the result, proved presently, that $\vdash A \sim \|A\|^\circ$ and $\vdash 1/A \sim \|A\|^\bullet$.

Lemma 15. *There exist maps $f(\cdot)$ and $g(\cdot)$ mapping any given A to joins witnessing $\vdash A \sim \|A\|^\circ$ and $\vdash 1/A \sim \|A\|^\bullet$ respectively.*

Proof. The (mutual) inductive definition of the desired maps is presented in parallel with the proof of their correctness. In the base cases, we set

$$
\begin{aligned}
f(p) &:= p & f(1) &:= 1 & f(0) &:= 0 \\
g(p) &:= 1/p & g(1) &:= 1 & g(0) &:= 1/0
\end{aligned}
$$

Correctness is nigh immediate, noting $\vdash 1/1 \to 1$ for $g(1)$. The diamond property is used for most of the inductive cases. To illustrate, consider $A = A_1 \oslash A_2$ and $A = A_2 \obslash A_1$, handled similarly. Starting with $f(\cdot)$, we have, by induction hypothesis, joins $f(A_1)$ and $g(A_2)$ for $\vdash A_1 \sim \|A_1\|^\circ$ and $\vdash (1/A_2) \sim \|A_2\|^\bullet$. Hence, by Lem.3(4), we have a join $f(A_1) \otimes (0 \otimes g(A_2))$ for $\vdash (A_1 \otimes (0 \otimes (1/A_2))) \sim (\|A_1\|^\circ \otimes (0 \otimes \|A_2\|^\bullet))$. In addition, we have joins

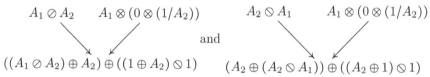

$$
\begin{array}{ccc}
A_1 \oslash A_2 \quad A_1 \otimes (0 \otimes (1/A_2)) & & A_2 \obslash A_1 \quad A_1 \otimes (0 \otimes (1/A_2)) \\
\searrow \qquad \swarrow & \text{and} & \searrow \qquad \swarrow \\
((A_1 \oslash A_2) \oplus A_2) \oplus ((1 \oplus A_2) \obslash 1) & & (A_2 \oplus (A_2 \obslash A_1)) \oplus ((A_2 \oplus 1) \obslash 1)
\end{array}
$$

We demonstrate the derivability claims found in the left diagram in Fig.3, those found in the right diagram being shown similarly. With these findings, we may now define $f(A_1 \oslash A_2)$ through the diamond property:

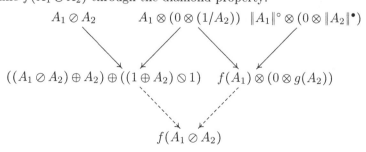

$$
\begin{array}{c}
A_1 \oslash A_2 \qquad A_1 \otimes (0 \otimes (1/A_2)) \quad \|A_1\|^\circ \otimes (0 \otimes \|A_2\|^\bullet) \\
\searrow \qquad \swarrow \qquad \searrow \qquad \swarrow \\
((A_1 \oslash A_2) \oplus A_2) \oplus ((1 \oplus A_2) \obslash 1) \qquad f(A_1) \otimes (0 \otimes g(A_2)) \\
\searrow \qquad \swarrow \\
f(A_1 \oslash A_2)
\end{array}
$$

$$\cfrac{\cfrac{A_1 \oslash A_2}{(A_1 \oslash A_2) \otimes \boxed{1}} \; 1}{\cfrac{(A_1 \oslash A_2) \otimes ((1 \oplus A_2) \oplus ((1 \oplus A_2) \oslash 1))}{\cfrac{\boxed{((A_1 \oslash A_2) \otimes (1 \oplus A_2))} \oplus ((1 \oplus A_2) \oslash 1)}{\cfrac{(\boxed{((A_1 \oslash A_2) \otimes 1)} \oplus A_2) \oplus ((1 \oplus A_2) \oslash 1)}{((A_1 \oslash A_2) \oplus A_2) \oplus ((1 \oplus A_2) \oslash 1)} \; 1^*} \; a^{\otimes \oplus}} \; a^{\otimes \oplus}} \; e^{\oslash}}$$

$$\cfrac{\cfrac{A_1 \otimes (0 \otimes \boxed{(1/A_2)})}{A_1 \otimes (0 \otimes ((1/A_2) \otimes \boxed{1}))} \; 1}{\cfrac{A_1 \otimes (0 \otimes \boxed{(1/A_2) \otimes ((1 \oplus A_2) \oplus ((1 \oplus A_2) \oslash 1))})}{\cfrac{A_1 \otimes (0 \otimes (\boxed{((1/A_2) \otimes (1 \oplus A_2))} \oplus ((1 \oplus A_2) \oslash 1)))}{\cfrac{A_1 \otimes (0 \otimes ((1 \oplus \boxed{((1/A_2) \otimes A_2)}) \oplus ((1 \oplus A_2) \oslash 1)))}{\cfrac{A_1 \otimes \boxed{0 \otimes ((1 \oplus 1) \oplus ((1 \oplus A_2) \oslash 1))}}{\cfrac{A_1 \otimes (\boxed{(0 \otimes (1 \oplus 1))} \oplus ((1 \oplus A_2) \oslash 1))}{\cfrac{A_1 \otimes ((\boxed{(0 \otimes 1)} \oplus 1) \oplus ((1 \oplus A_2) \oslash 1))}{\cfrac{A_1 \otimes (\boxed{(0 \oplus 1)} \oplus ((1 \oplus A_2) \oslash 1))}{\cfrac{A_1 \otimes (1 \oplus ((1 \oplus A_2) \oslash 1))}{\cfrac{\boxed{(A \otimes 1)} \oplus ((1 \oplus A_2) \oslash 1)}{\cfrac{\boxed{A_1} \oplus ((1 \oplus A_2) \oslash 1)}{((A_1 \oslash A_2) \oplus A_2) \oplus ((1 \oplus A_2) \oslash 1)} \; e^{\oslash}} \; 1^*} \; a^{\otimes \oplus}} \; 0} \; 1^*} \; a^{\otimes \oplus}} \; a^{\otimes \oplus}} \; e^{/}} \; c^{\otimes \oplus}} \; a^{\otimes \oplus}} \; e^{\oslash}}$$

Fig. 3. Showing $\mathbf{LG}_I \vdash A_1 \oslash A_2 \to ((A_1 \oslash A_2) \oplus A_2) \oplus ((1 \oplus A_2) \oslash 1)$ and $\mathbf{LG}_I \vdash A_1 \otimes (0 \otimes (1/A_2)) \to ((A_1 \oslash A_2) \oplus A_2) \oplus ((1 \oplus A_2) \oslash 1)$

while $f(A_2 \oslash A_1)$ is similarly defined

$$(A_2 \oplus (A_2 \oslash A_1)) \oplus ((A_2 \oplus 1) \oslash 1) \sqcup_{A \otimes (0 \otimes (1/A_2))} (f(A_1) \otimes (0 \otimes g(A_2)))$$

The same strategy is used to define $g(A_1 \oslash A_2)$ and $g(A_2 \oslash A_1)$, this time employing joins $(1 \oslash A_1) \oplus (1 \oplus (0 \oslash A_2))$ and $(1 \oslash A_1) \oplus ((A_2 \oslash 0) \oplus 1)$ witnessing $\vdash 1/(A_1 \oslash A_2) \sim (1/A_1) \otimes ((1/0) \otimes A_2)$ and $\vdash 1/(A_2 \oslash A_1) \sim (1/A_1) \otimes ((1/0) \otimes A_2)$. In the same vein, we can handle a significant portion of the remaining cases:

$$g(A_1 \otimes A_2) \coloneqq ((1 \oplus (A_2 \oslash 1)) \oplus (A_1 \oslash 1)) \sqcup_{(1/A_1) \otimes (1/A_2)} (g(A_1) \otimes g(A_2))$$
$$g(A_1/A_2) \coloneqq (1 \oplus ((A_1/A_2) \oslash 1)) \sqcup_{(1/A_1) \otimes A_2} (g(A_1) \otimes f(A_2))$$
$$g(A_2 \backslash A_1) \coloneqq (1 \oplus ((A_2 \backslash A_1) \oslash 1)) \sqcup_{(1/A_2) \otimes A_2} (g(A_1) \otimes f(A_2))$$
$$f(A_1 \oplus A_2) \coloneqq (A_1 \oplus (0 \oslash A_2)) \sqcup_{A_1 \otimes ((1/0) \otimes A_2)} f(A_1) \otimes ((1/0) \otimes f(A_2))$$
$$g(A_1 \oplus A_2) \coloneqq (1 \oplus ((A_1 \oplus A_2) \oslash 1)) \sqcup_{(1/A_1) \otimes ((1/A_2) \otimes 0)} (g(A_1) \otimes (g(A_2) \otimes 0))$$

To show $\vdash (1/A_1) \otimes ((1/A_2) \otimes 0) \to 1 \oplus ((A_1 \oplus A_2) \oslash 1)$ for the definition of $g(A_1 \oplus A_2)$ can be a bit tricky, so we give the derivation in Fig.4. We are left with the following cases, handled without use of the diamond property:

$$\dfrac{\dfrac{\dfrac{\dfrac{\dfrac{\dfrac{\dfrac{\dfrac{\dfrac{\dfrac{\dfrac{(1/A_1)\otimes(\boxed{(1/A_2)}\otimes 0)}{(1/A_1)\otimes(((1/A_2)\otimes\boxed{1})\otimes 0)}\;1}{(1/A_1)\otimes(\boxed{((1/A_2)\otimes((A_1\oplus A_2)\oplus((A_1\oplus A_2)\otimes 1)))}\otimes 0)}\;e^{\otimes}}{(1/A_1)\otimes((\boxed{((1/A_2)\otimes(A_1\oplus A_2))}\oplus((A_1\oplus A_2)\otimes 1))\otimes 0)}\;a^{\otimes\oplus}}{(1/A_1)\otimes(((A_1\oplus\boxed{(1/A_2)\otimes A_2)})\oplus((A_1\oplus A_2)\otimes 1))\otimes 0)}\;a^{\otimes\oplus}}{(1/A_1)\otimes\boxed{(((A_1\oplus 1)\oplus((A_1\oplus A_2)\otimes 1))\otimes 0)}}\;e^{/}}{(1/A_1)\otimes(\boxed{((A_1\oplus 1)\otimes 0)}\oplus((A_1\oplus A_2)\otimes 1))}\;c^{\oplus\otimes}}{(1/A_1)\otimes((A_1\oplus\boxed{(1\otimes 0)})\oplus((A_1\oplus A_2)\otimes 1))}\;a^{\oplus\otimes}}{(1/A_1)\otimes(\boxed{(A_1\oplus 0)}\oplus((A_1\oplus A_2)\otimes 1))}\;1^{*}}{(1/A_1)\otimes(A_1\oplus((A_1\oplus A_2)\otimes 1))}\;0}{\boxed{((1/A_1)\otimes A_1)}\oplus((A_1\oplus A_2)\otimes 1)}\;a^{\otimes\oplus}}{1\oplus((A_1\oplus A_2)\otimes 1)}\;e^{/}$$

Fig. 4. Showing $\mathbf{LG}_I \vdash (1/A_1)\otimes((1/A_2)\otimes 0)$

$$
\begin{aligned}
f(A_1\otimes A_2) &:= f(A_1)\otimes f(A_2)\\
f(A_1/A_2) &:= (f(A_1)\oplus(A_2\otimes g(A_2)))\oplus((1/A_2)\otimes 1)\\
f(A_2\backslash A_1) &:= (f(A_1)\oplus(A_2\otimes g(A_2)))\oplus((1/A_2)\otimes 1)
\end{aligned}
$$

Fig.5 shows well-definedness of $f(A_1/A_2)$, with $f(A_2\backslash A_1)$ handled similarly.

We shall decide type similarity by reference to the following invariants.

Definition 11. *For arbitrary p, we define by mutual inductions the functions $|\cdot|_p^+$ and $|\cdot|_p^-$ counting, respectively, the numbers of positive and negative occurrences of p inside their arguments. First, the positive count:*

$$
\begin{array}{lll}
|r|_p^+ := 1 \ \textit{iff } r = p & |A\otimes B|_p^+ := |A|_p^+ + |B|_p^+ & |A\oplus B|_p^+ := |A|_p^+ + |B|_p^+\\
|1|_p^+ := 0 & |A/B|_p^+ := |A|_p^+ + |B|_p^- & |B\otimes A|_p^+ := |A|_p^+ + |B|_p^-\\
|0|_p^+ := 0 & |B\backslash A|_p^+ := |A|_p^+ + |B|_p^- & |A\oslash B|_p^+ := |A|_p^+ + |B|_p^-
\end{array}
$$

and similarly, the negative count:

$$
\begin{array}{lll}
|r|_p := 0 & |A\otimes B|_p^- := |A|_p^- + |B|_p^- & |A\oplus B|_p^- := |A|_p^- + |B|_p^-\\
|1|_p := 0 & |A/B|_p^- := |A|_p^- + |B|_p^+ & |B\otimes A|_p^- := |A|_p^- + |B|_p^+\\
|0|_p := 0 & |B\backslash A|_p^- := |A|_p^- + |B|_p^+ & |A\oslash B|_p^- := |A|_p^- + |B|_p^+
\end{array}
$$

The atomic count $|A|_p$ for p is defined $|A|_p^+ - |A|_p^-$. In a similar fashion, we define positive and negative counts $|A|_0^+$ and $|A|_0^-$ for occurrences of the unit 0 inside A, and set $|A|_0 := |A|_0^+ - |A|_0^-$.

In practice, the previously defined counts shall prove only of interest with arguments of the form $\|A\|^{\circ}$. In the case of arbitrary formulas, we therefore define (by mutual induction) the positive and negative *operator counts* $|A|_\oplus^+$ and $|A|_\oplus^-$ (resembling, though slightly differing from, a concept of [9] bearing the same name), recording the values of $\|\|A\|^{\circ}|_0^+$ and $\|\|A\|^{\circ}|_0^-$ respectively.

$$
\frac{\|A_1\|^\circ \otimes \|A_2\|^\bullet}{\|A_1\|^\circ \otimes (\boxed{1} \otimes \|A_2\|^\bullet)}\ 1
$$

$$
\frac{\|A_1\|^\circ \otimes (\boxed{((1/A_2) \oplus ((1/A_2) \oslash 1)) \otimes \|A_2\|^\bullet})}{\|A_1\|^\circ \otimes (((1/A_2) \otimes \boxed{\|A_2\|^\bullet}) \oplus ((1/A_2) \oslash 1))}\ e^\circ
$$

$$
\frac{\|A_1\|^\circ \otimes (((1/A_2) \otimes \boxed{\|A_2\|^\bullet}) \oplus ((1/A_2) \oslash 1))}{\|A_1\|^\circ \otimes (((1/A_2) \otimes \boxed{g(A_2)}) \oplus ((1/A_2) \oslash 1))}\ c^{\oplus\otimes} \quad IH
$$

$$
\frac{\|A_1\|^\circ \otimes (\boxed{((1/A_2) \otimes (A_2 \oplus (A_2 \oslash g(A_2))))} \oplus ((1/A_2) \oslash 1))}{\|A_1\|^\circ \otimes ((\boxed{((1/A_2) \otimes A_2)} \oplus (A_2 \oslash g(A_2))) \oplus ((1/A_2) \oslash 1))}\ e^\circ
$$

$$
\frac{\|A_1\|^\circ \otimes ((1 \oplus (A_2 \oslash g(A_2))) \oplus ((1/A_2) \oslash 1))}{\boxed{(\|A_1\|^\circ \otimes (1 \oplus (A_2 \oslash g(A_2))))} \oplus ((1/A_2) \oslash 1)}\ a^{\otimes\oplus} \quad e^/
$$

$$
\frac{(\boxed{(\|A_1\|^\circ \otimes 1)} \oplus (A_2 \oslash g(A_2))) \oplus ((1/A_2) \oslash 1)}{(\boxed{\|A_1\|^\circ} \oplus (A_2 \oslash g(A_2))) \oplus ((1/A_2) \oslash 1)}\ a^{\otimes\oplus} \quad a^{\otimes\oplus} \quad 1^*
$$

$$
\frac{(\boxed{\|A_1\|^\circ} \oplus (A_2 \oslash g(A_2))) \oplus ((1/A_2) \oslash 1)}{(f(A_1) \oplus (A_2 \oslash g(A_2))) \oplus ((1/A_2) \oslash 1)}\ IH
$$

$$
\frac{A_1/A_2}{(A_1/A_2) \otimes \boxed{1}}\ 1
$$

$$
\frac{(A_1/A_2) \otimes ((1/A_2) \oplus ((1/A_2) \oslash 1))}{((A_1/A_2) \otimes \boxed{(1/A_2)}) \oplus ((1/A_2) \oslash 1)}\ e^\circ \quad a^{\otimes\oplus}
$$

$$
\frac{((A_1/A_2) \otimes \boxed{(1/A_2)}) \oplus ((1/A_2) \oslash 1)}{((A_1/A_2) \otimes \boxed{g(A_2)}) \oplus ((1/A_2) \oslash 1)}\ IH
$$

$$
\frac{\boxed{((A_1/A_2) \otimes (A_2 \oplus (A_2 \oslash g(A_2))))} \oplus ((1/A_2) \oslash 1)}{(\boxed{((A_1/A_2) \otimes A_2)} \oplus (A_2 \oslash g(A_2))) \oplus ((1/A_2) \oslash 1)}\ e^\circ \quad a^{\otimes\oplus}
$$

$$
\frac{(\boxed{A_1} \oplus (A_2 \oslash g(A_2))) \oplus ((1/A_2) \oslash 1)}{(f(A_1) \oplus (A_2 \oslash g(A_2))) \oplus ((1/A_2) \oslash 1)}\ e^/ \quad IH
$$

Fig. 5. Proving well-definedness of $f(A_1/A_2)$

Definition 12. *For arbitrary A, define the positive and negative operator counts $|A|^+_\oplus$ and $|A|^-_\oplus$ are defined by induction over A, as follows:*

$$
\begin{aligned}
&|p|^+_\oplus := 0 & &|A \otimes B|^+_\oplus := |A|^+_\oplus + |B|^+_\oplus & &|A \oplus B|^+_\oplus := |A|^+_\oplus + |B|^+_\oplus \\
&|1|^+_\oplus := 0 & &|A/B|^+_\oplus := |A|^+_\oplus + |B|^-_\oplus & &|B \oslash A|^+_\oplus := |A|^+_\oplus + |B|^+_\oplus + 1 \\
&|0|^+_\oplus := 1 & &|B\backslash A|^+_\oplus := |A|^+_\oplus + |B|^-_\oplus & &|A \oslash B|^+_\oplus := |A|^+_\oplus + |B|^+_\oplus + 1
\end{aligned}
$$

and

$$
\begin{aligned}
&|p|^-_\oplus := 0 & &|A \otimes B|^-_\oplus := |A|^-_\oplus + |B|^-_\oplus & &|A \oplus B|^-_\oplus := |A|^-_\oplus + |B|^-_\oplus + 1 \\
&|1|^-_\oplus := 0 & &|A/B|^-_\oplus := |A|^-_\oplus + |B|^+_\oplus & &|B \oslash A|^-_\oplus := |A|^-_\oplus + |B|^+_\oplus \\
&|0|^-_\oplus := 0 & &|B\backslash A|^-_\oplus := |A|^-_\oplus + |B|^+_\oplus & &|A \oslash B|^-_\oplus := |A|^-_\oplus + |B|^+_\oplus
\end{aligned}
$$

Finally, the operator count $|A|_\oplus$ is defined $|A|^+_\oplus - |A|^-_\oplus$.

Lemma 16. *For any A, $|A|^+_\oplus = \|A\|^\circ|^+_0 = \|A\|^\bullet|^-_0$ and $|A|^-_\oplus = \|A\|^\bullet|^+_0 = \|A\|^\circ|^-_0$.*

Lemma 17. *If $\vdash A \to B$, then $|A|_\oplus = |B|_\oplus$, and $|A|_p = |B|_p$ for all p_i.*

Corollary 1. *If* $\vdash A \sim B$, *then* $|A|_\oplus = |B|_\oplus$, *and* $|A|_p = |B|_p$ *for all* p_i.

We now prove the inverse of the above corollary. Our aim is to define a meet for $\|A\|^\circ$ and $\|B\|^\circ$, entering into the construction of a join for A and B through use of the diamond property along with $f(A)$ and $f(B)$. To this end, we first require a few more definitions and lemmas. The following is an easy observation.

Lemma 18. *Formulas* $\|C\|^\circ, \|C\|^\bullet$ *for any* C *are included in the proper subset of* $\mathscr{F}(Atom)$ *generated by the following grammar:*

$$\phi ::= 0 \mid p_i \mid (1/0) \mid (1/p_i)$$
$$A^{nf}, B^{nf} ::= 1 \mid \phi \mid (A^{nf} \otimes B^{nf})$$

Thus, positive and negative occurrences of 0 (p_i) take the forms 0 (p_i) and $1/0$ $(1/p_i)$, being glued together through $\otimes$ only. We next detail the corresponding notion of *context*. Through universal quantification over said concept in stating derivability of certain rules pertaining to the Grishin interactions (cf. Lem.19), we obtain the non-determinacy required for the construction of the desired meet.

Definition 13. *A (tensor) context* $A^\otimes[]$ *is a bracketing of a series of formulae connected through* $\otimes$, *containing a unique occurrence of a hole* $[]$:

$$A^\otimes[], B^\otimes[] ::= [] \mid (A^\otimes[] \otimes B) \mid (A \otimes B^\otimes[])$$

Given $A^\otimes[], B$, *let* $A^\otimes[B]$ *denote the substitution of* B *for* $[]$ *in* $A^\otimes[]$.

We next characterize (half of) the type I Grishin interaction using contexts.

Lemma 19. *If* $\vdash A^\otimes[B \oslash C] \to D$, *then* $\vdash B \oslash A^\otimes[C] \to D$.

Proof. Assuming $f : A^\otimes[B \oslash C] \to D$, we proceed by induction on $A^\otimes[]$. The base case being immediate, we check $A^\otimes[] = A_1^\otimes[] \otimes A_2$ and $A^\otimes[] = A_1 \otimes A_2^\otimes[]$:

$$\cfrac{\cfrac{\cfrac{B \oslash (A_1^\otimes[C] \otimes A_2)}{\underline{(B \oslash A_1^\otimes[C])} \otimes A_2} \; a^\oslash}{A_1^\otimes[B \oslash C] \otimes A_2} \; IH}{D} \; f \qquad \cfrac{\cfrac{\cfrac{B \oslash (A_1 \otimes A_2^\otimes[C])}{A_1 \otimes \underline{(B \oslash A_2^\otimes[C])}} \; IH}{A_1 \otimes A_2^\otimes[B \oslash C]} \; c^\oslash}{D} \; f$$

The nondeterminacy required for the construction of our desired meet is obtained through the liberty of choosing one's context in instantiating the above rules. In practice, we only require recourse to the following restricted form.

Corollary 2. *If* $\vdash A^\otimes[B] \to C$, *then* $\vdash (1 \oslash B) \oslash A^\otimes[1] \to C$.

Proof. Suppose $f : (1 \oslash B) \oslash A^\otimes[1] \to C$. We then proceed as follows:

$$\cfrac{\cfrac{\cfrac{(1 \oslash B) \oslash A^\otimes[1]}{A^\otimes[\underline{((1 \oslash B) \oslash 1)}]} \; Lem.19}{A^\otimes[B]} \; l^\oslash}{C} \; f$$

Theorem 3. $\vdash A \sim B$ *if* $|A|_\oplus = |B|_\oplus$ *and* $\|A\|^\circ|_p = \|B\|^\circ|_p$ *for all* p.

Proof. First, we require some notation. We shall write a (non-empty) *list* of formulas $[A_1, \ldots, A_n, B]$ to denote the right-associative bracketing of $A_1 \oslash \ldots A_n \oslash B$. Further, given $n \geq 0$, let A^n denote the list of n repetitions of A. Finally, we write ++ for list concatenation. Now let there be given an enumeration

$$p_1, p_2, \ldots p_n$$

of all the atoms occurring in A and B. Define

$$k := max(\|A\|^\circ|_0^+, \|B\|^\circ|_0^+) = max(|A|_\oplus^+, |B|_\oplus^+)$$
$$l := max(\|A\|^\circ|_0^-, \|B\|^\circ|_0^-) = max(|A|_\oplus^-, |B|_\oplus^-)$$
$$k(i) := max(\|A\|^\circ|_{p_i}^+, \|B\|^\circ|_{p_i}^+) \; (1 \leq i \leq n)$$
$$l(i) := max(\|A\|^\circ|_{p_i}^-, \|B\|^\circ|_{p_i}^-) \; (1 \leq i \leq n)$$

We now witness $\vdash \|A\|^\circ \sim \|B\|^\circ$ by a meet

$$D := (1 \oslash p_1)^{k(1)} \; ++ \; (1 \oslash (1/p_1))^{l(1)}$$
$$++ \ldots$$
$$++ \; (1 \oslash p_n)^{k(n)} \; ++ \; (1 \oslash (1/p_n))^{l(n)}$$
$$++ \; (1 \oslash 0)^k \qquad ++ \; (1 \oslash (1/0))^l \qquad ++ \; [1]$$

Since we know from Lem.15 that $\vdash A \sim \|A\|^\circ$ and $\vdash B \sim \|B\|^\circ$ with joins $f(A)$ and $f(B)$, we can construct a join $f(A) \sqcup_D f(B)$ witnessing $\vdash A \sim B$. Suffice it to show that D, as defined above, is indeed a meet for $\|A\|^\circ$ and $\|B\|^\circ$. W.l.o.g., we show $\vdash D \to \|A\|^\circ$, dividing our proof in three steps. We shall a running example for illustrating each step, considering the concrete case where $A = p_2 \otimes (p_1/p_2)$ and $B = p_3 \oslash (p_3 \oslash p_1)$. Then

$$\|A\|^\circ = p_2 \otimes (p_1 \otimes (1/p_2))$$
$$\|B\|^\circ = p_3 \otimes (0 \otimes ((1/p_3) \otimes ((1/0) \otimes p_1)))$$
$$D = [1 \oslash p_1, 1 \oslash p_2, 1 \oslash (1/p_2), 1 \oslash p_3, 1 \oslash (1/p_3), 1 \oslash 0, 1 \oslash (1/0), 1]$$

$$\begin{array}{llll} k(1) = 1 & k(2) = 1 & k(3) = 1 & k = 1 \\ l(1) = 0 & l(2) = 1 & l(3) = 1 & l = 1 \end{array}$$

1. First, note that we have

   If $f : E \to F$, also $(f \circ (1_{\otimes E}^* \circ (e_{1,G}' \otimes i_E))) : ((1/G) \otimes G) \otimes E \to F$ (*)

   Starting with $i_{\|A\|^\circ} : \|A\|^\circ \to \|A\|^\circ$, for $i = 1$ to n, recursively apply (*) $k(i) - \|A\|^\circ|_{p_i}^+ \; (= l(i) - \|A\|^\circ|_{p_i}^-)$, by $|A|_{p_i} = |B|_{p_i}$) times, instantiating G with p_i, followed by another $k - \|A\|^\circ|_\oplus^+ \; (= l - \|A\|^\circ|_\oplus^-)$, since $|A|_\oplus = |B|_\oplus)$ recursive applications, this time instantiating G by 0. In our example, we obtain the following arrows:

$$\cfrac{\cfrac{\cfrac{\boxed{((1/0) \otimes 0)} \otimes (((1/p_3) \otimes p_3) \otimes \|A\|^\circ)}{\boxed{1} \otimes (((1/p_3) \otimes p_3) \otimes \|A\|^\circ)} \, 1^*}{\cfrac{\boxed{((1/p_3) \otimes p_3)} \otimes \|A\|^\circ}{\boxed{1} \otimes \|A\|^\circ} \, e^l}}{\|A\|^\circ} \, 1^* \qquad e^l \qquad \cfrac{\cfrac{\boxed{((1/p_2) \otimes p_2)} \otimes \|B\|^\circ}{\boxed{1} \otimes \|B\|^\circ} \, e^l}{\|B\|^\circ} \, 1^*$$

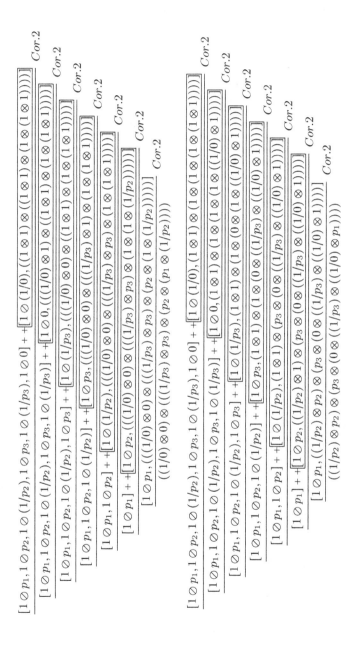

Fig. 6. Illustrating the proof of Thm. 3

Note that the antecedent of $\rightarrow$ now contains exactly $k(i)$ (k) and $l(i)$ (l) occurrences of p_i (0) and $1/p_i$ (1/0) respectively.

2. For $i = 1$ to n, apply the following procedure. Starting with the arrow constructed in the previous step, recursively apply Cor.2 $k(i)$ times, instantiating B with p_i, followed by another $l(i)$ applications where B is instantiated with $1/p_i$. Finally, we repeat the above procedure one last time with the positive and negative occurrences of 0. We continue with our example in Fig.6.

3. From D, we can derive the antecedent of the arrow in the previous step through repeated applications of 1, thus obtaining the desired result.

As a corollary of the above theorem, we can prove the decidability of the word problem in $\mathscr{A}(Atom)$. Lem.14 in turn implies decidability of type similarity in each of the variants of the Lambek-Grishin calculus discussed in this chapter.

Lemma 20. *For any expression ϕ in $\mathscr{A}(Atom)$, there exists a formula A in $\boldsymbol{LG}_I^{0,1}$ s.t. $[\![A]\!] = \phi$.*

Proof. We define the map $[\![\cdot]\!]^{-1}$ taking ϕ to a formula, as follows:

$$[\![p]\!]^{-1} := p \qquad\qquad [\![\phi^\perp]\!]^{-1} := 0/[\![\phi]\!]^{-1}$$
$$[\![\top]\!]^{-1} := 1 \qquad\qquad [\![\perp]\!]^{-1} := 0$$
$$[\![\phi \times \psi]\!]^{-1} := [\![\phi]\!]^{-1} \otimes [\![\psi]\!]^{-1} \quad [\![\phi + \psi]\!]^{-1} := [\![\phi]\!]^{-1} \oplus [\![\psi]\!]^{-1}$$

An easy induction ensures $[\![[\![\phi]\!]^{-1}]\!] = \phi$. To illustrate, consider the case $\phi^\perp$:
$[\![[\![\phi^\perp]\!]^{-1}]\!] = [\![0/[\![\phi]\!]^{-1}]\!] = \perp + [\![[\![\phi]\!]^{-1}]\!]^\perp = \perp + \phi^\perp = \phi^\perp$.

Lemma 21. *For any $\phi, \psi \in \mathscr{A}(Atom)$, we can decide whether or not $\phi = \psi$.*

Proof. By Thm.3, we can decide $\boldsymbol{LG}_I^{0,1} \vdash [\![\phi]\!]^{-1} \sim [\![\psi]\!]^{-1}$ through comparison of atomic- and operator counts. If affirmative, then also $[\![[\![\phi]\!]^{-1}]\!] = [\![[\![\psi]\!]^{-1}]\!]$ by Thm.1, i.e., $\phi = \psi$ by Lem.20. If instead $\boldsymbol{LG}_I^{0,1} \not\vdash [\![\phi]\!]^{-1} \sim [\![\psi]\!]^{-1}$, then also $[\![[\![\phi]\!]^{-1}]\!] \neq [\![[\![\psi]\!]^{-1}]\!]$, i.e., $\phi \neq \psi$ by Thm.2.

Theorem 4. *For any A, B, it is decidable whether $\boldsymbol{T} \vdash A \sim B$ for any $\boldsymbol{T} \in \{\boldsymbol{LG}_I, \boldsymbol{LG}_{IV}, \boldsymbol{LG}_I^{0,1}, \boldsymbol{LG}_{IV}^{0,1}\}$.*

Proof. Use Lem.21 to decide whether or not $[\![A]\!] = [\![B]\!]$ in $\mathscr{A}(Atom)$. If so, then $\vdash A \sim B$ by Lem.14. Otherwise, $\not\vdash A \sim B$ by Thm.1.

References

1. Brünnler, K., McKinley, R.: An algorithmic interpretation of a deep inference system. In: Cervesato, I., Veith, H., Voronkov, A. (eds.) LPAR 2008. LNCS, vol. 5330, pp. 482–496. Springer, Heidelberg (2008)
2. Cockett, J.R.B., Seely, R.A.G.: Weakly distributive categories. Journal of Pure and Applied Algebra, 45–65 (1991)

3. Foret, A.: On the computation of joins for non associative Lambek categorial grammars. In: Levy, J., Kohlhase, M., Niehren, J., Villaret, M. (eds.) Proceedings of the 17th International Workshop on Unification Valencia, UNIF 2003, Spain, June 8-9 (2003)
4. Grishin, V.N.: On a generalization of the Ajdukiewicz-Lambek system. In: Mikhailov, A.I. (ed.) Studies in Nonclassical Logics and Formal Systems, Nauka, Moscow, pp. 315–334 (1983)
5. Lambek, J.: On the calculus of syntactic types. In: Jakobson, R. (ed.) Structure of Language and its Mathematical Aspects, Proceedings of the Twelfth Symposium in Applied Mathematics (1961)
6. Lambek, J., Scott, P.J.: Introduction to Higher-Order Categorical Logic. Cambridge Studies in Advanced Mathematics. Cambridge University Press (1988)
7. Curien, P.L.: Categorical Combinators, Sequential Algorithms and Functional Programming. Progress in Theoretical Computer Science. Birkhäuser, Boston (1993)
8. Moortgat, M.: Symmetric categorial grammar. Journal of Philosophical Logic 38(6), 681–710 (2009)
9. Moortgat, M., Pentus, M.: Type similarity for the Lambek-Grishin calculus. In: Proceedings of the Twelfth Conference on Formal Grammar (2007)
10. Pentus, M.: The conjoinability relation in Lambek calculus and linear logic. ILLC Prepublication Series ML–93–03, Institute for Logic, Language and Computation, University of Amsterdam (1993)

NP-Completeness of Grammars
Based Upon Products of Free Pregroups

Denis Béchet

LINA - UMR 6241,
Université de Nantes & CNRS
2, rue de la Houssiniére - BP 92208
44322 Nantes Cedex 03 - France
Denis.Bechet@univ-nantes.fr

Abstract. Pregroup grammars are context-free lexicalized grammars based upon free pregroups which can describe parts of the syntax of natural languages. Some extensions are useful to model special constructions like agreements with complex features or non-projective relations or dependencies. A simple solution for these problems is given by lexicalized grammars based upon the product of free pregroups rather than on a single free pregroup. Such grammars are not necessarily context-free. However, the membership problem is NP-complete. To prove this theorem, the article defines a particular grammar built on the product of three free pregroups. This grammar is used to encode any SAT problem as a membership problem in the language corresponding to the grammar.

Keywords: Lambek Categorial Grammar, Pregroup Grammar, Free Pregroup, Product of Pregroups.

1 Introduction

Pregroup grammars [15] are a simplification of Lambek calculus [18] that can model parts of several natural languages: English [15], Italian [8], French [1], German [16,17], Japanese [7], Persian [19], etc. As with Lambek calculus, some extensions have been proposed for various constructions. For instance, in [4,3], simple type iterations are introduced into pregroup grammars for adjectival or adverbial phrases. [11] presents other extensions based upon modalities, product pregroup grammars and tupled pregroup grammars and applies these extensions to Polish.

In [12,13], the author proposes to use products of pregroups as a general construction to extend the generative power of pregroup grammars based upon free pregroups. For the author, this construction is interesting, for instance, for the Italian nominal and adjectival paradigm with binary valued features (masculine/feminine or singular/plural). With a product, every feature can be put in its own space giving a very simple solution for agreement. In [14], grammars based upon any pregroup (not only on a free pregroup) are proved to be Turing complete (the languages are all the ϵ-free recursively enumerable languages). The construction uses a grammar based upon the product of two free pregroups and its image through a string homomorphism.

C. Casadio et al. (Eds.): Lambek Festschrift, LNCS 8222, pp. 51–62, 2014.
© Springer-Verlag Berlin Heidelberg 2014

Products can also be used for long distance dependencies, in particular when the projective nature of free pregroup deductions limit axioms. For instance a pregroup analysis of "quand il l'avait ramené" (when he took her at home) needs a complex type for the clitic " l' " (her).

"l' " is assigned $\pi_3^r so^{ll}s^l\pi_3$ rather than o^{ll}. A better analysis would be:

quand il l' avait ramenée

For this kind of constructions, we need non-projective axioms. The product of free pregroups can be used for this purpose. The idea is used by Categorial Dependency Grammars[9] for non-projective dependencies even if in this case the polarities that define the ends of non-projective dependencies do not define complete categorial components[1].

Of course, we expect that such extensions preserve the interesting properties of pregroup grammars. One of these properties is that the membership problem is polynomial. This is no more the case with grammars based upon the products of at least 3 free pregroups: Here, we present a grammar based upon the product of 3 free pregroups that can code any SAT problem proving that the membership problem for this grammar is NP-hard.

The rest of the article begins by presenting pregroups, free pregroups and pregoup grammars (lexicalized grammars based on a free pregroup). Section 3 introduces the product of pregroups, pregroup product grammars (lexicalized grammars based on the product of several free pregroups) and gives some properties of the associated class of languages. Section 4 proves that the membership problem of the class of languages is NP-hard (in fact NP-complete). The last section concludes.

2 Background

Definition 1 (Pregroup). *A* pregroup *is a structure* $(P, \leq, \circ, l, r, 1)$ *such that* $(P, \leq, \circ, 1)$ *is a partially ordered monoid* [2] *and* l, r *are two unary operations on* P *that satisfy the inequalities* $x^l x \leq 1 \leq x x^l$ *and* $x x^r \leq 1 \leq x^r x$ *for all* $x \in P$.

---

[1] The types in a CDG can be defined as the set of the product of a categorial type and a list of signed integers.

[2] We briefly recall that a *monoid* is a structure $< M, \circ, 1 >$, such that $\circ$ is associative and has a neutral element 1 ($\forall x \in M : 1 \circ x = x \circ 1 = x$). A partially ordered monoid is a monoid $< M, \circ, 1 >$ with a partial order $\leq$ that satisfies $\forall a, b, c: a \leq b \Rightarrow c \circ a \leq c \circ b$ and $a \circ c \leq b \circ c$.

Definition 2 (Free Pregroup). *Let* $(P, \leq)$ *be a partially ordered set of basic types. We write* $\mathbb{Z}$ *for the set of signed integers.* $P^{(\mathbb{Z})} = \{\, p^{(i)} \mid p \in P,\ i \in \mathbb{Z} \,\}$ *is the set of simple types and* $T_{(P, \leq)} = \left(P^{(\mathbb{Z})} \right)^* = \{ p_1^{(i_1)} \cdots p_n^{(i_n)} \mid 0 \leq k \leq n,\ p_k \in P$ *and* $i_k \in \mathbb{Z} \}$ *is the set of types. The empty sequence in* $T_{(P, \leq)}$ *is denoted by* 1. *For* X *and* $Y \in T_{(P, \leq)}$, $X \leq Y$ *iff this relation is derivable in the following system where* $p, q \in P$, $n, k \in \mathbb{Z}$ *and* $X, Y, Z \in T_{(P, \leq)}$:

$$X \leq X \ \ (Id) \qquad\qquad \frac{X \leq Y \quad Y \leq Z}{X \leq Z} \ (Cut)$$

$$\frac{XY \leq Z}{Xp^{(n)}p^{(n+1)}Y \leq Z} \ (A_L) \qquad \frac{X \leq YZ}{X \leq Yp^{(n+1)}p^{(n)}Z} \ (A_R)$$

$$\frac{Xp^{(k)}Y \leq Z}{Xq^{(k)}Y \leq Z} \ (IND_L) \qquad \frac{X \leq Yq^{(k)}Z}{X \leq Yp^{(k)}Z} \ (IND_R)$$

(where $q \leq p$ *if* k *is even, and* $p \leq q$ *if* k *is odd)*

The construction, proposed by Buskowski [6], defines a pregroup that extends $\leq$ on basic types P to $T_{(P, \leq)}$ [3,4].

Cut Elimination. On the one hand, the cut rule is useful for clear and compact representation of derivations. On the other hand, it creates problems for derivation search because, due to this rule, one cannot in general bound the number of hypothetical premises needed in a derivation of a given inequality. Fortunately, this rule can be eliminated in pregroups without loss of generality, i.e. every derivable inequality has a cut-free derivation (see [5]) .

Definition 3 (Pregroup Grammar). *Let* $(P, \leq)$ *be a finite partially ordered set. A* pregroup grammar *based upon* $(P, \leq)$ *is a lexicalized* [5] *grammar* $G = (\Sigma, I, s)$ *on categories* $T_{(P, \leq)}$ *such that* $s \in P$. *G assigns a type* X *to a string* $v_1 \cdots v_n$ *of* Σ^* *iff for* $1 \leq i \leq n$, $\exists X_i \in I(v_i)$ *such that* $X_1 \cdots X_n \leq X$ *in the free pregroup* $T_{(P, \leq)}$. *The language* $\mathcal{L}(G)$ *is the set of strings in* Σ^* *that are assigned* s *by* G.

Example 1. Let us look at an analysis of a complete sentence from Marcel Proust (a part of it is shown in the introduction). The basic types used in this analysis are: $\pi_3, \overline{\pi}_3$: third person (subject) with $\pi_3 \leq \overline{\pi}_3$, p_2: past participle, ω: object, s: sentence, s_5: subjunctive clause, σ: complete subjunctive clause, τ: adverbial phrase.

---

[3] Left and right adjoints are defined by $(p^{(n)})^l = p^{(n-1)}$, $(p^{(n)})^r = p^{(n+1)}$, $(XY)^l = Y^l X^l$ and $(XY)^r = Y^r X^r$. p stands for $p^{(0)}$. The left and right adjoints of $X \in T_{(P, \leq)}$ are defined recursively: $X^{(0)} = X$, $X^{(n+1)} = (X^r)^{(n)}$ and $X^{(n-1)} = (X^l)^{(n)}$.

[4] $\leq$ is only a preorder. Thus, in fact, the pregroup is the quotient of $T_{(P, \leq)}$ under the equivalence relation $X \leq Y$ & $Y \leq X$.

[5] A lexicalized grammar is a triple (Σ, I, s): Σ is a finite alphabet, I assigns a finite set of categories (or types) to each $c \in \Sigma$, s is a category (or type) associated to correct strings.

| quand il | l' | avait | ramenée | chez-elle | il | fallait | qu' | il entrât |

$$\tau s^l \quad \pi_3 \quad \pi_3^r s \omega^{ll} s^l \pi_3 \quad \overline{\pi_3}^r s\, p_2^l \qquad p_2 \omega \lambda^l \qquad \lambda \qquad \pi_3 \quad \overline{\pi_3}^r \tau^r s\, \sigma^l \quad \sigma\, s_5^l \quad \pi_3 \quad \overline{\pi_3}^r s_5$$

Using only left rules (A_L) and (IND_L) and one (Id), we can prove that the product of the assigned types is less than or equal to s. The proof is schematically presented above. In this proof, each link corresponds to one application of (A_L) (eventually with a (IND_L) when the corresponding basic types are different).

3 Product of Pregroups

A natural idea to combine pregroups is to define a structure over the product of the corresponding monoids.

Definition 4 (Product of Pregroups). *For $N \geq 1$, let $P_i = (M_i, \leq_i, \circ_i, l_i, r_i, 1_i)$, $1 \leq i \leq N$, be N pregroups. We define $P_1 \times \cdots \times P_N$ as $(M_1 \times \cdots \times M_N, \leq , \circ, l, r, (1_1, \ldots, 1_N))$ where:*
- *$(x_1, \ldots, x_N) \leq (y_1, \ldots, y_N)$ iff $\forall i, 1 \leq i \leq N$, $x_i \leq_i y_i$,*
- *$(x_1, \ldots, x_N) \circ (y_1, \ldots, y_N) = (x_1 \circ_1 y_1, \ldots, x_N \circ_N y_N)$,*
- *$(x_1, \ldots, x_N)^l = (x_1^{l_1}, \ldots, x_N^{l_N})$ and $(x_1, \ldots, x_N)^r = (x_1^{r_1}, \ldots, x_N^{r_N})$.*

The product of several pregroups gives a structure that is also a pregroup[6].

3.1 Pregroup Product Grammars

Pregroup grammars are defined over a free pregroup. We relax this definition here and define grammars on any pregroup. In fact, we are interested only in the product of free pregroups.

Definition 5 (Pregroup Product Grammar). *Let $(P_1, \leq_1), \ldots, (P_N, \leq_N)$ be $N \geq 1$ finite partially ordered sets. A pregroup product grammar based upon $(P_1, \leq_1), \ldots, (P_N, \leq_N)$ is a lexicalized grammar $G = (\Sigma, I, s)$ on categories $T_{(P_1, \leq_1)} \times \cdots \times T_{(P_N, \leq_N)}$ such that $s \in P_1$. G assigns a type X to a string $v_1 \cdots v_n$ of Σ^* iff for $1 \leq i \leq n$, $\exists X_i \in I(v_i)$ such that $X_1 \circ \cdots \circ X_n \leq X$ in the product of the free pregroups $T_{(P_1, \leq_1)}, \ldots, T_{(P_N, \leq_N)}$ with $\circ$ as the binary operation of the product and $\leq$ as its partial order. The language $\mathcal{L}(G)$ is the set of strings in Σ^* that are assigned $(s, 1, \ldots, 1)$ by G.*

In the definition, when a string is assigned $(s, 1, \ldots, 1)$, the first component of the product must be less than or equal to s, a special basic type of the first free pregroup. The other components must be less than or equal to the unit of the corresponding free pregroup. It is possible to have a different definition, for instance by choosing that all components must be less than or equal to the unit

---

[6] The definition can also be extended to the empty product $(N = 0)$. In this case, the resulting structure is the monoid with a unique element which is also the unit element.

of this component and by adding a "wall" in Σ that is associated by the lexicon to the type $(s^r, 1, \ldots, 1)$.

3.2 Pregroup Product Grammars: Context Sensitive but NP Membership Problem

The membership problem of a string into the language associated to a pregroup grammar is polynomial in time either from the size of the string or from the size of the string plus the size of the pregroup grammar. In fact, the languages of pregroup grammars are the context-free languages [6,2].

For pregroup product grammars, the number $N \geq 1$ of free pregroups for the product is important. For $N = 1$, the product is equivalent to a free pregroup. Thus the same result can be proved on the membership problem and the expressive power (the membership problem is polynomial in time and the languages are context-free). This is completely different if $N > 1$. With regard to the expressive power, the article proves that for $k > 1$, $L_k = \{a_1^i a_2^i \cdots a_k^i \mid i \geq 1\}$ is generated by a pregroup product grammar based upon the product of $k - 1$ free pregroups.

Definition 6 (Pregroup Product Grammar for $\{a_1^i a_2^i \cdots a_k^i \mid i \geq 1\}$). *Let $P = (\{x_1, \ldots, x_k, z\}, =)$ a partially ordered set (the partial order on basic types is equality). We consider the product of the k free pregroups based upon k copies of P. Let $G_k = (\{a_1, \ldots, a_k\}, I_k, x_1)$ be the pregroup product grammar based upon the product and defined by the following lexicon:*

$$
\begin{aligned}
I_k(a_1) &= \{ (x_1 z x_1^l, 1, \ldots, 1), (x_1 z x_2^l, 1, \ldots, 1) \} \\
I_k(a_2) &= \{ (x_2 z^l x_2^l, z, 1, \ldots, 1), (x_2 z^l x_3^l, z, 1, \ldots, 1) \} \\
I_k(a_3) &= \{ (x_3 x_3^l, z^l, z, 1, \ldots, 1), (x_3 x_4^l, z^l, z, 1, \ldots, 1) \} \\
&\cdots \\
I_k(a_{k-1}) &= \{ (x_{k-1} x_{k-1}^l, 1, \ldots, 1, z^l, z), (x_{k-1} x_k^l, 1, \ldots, 1, z^l, z) \} \\
I_k(a_k) &= \{ (x_k x_k^l, 1, \ldots, 1, z), (x_k, \ldots, 1, z) \}
\end{aligned}
$$

Theorem 1. *For $k > 1$, $L_k = \{a_1^i a_2^i \cdots a_k^i \mid i \geq 1\} = \mathcal{L}(G_k)$.*

Proof. Firstly, it is easy to find a derivation in G_k corresponding to the string $A = a_1^i a_2^i \cdots a_k^i$ for $i > 0$: Using the lexicon I_k, we can associate the following expression to A:

$$\underbrace{(x_1 z^l x_1^l, 1, \ldots, 1) \circ \cdots \circ (x_1 z^l x_1^l, 1, \ldots, 1)}_{i-1} \circ (x_1 z^l x_2^l, 1, \ldots, 1) \circ$$

$$\underbrace{(x_2 z x_2^l, z^l, 1, \ldots, 1) \circ \cdots \circ (x_2 z x_2^l, z^l, 1, \ldots, 1)}_{i-1} \circ (x_2 z x_3^l, z^l, 1, \ldots, 1) \circ$$

$$\cdots$$

$$\underbrace{(x_{k-1} x_{k-1}^l, 1, \ldots, 1, z, z^l) \circ \cdots \circ (x_{k-1} x_{k-1}^l, 1, \ldots, 1, z, z^l)}_{i-1}$$

$$\circ (x_{k-1} x_k^l, 1, \ldots, 1, z, z^l) \circ$$

$$\underbrace{(x_k x_k^l, 1, \ldots, 1, z) \circ \cdots \circ (x_k x_k^l, 1, \ldots, 1, z)}_{i-1} \circ (x_k, 1, \ldots, 1, z)$$

For the first component:

$$\underbrace{(x_1z^lx_1^l)\cdots(x_1z^lx_1^l)}_{i-1}x_1z^lx_2^l\underbrace{(x_2zx_2^l)\cdots(x_2zx_2^l)}_{i-1}x_2zx_3^l$$

$$\underbrace{(x_3x_3^l)\cdots(x_3x_3^l)}_{i-1}x_3x_4^l\cdots\underbrace{(x_kx_k^l)\cdots(x_kx_k^l)}_{i-1}x_k\leq x_1$$

For the other components:

$$1\cdots1\underbrace{z^l\cdots z^l}_{i}\underbrace{z\cdots z}_{i}1\cdots1\leq1$$

Therefore $L_k\subseteq\mathcal{L}(G_k)$.

For the other direction, we prove that if $A\in\mathcal{L}(G_k)$ then for $1\leq i\leq k-1$, every occurrence of a_i in A must be before any occurrence of a_{i+1} and the number of a_i is the same as the number of a_{i+1}. The first property is given by the basic types $x_1,\ldots x_k$ of the first component of the derivation of A in G_k. Couples of x_i^l and x_i form a list from left to right leaving only one basic type x_1. For the second property, the number of a_i is the same as the number of a_{i+1}, because in the i-th component, basic type z is given by a_i as z^l and by a_{i+1} as z (each z on the right corresponds exactly to one z^l on the left).

In the Chomsky hierarchy, a pregroup product grammar can be simulated by a context-sensitive grammar using contextual rules. Intuitively, in the context-sensitive grammar, some contextual rules play the role of the free pregroup left rules (A_L) and (IND_L) of Definition 2. A second set of contextual rules performs local "permutations" of simple types that are not in the same component: A simple type in the i-th component permutes with a simple type in the j-th component if the first one is before the second one and if $i>j$.

In fact, the membership problem is clearly a NP problem because if we want to check that a string is in the language associated to a pregroup product grammar where the product has N components, we only have to produce an assignment for each symbol and prove that the N concatenations of each component of the types are less than or equal to s or 1 which are N polynomial problems.

The conclusion of this remark is that the languages of pregroup product grammars are contextual but most probably several context-sensitive language are not generated by a pregroup product grammar (the membership problem of the context-sensitive languages is PSPACE-complete). The next section proves that the membership problem is also NP-hard. Thus pregroup product grammars are not mildly context-sensitive [10].

4   Pregroup Product Grammars: NP-Hard

The section presents the main result of the paper: The membership problem for a particular pregroup product grammar is NP-hard. The proof is based upon an encoding of any SAT problem. The grammar is based upon the product of 3 free pregroups. As a consequence, the membership problem of pregroup product grammars is NP-complete at least for pregroup product grammars built with at least 3 free pregroups.

The proof uses the product of three copies of the free pregroup on $P_{SAT} = \{t, f\}$ with equality as the partial order on basic types. The set of elements of the pregroup is $T_{SAT} = T_{(P_{SAT},=)} \times T_{(P_{SAT},=)} \times T_{(P_{SAT},=)}$. The first component corresponds to the encoding of the formula that we want to satisfy. The two other components are used to propagate the boolean values of variables.

The formula is transformed into a string and it can be satisfied iff the string is included in the language generated by a fixed pregroup product grammar $\mathcal{G}_{SAT}$ based upon T_{SAT}.

Definition 7 (Formula Transformation $\mathcal{T}_n(F)$). *A boolean formula F that contains (at most) n variables $v_1, \ldots, v_n$, operators $\wedge$ (binary conjunction), $\vee$ (binary disjunction) and $\neg$ (negation) is transformed into a string $\mathcal{T}_n(F) \in \{a, b, c, d, e, \wedge, \vee, \neg\}^*$. $\mathcal{T}_n(F)$ and $[F]_n$ are defined as follows:*

- $\mathcal{T}_n(F) = \underbrace{a \cdots a}_{n}[F]_n \underbrace{e \cdots e}_{n}$
- $[v_i]_n = \underbrace{b \cdots b}_{i-1} c \underbrace{b \cdots b}_{n-i} \underbrace{d \cdots d}_{n}$
- $[F_1 \vee F_2]_n = \vee [F_1]_n [F_2]_n$
- $[F_1 \wedge F_2]_n = \wedge [F_1]_n [F_2]_n$
- $[\neg F_1]_n = \neg [F_1]_n$

Example 2. A boolean formula is transformed into a string using the prefix notation for operators. The transformations of $v_1 \wedge v_1$ and $v_1 \vee (v_1 \wedge v_2)$ are:

$$\mathcal{T}_1(v_1 \wedge v_1) \qquad = a \wedge \underbrace{cd}_{\text{for } v_1} \underbrace{cd}_{\text{for } v_1} e$$

$$\mathcal{T}_2(v_1 \vee (v_1 \wedge v_2)) = aa \vee \underbrace{cbdd}_{\text{for } v_1} \wedge \underbrace{cbdd}_{\text{for } v_1} \underbrace{bcdd}_{\text{for } v_2} ee$$

Definition 8 (Pregroup Product Grammar $\mathcal{G}_{SAT}$). *The pregroup product grammar $\mathcal{G}_{SAT} = (\{a, b, c, d, e, \wedge, \vee, \neg\}, I_{SAT}, t)$, based upon the product of three copies of the free pregroup on $(P_{SAT}, =)$ where $P_{SAT} = \{t, f\}$, is defined by the following lexicon:*

$$
\begin{aligned}
I_{SAT}(a) &= \{ (1, t^l, 1), (1, f^l, 1) \} \\
I_{SAT}(b) &= \{ (1, t, t^l), (1, f, f^l) \} \\
I_{SAT}(c) &= \{ (t, t, t^l), (f, f, f^l) \} \\
I_{SAT}(d) &= \{ (1, t^l, t), (1, f^l, f) \} \\
I_{SAT}(e) &= \{ (1, t, 1), (1, f, 1) \} \\
I_{SAT}(\wedge) &= \{ (tt^l t^l, 1, 1), (ff^l t^l, 1, 1), (ft^l f^l, 1, 1), (ff^l f^l, 1, 1) \} \\
I_{SAT}(\vee) &= \{ (tt^l t^l, 1, 1), (tf^l t^l, 1, 1), (tt^l f^l, 1, 1), (ff^l f^l, 1, 1) \} \\
I_{SAT}(\neg) &= \{ (tf^l, 1, 1), (ft^l, 1, 1) \}
\end{aligned}
$$

We write $\leq_{T_{(P_{SAT},=)}}$ for the partial order of the free pregroup on $(P_{SAT}, =)$ and $\leq_{SAT}$ for the partial order of the product of the three free pregroups based on $(P_{SAT}, =)$. The types assigned to the strings of $\mathcal{L}(\mathcal{G}_{SAT})$ are $\leq_{SAT} (t, 1, 1)$

Example 3. The formula $v_1 \wedge v_1$ can be satisfied for $v_1 = true$. There exists a type assignment of the symbols of $\mathcal{T}_1(v_1 \wedge v_1) = a \wedge cd\ cd\ e$ by $\mathcal{G}_{SAT}$ that is $\leq_{SAT} (t, 1, 1)$:

$$\underbrace{(1, t^l, 1)}_{for\ a} \circ \underbrace{(tt^l t^l, 1, 1)}_{for\ \wedge} \circ \underbrace{(t, t, t^l)}_{for\ c} \circ \underbrace{(1, t^l, t)}_{for\ d} \circ \underbrace{(t, t, t^l)}_{for\ c} \circ \underbrace{(1, t^l, t)}_{for\ d} \circ \underbrace{(1, t, 1)}_{for\ e}$$

$$\leq_{SAT} (t, 1, 1)$$

The formula $v_1 \wedge \neg v_2$ can be satisfied for $v_1 = true$ and $v_2 = false$. There exists a type assignment of the symbols of $\mathcal{T}_2(v_1 \wedge \neg v_2) = aa \wedge cbdd \neg bcdd\ ee$ by $\mathcal{G}_{SAT}$ that is $\leq_{SAT} (t, 1, 1)$:

$$\underbrace{(1, f^l, 1)}_{for\ a} \circ \underbrace{(1, t^l, 1)}_{for\ a} \circ \underbrace{(tt^l t^l, 1, 1)}_{for\ \wedge} \circ \underbrace{(t, t, t^l)}_{for\ c} \circ \underbrace{(1, f, f^l)}_{for\ b} \circ \underbrace{(1, f^l, f)}_{for\ d} \circ \underbrace{(1, t^l, t)}_{for\ d} \circ$$

$$\underbrace{(tf^l, 1, 1)}_{for\ \neg} \circ \underbrace{(1, t, t^l)}_{for\ b} \circ \underbrace{(f, f, f^l)}_{for\ c} \circ \underbrace{(1, f^l, f)}_{for\ d} \circ \underbrace{(1, t^l, t)}_{for\ d} \circ \underbrace{(1, t, 1)}_{for\ e} \circ \underbrace{(1, f, 1)}_{for\ e}$$

$$\leq_{SAT} (t, 1, 1)$$

Theorem 2. *A boolean formula F that contains (at most) n variables $v_1, \ldots, v_n$, operators $\wedge$ (binary conjunction), $\vee$ (binary disjunction) and $\neg$ (negation) can be satisfied iff $\mathcal{T}_n(F) \in \mathcal{L}(\mathcal{G}_{SAT})$*

Example 4. Example 3 shows two formulas that can be satisfied. Their transformations using $\mathcal{T}_n$ are in $\mathcal{L}(\mathcal{G}_{SAT})$. The formula $v_1 \wedge \neg v_1$ cannot be satisfied. A type assignment of $\mathcal{T}_1(v_1 \wedge \neg v_1) = a \wedge cd \neg cd\ e$ by $\mathcal{G}_{SAT}$ would produce the following type where for $1 \leq i \leq 11$, $x_i \in \{t, f\}$, $x_2 = x_3 \wedge x_4$ and $x_7 = \neg x_8$ (both equalities come from entries of $\wedge$ and $\neg$ of the lexicon I_{SAT} – we identify here true with t and false with f):

$$\underbrace{(1, x_1^l, 1)}_{a} \circ \underbrace{(x_2 x_3^l x_4^l, 1, 1)}_{\wedge} \circ \underbrace{(x_5, x_5, x_5^l)}_{c} \circ \underbrace{(1, x_6^l, x_6)}_{d} \circ$$

$$\underbrace{(x_7 x_8^l, 1, 1)}_{\neg} \circ \underbrace{(x_9, x_9, x_9^l)}_{c} \circ \underbrace{(1, x_{10}^l, x_{10})}_{d} \circ \underbrace{(1, x_{11}, 1)}_{e}$$

The type must be $\leq_{SAT} (t, 1, 1)$. Therefore, $x_2 x_3^l x_4^l x_5 x_7 x_8^l x_9 \leq_{T_{(P_{SAT}, =)}} t$, $x_1^l x_5 x_6^l x_9 x_{10}^l x_{11} \leq_{T_{(P_{SAT}, =)}} 1$ and $x_5^l x_6 x_9^l x_{10} \leq_{T_{(P_{SAT}, =)}} 1$. As a consequence, $x_2 = t$, $x_3 = x_7$, $x_4 = x_5$, $x_8 = x_9$, $x_1 = x_5$, $x_6 = x_9$, $x_{10} = x_{11}$, $x_5 = x_6$ and $x_9 = x_{10}$. There is no solution to all these equations: The transformation of the formula $v_1 \wedge \neg v_1$ through $\mathcal{T}_1$ is not in $\mathcal{L}(\mathcal{G}_{SAT})$.

Proof. Firstly, we prove that if a formula F on variables $v_1, \ldots, v_n$ can be satisfied, then $\mathcal{T}_n(F)$ is in $\mathcal{L}(\mathcal{G}_{SAT})$. Let $(x_1, \ldots, x_n) \in \{true, false\}^n$ be an assignment of variables $v_1, \ldots, v_n$ that satisfies F. Using the assignment, the occurrences of the variables and the occurrences of the operators of F can be annotated by boolean values that correspond to the value of the variable or the output value of the operator plus the input value for $\neg$ or both input values for $\vee$ and $\wedge$. Of course, the boolean values associated to an operator follow the truth table of the corresponding boolean operator. Now, we can assign a type in I_{SAT} to each symbol of $\mathcal{T}_n(F)$:

- The assignment of the i-th a in $\mathcal{T}_n(F) = \underbrace{a \cdots a}_{n}[F]\underbrace{e \cdots e}_{n}$ corresponds to the

  value x_{n+1-i} of the $(n+1-i)$-th boolean variable v_{n+1-i}. If x_{n+1-i} is true, the occurrence is assigned to $(1, t^l, 1)$, otherwise, it is assigned to $(1, f^l, 1)$.
- The assignment of the i-th e in $\mathcal{T}_n(F) = \underbrace{a \cdots a}_{n}[F]\underbrace{e \cdots e}_{n}$ corresponds to the

  value x_i of the i-th variable. If x_i is true, the occurrence is assigned to $(1, t, 1)$ otherwise to $(1, f, 1)$
- The i-th b or c in $[v_j]_n = \underbrace{b \cdots b}_{j-1} c \underbrace{b \cdots b}_{n-j} d \underbrace{\cdots d}_{n}$ corresponds to the value x_i of

  the i-th variable. If $i = j$, we have c. Then, if x_i is true, the occurrence is assigned to (t, t, t^l) otherwise to (f, f, f^l). If $i \neq j$, we have b. If x_i is true, the occurrence is assigned to $(1, t, t^l)$ otherwise to $(1, f, f^l)$.
- The i-th d in $[v_j]_n = \underbrace{b \cdots b}_{j-1} c \underbrace{b \cdots b}_{n-j} d \underbrace{\cdots d}_{n}$ corresponds to the value x_{n+1-i} of

  the $(n+1-i)$-th boolean variable v_{n+1-i}. If x_{n+1-i} is true, the occurrence is assigned to $(1, t^l, t)$, otherwise, it is assigned to $(1, f^l, f)$.
- For $\neg$ in $[\neg F_1]_n = \neg[F_1]_n$, the assignment of variables $v_1, \ldots, v_n$ that satisfies F induces a boolean value to the sub-formula F_1 that is either true or false. The output value of $\neg F_1$ is the opposite value (false for true and true for false). Thus, $\neg$ is assigned to $(t f^l, 1, 1)$ if the input is $false$ and the output is $true$ or to $(f t^l, 1, 1)$ if the input is true and the output is false.
- For $\wedge$ in $[F_1 \wedge F_2]_n = \wedge[F_1]_n[F_2]_n$, the assignment of variables $v_1, \ldots, v_n$ that satisfies F, induces a boolean value to each sub-formula F_1 and F_2. The output follows the truth table of the logical "and" operator. Following the input values, the assignment of $\wedge$ is given by the following table (the values of the inputs are reverse in the type because they appeared as left adjoints t^l or f^l):

$F_1(x_1, \ldots, x_n)$	$F_2(x_1, \ldots, x_n)$	$\wedge$
$true$	$true$	$(t t^l t^l, 1, 1)$
$true$	$false$	$(f f^l t^l, 1, 1)$
$false$	$true$	$(f t^l f^l, 1, 1)$
$false$	$false$	$(f f^l f^l, 1, 1)$

- $\vee$ is very similar to $\wedge$ except that we follow the truth table of the logical "or" operator:

$$\frac{F_1(x_1,\ldots,x_n)\ F_2(x_1,\ldots,x_n)\qquad\qquad\vee}{}$$

$true$	$true$	$(tt^lt^l,1,1)$
$true$	$false$	$(tf^lt^l,1,1)$
$false$	$true$	$(tt^lf^l,1,1)$
$false$	$false$	$(ff^lf^l,1,1)$

Now, we create three derivations (one for each component) that prove that the type assignment of $\mathcal{T}_n(F)$ (with the values $x_1,\ldots,x_n$ for the boolean variables $v_1,\ldots,v_n$) is $\leq_{SAT}$ $(t,1,1)$. The first component starts with $s \leq_{T_{(P_{SAT},=)}} t$, the other components with $1 \leq_{T_{(P_{SAT},=)}} 1$. The applications of (A_L) on the first component follow the syntactic tree of F written with the prefix notation for binary operators $\wedge$ and $\vee$. For this component, only the assignments of c, $\neg$, $\wedge$ and $\vee$ are important (the other symbols are assigned to 1 in the first component). The application of rule (A_L) between an occurrence of f^l (on the left) and an occurrence of f (on the right) corresponds to the link between the output of a variable or an operator that is false and one of the inputs of an operator. Similarly the application of rule (A_L) between an occurrence of t^l (on the left) and an occurrence of t (on the right) corresponds to the propagation of the true value. The basic type t that remains at the end is the value of the main operator or variable. It is t because F is true in this case. The two other components are used to synchronize the value given to each occurrence of the variables $v_1,\ldots,v_n$ (each c in $\mathcal{T}_n(F)$). For each occurrence of v_i, this is done on the complete vector of variables $v_1,\ldots,v_n$ but only one of the values (the value that corresponds to v_i) is copied into the first component. If we write $\overline{true}=t$ and $\overline{false}=f$ and if we only look at the second and third components, we have, for $[v_i]_n$, the type $\overline{x_1}\cdots\overline{x_n}\,\overline{x_n}^l\cdots\overline{x_1}^l$ for the second component and the type $\overline{x_1}^l\cdots\overline{x_n}^l\overline{x_n}\cdots\overline{x_1}$ for the third component. The n occurrences of a in $\mathcal{T}_n(F)$ give the type $\overline{x_n}^l\cdots\overline{x_1}^l$ for the second component and 1 for the third. The n occurrences of e give the type $\overline{x_1}\cdots\overline{x_n}$ for the second component and 1 for the third. Obviously, if we write $X = \overline{x_1}\cdots\overline{x_n}$, the global type of the second component is $\underbrace{X}_{\text{for }a}\underbrace{X^lX}_{\text{for }v_{i_1}}\cdots\underbrace{X^lX}_{\text{for }v_{i_m}}\underbrace{X^l}_{\text{for }e}$ which is $\leq_{T_{(P_{SAT},=)}}$ 1. For the third component, each variable corresponds to X^lX, which is $\leq_{T_{(P_{SAT},=)}}$ 1. Thus, the type assigned to $\mathcal{T}_n(F)$ using $x_1,\ldots,x_n$ for $v_1,\ldots,v_n$ is $\leq_{SAT}$ $(t,1,1)$ and $\mathcal{T}_n(F) \in \mathcal{L}(\mathcal{G}_{SAT})$.

The reverse inclusion proves that if F is a boolean function with n variables $v_1,\ldots,v_n$ and if $\mathcal{T}_n(F) \in \mathcal{L}(\mathcal{G}_{SAT})$ then F can be satisfied. The derivations of the three components that prove that a type assignment of $\mathcal{T}_n(F)$ by I_{SAT} is $\leq_{SAT}$ $(t,1,1)$ only use rule (A_L). The other rules (except for the applications of (Id) giving $t \leq_{T_{(P_{SAT},=)}} t$ for the first component and giving $1 \leq_{T_{(P_{SAT},=)}} 1$ for the second and third components and the cut rule) are never used in the system because the right part of the inequalities is either a basic type t for the first component or the unit for the other components and because the partial order on the simple types of the free pregroup is equality. Moreover, $\mathcal{G}_{SAT}$ only uses the four simple types t, t^l, f, and f^l and an assignment of each symbol of $\mathcal{T}_n(F)$ gives always the same formula when basic types t and f are identified. Thus, there exists at most one class of equivalent derivations of any type assignment of $\mathcal{T}_n(F)$ if we look at the set of applications of rule (A_L) in the three components of the type assignment. In a derivation, each application of (A_L) corresponds to an "axiom" between one t^l on the left and one t on the right or between one f^l on the left and one f on the right (as it is shown in Example 1) and all the "axioms" form a projective structure (like the couples of corresponding parentheses in a string of the Dyck language). The class of equivalent derivations (some applications of (A_L) can commute) must correspond to the construction shown above: The first component corresponds to the applications of rule (A_L) that propagate the output of variables and operators to the inputs of the corresponding operator in F. The remaining basic type of the first component (f or t) is the output of F. The second and the third components synchronize the variables in such a way that all the occurrences of the same variable have the same value. Now, if $\mathcal{T}_n(F) \in \mathcal{L}(\mathcal{G}_{SAT})$, the type assignment of the symbols of $\mathcal{T}_n(F)$ is such that the variable v_i has the value corresponding to the second component of the type assignment of the i-th e of $\mathcal{T}_n(F)$: if it is $(1, t, 1)$, v_i is set to true, if it is $(1, f, 1)$ v_i is set to false. For this set of values, the first component of the assignment of $\mathcal{T}_n(F)$ is $\leq_{T_{(P_{SAT},=)}} t$. This means that the value of F is true when the variables $v_1, \ldots, v_n$ are set to the values above. Thus F can be satisfied.

Of course because the membership problem in $\mathcal{L}(\mathcal{G}_{SAT})$ is a NP problem, this problem is NP-complete. As a consequence, the membership problem of $\mathcal{L}(G)$ when G is a pregroup product grammar is also NP-complete. The problem is still open for pregroup product grammar based of two free pregroups but this problem is most probably NP-complete.

5   Conclusion

The article introduces pregroup product grammars, grammars based of the product of free pregroups. It is shown that the class of languages is very expressive. For instance, $\{x_1^i \cdots x_N^i \mid i \geq 1\}$ for any $N \geq 1$ can be generated. However, the membership problem is NP-complete. Thus even if they are much more expressive, pregroup product grammars are less interesting than pregroup grammars with respect to the complexity of the membership problem.

References

1. Bargelli, D., Lambek, J.: An algebraic approach to french sentence structure. In: de Groote, P., Morrill, G., Retoré, C. (eds.) LACL 2001. LNCS (LNAI), vol. 2099, pp. 62–78. Springer, Heidelberg (2001)

2. Béchet, D.: Parsing pregroup grammars and Lambek calculus using partial composition. Studia Logica 87(2/3) (2007)
3. Béchet, D., Dikovsky, A., Foret, A., Garel, E.: Introduction of option and iteration into pregroup grammars. In: Casadio, C., Lambek, J. (eds.) Computational Algebraic Approaches to Natural Language, pp. 85–107. Polimetrica, Monza, Milan (2008), http://www.polimetrica.com
4. Béchet, D., Dikovsky, A., Foret, A., Garel, E.: Optional and iterated types for pregroup grammars. In: Martín-Vide, C., Otto, F., Fernau, H. (eds.) LATA 2008. LNCS, vol. 5196, pp. 88–100. Springer, Heidelberg (2008), http://grammars.grlmc.com/LATA2008
5. Buszkowski, W.: Cut elimination for the lambek calculus of adjoints. In: Abrusci, V., Casadio, C. (eds.) New Perspectives in Logic and Formal Linguisitics, Proceedings Vth ROMA Workshop. Bulzoni Editore (2001)
6. Buszkowski, W.: Lambek grammars based on pregroups. In: de Groote, P., Morrill, G., Retoré, C. (eds.) LACL 2001. LNCS (LNAI), vol. 2099, pp. 95–109. Springer, Heidelberg (2001)
7. Cardinal, K.: An algebraic study of Japanese grammar. Master's thesis, McGill University, Montreal (2002)
8. Casadio, C., Lambek, J.: An algebraic analysis of clitic pronouns in italian. In: de Groote, P., Morrill, G., Retoré, C. (eds.) LACL 2001. LNCS (LNAI), vol. 2099, pp. 110–124. Springer, Heidelberg (2001)
9. Dekhtyar, M., Dikovsky, A.: Categorial dependency grammars. In: Moortgat, M., Prince, V. (eds.) Proc. of Intern. Conf. on Categorial Grammars, pp. 76–91. Montpellier (2004)
10. Joshi, A., Vijay-Shanker, K., Weir, D.: The convergence of mildly context-sensitive grammar formalisms. In: Sells, P., Schieber, S., Wasow, T. (eds.) Fundational Issues in Natural Language Processing. MIT Press (1991)
11. Kiślak-Malinowska, A.: Extended pregroup grammars applied to natural languages. Logic and Logical Philosophy 21(3), 229–252 (2012)
12. Kobele, G.M.: Pregroups, products, and generative power. In: Proceedings of the Workshop on Pregroups and Linear Logic 2005, Chieti, Italy (May 2005)
13. Kobele, G.M.: Agreement bottlenecks in Italian. In: Casadio, C., Lambek, J. (eds.) Computational Algebraic Approaches to Natural Language, pp. 191–212. Polimetrica, Monza, Milan (2008), http://www.polimetrica.com
14. Kobele, G.M., Kracht, M.: On pregroups, freedom, and (virtual) conceptual necessity. In: Eilam, A., Scheffler, T., Tauberer, J. (eds.) Proceedings of the 29th Pennsylvania Linguistics Colloquium, vol. 12(1), pp. 189–198. University of Pennsylvania Working Papers in Linguistics (2006)
15. Lambek, J.: Type grammars revisited. In: Lecomte, A., Perrier, G., Lamarche, F. (eds.) LACL 1997. LNCS (LNAI), vol. 1582, pp. 1–27. Springer, Heidelberg (1999)
16. Lambek, J.: Type grammar meets german word order. Theoretical Linguistics 26, 19–30 (2000)
17. Lambek, J., Preller, A.: An algebraic approach to the german noun phrase. Linguistic Analysis 31, 3–4 (2003)
18. Lambek, J.: The mathematics of sentence structure. American Mathematical Monthly 65, 154–170 (1958)
19. Sadrzadeh, M.: Pregroup analysis of persian sentences (2007), http://eprints.ecs.soton.ac.uk/13970/

Distributional Semantics: A Montagovian View

Raffaella Bernardi

DISI, University of Trento

Abstract. This paper describes the current status of research in Distributional Semantics looking at the results from the Montagovian tradition stand point. It considers the main aspects of the Montagovian view as binoculars to observe those results, in particular: compositionality, syntax-semantics interface, logical words and entailment. To this end, it reviews some work that aims to tackle those issues within the Distributional Semantics Models and tries to highlight some open questions formal and distributional semanticists could address together.

Credits: Some of the material in the background section is based on distributional semantics talks by Marco Baroni, Stefan Evert, Alessandro Lenci and Roberto Zamparelli.

1 Introduction

This paper is not a research paper, no new results are reported. Its aim is to bridge two research communities working on related questions using different but compatible methods in order to profit of each other results. The main question they share is how we can formally capture natural language semantics. In other words, how can a computer processes linguistic expressions like *"Two men play a game"*, *"Some people play chess"* and *"Some people play music"* and realize that the second sentence is semantically similar to the first, but not to the last one – e.g. the first two sentences can be the descriptions of the same image whereas the last one describes a different event, even though it shares several words with the other sentences. To answer this question, formal semanticists employ a logic framework and exploit its reasoning apparatus, whereas distributional semanticists look at how natural language is used by inducing statistical based representations and exploiting vector semantic space tools. Of course, none of the two communities has reached a final answer, but both have discovered interesting aspects of natural language that can possibly converge within an integrated enterprise. To reach our aim, we will first briefly introduce the core concepts at the heart of the two approaches (Section 2) and then look at distributional semantics with the eyes of formal semanticists (Section 3).

2 Background

In this section, we describe our standing point by briefly introducing the core concepts of Logic and its application to natural language analysis. We will then look at Distributional Semantics from these formal semantics pillars.

C. Casadio et al. (Eds.): Lambek Festschrift, LNCS 8222, pp. 63–89, 2014.

2.1   From Logic to Language: The Montagovian Pillars

In Logic, the interpretation of a complex formula depends on the interpretation of the parts and of the logical operators connecting them (*compositionality*). The interpretation of the logical operators determines whether from a set of propositions a given proposition follows: $\{\psi_1, \ldots \psi_n\} \models \phi$. The entailment $\models$ is said to be *satisfiable* when there is at least one interpretation for which the premises and the conclusion are true; *falsifiable* when there is at least one interpretation for which the premises are true and the conclusion is false; and *valid* when the set of interpretations for which the premises are true is included in the set of interpretations for which the conclusion is true (*logical entailment.*) These two aspects have been used to formalize natural language meaning too. The starting point has been Frege's solution to the following puzzle: There is the star a called "venus", "morning star" and "evening star" that are represented in First Order Logic (FOL) by venus′, morningst′, eveningst′: $[\![$venus′$]\!] = a$, $[\![$morningst′$]\!] = a$ and $[\![$eveningst′$]\!] = a$. a is the meaning (*reference*) of these linguistic signs. Checking whether it is true that (i) "*the morning star is the morning star*" or that (ii) "*the morning star is the evening star*" ends up checking that (i) $[\![$morningst′$]\!] = [\![$morningst′$]\!]$ and (ii) $[\![$morningst′$]\!] = [\![$eveningst′$]\!]$, both of which reduce to checking $a = a$. But checking whether (i) "*the morning star is the morning star*" or that (ii) "*the morning star is the evening star*" cannot amount to the same operation since (ii) is cognitively more difficult than (i). Frege solved this puzzle by claiming that a linguistic sign consists of a *Bedeutung* (reference), the object that the expression refers to, and a *Sinn* (sense), mode of presentation of the reference. Moreover, he claimed that natural language meaning can be represented by a logical language.

Following Frege, formal semanticists' aim has been to obtain FOL representations of natural language expressions compositionaly. A crucial contribution to this research line has come from Montague [36], hence we can refer to the general framework as the Montagovian view. Formal semanticists have wondered what the meaning representation of the lexical words is, and which operation(s) put the lexical meaning representation together. The most largely shared view takes syntax to drive the order of composition. In particular, to assemble the syntactic structure Montague employed Categorial Grammar (CG) in which syntactic categories are seen as functions – $A \backslash B$ (or B/A), a function that wants an argument A on the left (resp., on the right) to return an expression of category B – and their composition as function application. The Categorial Grammar view has been further elaborated into a Logical Grammar by [30], the general framework is known as Type Logical Grammar [37,38]. In it the connection between syntax and semantics has been tied up at both lexical and grammatical level as we will better see in the sequel. In brief, the core components of the Montagovian framework are:

Compositionality. The meaning representation of a phrase depends on the meaning representation of its parts and the way they are put together.

Syntax-semantics Interface. The meaning representation assembly is guided by the derivational structure and a tight connection must be established between domains of interpretation and syntactic categories. This connection can be captured by defining a mapping between semantic types and syntactic categories.

Logical words and Entailment. Entailment between phrases consisting only of content words is model dependent (it corresponds to satisfiability), entailment between phrases consisting also of logical (grammatical) words is model independent (it corresponds to validity.)

In the following, first we introduce the general background at the heart of Distributional Semantics Models (DSMs), then we zoom into those models that account for compositionality in the light of the main issues summarized above

2.2   Distributional Semantics Models

As with any framework, in order to understand and appreciate the results achieved, the main research questions of the people working on the framework should be clear. For DSMs we can say the key questions have been the following ones: 1.) What is the sense of a given *word?*; 2.) how can the sense be induced and represented? and 3.) how do we relate word senses (synonyms, antonyms, hyperonym etc.)?[1] Well established answers supported by several evaluations are 1.) The sense of a word can be given by its use, viz. by the *contexts* in which it occurs; 2.) it can be induced from (either raw or parsed) corpora and can be represented by *vectors* (viz., tensors of order one); 3.) vector *cosine similarity* captures synonyms (as well as other semantic relations).

Today DSMs found their inspiration in ideas of the Fifties: First of all, [50] claims that word usage can reveal semantics flavor; [25] observed that words that occur in similar (linguistic) context tend to have similar meanings, [47] looked at the applied side of these ideas by considering co-occurrence frequency of the context words near a given target word to be important for word sense disambiguation in machine translation tasks; and the famous slogan of the framework "you shall know a word by the company it keeps" is due to [17]. Finally, [15] put these intuitions at work. To easily capture the main intuition behind Firth's slogan, we can consider the example by [33] who show how everyone can get the meaning of a made-up word like *wampimuk* by looking at the contexts in which it is used, for instance *"He filled the* wampimuk *with the substance, passed it around and we all drunk some"* and *"We found a little, hairy* wampimuk *sleeping behind the tree"* would suggest that *wampimuk* is a liquid in the first case and an animate thing in the second. Based on these kinds of observations, people have developed formal DSMs, implemented and evaluated them on several semantic tasks.

---

[1] The use of "sense" (as in Frege terminology) is not standard and may found opponents, but we believe it's useful to highlight the different perspective natural language is looked at within distributional and formal semantics models.

Definition. A *Distributional Semantics Model* is a quadruple $\langle B, A, V, S \rangle$, where: B is the set of "basis elements" – the dimensions of the space; A is a lexical association function that assigns co-occurrence frequency of target words to the dimensions; V is an optional transformation that reduces the dimensionality of the semantic space; and S is a similarity measure. The results of the model can be depicted for instance by the picture below taken from [35].

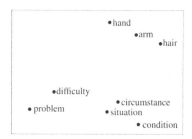

Toy example. To better understand the main points, let us take as toy example vectors in a 2 dimensional space, such that $B = \{shadow, shine\}$; $A=$ co-occurency frequency; and S the Euclidean distance. Let's take as target words: *moon, sun,* and *dog* and consider how often they co-occur with the basis elements:

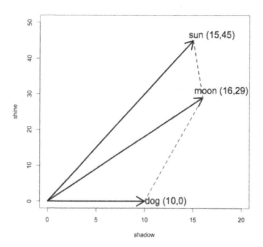

The Euclidean distance shows that *sun* is "closer" to *moon* than to *dog*. The two dimensional space representation give $\overrightarrow{moon}$=(16,29), $\overrightarrow{sun}$= (15,45), $\overrightarrow{dog}$=(10,0) that live together in a space representation (a matrix, dimensions × target-words):

$$\begin{bmatrix} 16 & 15 & 10 \\ 29 & 45 & 0 \end{bmatrix}$$

The most commonly used representation is the transpose matrix: target-words × dimensions:

	shine	shadow
$\overrightarrow{moon}$	16	29
$\overrightarrow{sun}$	15	45
$\overrightarrow{dog}$	10	0

The dimensions are also called "features" or "contexts".

Standard DSMs. In standard DSMs, words are taken to be all in the same space; the space dimensions are the *most k frequent words*, minus the "stop-words", viz. high-frequency words with relatively low information content, such as grammatical words (e.g. *of, the, and, them, . . .*). Hence, they may be around 2k-30K or even more; and they can be plain words, words with their part of speech (PoS), words with their syntactic relation. Hence, a text needs to be: tokenized, normalized (e.g., capitalization and stemming), annotated with PoS tags (N, J, etc.), and if required also parsed (to extract the syntactic relations). Instead of plain counts, the values can be more significant weights of the co-occurrence frequency: *tf-idf* (term frequency (tf) × inverse document frequency (idf)): an element gets a high weight when the corresponding term is frequent in the corresponding document (tf is high), but the term is rare in other documents of the corpus (df is low, idf is high.) [27]; or *PMI* (pointwise mutual information): measure how often two events x and y occur, compared with what we would expect if they were independent [7]. Finally, the many dimensions can be reduced so to obtain a matrix of a lower dimensionality (a matrix with less – linearly independent – dimensions) by either Singular Value Decomposition (SVD) that generalizes over sparser surface dimension by capturing "latent dimensions" or Random Indexing that improves efficiency by avoiding constructing too large matrices when not necessary.

Observation I. We can say that formal semanticists focused on "meaning" as *Bedeutung*, whereas distributional semanticists studied "meaning" as *Sinn*.

Observation II. We can say that formal semanticists have been interested in entailment as validity (entailment driven by logical words), whereas distributional semanticists have looked more to entailment as satisfiability (ISA-relation between content words).

2.3   Distributionality from the Montagavian View

Below we review some state-of-the art approaches to DSM beyond lexical meaning in the light of the Montagavian pillars and the type-logical view that has been developed based on it. In particular, we will briefly look at [35,12,43] and more in depth at [3]. The reader interested in a complete overview of DSMs is referred to [45].

Compositionality. All the work under consideration tackle the issue of compositionality and adopt the assumption that the meaning of the whole depends on the meaning representation of its parts. [35] take all constituents to be represented by vectors which combine together to produce a new vector and investigate possible vector composition operations; they focus their evaluation of such operations by looking at small phrases consisting of a head and a modifier or complement, and consider the class of additive and multiplicative operations to carry out the composition. As we will see below, the others instead use also tensors of order higher than vectors for capturing how a word can act on the word it composes with.

Syntax-Semantics Interface. The importance of taking the relation between the semantic composition and the syntactic structure into account is also discussed in all the work, though it is implemented in different ways and the strength of the connection varies from the very soft relation presented in [35] and [43] to the very tight one considered in [12,3].

[35] take syntax into account at least theoretically by presenting a flexible framework that covers under its umbrella several proposals. They consider the possibility of making the composition operation to depend on the syntactic relation. Formally, they define the result of the composition to be $\mathbf{p} = f(\mathbf{u}, \mathbf{v}, R, K)$ where R and K stand for the syntactic relation and the background knowledge, respectively. However, to simplify the implementation of the model, in practice they ignore K as well as the variety of function compositions based on the different syntactic relations. Moreover, they assume that the value vector $\mathbf{p}$ lies in the same space as $\mathbf{u}$ and $\mathbf{v}$. This essentially means that all syntactic categories correspond to semantic space of the same dimensionality. As the authors notice, the simplification may be too restrictive as it assumes that verbs, nouns, and adjectives are substantially similar enough to be represented in the same space, but it makes the implementation computationally simpler and the approach more feasible. Theoretically, they mention the possibility of considering the composition function to be asymmetric, for instance, as the action of a matrix, $\mathbf{U}$, representing one constituent, on a vector, $\mathbf{v}$, representing the other constituent: $\mathbf{p} = \mathbf{Cuv} = \mathbf{Uv}$ – as the authors notice, this is essentially [5]'s approach to adjective-noun composition to which we return below. Similarly, [43] take a softer approach to the syntax-semantics interface and consider all words to have the same type of representation: a matrix and a vector. The matrix component expresses the ability of (any) word to act on another when composing with it, each matrix word is composed with the lexical vector of the other word, the result of such composition is still a pair of a vector and a matrix; the vector is obtained by projecting the two matrix product results to the lexical vector space and the matrix is produced by projecting the pairing of matrices back to the matrix space. Hence, the role of the syntax is reduced to the minimum, it just provides the structure of the composition. We will look at how [12] and [3] handle the syntax-semantics interface in Section 3.1

Logical Words and Entailment. As emphasized by P. Bosch in his ESSLLI '08 talk, research on DSMs has traditionally focused on content words (open word class) whereas logical words (closed word class), like determiners, coordination, modals, prepositions have been neglected. However, if we aim to reach a compositional DSM able to capture the distributional meaning of sentences, we might need to encode the contribution of e.g. *every* and *no* in the sentences *"few dogs chase cats"* vs. *"no dog chases cats"*. Ed Hovey in his IWCS '11 talk discusses his intuitions regarding logical words, like negation of content words (*not hard*) and modal alteration of them (*possibly hard*) and claims that these expressions cannot be represented by tensors harvested from corpora but that they should be considered as operators: e.g. negation should negate the values of the tensors it composes with. Like Bosch and Hovey, we believe that researchers working on DSMs should go beyond lexical meaning and consider also phrases and sentences (a challenge that has been taken up by several research groups in the last few years), and we also believe that in this perspective it is time to consider grammatical words too (a challenge that is mostly overlooked); however, we raise doubts on Hovey's claim that considers grammatical words as pre-defined operators. In our view, the new question the DSM community might have to answer is whether from a distributional stand point there is a real distinction between grammatical and content words and if so to what extend. Do we really need to consider content words as given by their distribution and the grammatical words as pre-defined or do we instead need to change the distributional contexts to look at for capturing the meaning of the grammatical ones? We believe that a correct way to think of the issue should come from the observation of what leads a speaker to use for instance a determiner instead of another when expressing similar quantities (for instance, *few* vs. *a few, many* vs. *several.*) In the sequel (Section 3.2 and Section 3.3), we will review some preliminary work on this class of words within DSM.

3   The Montagovian Pillars within DSM

3.1   Syntax-Semantics Interface

Following the type logical view to the syntax-semantics interface, the connection between the two natural language levels needs to be captured by both the vocabulary and grammar rules; below we look at these two levels within DSMs.

Syntactic Categories and Semantic Types. In the type-logical view, a first step to establish the tight formal connection between syntax and semantics is achieved by defining a mapping between syntactic categories and semantic types, based on the assumption that expressions belonging to the same syntactic categories find their meaning in the same denotational domains and hence receive meaning representations of the same types. For instance, if one assumes that determiner phrases (category: DP) denote in the domain of entities (type: e), and sentences (category: S) denotes in the domain of truth values (type: t), viz. $\mathsf{Type}(DP) = e$ and $\mathsf{Type}(S) = t$, and that $\mathsf{Type}(A\backslash B) = \mathsf{Type}(B/A)$ $= \mathsf{Type}(A) \rightarrow \mathsf{Type}(B)$, then $\mathsf{Type}(DP\backslash S) = e \rightarrow t$, and $\mathsf{Type}((DP\backslash S)/DP) = e \rightarrow (e \rightarrow t)$.

This idea has been imported into the DSM realm in [9] (and in the extended version in [12]) where the authors assign to a lexical entry the product between the pregroup category and the vector in the tensor space, using a mathematical structure that unifies syntax and semantics. The use of pre-groups as grammar to analyse linguistic structure traces back again to Lambek [31,32]. [8] discusses the same framework in terms of multi-linear algebra providing a more concrete and intuitive view for those readers not familiar with category theory. At the level of lexical and phrasal interpretation, [9,12,8] import Frege's distinction into DSMs by representing "complete" and "incomplete" expressions as vectors and as higher-order tensors, respectively, and consider the syntax-semantics link established between syntactic categories and semantic types. For instance, a transitive verb has syntactic category $DP^r \cdot S \cdot DP^l$ (that corresponds to the functional CG category $(DP\backslash S)/DP$) and semantic type $N \otimes S \otimes N$, since expressions in DP and S are taken to live in the semantic space of type N and S, respectively, and the transitive verb relates these vector spaces via the tensor product ($\otimes$): its dimensions are combinations of those of the vectors it relates. As clearly explained in [8], the verb vector can be thought of as encoding all the ways in which the verb could interact with a subject and object in order to produce a sentence, and the composition (via inner product) with a particular subject and object reduces those possibilities to a single vector in the sentence space. Several implementations of this framework have been proposed, e.g., [21,22,13,28], but the connection between the syntactic categories and semantics types has been maintained only in [20].

The mapping between syntactic categories and semantic type is fully emphasised and employed in [3]. In the remaining of the paper, we will focus on this work. The authors generalize the distinction discussed in [5] between vectors (atomic categories, e.g., nouns) and matrices (one-argument function, e.g., adjectives) starting, as in the type-logical view, from defining a mapping from syntactic categories to semantic types, as specified below.[2]

$$\mathsf{Type}(a) \qquad\qquad = C_a \text{ (for } a \text{ atomic)}$$
$$\mathsf{Type}(A\backslash B) = \mathsf{Type}(B/A) = C_A \rightarrow C_B$$

In denotational semantics the semantic types indicate the type of the domain of denotation (for instance, *john* is of type e: it denotes in the domain of entities, D_e, whereas *walks* is of type $e \rightarrow t$ and denotes in the corresponding domains of functions from entities to truth values, $D_{e\rightarrow t}$); in distributional semantics [3] take types to stand for the semantics space in which the expression lives, namely the contexts or context transformations. Words that live in the space of vectors have an atomic type, whereas functional types are assigned to words that act as space mappings (context transformations): matrices (that is, second order tensors) have first order 1-argument functional types, third order tensors have

---

[2] [3] adopt the alternative notation: $\mathsf{Type}(B\backslash A) = \mathsf{Type}(B/A) = C_A \rightarrow C_B$.

first order 2-argument functional types, etc. In general, they assume that words of different syntactic categories live in different semantic spaces. As it is the case in formal semantics, where nouns and verb phrases are both functions from entities to truth values, one could decide that two different syntactic categories are mapped to the same semantic types – live in the same semantic space. [3] take as atomic categories N (noun), DP (determiner phrase) and S (sentence); their types are indexed to record the number of dimensions of the corresponding semantic space: $\text{Type}(N) = C_{n_i}, \text{Type}(DP) = C_{dp_j}, \text{Type}(S) = C_{s_k}$ – where C stands for context – whereas the types of the complex categories are obtained by the definition above.

Again following Montague, [3] consider a fragment of English that represents the variety of tensor composition the DSM should be able to cover both theoretically and practically. As vocabulary, they consider words in the syntactic categories listed in the table below. For sake of clarity, in the table next to the syntactic category we indicate also the corresponding semantic type as well as the order of the corresponding DS representation. In the sequel, following the standard practice, we will be using boldface lowercase letters, e.g., **a**, to represent a vector, boldface capital letters, e.g., **A**, to represent a matrix and Euler script letters, e.g., $\mathcal{X}$, to represent tensors of order higher than two.

Relative pronouns (RelPr) in subject or object positions should ideally receive the same syntactic category in CG. This can be done using other connectives besides the traditional functional ones (\ and /), but since the focus is on the syntax-semantics interface rather than about syntactic issues per se, the authors adopt the easiest CG solution and consider two syntactic categories: $(N\backslash N)/(DP\backslash S)$ for subject gap and $(N\backslash N)/(S/DP)$ for object gap, both mapping to the same semantic type.

Before going to look at how the relation between syntax and semantics is captured at the grammar rules level, we will still report some observation regarding CG categories within DSMs.

Table 1. Syntax-Semantics interface of an English Fragment

Lexicon			
Syn Cat	CG Cat	Semantic Type	Tensors
N	N	C_{n_i}	I vector (1st ord.)
NNS	DP	C_{dp_j}	J vector (1st ord.)
ADJ	N/N	$C_{n_i} \rightarrow C_{n_i}$	$I \times I$ matrix (2nd ord.)
DET	DP/N	$C_{n_i} \rightarrow C_{dp_j}$	$J \times I$ matrix (2nd ord.)
IV	$DP\backslash S$	$C_{dp_j} \rightarrow C_{s_k}$	$K \times J$ matrix (2nd ord.)
TV	$(DP\backslash S)/DP$	$C_{dp_j} \rightarrow (C_{dp_j} \rightarrow C_{s_k})$	$(K \times J) \times J$ (3rd ord.)
Pre	$(N\backslash N)/DP$	$C_{dp_j} \rightarrow (C_{n_i} \rightarrow C_{n_i})$	$(I \times I) \times J$ (3rd ord.)
CONJ	$(N\backslash N)/N$	$C_{n_i} \rightarrow (C_{n_i} \rightarrow C_{n_i})$	$(I \times I) \times I$ (3rd ord.)
CONJ	$(DP\backslash DP)/DP$	$C_{dp_j} \rightarrow (C_{dp_j} \rightarrow C_{dp_j})$	$(J \times J) \times J$ (3rd ord.)
RelPr	$(N\backslash N)/(DP\backslash S)$ $(N\backslash N)/(S/DP)$	$(C_{dp_j} \rightarrow C_{s_k}) \rightarrow (C_{n_i} \rightarrow C_{n_i})$	$(I \times I) \times (K \times J)$ (higher ord.)

Empirical Coverage. [39] has analysed two large corpora, Wikipedia and ukWaC[3] parsed with a Combinatorial Categorial Grammar (CCG) parser [10,26] aiming to understand which type of categories, hence tensors, are more frequent in natural language structures. From this analysis it results that in Wikipedia there are around 902M (ukWaC: around 1.8M) tokens belonging to an atomic category (vector); around 632M (ukWaC: around 1.4M) tokens belonging to a one-argument function category (matrices); around 114M (ukWaC: around 282M) tokens belonging to a two argument function category (3rd order tensor), and around 189M (uKaWac: 469M) tokens belonging to a tensor higher than 3; hence the large majority of tokens (around 90% in Wikipedia and 40% in ukaWaC) would be represented by a tensor of the order discussed in [3] and reviewed above.

Learning the Vocabulary. The vector representations of words belonging to atomic categories are obtained as explained above by harvesting the co-occurrence frequency and possibly converting them by means of some weighting schema. For the distributional functions, [5] propose to use regression methods. They look at adjective noun phrases, ADJ N, which again belong to the category of nouns and hence are represented by vectors as the modified noun. In other words, the distributional function is learned from examples of its input and output vectors extracted from the corpus; for instance, the matrices **RED** will be learned from vector pairs like (**army**, **RED army**), (**apple**, **RED apple**), etc. Standard machine learning methods are used to find the set of weights in the matrix that produces the best approximations to the corpus-extracted example output vectors when put together with[4] the corresponding input vectors. This method has been generalized to work with n-argument functions in [20]. In particular, when a function returns another function as output (e.g., it acts on a vector and generates a matrix) we need to apply a multiple-step regression learning method, inducing representations of example matrices in a first round of regressions, and then using regression again to learn the higher-order function. [20] have worked on transitive verbs. A transitive verb such as *eat* is a third-order tensor (e.g. $(2 \times 4) \times 4$ tensor, that takes an object, a DP represented by a 4-dimensional vector (e.g., *cake*) to return the corresponding VP (*"eat cake"*, a 2×4 matrix). To learn the weights of such tensor, [20] first use regression to obtain examples of matrices representing verb-object constructions with a specific verb. These matrices are estimated from corpus-extracted examples of <*subject, subject verb object*> vector pairs (picking subject-verb-object structures that occur with a certain frequency in the corpus, in order to be able to extract meaningful distributional vectors for them). After estimating a suitable number of such matrices for a variety of objects of the same verb (e.g., *"eat cake"*, *"eat meat"*, *"eat snacks"*), they use pairs of corpus-derived object vectors and the corresponding verb-object matrices estimated in the first step as

---

[3] Wikipedia English articles: around 820 million words, and ukWaC: around 2 billion words.

[4] We will see that the composition operation used is the product.

input-output examples in a second regression step, where the verb tensor components are determined.

CG Category Based DSMs. As we have mentioned above, the dimensions of standard DSMs have been taken to be words tagged with PoS tags or words labeled with dependency relations. Differently from this tradition, [39] exploits the richness of the CG categories to build a DSM model harvested from the large corpora parsed with the CCG parser mentioned above. We briefly report the results obtained. The model (CCG-DSM) has the 20K most frequent CG categories tagged words as dimensions, and the 10K most frequent nouns, 5K most frequent verbs, 5K most frequent adjectives as target words. The co-occurrence matrix harvested from the corpus has been converted by means of different weighting schema and reduced to 300 dimension by SVD. The model has been evaluated against a noun and verb clustering task as proposed in [4]. Interestingly, the CCG-DSM model outperforms both the one based on plain PoS-tagging and the one based on dependency relation-tagging in clustering verbs. The data-set, used for the evaluation, contains 45 verbs divided into five classes/clusters, viz. cognition: 10, motion: 15, body: 10, exchange: 5, change state: 5. The clustering has been done using CLUTO and evaluated with the standard clustering measures of entropy (clusters' level of disorder) and purity (proportion of the most frequent class in the cluster). The best performing results have been obtained with the Exponential Point-wise Mutual Information (epmi) weighting schema and the 2 window context (the 2 words on the left and the 2 words on the right of the target word). The measures are: entropy 0.305 (CCG-DSM) vs. 0.556 (dependency-DSM), purity 0.756 (CCG-DSM) vs. 0.667 (dependency-DSM). These results on the one hand confirm that the syntactic structure (encoded in the CG categories) plays a role in the distributional meaning of words, and on the other show that CG categories do carry important semantic information too.

Lambek's Lesson: Function Application and also Abstraction. As we explained earlier natural language expressions can correspond to first order or higher-order functions and can require one or more argument. Moreover, at the syntactic level, functions are directional ($A\backslash B$ vs. B/A), since in natural language function-argument order matters. Hence, CG and the type-logical grammar based on it consist of two function application rules: backward (when the argument is on the left of its function) and forward (when the argument is on the right of its function.)

Function application has been the main focus of several work aiming at combining CG-like syntax with DSMs. As mentioned above [9,12,5] have been among the pioneers of such enterprise. As anticipated earlier, [5] look at how an adjective modifies a noun by employing the matrix-by-vector product (see below) that allows a matrix (the adjectives, ADJ) to act on a vector (the noun, N) resulting in a new vector (a new noun, ADJ N). Interestingly, the authors show a great advantage of the DSM over the Formal Semantics one when dealing with composition of content words, namely they show that the same adjective modifies their argument differently accordingly to which is the noun it composes with (for instance, *"red apple"* vs. *"red army"*.) However, the authors, by focusing on the adjective-noun constructions, do not consider the variety of syntactic-semantics constructions natural language exhibits. [3] extend the approach to further cases generalizing the matrix-by-vector composition to handle n-argument functions and follow the type-logical framework exploiting the correspondence between Lambek and Lambda calculi. Again, we will report on this work and refer the reader to the cited bibliography for related work.

One of the earlier contribution of Lambek mathematical view to the natural language parsing problem is the discovery of the inverse rule of function application, namely *abstraction*. Lambek highlighted that if a structure can be composed, it can also be de-composed, in other words if one knows that $w_1 : B, w_2 : B \backslash A$ yields $w_1\ w_2 : A$ she also knows that e.g. $w_1 : A/(B \backslash A)$. Hence, the Lambek calculus, in the natural deduction format, consists of both elimination (function application – composition) and introduction (abstraction – de-composition) rules of the implicational operators ($\backslash$ and $/$). A restricted version of abstraction (type raising) is also present in the CG combinatory version, CCG [44] together with other function composition rules.[5]

In the type-logical framework, the syntactic trees (derivations) are labelled with lambda terms that record the operational steps and are therefore called "proof terms". Once the proof term of a parsed sentence is built, it can be replaced with the corresponding semantic representation of the lexical bits in the linguistic structure parsed. In a Montagovian view they will be replaced with λ-terms standing for the denotation of the words, in Continuation Semantics they would be replaced with λ-terms that take context into account (see [6,1]). In DSM, [3] propose to replace them with the corresponding tensors. Below we will see this system at work on some toy examples.

Function Application in DSM. [3] propose to use "generalized matrix-by-vector multiplication" to account for function application defined as below and explained by means of examples in the sequel. Given input $\mathcal{V}$ with shape $J_1 \times \ldots \times J_n$ and components denoted by $V_{j_1 \ldots j_n}$, and a linear transformation encoded in a tensor $\mathcal{M}$ with shape $(I_1 \times \ldots \times I_m) \times (J_1 \times \ldots \times J_n)$ and components

---

[5] From the empirical coverage study conducted in [39] it results that most of the sentences in the Wikipedia and ukWaC need only forward application (Wikipedia: around 3M and ukWaC: around 2.6M), backward application (Wikipedia: around 233K and ukWaC: around 391K), or combination of them: Wikipedia: 25M (uKWaC: 48M); hence totally around 28M sentences in Wikipedia (ukWaC: 51M) would require the generalized matrix-by-vector composition [3] in a rather straight-forward way; around 2.5M (ukWaC: 4.7M) sentences are parsed also with function composition (forward or backward) and around 5.8M (ukWac: 15M) sentences require also backward crossed composition. Furthermore, there are 18M sentences in Wikipedia and 40M in ukWac that require the conjunction rule, 149K sentences in Wikipedia and 494K sentences in ukWaC that require generalized backward crossed composition, and 800K sentences in Wikipedia and 3M sentences in ukWaC that require the type-raising rule. Of course these numbers are subject to possible mistakes of the parser.

denoted by $M_{i_1...i_m j_1...j_n}$, each component $W_{i_1...i_m}$ of the output tensor $\mathcal{W}$ (of shape $I_1 \times ... \times I_m$) is given by a weighted sum of all input components as follows:

$$W_{i_1...i_m} = \sum_{j_1=1}^{j_1=J_1} \cdots \sum_{j_n=1}^{j_n=J_n} M_{i_1...i_m j_1...j_n} V_{j_1...j_n}$$

The term of the operation is used by the authors to underline the fact that the general product operation they assume is equivalent to unfolding both the input and the output tensors into vectors, applying standard matrix-by-vector multiplication, and then re-indexing the components of the output to give it the appropriate shape. For example, to multiply a $(I \times J) \times (K \times L)$ fourth-order tensor by a $K \times L$ matrix, they treat the first as a matrix with $I \times J$ rows and $K \times L$ columns and the second as a vector with $K \times L$ components (e.g., a $(2 \times 3) \times (3 \times 3)$ tensor can be multiplied with a (3×3) matrix by treating the latter as a 9 component vector and the former as a 6×9 matrix). They perform matrix-by-vector multiplication and then rearrange the resulting $I \times J$-sized vector into a matrix of shape $I \times J$ (continuing the example, the values in the 6 component output vector are re-arranged into a 2×3 matrix). This is a straightforward way to apply linear transformations to tensors (indeed, there is a precise sense in which all tensors with the same shape constitute a "vector" space). The simple matrix-by-vector multiplication is used straight forwardly to apply a first-order function to an argument:

$$f(a) =_{def} \mathbf{F} \times \mathbf{a} = \mathbf{b}$$

where $\mathbf{F}$ is the matrix encoding function f as a linear transformation, $\mathbf{a}$ is the vector denoting the argument a and $\mathbf{b}$ is the vector output to the composition process. This is the rule used in [5] to account for the composition of an adjective with a noun. Let us assume that nouns live in a 2-dimensional space. Hence the adjective, as a function from nouns to nouns, is a 2×2 matrix (it multiplies with a 2 component vector to return another 2 component vector). See the toy example in Table 2: suppose *old* is associated to the toy matrix and applied to the *dog* vector, it returns the vector for *old dog*:

Table 2. The adjective *old* as the distributional function encoded in the matrix on the left. The function is applied to the noun *dog* via matrix-by-vector multiplication to obtain a compositional distributional representation of *old dog* (right).

OLD	runs	barks		dog		OLD(dog)
runs	0.5	0	$\times$ runs	1	= runs	$(0.5 \times 1) + (0 \times 5) = 0.5$
barks	0.3	1	barks	5	barks	$(0.3 \times 1) + (5 \times 1) = 5.3$

As observed in [3], in the case of *old*, we can imagine the adjective having a relatively small effect on the modified noun, not moving its vector too far from its original location (an *old dog* is still a *barking* creature). This will be reflected in a matrix that has values close to 1 on the diagonal cells (the ones whose weights govern the mapping between the same input and output components), and values close to 0 in the other cells (reflecting little "interference" from other features). On the other hand, an adjective such as *dead* that alters the nature of the noun it modifies more radically could have 0 or even negative values on the diagonal, and large negative or positive values in many non-diagonal cells, reflecting the stronger effect it has on the noun.

We can now look at the function application cases required by the fragment of English whose vocabulary is presented in Table 1.

(a) A matrix (2nd order tensor) composes with a vector (ADJ N e.g., *red dog*, DET N e.g., *the dog*, DP IV e.g., *the dog barks, dogs bark*);
(b) A 3rd order tensor composes with two vectors (DP TV DP, *dogs chase cats*, N Pre DP, *dog with tails*, DP CONJ DP *dogs and cats*)
(c) A higher-order tensor composes with a matrix ((c1) Rel IV, e.g., *which barks*, Rel TV DP *which chases cats*, and (c2) Rel DP TV, *which dogs chase*)

For instance, when parsing the expressions *dogs bark*, *dogs chase cats* and *which chase cats*, CG produces the structures and terms in the trees of Figure 1. To help reading the proof term, we use the @ symbol to indicate the application of a function to an argument ($f@a$).

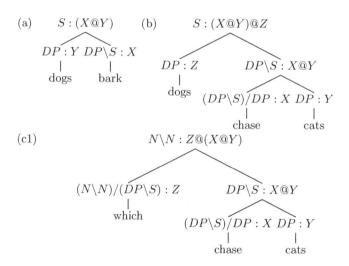

Fig. 1. Proof terms: Function application

[3] replace the variables with the corresponding DSM representations obtained as described above and compute the vectors representing the sentences. In particular, in the (a) tree X and Y are replaced with the matrix **BARK** and

the vector **dogs**, respectively, giving $\mathbf{BARK} \times \mathbf{dogs}$; whereas in the (b) tree X is replaced by the 3rd order tensor representing the meaning of *chase*, and Y and Z are replaced with the vectors representing the meaning of *dogs* and *cats*, respectively. Hence, we obtain $(\mathcal{CHASE} \times \mathbf{cats}) \times \mathbf{dogs}$. Similarly, for (c1) $\mathcal{WHICH} \times (\mathcal{CHASE} \times \mathbf{cats})$. Once we have built the meaning representation of the sentence, we can compute its meaning by means of generalized matrix-by-vector multiplication introduced above.

Below, we simplify the problem by using a toy scenario in which sentences live in a two dimensional space (C_{s_2}), determiner phrases in a four dimensional space (C_{dp_4}), and nouns into a three dimensional space (C_{n_3}). As said above, we have three cases to consider.

(a) Matrix Vector Composition. Matrix vector composition can be exemplified by the operation composing a determiner phrase and an intransitive verb, as in the sentence before *dogs bark*. The CG labeled syntactic tree of this sentence gives us $\mathbf{BARK}$ $\mathbf{dogs}$. Since in our toy semantic space scenario the semantic type of an intransitive verb is $C_{dp_4} \to C_{s_2}$ and of a determiner phrases is C_{dp_4}, we take $\mathbf{BARK}$ and $\mathbf{dog}$ to be a 2×4 matrix and a 4-dimensional vector, respectively; these terms are composed simply by function application which returns a 2-dimensional vector standing for the meaning of the whole sentence.

Below we represent the *dp* contexts as dp1, dp2, dp3, dp4 and similarly the two *s* contexts as s1, s2.

$DP\backslash S$: matrix

bark	dp1	dp2	dp3	dp4
s1	n_{11}	n_{12}	n_{13}	n_{14}
s2	n_{21}	n_{22}	n_{23}	n_{24}

DP: vector

	dogs
dp1	k_1
dp2	k_2
dp3	k_3
dp4	k_4

S: vector

	dogs bark
s1	$(n_{11}, n_{12}, n_{13}, n_{14}) \cdot (k_1, k_2, k_3, k_4)$
s2	$(n_{21}, n_{22}, n_{23}, n_{24}) \cdot (k_1, k_2, k_3, k_4)$

$=$

S: vector

	dogs bark
s1	$(n_{11} \times k_1) + \ldots + (n_{14} \times k_4)$
s2	$(n_{21} \times k_1) + \ldots + (n_{24} \times k_4)$

(b) 3rd Order Tensor Composed with Two Vectors. An example of this case is provided by the composition of a transitive verb with its object and subject. For instance, for the sentence *dogs chase cats*, CG produces the labeled syntactic tree seen above which gives us the DS representation $(\mathcal{CHASE} \times \mathbf{cats}) \times \mathbf{dogs}$. Hence, we need to apply step-wise the 3rd order tensor, the transitive verb, to two vectors, the object and the subject. In our toy example, the transitive verbs have semantic type $C_{dp_4} \to (C_{dp_4} \to C_{s_2})$. Hence, the DS meaning representation of *chase* is a $(2 \times 4) \times 4$ tensor; we can think of it as tensor of four slices of one 2×4 matrix each.

$$\text{chase}\;\frac{\begin{array}{l}\text{slice 1:}\\ |\text{dp1 dp2 dp3 dp4}\end{array}}{\begin{array}{l}s1|n_{11}^1\; n_{12}^1\; n_{13}^1\; n_{14}^1\\ s2|n_{21}^1\; n_{22}^1\; n_{23}^1\; n_{24}^1\end{array}}\;\cdots\;\frac{\begin{array}{l}\text{slice 4:}\\ |\text{dp1 dp2 dp3 dp4}\end{array}}{\begin{array}{l}s1|n_{11}^4\; n_{12}^4\; n_{13}^4\; n_{14}^4\\ s2|n_{21}^4\; n_{22}^4\; n_{23}^4\; n_{24}^4\end{array}}$$

The application of *chase* to *cats* gives the following 2×4 matrix:

chase cats	dp1	$\cdots$	dp4
$s1$	$(n_{11}^1, n_{12}^1, n_{13}^1, n_{14}^1) \cdot (k_1, k_2, k_3, k_4)$	$\cdots$	$(n_{11}^4, n_{12}^4, n_{13}^4, n_{14}^4) \cdot (k_1, k_2, k_3, k_4)$
$s2$	$(n_{21}^1, n_{22}^1, n_{23}^1, n_{24}^1) \cdot (k_1, k_2, k_3, k_4)$	$\cdots$	$(n_{21}^4, n_{22}^4, n_{23}^4, n_{24}^4) \cdot (k_1, k_2, k_3, k_4)$

which can then be applied to the 4 dimensional vector representing *dogs* yielding a 2 dimensional vector representing the whole sentence.

(c) Higher Order-Tensor Matrix Composition. The only higher-order tensor in Table 1 is the one representing a relative pronoun. From a formal semantic point of view, a relative pronoun creates the intersection of two properties. e.g. $[\![dog]\!] \cap [\![chase\ cats]\!]$; in distributional semantics, we can look at it as a 4th order tensor whose first argument is a verb phrase, hence a matrix, and its second argument is a noun, hence a vector, and it yields a modified noun, hence again a vector. In our toy example, it lives in a $(3 \times 3) \times (2 \times 4)$ space, it is of semantic type $(C_{dp_4} \to C_{s_2}) \to (C_{n_3} \to C_{n_3})$ and can be applied to a 2×4 matrix. For instance *which* can be applied to the matrix obtained above *chase cats*. As explained in [3], this operation can be reduced to the simpler one considered above, namely to the application of a 3rd order tensor to a vector. To this end, [3] apply the *unfolding* method that transforms a tensor into one of lower order by reordering its elements. There are several ways of reordering the elements, for instance, a $(2 \times 3) \times 4$ tensor can be arranged as a 6×4 matrix. Which mode is chosen is not important as far as across related calculations the same mode is used. Going back to our linguistic example, the relative pronoun and the VP-matrix could be unfolded into a $(3 \times 3) \times 8$ tensor and a 8 dimensional vector, respectively. To understand the unfolding method, let us look at how it could transform the 2×4 VP-matrix into a 8 dimensional vector and let us take as unfolding mode the concatenation of its elements as illustrated below. Let us assume the matrix representing *chase cats* represented abstractly above is instantiated as below; by unfolding we obtain the corresponding vector as following:

$$\frac{\text{chase cats}\,|\,\text{dp1 dp2 dp3 dp4}}{\begin{array}{l}s1|\;\;1\quad\;3\quad\;5\quad\;7\\ s2|\;\;2\quad\;4\quad\;6\quad\;8\end{array}}\;\rightsquigarrow_{\text{unfolding}}\;(1,2,3,4,5,6,7,8).$$

The evaluation carried out so far using generalized matrix-by-vector operation for function application has obtained encouraging results. See [46,20] for the evaluation of the compositional distributional models for adjective-noun and transitive verb-object composition, respectively and [41] for an evaluation of [3]'s approach at sentential level. No evaluation has been carried out yet on relative sentences, these constructions are going to be object of investigation of the COM-POSES project.[6]

Abstraction in DSMs. Abstraction is used mostly for two cases: long distance dependency and inverse scope. The latter is more a challenge for the formal grammar researchers than for the semanticists: once the grammar provides the right representations of an ambiguous sentence the semantic operations should be able to compute the proper meaning straight forwardly. Hence, in the following we will look only at long distance dependencies as instances of abstraction, and in particular at the case of relative pronoun that extracts the object of the relative clause sentence.

As far as we know, the first attempt to handle cases of long distance dependencies within the compositional distributional semantic framework is presented in [3] where the authors discuss the dependency of a main sentence subject from the transitive verb of a relative clause, e.g., "*A cat which dogs chase runs away*": the object of the relative clause is missing and its role is played by "*A cat*" thanks to the presence of the relative pronoun *which*. The lack of the object can be marked by a trace "*(A cat (which dogs chase ...)) runs away*". The type-logical view on the composition of this sentence can be represented by the tree below (Figure 2) that starts by assuming an hypothetical object (hyp), builds the sentence *dogs chase hyp* (Figure 2, tree on the left) and then withdraws the hypothesis building a tree without it (Figure 2, tree on the right) using abstraction. The application of abstraction is governed by the presence of the higher-order two-argument category $(N\backslash N)/(S/DP)$ assigned to the relative pronoun; it requires a sentence missing a DP on the rightmost position to return the category $N\backslash N$. Hence, the parser encounters a category mismatch: It has the task of composing $(N\backslash N)/(S/DP)$ (*which*) with the tree of category S corresponding to "*dogs chase hyp*". The tree of category S, however, contains an hypothesis of category DP—it would be a sentence if a DP had been provided. The parser can now withdraw the hypothetical DP and build the tree of category S/DP. The rule that allows this step is the one-branch rule encoding hypothetical reasoning. The derivation can then proceed by function application. The lambda calculus goes step by step with this hypothetical reasoning process. Besides the function application rules we have used so far, it consists of the abstraction rule that abstracts from the term $(Z@X)@Y$ (namely the term assigned to the S tree –hence, a term of type t), the variable X assigned to the hypothetical DP (hence, a term of type e), building the lambda term $\lambda X.(Z@X)@Y$ (a term of type $(e \rightarrow t)$). The next step is again the application of a function (W of type $(e \rightarrow t) \rightarrow ((e \rightarrow t) \rightarrow t)$) to an argument (the lambda term of type $(e \rightarrow t)$ we have just built).

Syntactically, these constructions challenge any formal grammars, hence they have attracted the attention of researchers and several solutions have been proposed within the CG framework. [3] build the syntactic tree in the way more

---

[6] http://clic.cimec.unitn.it/composes/

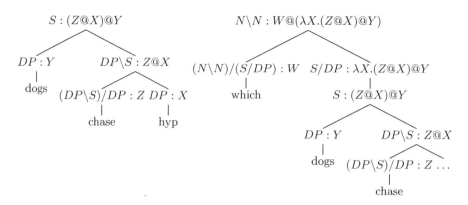

Fig. 2. Proof terms: function application and abstraction: hypothetical reasoning

straightforwardly linked to the distributional semantics analysis even though the rule employed (Figure 3), namely associativity, would cause over-generation problems.[7]

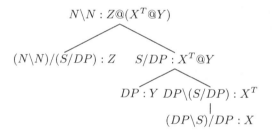

Fig. 3. Proof terms: associativity (syntax) and transformation (semantics)

The proof term consists of just function application of the kinds we have discussed above. The only step here to introduce is the one that transforms a transitive verb category from $(DP\backslash S)/DP$ into $DP\backslash(S/DP)$. Syntactically, this step corresponds to associativity, semantically it corresponds to the tensor transformation rule which establishes in general terms how the elements of a tensor can be switched:

$$(\mathcal{T} \times \mathbf{v}) \times \mathbf{w} = (\mathcal{T}^T \times \mathbf{w}) \times \mathbf{v}$$

---

[7] Its application could be controlled by employing for instance the multi-modal version of CG [37], but since our focus is on the composition of the distributional semantic representations involved in such constructions, we will overlook the syntactic issues. This semantic analysis or a variation of it could be connected to the different syntactic proposals in the literature.

This rule allows to transform a (pre-trained) transitive verb tensor that would normally be multiplied by an object and then a subject into the transposed form, that can take the subject first, and the object later, producing the same result. In the tree, this semantic rule is presented as taking the term X and yielding the term X^T. Now the proof term can be replaced with the actual distributional representation, obtaining $\mathcal{WHICH} \times (\mathcal{CHASE}^T \times \textbf{dogs})$, which can later modify the vector representing *cat*.

Let's now assume as toy example that a transitive verb is a $(3 \times 2) \times 2$ matrix; for instance, *chase* consists of the two slice tensor below; it can be transformed by switching the second and first column of the first and second slice, respectively. This guarantees that $(\mathcal{CHASE}\ \textbf{dogs})\ \textbf{cats}$ is equivalent to $(\mathcal{CHASE}^T\ \textbf{cats})\ \textbf{dogs}$ as the reader can test by herself.

$$
\mathcal{CHASE}\ \begin{array}{c} \text{slice 1:} \\ 1\ 4 \\ 2\ 5 \\ 3\ 6 \end{array}\ \begin{array}{c} \text{slice 2:} \\ 7\ 10 \\ 8\ 11 \\ 9\ 12 \end{array}\ \rightsquigarrow\ \mathcal{CHASE}^T\ \begin{array}{c} \text{slice 1:} \\ 1\ 7 \\ 2\ 8 \\ 3\ 9 \end{array}\ \begin{array}{c} \text{slice 2:} \\ 4\ 10 \\ 5\ 11 \\ 6\ 12 \end{array}
$$

The use of the tensor transformation rule above avoid having to use the counterpart of abstraction, and does not provide a way for handling structure decomposition, as a more general solution might require. Hence, we wonder whether as Lambek highlighted the possibility of extending CG with abstraction, DSM framework can be extended with an equivalent rule.

We conjecture that a solution to long distance dependency could come from the categorical view presented in [42], where the authors discuss the "eta maps" showing that they create Bell states that produce an extra space allowing for "teleportation", in other words the eta maps enable the information to flow between the quantum states that are not locally close. For instance, in the case of a negative transitive sentence *"John does not like Mary"* *does* and *not* are vectors that act as identity and as base swapper, respectively. The eta maps move the subject vector *john* to be the argument of the transitive verb so that the *does* and *not* vectors act on the representation of the positive transitive sentence swapping its bases, namely making it true if it was false, and vice versa.

Alternative solutions could come from the regression method discussed above. Following Montague's intuition regarding type lifting, namely that an expression like a personal name can be represented either as an object (a constant of type e) or as a set of those properties that hold for that object (a second order function of type $((e \to t) \to t)$), we could assume that an expression of syntactic category DP can be represented semantically either by a vector or by a higher order tensor obtained from the corresponding vector by means of some tensor transformation rule. This rule could be learned by regression: we could induce from the corpus the vectors of a quantifier phrase (DET N, e.g., *"some dog"*), learn the tensor representing the same expression by regression, e.g. given the input pair (*runs*, *"some dog runs"*), learn the function representing *"some dog"*, and then learn the transformation rule from the induced vector representing *"some dog"* to its higher order tensor representation learned by regression; or we could find ways to exploit the CCG parsed corpus for learning the type-raising rule from those structures in which it has been used.

Question. In short, the challenge we are posing to the compositional distributional semantic community is to handle the semantic composition of not-juxtaposed expressions.

3.2   From Logical Entailment to Entailment in DSM

In the brief review of the core logical concepts behind formal semantics we have highlighted the difference between satisfiability (an entailment that holds in a particular interpretation) and validity (an entailment that holds across all the interpretation). The latter is due to the presence of logical words, like determiners, whose meaning is independent from a specific interpretation. Hence, in the following before reviewing the work on entailment within DSMs, we are going to present DSMs views on logical (grammatical) words.

Grammatical Words as Predefined Logical Operators. [18] proposes an interesting hybrid framework combining the complementary strengths of FOL and DSM, namely the expressivity of the former and the flexibility of the latter. The authors use FOL for representing the logical structure of a sentence, and DSM for capturing the content words meaning and project lexical inferences from the vector space to logical form. In a nutshell, based on the idea that distribution similarity between expressions A and B corresponds to substitutability of B in the place of A, they generate inference projection rules like $\forall x.\mathtt{Car}(x) \rightarrow \mathtt{Vehicle}(x)$ using WordNet as a filter to validate such axioms. Then, FOL sentence representations are obtained using Boxer [14]. Given a pair of sentences (Prem, Hyp), they check whether Hyp could follow from Prem (Prem $\rightarrow$ Hyp) from a DS view by checking (i) if there are inference projection rules between the words occurring in the two sentences and (ii) contextualising such rules by assigning them a weight: given a rule that relates a word w_1 in Hyp with a word w_2 in Prem, the weight is obtained by computing the similarity of w_1 with the sum of the vectors of the words co-occurring with w_2 in Prem (the context of w_2.) In the simplified setting described in [18], the authors generate a potential alignment between any pair of words, within the two sentences, that are related (synonymous or hyponym/hypernym up to a certain distance) in WordNet, which also means that they have to be of the same syntactic category. Moreover, they currently only deal with single-word paraphrases. Finally, they use Markov Logic Networks for reasoning on FOL weighted clauses. For instance, given the projection rule (1) $\forall x.\mathtt{Car}(x) \rightarrow \mathtt{Vehicle}(x)$, by FOL they can infer that given $\neg\exists x.\mathtt{Vehicle}(x) \wedge \mathtt{Own}(x)$ and the rule (1) above, then $\neg\exists x.\mathtt{Car}(x) \wedge \mathtt{Own}(x)$. The rule (1) receives a high weight in the given contexts since *car* is similar to *own* (the context of *vehicle*). Following Natural Logic, to handle inferences involving sentences containing nested propositions, they mark the polarity of the position of the embedded proposition.

The view of considering logical words, in particular negation, as pre-defined and close to their truth-value meaning is present in other work too (see [9,49] among others), though in a full vector space model approach. [9] take sentences to be in the space spanned by a single vector (**1**), identified with "true" and the origin to be "false" (**0**). So a sentence like "*John likes Mary*" is represented by the vector **1** if the sentence is "true" and by **0** otherwise, moreover the authors leave open the possibility of considering degree of sentence meaning instead of just the two truth values. Within the DSM framework of [9], [42] takes negation to be a base swapper operator. The details of this approach are further explained in [12]. Similarly, [49] starts from the intuition that unrelated meanings should be orthogonal to one another, which is to say that they should have no features in common at all. Hence, he takes negation to generate a vector representation that is completely orthogonal to the negated term. In the following, we will report on some preliminary results carried out on grammatical words, more in particular on determiners and determiner phrases, that suggest the possibility of undertaking a truly distributional analysis of these words too.

Grammatical Words as Tensors Learned from Their Use. [34] studies the distributional behavior of 50 determiners (articles, quantifiers, and cardinals). First of all, the author aims to check how the distributional context changes if at the same nouns are applied different determiners and if similar determiners occur in similar contexts. To this end, he builds two DSMs using a large corpus:[8] (a) one with lemmatized content words as dimension (LDSM), and (b) a second one with inflected content and grammatical words as dimensions (IDSM). For each of the studied determiner, he extracts determiner phrases (DPs) from the corpus choosing the most frequent 20K nouns and their vector representation in the two DSMs mentioned above, and extract the closest neighbour of the DPs vectors. The experiment shows that in the DP vector representations of LDSM the meaning of the nouns emerges over the one of the determiner contrary to what happens in IDSM. Moreover, the use of a noun seems to change according to the determiner used: for instance, whereas *every dog* tends to co-occur with general concepts, usually attributed to dogs in general – *animal, tail, pet, love, cleaver* and *friend* – "*that dog*" occurs with more familiar words, usually associated to a single dog, single episodes or everyday situation – *bite, owner, bad, police, kill* or *bloody*. Interestingly, the author conjectures that the determiner *that* is usually preferred for describing negative events, creating a distance between the dog and the speaker, whereas other determiners like *this* are used in positive contexts, occurring with words as *rescue, show, wonderful, loving* or *companion*. All in all, the experiment brings evidence to the claim that using a determiner rather than another affects the context in which the DP occurs. The result is confirmed by a second experiment carried out in [34] based on clustering. In this case, the vector representation of the determiner is computed out of the DPs by

---

[8] ukWaC, a 2 billion word corpus crawled from the web, British National Corpus, a 100 million word corpus, Wikipedia, about 1.6 billion tokens. The three corpora have been tokenzied and tagged with Treetagger.

calculating their average vector representation. Interesting clusters have indeed emerged (e.g. {*too few, too many, too much*}, and {*four, three, two, several*} etc.), but further studies in this direction are necessary since no clustering method alone has succeed in the task. The experiment shows that also determiners seem to be characterized by their distributional context, but a DSM more suitable to their role should be built for them.

Finally, [34] reports on a third experiment in which pairs of DPs are studied. The attention is focused on nine determiners (*all, every, four, that, these, this, those, three* and *two.*) A classifier is used for recognizing similar vs. dissimilar DP pairs.[9] The author has carried out the task using different classifiers, the best results have been obtained with a polynomial super vector machine (SVM)[10] that has obtained the following weighted average of the precision and recall (F-measures): 0.8% (LDSM, with raw co-occurrence values) and 0.81 % (inflected IDSM, with lmi weight.) The same experiment has also be tried against unseen determiners, viz., the testing dataset contains one determiner more than the training dataset, but the same nouns already used for the training. The SVM was able to correctly classify up to 68.9% of the 1226 never seen instances.

Before concluding this section, we would like to draw the reader attention on some interesting studies on determiner phrases carried out within the psycholinguistic community. As it has been emphasized in [40] quantifiers have been studied in details from the formal semantics angle, but they have been mostly ignored by the empirical based community which has focused on content words. Interestingly, they have been studied in Pragmatics and Psycholinguistics. In Pragmatics, it has been claimed that quantifiers like *no, few, some* and *all* are scalar expressions: they can be ordered on a scale with respect to the strengths of the information that they convey. As it is well known, their use involves pragmatic inferences called scalar implicature [23] ("the participants in a conversation expect that each will tailor their contribution to be as informative as required but no more informative than is required"). Though, in formal semantics, for instance *some* has just one meaning, in practice it can be used in different ways, see for instance the example below taken from [40]

- R: if you ate some of the cookies, then I won't have enough for the party.
- M: I ate some of the cookies. In fact, I ate all of them.   [Meaning: "some and possibly all"]
- R: Where are the apples that I bought?
- M. I ate some of them   [Meaning: "some but not all")

---

[9] Similar are DPs that share the determiners or the noun (e.g., *four countries-four states*, or that have similar determiners and similar nouns (e.g., *two boats-four boats*) or have similar determiners and similar nouns (e.g., *this shirt-that coat*); whereas dissimilar DPs are such that they have different determiners and different nouns (e.g., *this village-every cat*), or different determiners and similar noun (e.g., *two musicians-those painters*) (or viceversa, e.g., *two artists-four ducks*)

[10] The other classifiers used are Naive, J48, SVM Radial Kernel. Interestingly, the need of a polynomial SVM classifier for classifying relations between DPs was shown in an other internal project too on DP entailment.

Moreover, [40] shows that quantifiers can have different "polarity" even when denoting the same vague quantity: Quantifiers with positive (negative) polarity, e.g., *a few*, *quite a few*, *many*, (resp. *few*, *very few*, *not many*) are used to encourage (resp., discourage) the speaker to do something. The distinction between positive vs. negative polarity QPs is reinforced by their different behaviour with respect to the set of discourse entities they refer to and they make accessible via anaphora. To this end, the authors distinguish between the "reference set", viz. the set of entities the quantifier operates upon, and the "complement set", viz., the complement of the reference set.[11] Positive polarity QPs put the focus on the reference set while negative polarity QPs put the focus on the complement set. Moreover, the reference set is available for anaphora. Example:

> "(a) A few/(b)Few of the students attended the lecture. They ...."

people continue (a) by speaking of properties of the reference set (e.g., "*They listen carefully and took notes*") and (b) by speaking of the complement set (e.g., "*They decided to stay at home instead*")

Question. The psycholinguistics results on determiners reported above seem to confirm the possibility of studying these words (and maybe the class of logical words in general) from a distributional view. We wonder whether they also suggest that the relevant part of the context of use could be of a different nature than the one considered within DSMs for content words. For instance, instead of just looking at co-occurency frequency within a sentence, we might need to consider the discourse level (see the comment above on anaphora), or we might need to consider instead of the words in isolation, the semantic relation holding within the words in the observed context (see the comment on the choice of the verb phrase above.)

3.3   Entailment in DSM

[11] studies the algebraic properties a vector space used for representing natural language meaning needs to have and identifies possible directions to account for degree of entailment between distributional representations proposing to use the partial order of the defined algebraic structure. However, he does not describe the idea in details and does not evaluate it on any empirical ground. Implementations and interesting evaluation results have been carried out at lexical level. For instance, [16] suggests that it may not be possible to induce hyponymy information from a vector space representation, but it is possible to encode the relation in this space after it has been obtained through some other means. On the other hand, recent studies [19,29,48] have pursued the intuition that entailment is the ability of one term to "substitute" for another. For example, *baseball* contexts are also *sport* contexts but not *vice versa*, hence *baseball* is "narrower"

---

[11] Example: "*many of the students attended the lecture*". The reference set is the set of all the students who were present at the lecture, the complement set is the set of all the students who were absent.

than *sport* (*baseball* $\models$ *sport*). On this view, entailment between vectors corresponds to inclusion of contexts or features, and can be captured by asymmetric measures of distribution similarity. In particular, [29] carefully crafted the *balAPinc* measure for lexical entailment. In brief, the balAPinc score is higher if many features are present and the included features are ranked high. [2] look at a similar issue but from a different perspective and by going beyond lexical level. The authors do not use a hand-crafted measure, but rather a machine learning based classifier. They use a SVM and show that it can learn the entailment relation between phrases of the same syntactic categories: from a training set of noun pairs it learns the entailment relation between expressions of this category ($\|=_N$) e.g., from training examples like *big dog* $\|=$ *dog*, it learns *dog* $\|=$ *animal*, and from a training set of quantifier phrases, e.g. *all dog* $\|=$ *some dog*, it learns the $\|=_{QP}$ even when the testing data set contains QPs never seen in the training data set. The entailment is specific to the syntactic category and does not generalize across the categories (if the SVM is trained on $\|=_N$ it will obtain bad performance on a $\|=_{QP}$ test dataset, and viceversa if trained on $\|=_{QP}$ it will obtain bad performance on a $\|=_N$ test dataset.)[12] The results reported in [2] had been obtained with a cubic polynomial kernel; interestingly, [24] shows that a linear classifier will obtain worse results and that a two degree classifier (either homogeneous or inhomogeneous) would perform equally well than the cubic one. These latter results confirm the involvement of features interaction, rather than purely inclusion, in the entailment relation of DSM representations.

4   Conclusions

By reviewing the core aspects of formal and distributional semantics models and by presenting the more recent results obtained within DSMs beyond lexical level adopting the formal semantics binoculars, we have highlighted some open issues. First of all, we have underlined the importance of considering the correspondence between syntax and semantic both as expressed between syntactic categories and semantic types (types of semantic space) and as captured by the composition rules. As for the latter, we have raised the question of how structures with gaps can be handled with DSMs for which researchers so far have focused only on function application of juxtaposed function-arguments. Moreover, by introducing the concepts of satisfiability and validity, we have focused on the role the logical (grammatical) words play in natural language reasoning from a logic view and compared their role when observed from the language of use perspective. More research needs to be carried out on this class of words to understand whether there is the need of an hybrid system that combines logic and distributional relations or whether the integration of the two approaches would be needed only to take into account the two fregean aspects of meaning, reference and sense.

---

[12] This second half of the experiment, training on QPs and testing on Ns has been carried out by [24].

References

1. Barker, C., Shan, C.-C.: Types as graphs: Continuations in type logical grammar. Journal of Logic, Language and Information 15, 331–370 (2006)
2. Baroni, M., Bernardi, R., Shan, C.-C., Do, N.Q.: Entailment above the word level in distributional semantics. In: Proceedings of EACL (2012)
3. Baroni, M., Bernardi, R., Zamparelli, R.: Frege in space: A program for compositional distributional semantics. Linguistic Issues in Language Technology; Special issues on Semantics for textual inference (in press)
4. Baroni, M., Evert, S., Lenci, A. (eds.): Proceedings of the ESSLLI Workshop on Distributional Lexical Semantics, ESSLLI 2008 (2008)
5. Baroni, M., Zamparelli, R.: Nouns are vectors, adjectives are matrices: Representing adjective-noun constructions in semantic space. In: Proceedings of EMNLP, Boston, MA, pp. 1183–1193 (2010)
6. Bernardi, R., Moortgat, M.: Continuation semantics for the Lambek-Grishin calculus. Information and Computation 208(5), 397–416 (2010)
7. Church, K.W., Hanks, P.: Word association norms, mutual information and lexicography. Proceedings of the 27th Annual Conference of the Association of Computational Linguistics (ACL 1989) (1989)
8. Clark, S.: Type Driven Syntax and Semantics for Composing Meaning Vectors. In: Quantum Physics and Linguistics: A Compositional, Diagrammatic Discourse. Oxford University Press (2012) (in press)
9. Clark, S., Coecke, B., Sadrzadeh, M.: A compositional distributional model of meaning. In: Bruza, P., Lawless, W., van Rijsbergen, K., Coecke, B., Sofge, D., Clark, S. (eds.) Proceedings of the Second Symposium on Quantum Interaction, pp. 133–140 (2008)
10. Clark, S., Curran, J.: Wide-coverage efficient statistical parsing with CCG and log-linear models. Computational Linguistics 33(4), 493–552 (2007)
11. Clarke, D.: A context-theoretic framework for compositionality in distributional semantics. Computational Linguistics 1(54) (2011)
12. Coecke, B., Sadrzadeh, M., Clark, S.: Mathematical foundations for a compositional distributed model of meaning. Lambek Festschirft, Linguistic Analysis 36, 36 (2010)
13. Coecke, B., Grefenstette, E., Sadrzadeh, M.: Lambek vs. Lambek: Vector space semantics and string diagrams for lambek calculus (2012) (unpublished)
14. Curran, J., Clark, S., Bos, J.: Linguistically motivated large-scale nlp with c&c and boxer. In: Proceedings of ACL (Demo and Poster Sessions), Prague, Czech Republic, pp. 33–36 (2007)
15. Deerwster, S., Dumai, S.T., Furnas, G.W., Landauer, T.K., Harshman, R.: Indexing by latent semantic analysis. In: J. Am. Soc. Information Science and Technology (1990)
16. Erk, K.: Supporting inferences in semantic space: representing words as regions. In: Proceedings of IWCS, Tilburg, Netherlands, pp. 104–115 (2009)
17. Firth, J.R.: A synopsis of linguistic theory 1930-1955. Studies in Linguistic Analysis (1957); Reprinted in Palmer, F.R. (ed.) Selected Papers of J.R. Firth 1952-1959. Longman, London (1968)
18. Garrette, D., Erk, K., Mooney, R.: Integrating logical representations with probabilistic information using markov logic. In: Proceedings of the International Conference on Computational Semantics, Oxford, England, pp. 105–114 (January 2011)
19. Geffet, M., Dagan, I.: The distributional inclusion hypotheses and lexical entailment. In: Proceedings of the 43rd Annual Meeting of the ACL, pp. 107–114 (2005)

20. Grefenstette, E., Dinu, G., Zhang, Y.-Z., Sadrzadeh, M., Baroni, M.: Multi-step regression learning for compositional distributional semantics. In: Proceedings of IWCS, Potsdam, Germany (2013)
21. Grefenstette, E., Sadrzadeh, M.: Experimental support for a categorical compositional distributional model of meaning. In: Proceedings of EMNLP, Edinburgh, UK, pp. 1394–1404 (2011)
22. Grefenstette, E., Sadrzadeh, M.: Experimenting with transitive verbs in a DisCoCat. In: Proceedings of GEMS, Edinburgh, UK, pp. 62–66 (2011)
23. Grice, H.P.: Logic and conversation. Syntax and Semantics (1975); Reprinted in Studies in the Way of Words Grice, H.P. (ed.), pp. 22–40. Harvard University Press, Cambridge (1989)
24. Vasques, M.X.G.: Quantifying determiners from the distributional semantics view. Master's thesis, Free University of Bozen-Bolzano. Erasmus Mundus European Master Program in Langauge and Communication Technologies (2012)
25. Harris, Z.: Distributional structure. Word 10(23), 146–162 (1954)
26. Honnibal, M., Curran, J.R., Bos, J.: Rebanking CCGBank for improved interpretation. In: Proceedings of the 48th Annual Meeting of ACuL, pp. 207–215 (2007)
27. Jones, K.S.: A statistical interpretation of term specificity and its application in retrieval. Journal of Documentation 28(11) (1972)
28. Kartsaklis, D., Sadrzadeh, M., Pulman, S., Coecke, B.: Reasoning about meaning in natural language with compact closed categories and frobenius algebras (2012) (unpublished)
29. Kotlerman, L., Dagan, I., Szpektor, I., Zhitomirsky-Geffet, M.: Directional distributional similarity for lexical inference. Natural Language Engineering (2010)
30. Lambek, J.: The mathematics of sentence structure. American Mathematical Monthly 65, 154–170 (1958)
31. Lambek, J.: Type grammars revisited. In: Lecomte, A., Perrier, G., Lamarche, F. (eds.) LACL 1997. LNCS (LNAI), vol. 1582, pp. 1–27. Springer, Heidelberg (1999)
32. Lambek, J.: A computational algebraic approach to english grammar. Syntax (2004)
33. McDonald, S., Ramscar, M.: Testing the distributional hypothesis: The influence of context on judgements of semantic similarity. In: Proceedings of CogSci, pp. 611–616 (2001)
34. Menini, S.: A distributional approach to determiners. Master's thesis, University of Trento (2012)
35. Mitchell, J., Lapata, M.: Composition in distributional models of semantics. Cognitive Science 34(8), 1388–1429 (2010)
36. Montague, R.: English as a formal language. Linguaggi Nella Società e Nella Tecnica, 189–224 (1970)
37. Moortgat, M.: Categorial Type Logics. In: van Benthem, J., ter Meulen, A. (eds.) Handbook of Logic and Language, pp. 93–178. The MIT Press, Cambridge (1997)
38. Morrill, G.: Type Logical Grammar. Kluwer, Dordrecht (1994)
39. Paramita: Exploiting ccg derivations within distributional semantic models. Master's thesis, Free University of Bozen-Bolzano. Erasmus Mundus European Master Program in Langauge and Communication Technologies (2012)
40. Paterson, K., Filik, R., Moxey, L.: Quantifiers and discourse processing. Language and Linguistics Compass, 1–29 (2011)
41. Pham, N., Bernardi, R., Zhang, Y.Z., Baroni, M.: Sentence paraphrase detection: When determiners and word order make the difference. In: Proceedings of the IWCS 2013 Workshop: Towards a Formal Distributional Semantics (2013)

42. Preller, A., Sadrzadeh, M.: Bell states as negation in natural languages. ENTCS. QPL, University of Oxford (2009)
43. Socher, R., Huval, B., Manning, C., Ng, A.: Semantic compositionality through recursive matrix-vector spaces. In: Proceedings of EMNLP, Jeju Island, Korea, pp. 1201–1211 (2012)
44. Steedman, M.: The Syntactic Process. MIT Press, Cambridge (2000)
45. Turney, P., Pantel, P.: From frequency to meaning: Vector space models of semantics. Journal of Artificial Intelligence Research 37, 141–188 (2010)
46. Vecchi, E.M., Baroni, M., Zamparelli, R.: (linear) maps of the impossible: Capturing semantic anomalies in distributional space. In: Proceedings of the ACL DISCo Workshop, pp. 1–9 (2011)
47. Weaver: Translation. In: Machine Translation of Languages. MIT Press, Cambridge (1955)
48. Weeds, J., Weir, D., McCarthy, D.: Characterising measures of lexical distributional similarity. In: Proceedings of the 20th International Conference of Computational Linguistics, COLING 2004, pp. 1015–1021 (2004)
49. Widdows, D.: Orthogonal negation in vector spaces for modelling word-meanings and document retrieval. In: Proceedings of the 41st Annual Meeting on Association for Computational Linguistics, ACL 2003, vol. 1, pp. 136–143. Association for Computational Linguistics (2003)
50. Wittgenstein, L.: Philosophical Investigations. Blackwell, Oxford (1953); Translated by G.E.M. Anscombe

A Logical Basis for Quantum Evolution and Entanglement

Richard F. Blute[1], Alessio Guglielmi[2], Ivan T. Ivanov[3],
Prakash Panangaden[4], and Lutz Straßburger[5]

[1] Department of Mathematics and Statistics, University of Ottawa
[2] Department of Computer Science, University of Bath
[3] Department of Mathematics, Vanier College
[4] School of Computer Science, McGill University
[5] INRIA Saclay, École Polytechnique

Abstract. We reconsider *discrete quantum causal dynamics* where quantum systems are viewed as discrete structures, namely directed acyclic graphs. In such a graph, events are considered as vertices and edges depict propagation between events. Evolution is described as happening between a special family of spacelike slices, which were referred to as *locative slices*. Such slices are not so large as to result in acausal influences, but large enough to capture nonlocal correlations.

In our logical interpretation, edges are assigned logical formulas in a special logical system, called BV, an instance of a *deep inference system*. We demonstrate that BV, with its mix of commutative and noncommutative connectives, is precisely the right logic for such analysis. We show that the commutative tensor encodes (possible) entanglement, and the noncommutative seq encodes causal precedence. With this interpretation, the locative slices are precisely the derivable strings of formulas. Several new technical results about BV are developed as part of this analysis.

Dedicated to Jim Lambek on the occasion of his 90th birthday.

1 Introduction

The subject of this paper is the analysis of the evolution of quantum systems. Such systems may be protocols such as *quantum teleportation* [1]. But we have a more general notion of system in mind. Of course the key to the success of the teleportation protocol is the possibility of entanglement of particles. Our analysis will provide a syntactic way of describing and analyzing such entanglements, and their evolution in time.

This subject started with the idea that since the monoidal structure of the category of Hilbert spaces, i.e. the tensor product, provides a basis for understanding entanglement, the more general theory of monoidal categories could provide a more abstract and general setting. The idea of using general monoidal categories in place of the specific category of Hilbert spaces can be found in

C. Casadio et al. (Eds.): Lambek Festschrift, LNCS 8222, pp. 90–107, 2014.

a number of sources, most notably [2], where it is shown that the notion of a symmetric compact closed dagger monoidal category is the correct level of abstraction to encode and prove the correctness of protocols. Subsequent work in this area can be found in [3], and the references therein.

A natural step in this program is to use the logic underlying monoidal categories as a syntactic framework for analyzing such quantum systems. But more than that is possible. While a logic does come with a syntax, it also has a built-in notion of dynamics, given by the cut-elimination procedure. In intuitionistic logic, the syntax is given by simply-typed λ-calculus, and dynamics is then given by β-reduction [4]. In linear logic, the syntax for specifying proofs is given by *proof nets* [5]. Cut-elimination takes the form of a local graph rewriting system.

In [6], it is shown that causal evolution in a discrete system can be modelled using monoidal categories. The details are given in the next section, but one begins with a directed, acyclic graph, called a *causal graph*. The nodes of the graph represent events, while the edges represent flow of particles between events. The dynamics is represented by assigning to each edge an object in a monoidal category and each vertex a morphism with domain the tensor of the incoming edges and codomain the tensor of the outgoing edges. Evolution is described as happening between a special family of spacelike slices, which were referred to as *locative slices*. Locative slices differ from the *maximal slices* of Markopolou [7]. Locative slices are not so large as to result in acausal influences, but large enough to capture nonlocal correlations.

In a longer unpublished version of [6], see [8], a first logical interpretation of this semantics is given. We assign to each edge a (linear) logical formula, typically an atomic formula. Then a vertex is assigned a sequent, saying that the conjunction (linear tensor) of the incoming edges entails the disjunction (linear par) of the outgoing edges. One uses logical deduction via the cut-rule to model the evolution of the system. There are several advantages to this logical approach. Having two connectives, as opposed to the single tensor, allows for more subtle encoding. We can use the linear par to indicate that two particles are (potentially) entangled, while linear tensor indicates two unentangled particles. Application of the cut-rule is a purely local phenomenon, so this logical approach seems to capture quite nicely the interaction between the local nature of events and the nonlocal nature of entanglement. But the earlier work ran into the problem that it could not handle all possible examples of evolution. Several specific examples were given. The problem was that over the course of a system evolving, two particles which had been unentangled can become entangled due to an event that is nonlocal to either. The simple linear logic calculus had no effective way to encode this situation. A solution was proposed, using something the authors called *entanglement update*, but it was felt at the time that more subtle encoding, using more connectives, should be possible.

Thus enters the new system of logics which go under the general name *deep inference*. Deep inference is a new methodology in proof theory, introduced in [9] for expressing the logic BV, and subsequently developed to the point that all major logics can be expressed with deep-inference proof systems (see [10] for a complete overview). Deep inference is more general than traditional Gentzen proof theory because proofs can be freely composed by the logical operators, instead of having a rigid formula-directed tree structure. This induces a new symmetry, which can be exploited for achieving locality of inference rules, and which is not generally achievable with Gentzen methods. Locality, in turn, makes it possible to use new methods, often with a geometric flavour, in the normalisation theory of proof systems.

Remarkably, the additional expressive power of deep inference turns out to be precisely what is needed to fully encode the sort of discrete quantum evolution that the first paper attempted to describe. The key is the noncommutativity of the added connective seq. This gives a method of encoding causal precedence directly into the syntax in a way that the original encoding of [6] using only linear logic lacked. This is the content of Theorem 4, which asserts that there is a precise correspondence between locative slices and derivable strings of formulas in the BV logic. This technical result is of independent interest beyond its use here.

2   Evolving Quantum Systems along Directed Acyclic Graphs

In earlier work [6], the basis of the representation of quantum evolution was the graph of events and causal links between them. An event could be one of the following: a unitary evolution of some subsystem, an interaction of a subsystem with a classical device (a measurement) or perhaps just the coming together or splitting apart of several spatially separated subsystems. Events will be depicted as vertices of a directed graph. The edges of the graph will represent a physical flow between the different events. The vertices of the graph are then naturally labelled with operators representing the corresponding events. We assume that there are no causal cycles; the underlying graph has to be a directed acyclic graph (DAG).

A typical dag is shown in Fig 1. The square boxes, the vertices of the dag, are events where interaction occurs. The labelled edges represent fragments of the system under scrutiny moving through spacetime. At vertex 3, for example, the components c and d come together, interact and fly apart as g and h. Each labelled edge has associated with it a Hilbert space and the state of the subsystem is represented by some density matrix. Each edge thus corresponds to a density matrix and each vertex to a physical interaction.

These dags of events could be thought of as *causal graphs* as they are an evident generalization of the causal sets of Sorkin [11]. A causal set is simply a poset, with the partial order representing causal precedence. A causal graph encodes much richer structure. So in a causal graph, we ask: What are the allowed physical effects? On physical grounds, the most general transformation of density matrices is a *completely positive, trace non-increasing map* or *superoperator* for short; see, for example, Chapter 8 of [1].

Density matrices are not just associated with edges, they are associated with larger, more distributed, subsystems as well. We need some basic terminology

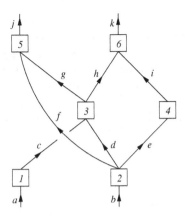

Fig. 1. A dag of events

associated with dags which brings out the causal structure more explicitly. We say that an edge e *immediately precedes* f if the target vertex of e is the source vertex of f. We say that e *precedes* f, written $e \preceq f$ if there is a chain of immediate precedence relations linking e and f, in short, "precedes" is the transitive closure of "immediately precedes". This is not quite a partial order, because we have left out reflexivity, but concepts like chain (a totally ordered subset) and antichain (a completely unordered subset) work as in partial orders.

We use the word "slice" for an antichain in the precedence order. The word is supposed to be evocative of "spacelike slice" as used in relativity, and has exactly the same significance.

A density matrix is a description of a part of a system. Thus it makes sense to ask about the density matrix associated with a part of a system that is not localized at a single event. In our dag of figure 1 we can, for example, ask about the density matrix of the portion of the system associated with the edges d, e and f. Thus density matrices can be associated with arbitrary slices. Note that it makes no sense to ask for the density matrix associated with a subset of edges that is not a slice.

The Hilbert space associated with a slice is the tensor product of the Hilbert spaces associated with the edges. Given a density matrix, say ρ, associated with, for example, the slice d, e, f, we get the density matrix for the subslice d, e by taking the partial trace over the dimensions associated with the Hilbert space f.

One can now consider a framework for evolution. One possibility, considered in [7], is to associate data with *maximal* slices and propagate from one slice to the next. Here, maximal means that to add any other vertex would destroy the antichain property. One then has to prove by examining the details of each dynamical law that the evolution is indeed causal. For example, one would like to show that the event at vertex 4 does not affect the density matrix at edge j. With data being propagated on maximal slices this does not follow automatically. One can instead work with local propagation; one keeps track of the density matrices on the individual edges only. This is indeed guaranteed to be causal, unfortunately it loses some essential nonlocal correlations. For example, the density matrices associated with the edges h and i will not reflect the fact that there might be nonlocal correlation or "entanglement" due to their common origin in the event at vertex 2. One needs to keep track of the density matrix on the slice i, h and earlier on d, e.

The main contribution of [6] was to identify a class of slices, called *locative* slices, that were large enough to keep track of all non-local correlations but "small enough" to guarantee causality.

Definition 1. *A* locative slice *is obtained as the result of taking any subset of the initial edges (all of which are assumed to be independent) and then propagating through edges without ever discarding an edge.*

In our running example, the initial slices are $\{a\}$, $\{b\}$ and $\{a, b\}$,. Just choosing for example the initial edge a as initial slice, and propagating from there gives the locatives slices $\{a\}$, $\{c\}$, $\{g, h\}$, $\{j, h\}$, $\{g, k\}$, and $\{j, h\}$.

In fact, the following is a convenient way of presenting the locative slices and their evolution[1].

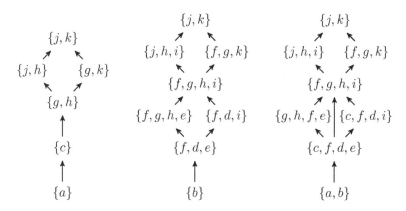

Examples of non-locative slices are c, d, e and g, h, i and g, k. The intuition behind the concept of locativity is that one never discards information (by computing partial traces) when tracking the density matrices on locative slices. This is what allows them to capture all the non-local correlations.

The prescription for computing the density matrix on a given slice, say e, given the density matrices on the incoming slices and the superoperators at the vertices is to evolve from the minimal locative slice in the past of e to the minimal locative slice containing e. Any choice of locative slices in between may be used. The main results that we proved in [6] were that the density matrix so computed is (a) independent of the choice of the slicing (covariance) and (b) only events to the causal past can affect the density matrix at e (causality). Thus the dag and the slices form the geometrical structure and the density matrices and superoperators form the dynamics.

---

[1] We thank an anonymous referee for this presentation.

3   A First Logical View of Quantum Causal Evolution

3.1   The Logic of Directed Acyclic Graphs

One of the common interpretations of a dag is as generating a simple logic. (For readers not familiar with the approach to logic discussed here, we recommend [12].) The nodes of the dag are interpreted as logical sequents of the form:

$$A_1, A_2, \ldots, A_n \vdash B_1, B_2, \ldots, B_m$$

Here $\vdash$ is the logical entailment relation. Our system will have only one inference rule, called the *Cut rule*, which states:

$$\frac{\Gamma \vdash \Delta, A \quad A, \Gamma' \vdash \Delta'}{\Gamma, \Gamma' \vdash \Delta, \Delta'}$$

Sequent rules should be interpreted as saying that if one has derived the two sequents above the line, then one can infer the sequent below the line. Proofs in the system always begin with *axioms*. Axioms are of the form $A_1, A_2, \ldots, A_n \vdash B_1, B_2, \ldots, B_m$, where $A_1, A_2, \ldots, A_n$ are the incoming edges of some vertex in our dag, and $B_1, B_2, \ldots, B_m$ will be the outgoing edges. There will be one such axiom for each vertex in our dag. For example, consider Figure 1. Then we will have the following axioms:

$$a \overset{1}{\vdash} c \quad b \overset{2}{\vdash} d, e, f \quad c, d \overset{3}{\vdash} g, h \quad e \overset{4}{\vdash} i \quad f, g \overset{5}{\vdash} j \quad h, i \overset{6}{\vdash} k$$

where we have labelled each entailment symbol with the name of the corresponding vertex. The following is an example of a deduction in this system of the sequent $a, b \vdash f, g, h, i$.

$$\frac{\dfrac{b \vdash d, e, f \quad \dfrac{a \vdash c \quad c, d \vdash g, h}{a, d \vdash g, h}}{a, b \vdash e, f, g, h} \quad e \vdash i}{a, b \vdash f, g, h, i}$$

Categorically, one can show that a dag canonically generates a free *polycategory* [13], which can be used to present an alternative formulation of the structures considered here.

3.2   The Logic of Evolution

We need to make the link between derivability in our logic and locativity. This is not completely trivial. One could, naively, define a set Δ of edges to be *derivable* if there is a deduction in the logic generated by G of $\Gamma \vdash \Delta$ where Γ is a set of initial edges. But this fails to capture some crucial examples. For example, consider the dag underlying the system in Figure 2. Corresponding to this dag, we get the following basic morphisms (axioms):

$$a \vdash b, c \quad b \vdash d \quad c \vdash e \quad d, e \vdash f.$$

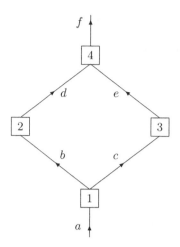

Fig. 2.

Evidently, the set $\{f\}$ is a locative slice, and yet the sequent $a \vdash f$ is not derivable. The sequent $a \vdash d, e$ is derivable, and one would like to cut it against $d, e \vdash f$, but one is only allowed to cut a single formula. Such "multicuts" are expressly forbidden, as they lead to undesirable logical properties [14].

Physically, the reason for this problem is that the sequent $d, e \vdash f$ does not encode the information that the two states at d and e are correlated. It is precisely the fact that they are correlated that implies that one would need to use a multicut. To avoid this problem, one must introduce some notation, specifically a syntax for specifying such correlations. We will use the logical connectives of the multiplicative fragment of linear logic to this end [5]. The multiplicative disjunction of linear logic, denoted $\otimes$ and called the *par* connective, will express such nonlocal correlations.

In our example, we will write the sequent corresponding to vertex 4 as $d \otimes e \vdash f$ to express the fact that the subsystems associated with these two edges are possibly entangled through interactions in their common past.

Note that whenever two (or more) subsystems emerge from an interaction, they are correlated. In linear logic, this is reflected by the following rule called the (right) *Par rule*:

$$\frac{\Gamma \vdash \Delta, A, B}{\Gamma \vdash \Delta, A \otimes B}$$

Thus we can always introduce the symbol for correlation in the right hand side of the sequent.

Notice that we can cut along a compound formula without violating any logical rules. So in the present setting, we would have the following deduction:

$$\frac{\dfrac{a \vdash b,c \quad b \vdash d}{a \vdash c,d} \quad c \vdash e}{\dfrac{\dfrac{a \vdash d,e}{a \vdash d \,\invamp\, e} \quad d \,\invamp\, e \vdash f}{a \vdash f}}$$

All the cuts in this deduction are legitimate; instead of a multicut we are cutting along a compound formula in the last step. So the first step in modifying our general prescription is to extend our dag logic, which originally contained only the cut rule, to include the connective rules of linear logic.

The above logical rule determines how one introduces a par connective on the righthand side of a sequent. For the lefthand side, one introduces pars in the axioms by the following general prescription.

Given a vertex in a multigraph, we suppose that it has incoming edges $a_1, a_2, \ldots, a_n$ and outgoing edges $b_1, b_2, \ldots, b_m$. In the previous formulation, this vertex would have been labelled with the axiom $\Gamma = a_1, a_2, \ldots, a_n \vdash b_1, b_2, \ldots, b_m$. We will now introduce several pars ($\invamp$) on the lefthand side to indicate entanglements of the sort described above. Begin by defining a relation $\sim$ by saying $a_i \sim a_j$ if there is an initial edge c and directed paths from c to a_i and from c to a_j. This is not an equivalence relation, but one takes the equivalence relation generated by the relation $\sim$. Call this new relation $\cong$. This relation partitions the set Γ into a set of equivalence classes. One then "pars" together the elements of each equivalence class, and this determines the structure of the lefthand side of our axiom. For example, consider vertices 5 and 6 in Figure 1. Vertex 5 would be labelled by $f \invamp g \vdash j$ and vertex 6 would be labelled by $h \invamp i \vdash k$. On the other hand, vertex 3 would be labelled by $c, d \vdash g, h$.

Just as the par connective indicates the existence of past correlations, we use the more familiar tensor symbol $\otimes$, which is also a connective of linear logic, to indicate the lack of nonlocal correlation. This connective also has a logical rule:

$$\frac{\Gamma \vdash \Delta, A \quad \Gamma' \vdash \Delta', B}{\Gamma, \Gamma' \vdash \Delta, \Delta', A \otimes B}$$

But we note that unlike in ordinary logic, this rule can only be applied in situations that are physically meaningful.

Definition 2. $\pi \colon \Gamma \vdash \Delta$ and $\pi' \colon \Gamma' \vdash \Delta'$ are spacelike separated *if the following two conditions are satisfied:*

- Γ and Γ' are disjoint subsets of the set of initial edges.
- The edges which make up Δ and Δ' are pairwise spacelike separated.

In our extended dag logic, we will only allow the tensor rule to be applied when the two deductions are space like separated.

Summarizing, to every dag G we associate its "logic", namely the edges are considered as formulas and vertices are axioms. We have the usual linear logical

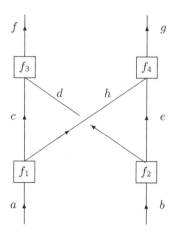

Fig. 3.

connective rules, including the cut rule which in our setting is interpreted physically as propagation. The par connective denotes correlation, and the tensor
lack of correlation. Note that every deduction in our system will conclude with
a sequent of the form $\Gamma \vdash \Delta$, where Γ is a set of initial edges.

Now one would like to modify the definition of derivability to say that a set
of edges Δ is *derivable* if in our extended dag logic, one can derive a sequent
$\Gamma \vdash \hat{\Delta}$ such that the list of edges appearing in $\hat{\Delta}$ was precisely Δ, and Γ is a set
of initial edges. However this is still not sufficient as an axiomatic approach to
capturing all locative slices. We note the example in Figure 3.

Evidently the slice $\{f, g\}$ is locative, but we claim that it cannot be derived
even in our extended logic. To this directed graph, we would associate the following axioms:

$$a \vdash c, h \quad b \vdash d, e \quad c, d \vdash f \quad h, e \vdash g$$

Note that there are no correlations between c and d or between h and e. Thus
no $\otimes$-combinations can be introduced. Now if one attempts to derive $a, b \vdash f, g$,
we proceed as follows:

$$\frac{\dfrac{a \vdash c, h \quad b \vdash d, e}{a, b \vdash c \otimes d, h, e} \quad \dfrac{c, d \vdash f}{c \otimes d \vdash f}}{a, b \vdash h, e, f}$$

At this point, we are unable to proceed. Had we attempted the symmetric approach tensoring h and e together, we would have encountered the same problem.

The problem is that our logical system is still missing one crucial aspect, and
that is that correlations develop dynamically as the system evolves, or equivalently as the deduction proceeds. We note that this logical phenomenon is
reflected in physically occurring situations. But a consequence is that our axioms must change dynamically as well. This seems to be a genuinely new logical
principle.

We give the following definition.

Definition 3. *Suppose we have a deduction π of the sequent $\Gamma \vdash \Delta$ in the logic associated to the dag G, and that T is a vertex in G to the future or acausal to the edges of the set Δ with a and b among the incoming edges of T. Then a and b are* correlated *with respect to π if there exist outgoing edges c and d of the proof π and directed paths from c to a and from d to b.*

So the point here is that when performing a deduction, one does not assign an axiom to a given vertex until it is necessary to use that axiom in the proof. Then one assigns that axiom using this new notion of correlation and the equivalence relation defined above. This prescription reflects the physical reality that entanglement of local quantum subsystems could develop as a result of a distant interaction between some other subsystems of the same quantum system. We are finally able to give the following crucial definition:

Definition 4. *A set Δ of edges in a dag G is said to be* derivable *if there is a deduction in the logic associated to G of $\Gamma \vdash \hat{\Delta}$ where $\hat{\Delta}$ is a sequence of formulas whose underlying set of edges is precisely Δ and where Γ is a set of initial edges, in fact the set of initial edges to the past of Δ.*

Theorem 1. *A set of edges is derivable if and only if it is locative. More specifically, if there is a deduction of $\Gamma \vdash \hat{\Delta}$ as described above, then Δ is necessarily locative. Conversely, given any locative slice, one can find such a deduction.*

Proof. Recall that a locative slice L is obtained from the set of initial edges in its past by an inductive procedure. At each step, we choose arbitrarily a minimal vertex u in the past of L, remove the incoming edges of u and add the outgoing edges. This step corresponds to the application of a cut rule, and the method we have used of assigning the par connective to the lefthand side of an axiom ensures that it is always a legal cut. The tensor rule is necessary in order to combine spacelike separated subsystems in order to prepare for the application of the cut rule.

Thus we have successfully given an axiomatic logic-based approach to describing evolution. In summary, to find the density matrix associated to a locative slice Δ, one finds a set of linear logic formulas whose underlying set of atoms is Δ and a deduction of $\Gamma \vdash \hat{\Delta}$ where Γ is as above.

4   Using Deep Inference to Capture Locativity

In the previous sections we explained the approach of [6], using as key unit of deduction a sequent $a_1, \ldots, a_k \vdash b_1, \ldots, b_l$ meaning that the slice $\{b_1, \ldots, b_l\}$ is *reachable* from $\{a_1, \ldots, a_k\}$ by firing a number of events (vertices). However, this approach is not able to entirely capture the notion of locative slices, because correlations develop dynamically as the system evolves, or equivalently, as the deduction proceeds. Thus, we had to let axioms evolve dynamically.

The deep reason behind this problem is that the underlying logic is multiplicative linear logic (MLL): The sequent above represents the formula $a_1 \otimes \cdots \otimes a_k \multimap b_1 \mathbin{\bindnasrepma} \cdots \mathbin{\bindnasrepma} b_l$ or equivalently $a_1^\perp \mathbin{\bindnasrepma} \cdots \mathbin{\bindnasrepma} a_k^\perp \mathbin{\bindnasrepma} b_1 \mathbin{\bindnasrepma} \cdots \mathbin{\bindnasrepma} b_l$, i.e., the logic is not able to see the aspect of *time* in the causality. For this reason we propose to use the logic BV, which is essentially MLL (with mix) enhanced by a third binary connective $\lhd$ (called *seq* or *before*) which is associative and non-commutative and self-dual, i.e., the negation of $A \lhd B$ is $A^\perp \lhd B^\perp$. It is this non-commutative connective, which allows us to properly capture quantum causality.

Of course, we are interested in expressing our logic in a deductive system that admits a complete cut-free presentation. In this case, as we briefly argue in the following, the adoption of deep inference is necessary to deal with a self-dual non-commutative logical operator.

4.1   Review of BV and Deep Inference

The significance of deep inference systems was discussed in the introduction. We note now that within the range of the deep-inference methodology, we can define several formalisms, *i.e.* general prescriptions (like the sequent calculus or natural deduction) on how to design proof systems. The first, and conceptually simplest, formalism that has been defined in deep inference is called the *calculus of structures*, or *CoS*, and this is what we adopt in this paper and call "deep inference". In fact, the fine proof-theoretic points about the various deep inference formalisms are not relevant to this paper.

The proof theory of deep inference is now well developed for classical [15], intuitionistic [16,17], linear [18,19] and modal [20,21] logics. More relevant to us, there is an extensive literature on BV and commutative/non-commutative linear logics containing BV. We cannot here provide a tutorial on BV, so we refer to its literature. In particular, [9] provides the semantic motivation and intuition behind BV, together with examples of its use. In [22], Tiu shows that deep inference is necessary for giving a cut-free deductive system for the logic BV. Kahramanoğulları proves that System BV is NP-complete [23].

We now proceed to define system BV, quickly and informally. The inference rules are:

$$\mathsf{ai}{\downarrow}\,\frac{F\{\circ\}}{F\{a \mathbin{\bindnasrepma} a^\perp\}} \qquad \mathsf{s}\,\frac{F\{A \otimes [B \mathbin{\bindnasrepma} C]\}}{F\{(A \otimes B) \mathbin{\bindnasrepma} C\}} \qquad \mathsf{ai}{\uparrow}\,\frac{F\{a \otimes a^\perp\}}{F\{\circ\}}$$

$$\mathsf{q}{\downarrow}\,\frac{F\{[A \mathbin{\bindnasrepma} C] \lhd [B \mathbin{\bindnasrepma} D]\}}{F\{\langle A \lhd B \rangle \mathbin{\bindnasrepma} \langle C \lhd D \rangle\}} \qquad \mathsf{q}{\uparrow}\,\frac{F\{\langle A \lhd B \rangle \otimes \langle C \lhd D \rangle\}}{F\{(A \otimes C) \lhd (B \otimes D)\}}$$

They have to be read as ordinary rewrite rules acting on the formulas inside arbitrary contexts $F\{\ \}$. Note that we push negation via DeMorgan equalities to the atoms, and thus, all contexts are positive. The letters A, B, C, D stand for arbitrary formulas and a is an arbitrary atom. Formulas are considered equal modulo the associativity of all three connectives $\mathbin{\bindnasrepma}$, $\lhd$, and $\otimes$, the commutativity of the two connectives $\mathbin{\bindnasrepma}$ and $\otimes$, and the unit laws for $\circ$, which is unit to all three connectives, i.e., $A = A \mathbin{\bindnasrepma} \circ = A \otimes \circ = A \lhd \circ = \circ \lhd A$.

Since, in our experience, working modulo equality is a sticky point of deep inference, we invite the reader to meditate on the following examples which are some of the possible instances of the $q\downarrow$ rule:

$$q\downarrow\frac{\langle[a\,\eightpointstar\,c]\triangleleft[b\,\eightpointstar\,d]\rangle\,\eightpointstar\,e}{\langle a\triangleleft b\rangle\,\eightpointstar\,\langle c\triangleleft d\rangle\,\eightpointstar\,e},\qquad q\downarrow\frac{[\langle a\triangleleft b\rangle\,\eightpointstar\,c\,\eightpointstar\,e]\triangleleft d}{\langle a\triangleleft b\rangle\,\eightpointstar\,\langle c\triangleleft d\rangle\,\eightpointstar\,e},\qquad q\downarrow\frac{\langle c\triangleleft d\triangleleft a\triangleleft b\rangle\,\eightpointstar\,e}{\langle a\triangleleft b\rangle\,\eightpointstar\,\langle c\triangleleft d\rangle\,\eightpointstar\,e}.$$

By referring to the previously defined $q\downarrow$ rule scheme, we can see that the second instance above is produced by taking $F\{\ \}=\{\ \}$, $A=\langle a\triangleleft b\rangle\,\eightpointstar\,e$, $B=\circ$, $C=c$ and $D=d$, and the third instance is produced by taking $F\{\ \}=\{\ \}\,\eightpointstar\,e$, $A=c\triangleleft d$, $B=\circ$, $C=\circ$ and $D=a\triangleleft b$. The best way to understand the rules of BV is to learn their intuitive meaning, which is explained by an intuitive "space-temporal" metaphor in [9].

The set of rules $\{ai\downarrow,ai\uparrow,s,q\downarrow,q\uparrow\}$ is called SBV, and the set $\{ai\downarrow,s,q\downarrow\}$ is called BV. We write

$$
\begin{array}{c}
A\\
\Delta\,\|\,\text{SBV}\\
B
\end{array}
$$

to denote a derivation Δ from premise A to conclusion B using SBV, and we do analogously for BV.

Much like in the sequent calculus, we can consider BV a cut-free system, while SBV is essentially BV plus a cut rule. The two are related by the following theorem.

Theorem 2. *For all formulas A and B, we have*

$$
\begin{array}{c}
A\\
\|\,\text{SBV}\\
B
\end{array}
\qquad\text{\emph{if and only if}}\qquad
\begin{array}{c}
\circ\\
\|\,\text{BV}\\
A^{\perp}\,\eightpointstar\,B
\end{array}.
$$

Again, all the details are explained in [9]. Let us here only mention that the usual cut elimination is a special case of Theorem 2, for $A=\circ$. Then it says that a formula B is provable in BV iff it is provable in SBV.

Observation 3. *If a formula A is provable in BV, then every atom a occurs as often in A as $a^{\perp}$. This is easy to see: the only possibility for an atom a to disappear is in an instance of $ai\downarrow$; but then at the same time an atom $a^{\perp}$ disappears.*

Definition 5. *A BV formula Q is called a negation cycle if there is a nonempty set of atoms $\mathscr{P}=\{a_0,a_2,\ldots,a_{n-1}\}$, such that no two atoms in $\mathscr{P}$ are dual, $i\neq j$ implies $a_i\neq a_j$, and such that $Q=Z_0\,\eightpointstar\cdots\,\eightpointstar\,Z_{n-1}$, where, for every $j=0,\ldots,n-1$, we have $Z_j=a_j\otimes a_{j+1\ (\mathrm{mod}\ n)}^{\perp}$ or $Z_j=a_j\triangleleft a_{j+1\ (\mathrm{mod}\ n)}^{\perp}$. We say that a formula P contains a negation cycle if there is a negation cycle Q such that*

- *Q can be obtained from P by replacing some atoms in P by $\circ$, and*
- *all the atoms that occur in Q occur only once in P.*

Example 1. The formula $(a \otimes c \otimes [d^\perp \mathbin{\rotatebox[origin=c]{180}{\&}} b]) \mathbin{\rotatebox[origin=c]{180}{\&}} c^\perp \mathbin{\rotatebox[origin=c]{180}{\&}} \langle b^\perp \vartriangleleft [a^\perp \mathbin{\rotatebox[origin=c]{180}{\&}} d] \rangle$ contains a negation cycle $(a \otimes b) \mathbin{\rotatebox[origin=c]{180}{\&}} \langle b^\perp \vartriangleleft a^\perp \rangle = (a \otimes \circ \otimes [\circ \mathbin{\rotatebox[origin=c]{180}{\&}} b]) \mathbin{\rotatebox[origin=c]{180}{\&}} \circ \mathbin{\rotatebox[origin=c]{180}{\&}} \langle b^\perp \vartriangleleft [a^\perp \mathbin{\rotatebox[origin=c]{180}{\&}} \circ] \rangle$.

Proposition 1. *Let A be a* BV *formula. If P contains a negation cycle, then P is not provable in* BV.

A proof of this propostion can be found in [24, Proposition 7.4.30]. A symmetric version of this proposition has been shown for SBV in [25, Lemma 5.20].

4.2   Locativity via BV

Let us now come back to dags. A vertex $v \in \mathcal{V}$ in such a graph $\mathcal{G} = (\mathcal{V}, \mathcal{E})$ is now encoded by the formula

$$V = (a_1^\perp \otimes \cdots \otimes a_k^\perp) \vartriangleleft [b_1 \mathbin{\rotatebox[origin=c]{180}{\&}} \cdots \mathbin{\rotatebox[origin=c]{180}{\&}} b_l]$$

where $\{a_1, \ldots, a_k\} = \mathsf{target}^{-1}(v)$ is the set of edges having their target in v, and $\{b_1, \ldots, b_l\} = \mathsf{source}^{-1}(v)$ is the set of edges having their source in v. For a slice $\mathcal{S} = \{e_1, \ldots, e_n\} \subseteq \mathcal{E}$ we define its encoding to be the formula $S = e_1 \mathbin{\rotatebox[origin=c]{180}{\&}} \cdots \mathbin{\rotatebox[origin=c]{180}{\&}} e_n$.

Lemma 1. *Let $(\mathcal{V}, \mathcal{E})$ be a dag, let $\mathcal{S} \subseteq \mathcal{E}$ be a slice, let $v \in \mathcal{V}$ be such that $\mathsf{target}^{-1}(v) \subseteq \mathcal{S}$, and let $\mathcal{S}'$ be the propagation of $\mathcal{S}$ through v. Then there is a derivation*

$$
\begin{array}{c}
S \otimes V \\
\Big\| \, \mathsf{SBV} \\
S'
\end{array}
\tag{1}
$$

where V, S, and S' are the encodings of v, $\mathcal{S}$, and $\mathcal{S}'$, respectively.

Proof. Assume $\mathsf{source}^{-1}(v) = \{b_1, \ldots, b_l\}$ and $\mathsf{target}^{-1}(v) = \{a_1, \ldots, a_k\}$ and $\mathcal{S} = \{e_1, \ldots, e_m, a_1, \ldots, a_k\}$. Then $\mathcal{S}' = \{e_1, \ldots, e_m, b_1, \ldots, b_l\}$. Now we can construct

$$
\begin{array}{c}
[e_1 \mathbin{\rotatebox[origin=c]{180}{\&}} \cdots \mathbin{\rotatebox[origin=c]{180}{\&}} e_m \mathbin{\rotatebox[origin=c]{180}{\&}} a_1 \mathbin{\rotatebox[origin=c]{180}{\&}} \cdots \mathbin{\rotatebox[origin=c]{180}{\&}} a_k] \otimes \langle (a_1^\perp \otimes \cdots \otimes a_k^\perp) \vartriangleleft [b_1 \mathbin{\rotatebox[origin=c]{180}{\&}} \cdots \mathbin{\rotatebox[origin=c]{180}{\&}} b_l] \rangle \\
\hline
{\scriptstyle s} \quad e_1 \mathbin{\rotatebox[origin=c]{180}{\&}} \cdots \mathbin{\rotatebox[origin=c]{180}{\&}} e_m \mathbin{\rotatebox[origin=c]{180}{\&}} ([a_1 \mathbin{\rotatebox[origin=c]{180}{\&}} \cdots \mathbin{\rotatebox[origin=c]{180}{\&}} a_k] \otimes \langle (a_1^\perp \otimes \cdots \otimes a_k^\perp) \vartriangleleft [b_1 \mathbin{\rotatebox[origin=c]{180}{\&}} \cdots \mathbin{\rotatebox[origin=c]{180}{\&}} b_l] \rangle) \\
\hline
{\scriptstyle q\uparrow} \quad e_1 \mathbin{\rotatebox[origin=c]{180}{\&}} \cdots \mathbin{\rotatebox[origin=c]{180}{\&}} e_m \mathbin{\rotatebox[origin=c]{180}{\&}} \langle ([a_1 \mathbin{\rotatebox[origin=c]{180}{\&}} \cdots \mathbin{\rotatebox[origin=c]{180}{\&}} a_k] \otimes a_1^\perp \otimes \cdots \otimes a_k^\perp) \vartriangleleft [b_1 \mathbin{\rotatebox[origin=c]{180}{\&}} \cdots \mathbin{\rotatebox[origin=c]{180}{\&}} b_l] \rangle \\
\hline
{\scriptstyle s} \quad e_1 \mathbin{\rotatebox[origin=c]{180}{\&}} \cdots \mathbin{\rotatebox[origin=c]{180}{\&}} e_m \mathbin{\rotatebox[origin=c]{180}{\&}} \langle ([(a_1 \otimes a_1^\perp) \mathbin{\rotatebox[origin=c]{180}{\&}} a_2 \mathbin{\rotatebox[origin=c]{180}{\&}} \cdots \mathbin{\rotatebox[origin=c]{180}{\&}} a_k] \otimes \cdots \otimes a_k^\perp) \vartriangleleft [b_1 \mathbin{\rotatebox[origin=c]{180}{\&}} \cdots \mathbin{\rotatebox[origin=c]{180}{\&}} b_l] \rangle \\
\hline
{\scriptstyle ai\uparrow} \quad e_1 \mathbin{\rotatebox[origin=c]{180}{\&}} \cdots \mathbin{\rotatebox[origin=c]{180}{\&}} e_m \mathbin{\rotatebox[origin=c]{180}{\&}} \langle ([a_2 \mathbin{\rotatebox[origin=c]{180}{\&}} \cdots \mathbin{\rotatebox[origin=c]{180}{\&}} a_k] \otimes a_2^\perp \otimes \cdots \otimes a_k^\perp) \vartriangleleft [b_1 \mathbin{\rotatebox[origin=c]{180}{\&}} \cdots \mathbin{\rotatebox[origin=c]{180}{\&}} b_l] \rangle \\
{\scriptstyle s} \\
\vdots
\end{array}
$$

$$
\begin{array}{c}
{\scriptstyle ai\uparrow} \quad e_1 \mathbin{\rotatebox[origin=c]{180}{\&}} \cdots \mathbin{\rotatebox[origin=c]{180}{\&}} e_m \mathbin{\rotatebox[origin=c]{180}{\&}} \langle ([a_{k-1} \mathbin{\rotatebox[origin=c]{180}{\&}} a_k] \otimes a_{k-1}^\perp \otimes a_k^\perp) \vartriangleleft [b_1 \mathbin{\rotatebox[origin=c]{180}{\&}} \cdots \mathbin{\rotatebox[origin=c]{180}{\&}} b_l] \rangle \\
\hline
{\scriptstyle s} \quad e_1 \mathbin{\rotatebox[origin=c]{180}{\&}} \cdots \mathbin{\rotatebox[origin=c]{180}{\&}} e_m \mathbin{\rotatebox[origin=c]{180}{\&}} \langle ([(a_{k-1} \otimes a_{k-1}^\perp) \mathbin{\rotatebox[origin=c]{180}{\&}} a_k] \otimes a_k^\perp) \vartriangleleft [b_1 \mathbin{\rotatebox[origin=c]{180}{\&}} \cdots \mathbin{\rotatebox[origin=c]{180}{\&}} b_l] \rangle \\
\hline
{\scriptstyle ai\uparrow} \quad e_1 \mathbin{\rotatebox[origin=c]{180}{\&}} \cdots \mathbin{\rotatebox[origin=c]{180}{\&}} e_m \mathbin{\rotatebox[origin=c]{180}{\&}} \langle (a_k \otimes a_k^\perp) \vartriangleleft [b_1 \mathbin{\rotatebox[origin=c]{180}{\&}} \cdots \mathbin{\rotatebox[origin=c]{180}{\&}} b_l] \rangle \\
\hline
{\scriptstyle ai\uparrow} \quad e_1 \mathbin{\rotatebox[origin=c]{180}{\&}} \cdots \mathbin{\rotatebox[origin=c]{180}{\&}} e_m \mathbin{\rotatebox[origin=c]{180}{\&}} \langle \circ \vartriangleleft [b_1 \mathbin{\rotatebox[origin=c]{180}{\&}} \cdots \mathbin{\rotatebox[origin=c]{180}{\&}} b_l] \rangle \\
= \overline{\quad e_1 \mathbin{\rotatebox[origin=c]{180}{\&}} \cdots \mathbin{\rotatebox[origin=c]{180}{\&}} e_m \mathbin{\rotatebox[origin=c]{180}{\&}} b_1 \mathbin{\rotatebox[origin=c]{180}{\&}} \cdots \mathbin{\rotatebox[origin=c]{180}{\&}} b_l \quad}
\end{array} \, ,
$$

as desired.

Lemma 2. *Let $(\mathcal{V}, \mathcal{E})$ be a dag, let $\mathcal{S}, \mathcal{S}' \subseteq \mathcal{E}$ be slices, such that $\mathcal{S}'$ is reachable from $\mathcal{S}$ by firing a number of events (vertices). Then there is a derivation*

$$
\begin{array}{c}
S \otimes V_1 \otimes \cdots \otimes V_n \\
\left\| \, \mathsf{SBV} \right. \\
S'
\end{array}
\qquad (2)
$$

where $V_1, \ldots, V_n$ encode $v_1, \ldots, v_n \in \mathcal{V}$ (namely, the vertices through which the slices are propagated), and S, S' encode $\mathcal{S}$, $\mathcal{S}'$.

Proof. If $\mathcal{S}'$ is reachable from $\mathcal{S}$ then there is an $n \geq 0$ and slices $\mathcal{S}_0, \ldots, \mathcal{S}_n \subseteq \mathcal{E}$ and vertices $v_1, \ldots, v_n \in \mathcal{V}$ such that for all $i \in \{1, \ldots, n\}$ we have that $\mathcal{S}_i$ is the propagation of $\mathcal{S}_{i-1}$ through v_i, and $\mathcal{S} = \mathcal{S}_0$ and $\mathcal{S}' = \mathcal{S}_n$. Now we can apply Lemma 1 n times to get the derivation (2).

Lemma 3. *Let $(\mathcal{V}, \mathcal{E})$ be a dag, let S and S' be the encodings of $\mathcal{S}, \mathcal{S}' \subseteq \mathcal{E}$, where $\mathcal{S}$ is a slice. Further, let $V_1, \ldots, V_n$ be the encodings of $v_1, \ldots, v_n \in \mathcal{V}$. If there is a proof*

$$
\begin{array}{c}
\Pi \left\| \, \mathsf{BV} \right. \\
V_1^\perp \parr \cdots \parr V_n^\perp \parr S^\perp \parr S'
\end{array}
$$

then $\mathcal{S}'$ is a slice reachable from $\mathcal{S}$ and $v_1, \ldots, v_n$ are the vertices through which it is propagated.

Proof. By induction on n. If $n = 0$, we have a proof of $S^\perp \parr S'$. Since $S^\perp$ contains only negated propositional variables, and S' only non-negated ones, we have that every atom in S' has its killer in $S^\perp$. Therefore $\mathcal{S}' = \mathcal{S}$. Let now $n \geq 1$. We can assume that $S' = e_1 \parr \cdots \parr e_m$, and that for every $i \in \{1, \ldots, n\}$ we have $V_i^\perp = [a_{i1} \parr \cdots \parr a_{ik_i}] \lhd (b_{i1}^\perp \otimes \cdots \otimes b_{il_i}^\perp)$. i.e., $\mathsf{target}^{-1}(v_i) = \{a_{i1}, \ldots, a_{ik_i}\}$ and $\mathsf{source}^{-1}(v_i) = b_{i1}, \ldots, b_{il_i}$. Now we claim that there is an $i \in \{1, \ldots, n\}$ such that $\{b_{i1}, \ldots, b_{il_i}\} \subseteq \{e_1, \ldots, e_m\}$. In other words, there is a vertex among the $v_1, \ldots, v_n$, such that all its outgoing edges are in $\mathcal{S}'$. For showing this claim assume by way of contradiction that every vertex among $v_1, \ldots, v_n$ has an outgoing edge that does not appear in $\mathcal{S}'$, i.e., for all $i \in \{1, \ldots, n\}$, there is an $s_i \in 1, \ldots, l_i$ with $b_{is_i} \notin \{e_1, \ldots, e_m\}$. By Observation 3, we must have that for every $i \in \{1, \ldots, n\}$ there is a $j \in \{1, \ldots, n\}$ with $b_{is_i} \in \{a_{j1}, \ldots, a_{jk_j}\}$, i.e., the killer of b_{is_i} occurs as incoming edge of some vertex v_j. Let $\mathsf{jump} \colon \{1, \ldots, n\} \to \{1, \ldots, n\}$ be a function that assigns to every i such a j (there might be many of them, but we pick just one). Now let $i_1 = 1$, $i_2 = \mathsf{jump}(i_1)$, $i_3 = \mathsf{jump}(i_2)$, and so on. Since there are only finitely many V_i, we have an p and q with $p \leq q$ and $i_{q+1} = i_p$. Let us take the minimal such q, i.e., $i_p, \ldots, i_q$ are all different. Inside the proof Π above, we now replace everywhere all atoms by $\circ$, except for $b_{i_p}, b_{i_p}^\perp, \ldots, b_{i_q}, b_{i_q}^\perp$. By this, the proof remains valid and has conclusion

$$
\langle b_{i_q} \lhd b_{i_p}^\perp \rangle \parr \langle b_{i_p} \lhd b_{i_{p+1}}^\perp \rangle \parr \cdots \parr \langle b_{i_{q-1}} \lhd b_{i_q}^\perp \rangle \quad ,
$$

which is a contradiction to Proposition 1. This finishes the proof of the claim.

Now we can, without loss of generality, assume that v_n is the vertex with all its outgoing edges in $\mathscr{S}'$, i.e., $\{b_{n1}, \ldots, b_{nl_n}\} \subseteq \{e_1, \ldots, e_m\}$, and (again without loss of generality) $e_1 = b_{n1}, \ldots, e_{l_n} = b_{nl_n}$. Our proof Π looks therefore as follows:

$$V_1^\perp \,\mathord{\otimes}\, \cdots \,\mathord{\otimes}\, V_{n-1}^\perp \,\mathord{\otimes}\, S^\perp \,\mathord{\otimes}\, \underbrace{\langle [a_{n1} \,\mathord{\otimes}\, \cdots \,\mathord{\otimes}\, a_{nk_n}] \vartriangleleft (b_{n1}^\perp \otimes \cdots \otimes b_{nl_n}^\perp) \rangle}_{V_n^\perp} \,\mathord{\otimes}\, S'$$

$$\left\| \begin{array}{c} \Pi \\ \end{array} \mathsf{BV} \right.$$

where $S' = b_{n1} \,\mathord{\otimes}\, \cdots \,\mathord{\otimes}\, b_{nl_n} \,\mathord{\otimes}\, e_{l_n+1} \,\mathord{\otimes}\, \cdots \,\mathord{\otimes}\, e_m$. In Π we can now replace the atoms $b_{n1}, b_{n1}^\perp, \ldots, b_{nl_n}, b_{nl_n}^\perp$ everywhere by $\circ$. This yields a valid proof

$$V_1^\perp \,\mathord{\otimes}\, \cdots \,\mathord{\otimes}\, V_{n-1}^\perp \,\mathord{\otimes}\, S^\perp \,\mathord{\otimes}\, a_{n1} \,\mathord{\otimes}\, \cdots \,\mathord{\otimes}\, a_{nk_n} \,\mathord{\otimes}\, e_{l_n+1} \,\mathord{\otimes}\, \cdots \,\mathord{\otimes}\, e_m$$

$$\left\| \begin{array}{c} \Pi' \\ \end{array} \mathsf{BV} \right.$$

to which we can apply the induction hypothesis, from which we can conclude that

$$\mathscr{S}'' = \{a_{n1}, \ldots, a_{nk_n}, e_{l_n+1}, \ldots, e_m\}$$

is a slice that is reachable from S. Clearly $\mathscr{S}'$ is the propagation of $\mathscr{S}''$ through v_n, and therefore it is a slice and reachable from $\mathscr{S}$.

Theorem 4. *Let* $\mathscr{G} = (\mathscr{V}, \mathscr{E})$ *be a dag. A subset* $\mathscr{S} \subseteq \mathscr{E}$ *is a locative slice if and only if there is a derivation*

$$I \otimes V_1 \otimes \ldots \otimes V_n$$
$$\left\| \mathsf{SBV} \right.$$
$$S$$

,

where S *is the encoding of* $\mathscr{S}$, *and* I *is the encoding of a subset of the initial edges, and* $V_1, \ldots, V_n$ *encode* $v_1, \ldots, v_n \in \mathscr{V}$.

Proof. The "only if" direction follows immediately from Lemma 2. For the "if" direction, we first apply Theorem 2, and then Lemma 3.

5 Conclusion

Having a logical syntax also leads to the possibility of discussing semantics; this would be a mathematical universe in which the logical structure can be interpreted. This has the potential to be of great interest in the physical systems we are considering here, where one would want to calculate such things as expectation values. As in any categorical interpretation of a logic, one needs a category with appropriate structure to support the logical connectives and model the inference rules. The additional logical connectives of BV allows for more subtle encodings than can be expressed in a compact closed category.

The structure of BV leads to interesting category-theoretic considerations [26]. One must find a category with the following structure:

- *-autonomous structure, i.e. the category must be symmetric, monoidal closed and self-dual.
- an additional (noncommutative) monoidal structure commuting with the above duality.
- coherence isomorphisms necessary to interpret the logic, describing the interaction of the various tensors.

Such categories are called BV-*categories* in [26]. Of course, trivial examples abound. One can take the category Rel of sets and relations, modelling all three monoidal structures as one. Similarly the category of (finite-dimensional) Hilbert spaces, or any symmetric compact closed category would suffice. But what is wanted is a category in which the third monoidal structure is genuinely noncommutative.

While this already poses a significant challenge, we are here faced with the added difficulty that we would like the category to have some physical significance, to be able to interpret the quantum events described in this paper. Fortunately, work along these lines has already been done. See [26].

That paper considers the category of Girard's *probabilistic coherence spaces* PCS, introduced in [27]. While Girard demonstrates the *-autonomous structure, the paper [26] shows that the category properly models the additional noncommutative tensor of BV. We note that the paper [27] also has a notion of *quantum coherence space*, where analogous structure can be found.

Roughly, a probabilistic coherence space is a set X equipped with a set of generalized measures, i.e. functions to the set of nonnegative reals. These are called the *allowable* generalized measures. The set must be closed with respect to the double dual operation, where duality is determined by *polarity*, where we say that two generalized measures on X are polar, written $f \perp g$, if

$$\sum_{x \in X} f(x)g(x) \leq 1$$

The noncommutative connective is then modelled by the formula:

$$A \oslash B = \{ \, \textstyle\sum_{i=1}^{n} f_i \otimes g_i \mid f_i \text{ is an allowable measure on } A \text{ and }$$
$$\textstyle\sum_{i=1}^{n} g_i \text{ is an allowable measure on } B \, \}$$

Note the lack of symmetry in the definition. Both the categories of probabilistic and quantum coherence spaces will likely provide physically interesting semantics of the discrete quantum dynamics presented here. We hope to explore this in future work.

Acknowledgements. Research supported in part by NSERC, by the ANR project "INFER" and the INRIA ARC "Redo".

References

1. Nielsen, M., Chuang, I.: Quantum Computation and Quantum Information. Cambridge University Press (2000)
2. Abramsky, S., Coecke, B.: A categorical semantics of quantum protocols. In: Proceedings of the 19th Annual IEEE Symposium on Logic in Computer Science 2004, pp. 415–425. IEEE Computer Society (2004)
3. Abramsky, S., Coecke, B.: Physics from computer science. International Journal of Unconventional Computing 3, 179–197 (2007)
4. Lambek, J., Scott, P.: Introduction to Higher-order Categorical Logic. Cambridge Univ. Press (1986)
5. Girard, J.Y.: Linear logic: Its syntax and semantics. In: Girard, J.Y., Lafont, Y., Regnier, L. (eds.) Advances in Linear Logic. LMS Lecture Note Series, vol. 222, pp. 1–42. Cambridge University Press (1995)
6. Blute, R., Ivanov, I., Panangaden, P.: Discrete quantum causal dynamics. International Journal of Theoretical Phsysics 42, 2025–2041 (2003)
7. Markopoulou, F.: Quantum causal histories. Classical and Quantum Gravity 17, 2059–2077 (2000)
8. Blute, R.F., Ivanov, I.T., Panangaden, P.: Discrete quantum causal dynamics (2001), http://lanl.arxiv.org/abs/gr-qc/0109053
9. Guglielmi, A.: A system of interaction and structure. ACM Transactions on Computational Logic 8(1), 1–64 (2007), http://cs.bath.ac.uk/ag/p/SystIntStr.pdf
10. Guglielmi, A.: Deep inference, http://alessio.guglielmi.name/res/cos
11. Sorkin, R.: Spacetime and causal sets. In: D'Olivo, J., et al. (eds.) Relativity and Gravitation: Classical and Quantum. World Scientific (1991)
12. Girard, J.Y., Lafont, Y., Taylor, P.: Proofs and Types. Cambridge Tracts in Theoretical Computer Science. Cambridge University Press (1989)
13. Szabo, M.: Polycategories. Communications in Algebra 3, 663–689 (1975)
14. Blute, R., Cockett, J., Seely, R., Trimble, T.: Natural deduction and coherence for weakly distributive categories. Journal of Pure and Applied Algebra 113, 229–296 (1996)
15. Brünnler, K.: Locality for classical logic. Notre Dame Journal of Formal Logic 47(4), 557–580 (2006),
    http://www.iam.unibe.ch/~kai/Papers/LocalityClassical.pdf
16. Tiu, A.F.: A local system for intuitionistic logic. In: Hermann, M., Voronkov, A. (eds.) LPAR 2006. LNCS (LNAI), vol. 4246, pp. 242–256. Springer, Heidelberg (2006)
17. Guenot, N.: Nested Deduction in Logical Foundations for Computation. Phd thesis, Ecole Polytechnique (2013)
18. Straßburger, L.: A local system for linear logic. In: Baaz, M., Voronkov, A. (eds.) LPAR 2002. LNCS (LNAI), vol. 2514, pp. 388–402. Springer, Heidelberg (2002)
19. Straßburger, L.: MELL in the calculus of structures. Theoretical Computer Science 309, 213–285 (2003),
    http://www.lix.polytechnique.fr/~lutz/papers/els.pdf
20. Brünnler, K.: Deep sequent systems for modal logic. In: Governatori, G., Hodkinson, I., Venema, Y. (eds.) Advances in Modal Logic, vol. 6, pp. 107–119. College Publications (2006), http://www.aiml.net/volumes/volume6/Bruennler.ps
21. Straßburger, L.: Cut elimination in nested sequents for intuitionistic modal logics. In: Pfenning, F. (ed.) FOSSACS 2013. LNCS, vol. 7794, pp. 209–224. Springer, Heidelberg (2013)

22. Tiu, A.: A system of interaction and structure II: The need for deep inference. Logical Methods in Computer Science 2(2:4), 1–24 (2006), http://arxiv.org/pdf/cs.LO/0512036
23. Kahramanoğulları, O.: System BV is NP-complete. Annals of Pure and Applied Logic 152(1-3), 107–121 (2007), http://dx.doi.org/10.1016/j.apal.2007.11.005
24. Straßburger, L.: Linear Logic and Noncommutativity in the Calculus of Structures. PhD thesis, Technische Universität Dresden (2003)
25. Straßburger, L., Guglielmi, A.: A system of interaction and structure IV: The exponentials and decomposition. ACM Trans. Comput. Log. 12(4), 23 (2011)
26. Blute, R.F., Panangaden, P., Slavnov, S.: Deep inference and probabilistic coherence spaces. Applied Categorical Structures 20, 209–228 (2012)
27. Girard, J.Y.: Between Logic and Quantic: A Tract. LMS Lecture Note Series, vol. 316, ch. 10, pp. 346–381. Cambridge University Press (2004)

Learning Lambek Grammars from Proof Frames

Roberto Bonato[1,*] and Christian Retoré[2,**]

[1] Questel SAS, Sophia Antipolis, France
[2] IRIT, Toulouse, France & Univ. Bordeaux, France

Abstract. In addition to their limpid interface with semantics, categorial grammars enjoy another important property: learnability. This was first noticed by Buszkowski and Penn and further studied by Kanazawa, for Bar-Hillel categorial grammars.

What about Lambek categorial grammars? In a previous paper we showed that product free Lambek grammars are learnable from structured sentences, the structures being incomplete natural deductions. Although these grammars were shown to be unlearnable from strings by Foret ad Le Nir, in the present paper, we show that Lambek grammars, possibly with product, are learnable from proof frames i.e. incomplete proof nets.

After a short reminder on grammatical inference à la Gold, we provide an algorithm that learns Lambek grammars with product from proof frames and we prove its convergence. We do so for 1-valued "(also known as rigid) Lambek grammars with product, since standard techniques can extend our result to k-valued grammars. Because of the correspondence between cut-free proof nets and normal natural deductions, our initial result on product free Lambek grammars can be recovered.[1]

*We are glad to dedicate the present paper to
Jim Lambek for his 90th birthday: he is the living proof that research is
an eternal learning process.*

1 Presentation

Generative grammar exhibited two characteristic properties of the syntax of human languages that distinguish them from other formal languages:

1. Sentences should be easily parsed and generated, since we speak and understand each other in real time.
2. Any human language should be easily learnable, preferably from not so many positive examples, as first language acquisition shows.

---

[*] I am deeply indebted to my co-author for having taken up again after so many years our early work on learnability for k-valued Lambek grammars, extended and coherently integrated it into the framework of learnability from proof frames.

[**] Thanks to IRIT-CNRS for hosting me during my sabbatical, to the Loci ANR project for its intellectual and financial support, to C. Casadio, M. Moortgat for their encouragement and to A. Foret for her helpful remarks.

[1] At the turn of the millenium, our initial work benefited from a number of valuable discussions with Philippe Darondeau. We are very sorry to learn of his premature passing. Adieu, Philippe.

C. Casadio et al. (Eds.): Lambek Festschrift, LNCS 8222, pp. 108–135, 2014.

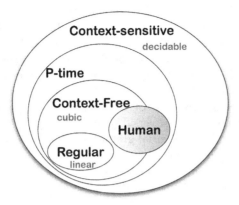

Fig. 1. Human languages and the classes of the Chomsky hierarchy (with parsing complexity)

Formally, the first point did receive a lot of attention, leading to the class of mildly context sensitive languages [20]: they enjoy polynomial parsing but are rich enough to describe natural language syntax. A formal account of learnability was harder to find. Furthermore, as soon as a notion of formal learnability was proposed, the first results seemed so negative that the learnability criterion was left out of the design of syntactical formalisms. This negative result stated that whenever a class of languages contains all the regular languages it cannot be learnt.

By that time, languages were viewed through the Chomsky hierarchy (see figure 1) and given that regular languages are the simplest and that human languages were known to include non regular languages, an algorithm that learns the syntax of a human language from positive examples was considered as impossible. This pessimistic viewpoint was erroneous for at least two reasons:

- The class of human languages does not include all regular languages and it is likely that it does to even include a single regular language, see on figure 1 the present hypothesis on human languages.
- The positive examples were thought to be sequences of words, while it has been shown long ago that grammatical rules operate on *structured* sentences and phrases (that are rather trees or graphs), see e.g. [8] for a recent account.

Gold's notion of *learning a class of languages* generated by a class of grammars $\mathcal{G}$ — that we shall recall in the first section of the present paper — is that a learning function ϕ maps a sequence of sentences $e_1, \ldots, e_n$ to a grammar $G_n = \phi(e_1, \ldots, e_n)$ in the class in such a way that, when the examples enumerate a language $\mathcal{L}(G)$ in the class $\mathcal{G}$, there exists an integer N such that if $n > N$ the G_n are constantly equal to G_N generating the same language i.e. $\mathcal{L}(G_n) = \mathcal{L}(G)$. The fundamental point is that the function learns a *class* of languages: the algorithm eventually finds out that the enumerated language cannot be any other language in the class. Therefore the very same language can be *learnable* as a member of a learnable class of languages, and *unlearnable* as the member of another class of languages. Although surprising at first sight, this notion according to which one *learns in a predefined class of languages* is rather compatible with our knowledge of first language acquisition.

Overtaking the pessimistic view of Gold's theorem, Angluin established in the 80s that some large but transversal classes of languages were learnable in Gold's sense. [4] Regarding categorial grammars, Buszkowski and Penn defined in late 80s [12,11] an algorithm that learns basic categorical grammars from structured sentences, functor-argument structures, and Kanazawa proved in 1994 that their algorithm converges: it actually learns categorial grammar in Gold's sense. [22,21]

The result in the present paper is much in the same vein as Buszkowski, Penn and Kanazawa.

Section 2. We first recall the Gold learning paradigm, identification in the limit from positive examples.

Sections 3, 4. Next we briefly present Lambek categorial grammars, and define the parsing of Lambek categorial grammar with product as cut-free proof nets construction and introduce the *proof frames*, that will be the structures we shall learn from. Informally, proof frames are name free parse structures, just like functor argument structures that are commonly used for learning basic categorial grammars. Such grammars ought to be learnt from *structured sentences* since Foret and Le Nir established that they cannot be learnt from strings [14].

Sections 5,6,7. After a reminder on unification and categorial grammars, we present our algorithm that learns rigid Lambek categorial grammars with product from proof frames and perform it on sample data involving introduction rules that are not in basic categorial grammars and product rules that are not in Lambek grammars. We then prove the convergence of this algorithm.

Section 8. We show that the present result strictly encompasses our initial result [10] that learns rigid product-free Lambek grammars from name-free natural deductions. To do so, we give the bijective correspondence between *cut-free* proof nets for the product-free Lambek calculus and *normal* natural deduction that are commonly used as parse structures.

In the conclusion, we discuss the merits and limits of the present work. We briefly explain how it can generalise to k-valued Lambek grammars with product and suggest direction for obtaining corpora with proof frame annotations from dependency-annotated corpora.

2   Exact Learning à la Gold: A Brief Reminder

We shall just give a brief overview of the Gold learning model of [17], with some comments, and explain why his famous unlearnability theorem of [17] (theorem 1 below) is not as negative as it may seem — as [4] or the present article shows.

The principles of first language acquisition as advocated by Chomsky [31] and more recently by Pinker [32,33] can be very roughly summarised as follows:

1. One *learns from positive examples only*: an argument says that in certain civilisations children uttering ungrammatical sentences are never corrected although they learn the grammar just as fast as ours — this can be discussed, since the absence of reaction might be considered as negative evidence, as well as the absence of some sentences in the input.
2. The target language is reached by *specialisation* more precisely by restricting word order from languages with a freer word order: rare are the learning algorithms for natural language that proceed by specialisation although, when starting from semantics, there are some, like the work of Tellier [39]
3. *Root meaning* is known first, hence the argumental structure or valencies are correct before the grammar learning process starts. This implies that all needed words are there, possibly in a non correct order, hence enforcing the idea of learning by specialisation — the afore mentioned work by Tellier proceeds from argument structures [39]
4. The examples that the child is exposed to are not so many: this is known as the *Poverty Of Stimulus* argument. It has been widely discussed since 2000 in particular for supporting quantitative methods. [31,34,35,8]

In his seminal 1967 paper, Gold introduced a formalisation of the process of the acquisition of one's first language grammar, which follows the first principle stated above, which is the easiest to formalise: the formal question he addressed could be more generally stated as *grammatical inference from positive examples*. It also should be said that this notion of learning may be used for other purposes every time one wants to extract some regularity out of sequences observations other fields being genomics (what would be a grammar of strings issued from ADN sequences) and diagnosis (what are the regular behaviours of system, what would be a grammar generating the sequences of normal observations provided by captors for detecting abnormal behaviours).

We shall provide only a minimum of information on formal languages and grammars. Let us just say that a language is a subset of inductive class $\mathcal{U}$. Elements of $\mathcal{U}$ usually are finite sequences (a.k.a. strings) of words, trees whose leaves are labelled by words, or graphs whose vertices are words — we here say "words" because they are linguistic words, while other say "letters" or "terminals," and we say "sentences" for sequences of words where others say "words" for sequences of "letters" or "terminals". A grammar G is a process generating the objects of a language $\mathcal{L}(G) \subset \mathcal{U}$. The membership question is said to be decidable for a grammar G when the characteristic function of $L(G)$ in U is computable. The most standard example of U is Σ^* the set of finite sequences over some set of symbols (e.g. words) Σ. The phrase structure grammars of Chomsky-Schutzenberger are the most famous grammars producing languages that are parts of Σ^*. Lambek categorial grammars and basic categorial grammars are an alternative way to generate sentences as elements of Σ^*: they produce the same languages as context–free languages [7,30,26, chapters 2, 3]. Finite labeled trees also are a possible class of object. A regular tree grammar produces such a tree language, and the yields of the trees in $\mathcal{L}(G)$ define a context free string language. In the formal study of human languages, $\mathcal{U}$ usually consists in strings of words or in trees.

Definition 1 (Gold, 1967, [17]). *A learning function for a class of grammars $\mathcal{G}$ producing $\mathcal{U}$-objects ($\mathcal{L}(G) \subset \mathcal{U}$) is a partial function ϕ that maps any finite sequence of positive examples $e_1, e_2, \ldots, e_k$ with $e_i \in \mathcal{U}$ to a grammar in the class $\phi(e_1, e_2, \ldots, e_k) \in \mathcal{G}$ such that:*

if *$(e_i)_{i \in I}$ is any enumeration of a language $\mathcal{L}(G) \subset \mathcal{U}$ with $G \in \mathcal{G}$,*
then *there exists an integer N such that, calling $G_i = \phi(e_1, \ldots, e_i)$:*
  - *$G_P = G_N$ for all $P \geq N$.*
  - *$\mathcal{L}(G_N) = \mathcal{L}(G)$.*

Several interesting properties of learning functions have been considered:

Definition 2. *A learning function ϕ is said to be*

  - *effective or computable when ϕ is recursive. In this case one often speaks of a learning algorithm. We shall only consider effective learning functions: this is consistent both with language being viewed as a computational process and with applications to computational linguistics. Observe that the learning function does not have to be a total recursive function: it may well be undefined for some sequences of sentences and still be a learning function.*
  - *conservative if $\phi(e_1, \ldots, e_p, e_{p+1}) = \phi(e_1, \ldots, e_p)$ whenever $e_{p+1} \in \mathcal{L}(\phi(e_1, \ldots, e_p))$.*
  - *consistent if $\{e_1, \ldots, e_p\} \subset \mathcal{L}(\phi(e_1, \ldots, e_p))$ whenever $\phi(e_1, \ldots, e_p)$ is defined.*
  - *set driven if $\phi(e_1, \ldots, e_p) = \phi(e_1', \ldots, e_q')$ whenever $\{e_1, \ldots, e_p\} = \{e_1', \ldots, e_q'\}$ — neither the order of the examples nor their repetitions matters.*
  - *incremental if there exists a binary function Ψ such that*
    $\phi(e_1, \ldots, e_p, e_{p+1}) = \Psi(\phi(e_1, \ldots, e_p), e_{p+1})$
  - *responsive if the image $\phi(e_1, \ldots, e_p)$ is defined whenever there exists L in the class with $\{e_1, \ldots, e_p\} \subset L$*
  - *monotone increasing when $\phi(e_1, \ldots, e_p, e_{p+1}) \subset \phi(e_1, \ldots, e_p)$*

In this paper the algorithm for learning Lambek grammars enjoys all those properties. They all seem to be sensible with respect to first language acquisition but the last one: indeed, as said above, children rather learn by specialisation.

It should be observed that the learning algorithm applies to a *class of languages*. So it is fairly possible that a given language L which both belongs to the classes $\mathcal{G}_1$ and $\mathcal{G}_2$ can be identified as a member of $\mathcal{G}_1$ and not as a member of $\mathcal{G}_2$. Learning L in such a setting is nothing more than to be sure, given the examples seen so far, that the language is not any other language in the class.

The classical result from the same 1967 paper by Gold [17] that has be over interpreted see e.g. [5,19] can be stated as follows:

Theorem 1 (Gold, 1967, [17]). *If a class $\mathcal{G}_r$ of grammars generates*

- *languages $(L_i)_i \in \mathbb{N}$ with $L_i \in \mathbb{N}$ which are strictly embedded that is $L_i \subsetneq L_{i+1}$ for all $i \in \mathbb{N}$*
- *together with the union of all these languages $\cup_{i\in\mathbb{N}} L_i \in \mathcal{G}_r$*

then no function may learn $\mathcal{G}_r$.

Proof. From the definition, we see that a learning function should have guessed the grammar of a language $\mathcal{L}(G)$ with $G \in \mathcal{G}$ after a finite number of examples in the enumeration of $\mathcal{L}(G)$. Consequently, for any enumeration of any language in the class,

(1)   the learning function may only change its mind finitely many times.

Assume that is a learning function ϕ for the class $\mathcal{G}_r$. Since the L_i are nested as stated, we can provide an enumeration of $L = \cup L_i$ according to which we firstly see examples $x_0^1, \cdots, x_0^{p_0}$ from L_0 until ϕ proposes G_0 with $\mathcal{L}G_0 = L_0$, then we see examples $x_1^1, \cdots, x_1^p$ in L_1 until ϕ proposes G_1 with $\mathcal{L}G_1 = L_1$, then we see examples $x_2^1, \cdots, x_2^p$ in L_2 until ϕ proposes G_2 with $\mathcal{L}G_2 = L_2$, etc. In such an enumeration of L the learning function changes its mind infinitely many times, conflicting with (1). Thus there cannot exists a learning function for the class $\mathcal{G}_r$.

Gold's theorem above has an easy consequence that was interpreted quite negatively:

Corollary 1. *No class containing the regular languages can be learnt.*

Indeed, by that time the Chomsky hierarchy was so present that no one thought that transverse classes could be of any interest let alone learnable. Nowadays, it is assumed that the syntax of human languages contains no regular languages and goes a bit beyond context free languages as can be seen in figure 1. It does not seem likely that human languages contain a series of strictly embedded languages as well as their unions. Hence Gold's theorem does not prevent large and interesting classes of languages from being learnt. For instance Angluin showed that pattern languages, a transversal class can be learnt by identification in the limit [4] and she also provided a criterion for learnability base on telltale sets:

Theorem 2 (Angluin, 1980, [5]). *An enumerable family of languages L_i with a decidable membership problem is effectively learnable whenever for each i there is a computable finite $T_i \subset_f L_i$ such that if $T_i \subset L_j$ then there exists $w \in (L_j \smallsetminus L_i)$*

As a proof that some interesting classes are learnable, we shall define particular grammars, Lambek categorial grammars with product, and their associated structure languages, before proving that they can be learnt from these structures, named proof frames.

3 Categorial Grammars and the LCGp Class

Given a finite set of words Σ and an inductively defined set of categories $\mathcal{C}$ including a special category s and an inductively defined set of derivable sequents $\vdash \subset (\mathcal{C}^* \times \mathcal{C})$ (each of them being written $t_1, \ldots, t_n \vdash t$) a categorial grammar G is defined as a map lex_G from words to finite sets of categories. An important property, as far as learnability is concerned, is the maximal number of categories per word i.e. $\max_{w \in \Sigma} |\text{lex}_G(w)|$. When it is less than k, the categorial grammar G is said to be k-valued and 1-valued categorial grammars are said to be *rigid*.

Some standard family of categorial grammars are:

1. *Basic categorial grammars BCG* also known as AB grammars have their categories in $\mathcal{C} ::= \text{s} \mid B \mid \mathcal{C} \backslash \mathcal{C} \mid \mathcal{C} / \mathcal{C}$ and the derivable sequents are the ones that are derivable in the Lambek calculus with elimination rules only $\Delta \vdash A$ and $\Gamma \vdash B / A$ (respectively $\Gamma \vdash A \backslash B$) yields $\Gamma, \Delta \vdash B$ (respectively $\Delta, \Gamma \vdash B$) — in such a setting the empty sequence is naturally prohibited even without saying so. [6]

2. The original *Lambek grammars* [23] also have their categories in the same inductive set $\mathcal{C} ::= \text{s} \mid B \mid \mathcal{C} \backslash \mathcal{C} \mid \mathcal{C} / \mathcal{C}$ and the derivable sequents are the ones that are derivable in the Lambek calculus without empty antecedent, i.e. with rules of figure 3 except $\otimes_i$ and $\otimes_h$ — a variant allows empty antecedents.

3. *Lambek grammars with product* (LCGp) have their categories in $\mathcal{C}_\otimes ::= \text{s} \mid B \mid \mathcal{C}_\otimes \backslash \mathcal{C}_\otimes \mid \mathcal{C}_\otimes / \mathcal{C}_\otimes \mid \mathcal{C}_\otimes \otimes \mathcal{C}_\otimes$ and the derivable sequents are the ones that are derivable in the Lambek calculus with product without empty antecedent i.e. with all the rules of figure 3 — a variant allows empty antecedents.

A phrase, that is a sequence of words $w_1 \cdots w_n$, is said to be of category C according to G when, for every i between 1 and p there exists $t_i \in \text{lex}_G(w_i)$ such that $t_1, \ldots, t_n \vdash C$ is a derivable sequent. When C is s the phrase is said to be a *sentence* according to G. The string language generated by a categorial grammar is the subset of Σ^* consisting in strings that are of category s i.e. sentences. Any language generated by a grammar in one of the aforementioned classes of categorial grammars is context free.

In this paper we focus on Lambek grammars with product (LCGp). The explicit use of product categories in Lambek grammars is not so common. Category like $(a \otimes b) \backslash c$ can be viewed as $b \backslash (a \backslash c)$ so they do not really involve a product. The comma in the left-hand side of the sequent, as well as the separation between words are implicit products, but grammar and parsing can be defined without explicitly using the product. Nevertheless, there are cases when the product is appreciated.

- For analysing the French Treebank, Moot in [25] assigns the category $((np \otimes pp) \backslash (np \otimes pp)) / (np \otimes pp)$ to "*et*" ("*and*") for sentences like:

  (2) Jean donne un livre à Marie et une fleur à Anne.

- According to Glyn Morrill [28,27] past participles like *raced* should be assigned the category $((CN \backslash CN)/(N \backslash (N \backslash \text{s--}))) \otimes (N \backslash (N \backslash \text{s--}))$ where s-- is an untensed sentence in sentences like:

  (3) The horse raced past the barn fell.

The derivable sequents of the Lambek syntactic calculus with product are obtained form the axiom $C \vdash C$ for any category C and the rules are given below, where A, B are categories and Γ, Δ finite sequences of categories:

$$\frac{\Gamma, B, \Gamma' \vdash C \quad \Delta \vdash A}{\Gamma, \Delta, A \backslash B, \Gamma' \vdash C} \backslash_h \qquad \frac{A, \Gamma \vdash C}{\Gamma \vdash A \backslash C} \backslash_i \quad \Gamma \neq \varnothing$$

$$\frac{\Gamma, B, \Gamma' \vdash C \quad \Delta \vdash A}{\Gamma, B / A, \Delta, \Gamma' \vdash C} /_h \qquad \frac{\Gamma, A \vdash C}{\Gamma \vdash C / A} /_i \quad \Gamma \neq \varnothing$$

$$\frac{\Gamma, A, B, \Gamma' \vdash C}{\Gamma, A \otimes B, \Gamma' \vdash C} \otimes_h \qquad \frac{\Delta \vdash A \quad \Gamma \vdash B}{\Delta, \Gamma \vdash A \otimes B} \otimes_i$$

Fig. 2. Sequent calculus rule for the Lambek calculus

4 Categorial Grammars Generating Proof Frames

The classes of languages that we wish to learn include some proper context free languages [7], hence they might be difficult to learn. So we shall learn them from *structured sentences*, and this section is devoted to present the proof frames that we shall use as structured sentences.

A neat natural deduction system for Lambek calculus with product is rather intricate [3,1], mainly because the product elimination rules have to be carefully commuted for having a unique normal form. Cut-free sequent calculus proofs are also not so good structures because they are quite redundant and some of their rules can be swapped. As explained in [26, chapter 6] proof nets provide perfect parse structure for Lambek grammars even if they use the product. When the product is not used, cut-free proof nets and normal natural deduction are iso-morphic, as we shall show in subsection 8.1. Consequently the structures that we used for learning will be proof frames that are proof nets with missing informations. Let us see how categorial grammars generate such structures, and first let us recall the correspondence between polarised formulae of linear logic and Lambek categories.

4.1 Polarised Linear Formulae and Lambek Categories

A Lambek grammar is better described with the usual Lambek categories, while proof nets are better described with linear logic formulae. Hence we need to re-call the correspondence between these two languages as done in [26, chapter 6]. Lambek categories (with product) are $\mathcal{C}_\otimes$ defined in the previous section 3. Linear formula L are defined by:

$$\mathsf{L} ::= \mathsf{P} \mid \mathsf{P}^{\perp} \mid (\mathsf{L} \otimes \mathsf{L}) \mid (\mathsf{L} \wp \mathsf{L})$$

the negation of linear logic $\_)^{\perp}$ is only used on propositional variables from P as the De Morgan laws allow:

$$(A^{\perp})^{\perp} \equiv A \qquad (A \wp B)^{\perp} \equiv (B^{\perp} \otimes A^{\perp}) \qquad (A \otimes B)^{\perp} \equiv (B^{\perp} \wp A^{\perp})$$

To translate Lambek categories into linear logic formulae, one has to distinguish the polarised formulae, the output or positive ones L° and the input or negative ones from $\mathsf{L}^{\bullet}$ with $F \in \mathsf{L}^{\circ} \iff F^{\perp} \in \mathsf{L}^{\bullet}$ and $(\mathsf{L}^{\circ} \cup \mathsf{L}^{\bullet}) \not\subseteq \mathsf{L}$:

$$\begin{cases} \mathsf{L}^{\circ} ::= \mathsf{P} \mid (\mathsf{L}^{\circ} \otimes \mathsf{L}^{\circ}) \mid (\mathsf{L}^{\bullet} \wp \mathsf{L}^{\circ}) \mid (\mathsf{L}^{\circ} \wp \mathsf{L}^{\bullet}) \\ \mathsf{L}^{\bullet} ::= \mathsf{P}^{\perp} \mid (\mathsf{L}^{\bullet} \wp \mathsf{L}^{\bullet}) \mid (\mathsf{L}^{\circ} \otimes \mathsf{L}^{\bullet}) \mid (\mathsf{L}^{\bullet} \otimes \mathsf{L}^{\circ}) \end{cases}$$

Any formula of the Lambek L calculus can be translated as an output formula $+L$ of multiplicative linear logic and its negation can be translated as an input linear logic formulae $-L$ as follows:

L	$\alpha \in P$	$L = M \otimes N$	$L = M \backslash N$	$L = N / M$
$+L$	α	$+M \otimes +N$	$-M \wp +N$	$+N \wp -M$
$-L$	$\alpha^{\perp}$	$-N \wp -M$	$-N \otimes +M$	$+M \otimes -N$

Conversely any output formula of linear logic is the translation of a Lambek formula and any input formula of linear logic is the negation of the translation of a Lambek formula. Let $(\ldots)^{\circ}_{\mathsf{Lp}}$ denotes the inverse bijection of "+", from L° to Lp and $(\ldots)^{\bullet}_{\mathsf{Lp}}$ denotes the inverse bijection of "−" from $\mathsf{L}^{\bullet}$ to Lp. These two maps are inductively defined as follows:

$F \in \mathsf{L}^{\circ}$	$\alpha \in P$	$(G \in \mathsf{L}^{\circ}) \otimes (H \in \mathsf{L}^{\circ})$	$(G \in \mathsf{L}^{\bullet}) \wp (H \in \mathsf{L}^{\circ})$	$(G \in \mathsf{L}^{\circ}) \wp (H \in \mathsf{L}^{\bullet})$
F°_{Lp}	α	$G^{\circ}_{\mathsf{Lp}} \otimes H^{\circ}_{\mathsf{Lp}}$	$G^{\bullet}_{\mathsf{Lp}} \backslash H^{\circ}_{\mathsf{Lp}}$	$G^{\circ}_{\mathsf{Lp}} / H^{\bullet}_{\mathsf{Lp}}$

$F \in \mathsf{L}^{\bullet}$	$\alpha^{\perp} \in P^{\perp}$	$(G \in \mathsf{L}^{\bullet}) \wp (H \in \mathsf{L}^{\bullet})$	$(G \in \mathsf{L}^{\circ}) \otimes (H \in \mathsf{L}^{\bullet})$	$(G \in \mathsf{L}^{\bullet}) \otimes (H \in \mathsf{L}^{\circ})$
$F^{\bullet}_{\mathsf{Lp}}$	α	$H^{\bullet}_{\mathsf{Lp}} \otimes G^{\bullet}_{\mathsf{Lp}}$	$H^{\bullet}_{\mathsf{Lp}} / G^{\circ}_{\mathsf{Lp}}$	$H^{\circ}_{\mathsf{Lp}} \backslash G^{\bullet}_{\mathsf{Lp}}$

4.2 Proof Nets

A proof net is a graphical representation of a proof which identifies inessentially different proofs. A cut-free proof net has several conclusions, and it consists of

- the subformula trees of its conclusions, that possibly stops on a sub formula which is not necessarily a propositional variable (axioms involving complex formulae simplify the learning process).
- a cyclic order on these sub formula trees
- axioms that links two dual leaves F and $F^{\perp}$ of these formula subtrees.

Such a structure can be represented by a sequence of terms — admittedly easier to type than a graph — with indices for axioms. Each index appears exactly twice, once on a formula F (not necessarily a propositional variable) and one on $F^{\perp}$. Here are two proof nets with the same conclusions:

(4) $s^{\perp^1} \otimes (s^2 \wp np^{\perp^3}), np^3 \otimes (s^{\perp} \otimes np)^7, (np^{\perp} \wp s)^7 \otimes s^{\perp^2}, s^1$

(5) $s^{\perp^1} \otimes (s^2 \wp np^{\perp^3}), np^3 \otimes (s^{\perp^4} \otimes np^5), (np^{\perp^5} \wp s^4) \otimes s^{\perp^2}, s^1$

The second one is obtained from the first one by expansing the complex axiom $(s^{\perp} \otimes np)^7, (np^{\perp} \wp s)^7$ into two axioms: $(s^{\perp^4} \otimes np^5), (np^{\perp^5} \wp s^4)$. Complex axioms always can be expansed into atomic axioms — this is known as η-expansion. This is the reason why proof nets are often presented with atomic axioms. Nevertheless as we shall substitute propositional variables with complex formula during the learning process we need to consider complex axioms as well — see the processing of example (9) in section 6.

No any such structure does correspond to a proof:

Definition 3. *A proof structure with conclusions $C^1, I_1^1, \ldots, I_n^1$ is said to be a proof net of the Lambek calculus when it enjoys the correctness criterion defined by the following properties:*

1. *Acyclic: any cycle contains the two branches of a $\wp$ link*
2. *Intuitionistic: exactly one conclusion is an output formula of L°, all other conclusions are input formulae of $L^\bullet$*
3. *Non commutative: no two axioms cross each other*
4. *Without empty antecedent: there is no sub proof net with a single conclusion*

The first point in this definition is not stated precisely but, given that we learn from correct structured sentences, we shall not need a precise definition. The reader interested in the details can read [26, chapter 6]. Some papers require a form of connectedness but it is not actually needed since this connectedness is a consequence of the first two points see [18] or [26, section 6.4.8 pages 225–227].

Definition 4. *Proof nets for the Lambek calculus can be defined inductively as follows (observe that they contain exactly one output conclusion):*

- *given an output formula F an axiom $F, F^{\perp}$ is a proof net with two conclusions F and $F^{\perp}$ — we do no require that F is a propositional variable.*
- *given a proof net π^1 with conclusions $O^1, I_1^1, \ldots, I_n^1$ and a proof net π^2 with conclusions $O^2, I_1^2, \ldots, I_p^2$ where O^1 and O^2 are the output conclusions, one can add a $\otimes$-link between a conclusion of one and a conclusion of the other, at least one of the two being an output conclusion. We thus can obtain a proof net whose conclusions are:*
  - $O^1 \otimes I_k^2, I_{k+1}^2, \ldots, I_p^2, O^2, I_1^2, I_{k-1}^2, I_1^1, \ldots, I_n^1 - O^2$ *being the output conclusion*
  - $I_l^1 \otimes O^2, I_1^2, \ldots, I_p^2, I_{l+1}^1, \ldots, I_n^1, O^1, I_1^1, \ldots, I_{l-1}^1, - O^1$ *being the output conclusion*
  - $O^1 \otimes O^2, I_1^2, \ldots, I_p^2, I_1^1, \ldots, I_n^1 - O^1 \otimes O^2$ *being the output conclusion.*
- *given a proof net π^1 with conclusions $O^1, I_1^1, \ldots, I_n^1$ one can add a $\wp$ link between any two consecutive conclusions, thus obtaining a proof nets with conclusions:*

- $O^1, I_1^1, \ldots, I_i \wp I_{i+1}, \ldots, I_n^1 - O^1$ being the output conclusion
- $O^1 \wp I_1^1, I_2^1 \ldots, I_n^1 - O^1 \wp I_1^1$ being the output conclusion
- $I_n^1 \wp O^1, I_1^1 \ldots, I_{n-1}^1 - O^1 \wp I_1^1$ being the output conclusion

A key result is that:

Theorem 3. *The inductively defined proof nets of definition 4, i.e. proofs, exactly correspond to the proof nets defined as graphs enjoying the universal properties of the criterion 3*

A parse structure for a sentence $w^1, \ldots, w^p$ generated by a Lambek grammar G defined by a lexicon lex_G is a proof net with conclusions $(c^n)^-, \ldots, (c^1)^-, \mathsf{s}^+$ with $c^i \in \text{lex}(w^i)$. This replaces the definition of parse structure as normal natural deductions [40] which does not work well when the product is used [3,1].

4.3   Structured Sentences to Learn from: s Proof Frames

An s proof frame (sPF) is simple a parse structure of a Lambek grammar i.e. a proof net whose formula names have been erased, except the s on the output conclusion. Regarding axioms, their positive and negative tips are also kept. Such a structure is the analogous of a functor argument structure for AB grammars or of a name free normal natural deduction for Lambek grammars used in [12,11,10] and it can be defined inductively as we did in 4, or by the conditions in definition 3.

Definition 5 (Proof frames, sPF). *An s proof frame (sPF) is a normal proof net π such that:*

- *The output of π is labelled with the propositional constant s — which is necessarily the conclusion of an axiom, the input conclusion of this axiom being labelled $\mathsf{s}^\perp$.*
- *The output conclusion of any other axiom in π is O its input conclusion being $O^\perp = I$.*

Given an s proof net π its associated s proof frame π_f is obtained by replacing in π the output of any axiom by O (and its dual by $I = O^\perp$) except the s that is the output of π itself which is left unchanged.

A given Lambek grammar G is said to generate an sPF ρ whenever there exists a proof net π generated by G such that $\rho = \pi^{IO}$. In such a case we write $\rho \in \text{sPF}(G)$.

The sPF associated with the two proof nets 4 and 5 above are:

(6) $\mathsf{s}^{\perp 1} \otimes (O^2 \wp I^3), O^3 \otimes (I^4 \otimes O^5), (O^5 \wp O^4) \otimes I^2, \mathsf{s}$

(7) $\mathsf{s}^{\perp 1} \otimes (O^2 \wp I^3), O^3 \otimes I^7, O^7 \otimes I^2, \mathsf{s}$

5   Unification, Proof Frames and Categorial Grammars

Our learning algorithm makes a crucial use of category-unification, and this kind of technique is quite common in grammatical inference [29], so let us briefly define unification of categorial grammars.

As said in paragraph 3, a categorial grammar is defined from a lexicon that maps every word w to a finite set of categories $\text{lex}_G(w)$. Categories are usually defined from a finite set B of base categories that includes a special base category s. Here we shall consider simultaneously many different categorial grammars and to do so we shall have an infinite set B whose members will be s and infinitely many category variables denoted by x, y, x_1, x_2, $\ldots$, y_1, y_2, $\ldots$ In other words, $B = \{\mathsf{s}\} \cup V$, $\mathsf{s} \notin V$, V being an infinite set of category variables. The categories arising from B are defined as usual by $\mathcal{V} ::= \mathsf{s} \mid V \mid \mathcal{V} \backslash \mathcal{V} \mid \mathcal{V}/\mathcal{V} \mid \mathcal{V} \otimes \mathcal{V}$. This infinite set of base categories does not change much categorial grammars: since there are finitely many words each of them being associated with finitely many categories, the lexicon is *finite* and a given categorial grammar only makes use of finitely many base categories. Choosing an infinite language is rather important, as we shall substitute a category variable with a complex category using fresh variables, thus turning a categorial grammar into another one, and considering families of grammars over the same base categories.

A *substitution* σ is a function from categories $\mathcal{V}$ to categories $\mathcal{V}$ that is generated by a mapping σ_V of finitely many variables $x_{i_1}, \cdots, x_{i_p}$ in V to categories of $\mathcal{V}$:

$$\sigma(\mathsf{s}) = \mathsf{s}$$
$$\text{given } x \in V, \quad \sigma(x) = \begin{cases} \sigma_V(x) \text{ if } x = x_{i_k} \text{ for some } k \\ x \qquad\quad \text{otherwise} \end{cases}$$
$$\sigma(A \backslash B) = \sigma(A) \backslash \sigma(B)$$
$$\sigma(B / A) = \sigma(B) / \sigma(A)$$

The substitution σ is said to be a *renaming* when σ_V is a bijective mapping from V to V — otherwise stated σ_V is a permutation of the $x_{i_1}, \cdots, x_{i_p}$).

As usual, substitutions may be extended to sets of categories by stipulating $\sigma(A) = \{\sigma(a) | a \in A\}$. Observe that $\sigma(A)$ can be a singleton while A is not: $\{(a/(b\backslash c)), (a/u)\}[u \mapsto (b\backslash c] = \{a/(b\backslash c)\}$. A substitution can also be applied to a categorial grammar: $\sigma(G) = G'$ with $\text{lex}_{G'}(w) = \sigma(\text{lex}_G(w))$ for any word w, and observe that a substation turns a k-valued categorical grammar into a k'-valued categorical grammar with $k' \leq k$, and possibly into a rigid (or 1-valued) categorial grammar (cf. section 3).

A substitution σ on Lambek categories (defined by mapping finitely many category variables x_i to Lambek categories L_i, $x_i \mapsto L_i$) clearly defines a substitution on linear formulae σ^ℓ (by $x_i \mapsto L_i^+$), which preserves the polarities $\sigma^\ell(F)$ is positive(respectively negative) if and only if F is. Conversely, a substitution ρ on linear formulae defined by mapping variables to positive linear formulae $(x_i \mapsto F_i)$ defines a substitution on Lambek categories ρ^L with the mapping $x_i \mapsto F_{\mathsf{Lp}}^\circ$. One has: $\sigma(L) = (\sigma^\ell(L+))_{\mathsf{Lp}}^\circ$ and $\rho(F) = (\rho^L(F_{\mathsf{Lp}}^\circ))+$ if $F \in L^\circ$ and

$\rho(F) = (\rho^L(F_{\mathsf{Lp}}^{\bullet}))-$. Roughly speaking as far as we use only polarised linear formulae and substitution that preserve polarities, it does not make any difference to perform substitutions on linear formulae or on Lambek categories.

Substitution preserving polarities (or Lambek substitutions) can also be applied to proof nets: $\sigma(\pi)$ is obtained by applying the substitution to any formula in π, and they turn an s Lambek proof net into an s Lambek proof net – this is a good reason for considering axioms on complex formulae.

Proposition 1. *If σ is a substitution preserving polarities and π a proof net generated by a Lambek grammar G, then $\sigma(\pi)$ is generated by $\sigma(G)$ and $\sigma(\pi)$ have the same associated s proof frame: $\sigma(\pi)_f = \pi_f$*

Two grammars G_1 and G_2 with their categories in $\mathcal{V}$ are said to be *equal* whenever there is renaming ν such that $\nu(G_1) = G_2$.

A substitution σ is said to unify two categories A, B if one has $\sigma(A) = \sigma(B)$. A substitution is said to unify a set of categories T or to be a unifier for T if for all categories A, B in T one has $\sigma(A) = \sigma(B)$ — in other words, $\sigma(T)$ is a singleton.

A substitution σ is said to unify a categorial grammar G or to be a unifier of G whenever, for every word in the lexicon σ unifies $\mathrm{lex}_G(w)$, i.e. for any word w in the lexicon $\mathrm{lex}_{\sigma(G)}(w)$ has a unique category — in other words $\sigma(G)$ is rigid.

A unifier does not necessarily exists, but when it does, there exists a *most general unifier (mgu)* that is a unifier σ_u such for every unifier τ there exists a substitution σ_τ such that $\tau = \sigma_\tau \circ \sigma_u$. This most general unifier is unique up to renaming. This result also holds for unifiers that unify a set of categories and even for unifiers that unify a categorial grammar. [22]

Definition 6. *Let π be an s proof net whose associated sPF is π_f. If all the axioms in π but the $s, s^\perp$ whose s is π's main output are $\alpha_i, \alpha_i^\perp$ with $\alpha_i \neq \alpha_j$ when $i \neq j$, π is said to be a most general labelling of π_f. If π_f is the associated sPF of an s proof net π and π_v one of the most general labelling of π_f, then π_v is also said to be a most general labelling of π. The most general labelling of an s proof net is unique up to renaming.*

We have the following obvious but important property:

Proposition 2. *Let π_v is a most general labelling of an s proof net π, then there exists a substitution σ such that $\pi = \sigma(\pi_v)$.*

6   An RG-Like Algorithm for Learning Lambek Categorial Grammars from Proof Frames

Assume we are defining a *consistent* learning function from positive examples for a class of categorial grammar (see definition 2). Assume that we already mapped $e_1, \ldots, e_n$ to a grammar G_n with $e_1, \ldots, e_n \subset \mathcal{L}(G_n)$ and $e_{n+1} \notin \mathcal{L}(G_n)$. This means that for some word w in the sentence e_{n+1} no category of $\mathrm{lex}_{G_n}(w)$

The algorithm for unifying two categories C_1 and C_2 can be done by processing a finite multi-set E of potential equations on terms, until it fails or reaches a set of equations whose left hand side are variables, each of which appears in a unique such equation — a measure consisting in triple of integers ordered ensures that this algorithm always stops. This set of equations $x_i = t_i$ defines a substitution by setting $\nu(x_i) = t_i$. Initially $E = \{C_1 = C_2\}$. In the procedure below, upper case letters stand for categories, whatever they might be, x for a variable, $*$ and $\diamond$ stand for binary connectives among $\backslash, /, \otimes$. Equivalently, unifications could be performed on linear formulae, as said in the main text. The most general unifier of n categories can be performed by iterating binary unification, the resulting most general unifier does not depend on the way one proceeds.

$$E \cup \{C=C\} \longrightarrow E$$
$$E \cup \{A_1 * B_1 = A_2 * B_2\} \longrightarrow E \cup \{A_1 = A_2, B_1 = B_2\}$$
$$E \cup \{C=x\} \longrightarrow E \cup \{x=C\}$$
$$\text{if } x \in Var(C) \quad E \cup \{x=C\} \longrightarrow \bot$$
$$\text{if } x \notin Var(C) \wedge x \in Var(E) \quad E \cup \{x=C\} \longrightarrow E[x := C] \cup \{x=C\}$$
$$\text{if } \diamond \neq * \quad E \cup \{A_1 * B_1 = A_2 \diamond B_2\} \longrightarrow \bot$$
$$E \cup \{s = A_2 * B_2\} \longrightarrow \bot$$
$$E \cup \{A_1 * B_1 = s\} \longrightarrow \bot$$

Fig. 3. The unification algorithm for unifying two categories

is able to account for the behaviour of w in e_{n+1}. A natural but misleading idea would be to say: if word w^k needs category c_{n+1}^k in example e_{n+1}, let us add c^k to $\text{lex}_{G_n}(w^k)$ to define $\text{lex}_{G_{n+1}}(w^k)$. Doing so for every occurrence of problematic words in e_{n+1} we will have $e_1, \ldots, e_n, e_{n+1} \subset \mathcal{L}(G_{n+1})$ and in the limit we obtained the smallest grammar G_∞ such that $\forall i \ e_1, \ldots, e_i \in \mathcal{L}G_\infty$ which should be reached at some point. Doing so, there is little hope to identify a language in the limit in Gold sense. Indeed, nothing guarantees that the process will stop, and a categorial grammar with infinitely many types for some word is not even a grammar, that is a finite description of a possibly infinite language. Thus, an important guideline for learning categorial grammars is to bound the number of categories per word. That is the reason why we introduced in section 3 the notion of k-*valued* categorial grammars, which endow every word with at most k categories, and we shall start by learning *rigid* (1-valued) categorial grammars as the k-valued case derives from the rigid case.

Our algorithm can be viewed as an extension to Lambek grammars with product of the RG algorithm (learning Rigid Grammars) introduced by Buszkowski and Penn in [11,12] initially designed for rigid AB grammars. A difference from their seminal work is that the data ones learns from were functor argument trees while here they are proof frames (or natural deduction frames when the product is not used [10], see section 8). Proof frames may seem less natural than natural deduction, but we have two good reasons for using them:

- The first one is that product is of interest for some grammatical constructions as examples 2 and 3 show while there is no fully satisfying natural deduction for Lambek calculus with product. [3,1]
- The second one is that they resemble dependency structures, since an axiom between the two conclusions corresponding to two words expresses a dependency between these two words.

To illustrate our learning algorithm we shall proceed with the three examples below, whose corresponding s proof frames are given in figure 4. As their sPF structures shows, the middle one (9) involves a positive product in the (the $I \wp I$ in the category of "*and*") and the last one (10) involves an introduction rule (the $O \wp I$ in the category of "*that*").

(8)   Sophie gave a kiss to Christian

(9)   Christian gave a book to Anne and a kiss to Sophie

(10)   Sophie liked a book that Christian liked.

Unfortunately the use for proof nets is to use a reverse word order, for having conclusions only, and these conclusions are linear formulae, the dual of Lambek categories as explained in section 4 — in some papers by Glynn Morrill e.g. [28] the order is not reversed, but then the linear formulae and the proof net structure are less visible. One solution that will make the supporters of either notation happy is to write the sentences vertically as we do in figure 4.

Definition 7 (RG like algorithm for sPF). *Let* $D = (\pi_f^k)_{1 \le k \le n}$ *be the* s *proof frames associated with the examples* $(e_f^k) 1 \le k \le n$, *and let* (π^k) *be most general labelings of the* $(\pi_f^k)_{1 \le k \le n}$. *We can assume that they have no common variables — this is possible because the set of variables is infinite and because most general labelings are defined up to renaming. If example* e^k *contains n words* $w_1^k, \dots, w_n^k$ *then* π^k *has n conclusions* $(c_n^k)-, \dots, (w_1^k)-$, s, *where all the* c_i^k *are Lambek categories.*

Let $GF(D)$ *be the non necessarily rigid grammar defined by the assignments* $w_i^k : c_i^k$ — *observe that a for a given word w there may exist several i and k such that* $w = w_i^k$.

Let $RG(D)$ *be the rigid grammar defined as the most general unifier of the categories* lex(w) *for each word in the lexicon when such a most general unifier exists.*

Define $\phi(D)$ *as* $RG(D)$. *When unification fails, the grammar can be defined by* lex$(w) = \varnothing$ *for those words whose categories do not unify.*

With the sPF of our examples in sPF yields the following type assignments where the variable x_n corresponds to the axiom number n in the examples, they are all different as expected — remember that s is not a category variable but a constant.

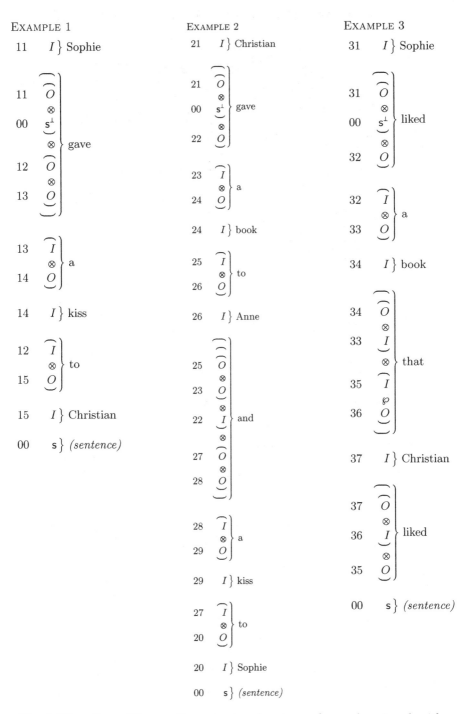

Fig. 4. Three S proof frames: three structured sentences for our learning algorithm

word	category (Lambek)	category$^{\perp}$ (linear logic)
and	$(((x_{23} \otimes x_{25}) \backslash x_{22})...$ $.../ (x_{28} \otimes x_{27}))$	$((x_{28} \otimes x_{27}) \otimes ...$ $...(x_{22} \otimes (x_{23} \otimes x_{25})))$
that	$((x_{34} \backslash x_{33}) / (x_{36} / x_{35}))$	$((x_{36} \wp x_{35}^{\perp}) \otimes (x_{33}^{\perp} \otimes x_{34}))$
liked	$(x_{31} \backslash \mathsf{s}) / x_{32}$	$x_{32} \otimes (\mathsf{s} \otimes x_{31})$
	$(x_{37} \backslash x_{36}) / x_{35}$	$x_{35} \otimes (x_{36} \otimes x_{37})$
gave	$((x_{11} \backslash \mathsf{s}) / (x_{13} \otimes x_{12}))$	$(x_{13} \otimes x_{12}) \otimes (\mathsf{s} \otimes x_{11})$
	$((x_{21} \backslash \mathsf{s}) / x_{22})$	$x_{22} \otimes (\mathsf{s} \otimes x_{21})$
to	x_{12} / x_{15}	$x_{15} \otimes x_{12}^{\perp}$
	x_{25} / x_{26}	$x_{26} \otimes x_{25}^{\perp}$
	x_{27} / x_{20}	$x_{20} \otimes x_{27}^{\perp}$
a	x_{13} / x_{14}	$x_{14} \otimes x_{13}^{\perp}$
	x_{23} / x_{24}	$x_{24} \otimes x_{23}^{\perp}$
	x_{28} / x_{29}	$x_{29} \otimes x_{28}^{\perp}$
	x_{32} / x_{33}	$x_{33} \otimes x_{32}^{\perp}$
Anne	x_{26}	$x_{26}^{\perp}$
Sophie	x_{11}	$x_{11}^{\perp}$
	x_{20}	$x_{20}^{\perp}$
	x_{31}	$x_{31}^{\perp}$
Christian	x_{15}	$x_{15}^{\perp}$
	x_{21}	$x_{21}^{\perp}$
	x_{37}	$x_{37}^{\perp}$
book	x_{24}	$x_{24}^{\perp}$
	x_{34}	$x_{34}^{\perp}$
kiss	x_{14}	$x_{14}^{\perp}$
	x_{29}	$x_{29}^{\perp}$

Unifications either performed on Lambek categories c_i^k or on the corresponding linear formulae (the (c_i^k)– that appear in the second column) yield the following equations:

liked
$$x_{31} = x_{37}$$
$$x_{36} = \mathsf{s}$$
$$x_{32} = x_{35}$$

gave
$$x_{11} = x_{21}$$
$$x_{22} = x_{13} \otimes x_{12}$$

to
$$x_{12} = x_{25} = x_{27}$$
$$x_{15} = x_{26} = x_{20}$$

a
$$x_{13} = x_{23} = x_{28} = x_{32}$$
$$x_{14} = x_{24} = x_{29} = x_{33}$$

Sophie
$$x_{11} = x_{20} = x_{31}$$

Christian
$$x_{15} = x_{21} = x_{37}$$

kiss
$$x_{14} = x_{29}$$

book
$$x_{24} = x_{34}$$

These unification equations can be solved by setting:

$$x_{36} = \mathsf{s}$$
$$x_{22} = x_{13} \otimes x_{12} = np \otimes pp$$
$$x_{12} = x_{25} = x_{27} = pp \qquad \text{prepositional phrase introduced by ``}to\text{''}$$
$$x_{13} = x_{23} = x_{28} = x_{32} = x_{35} = np \qquad \text{noun phrase}$$
$$x_{14} = x_{24} = x_{29} = x_{33} = x_{34} = cn \qquad \text{common noun}$$
$$x_{11} = x_{20} = x_{31} = x_{15} = x_{21} = x_{37} = x_{15} = x_{26} = pn \qquad \text{proper name}$$

The grammar can be unified into a rigid grammar G_r , namely:

word	category (Lambek)	category$^{\perp}$ (linearlogic)
and	$(((np \otimes pp) \setminus (np \otimes pp)...$ $... / (np \otimes pp))$	$((np \otimes pp) \otimes ...$ $...((np \otimes pp)^{\perp} \otimes (np \otimes pp)))$
that	$((n \setminus n) / (\mathsf{s} / np))$	$((\mathsf{s} \wp np^{\perp}) \otimes (n^{\perp} \otimes n))$
liked	$(pp \setminus \mathsf{s}) / np$	$np \otimes (\mathsf{s} \otimes pn)$
gave	$(pp \setminus \mathsf{s}) / (pp \otimes np))$	$(np \otimes pp) \otimes (\mathsf{s} \otimes pn)$
to	np / pn	$pn \otimes np^{\perp}$
a	np / cn	$cn \otimes pp^{\perp}$
Anne	pn	$pn^{\perp}$
Sophie	pn	$pn^{\perp}$
Christian	pn	$pn^{\perp}$
book	cn	$cn^{\perp}$
kiss	cn	$cn^{\perp}$

As stated in proposition 1, one easily observes that the sPF are indeed produced by the rigid grammar G_r.

Earlier on, in the definition of sPF , we allowed non atomic axioms, and we can now precisely see why: the axiom 22 could be instantiated by the single variable x_{22} but, when performing unification, it got finally instantiated with $x_{13} \otimes x_{12}$. Thus, if we would have forced axioms to always be on propositional variables, the sPF of example 2 would not have been generated by the G_r: instead, G_r would not have generated exactly the example 2 but only the sPF with the axioms $x_{13}, x_{13}^{\perp}$ and $x_{12}^{\perp}, x_{12}$ linked by an $\otimes$ link $x_{13}^{\perp} \otimes x_{12}$ and by a $\wp$ link $x_{12}^{\perp} \wp x_{13}^{\perp}$.

7   Convergence of the Learning Algorithm

This algorithm converges in the sense defined by Gold [17], see definition 1. The first proof of convergence of a learning algorithm for categorial grammars is the proof by Kanazawa [21] of the algorithm of Buszkowski and Penn [12] for learning rigid AB grammars from functor argument structures (name free proofs os this calculus with elimination rules only). We shall do something similar, but we learn a different class of grammars from different structures, and the proof follows [9] that is a simplification of [22].

The proof of convergence makes use of the following notions and notations:

$G \sqsubset G'$ This reflexive relation between G and G' holds whenever every lexical category assignment $a : T$ in G is in G' as well — in particular when G' is rigid, so is G, and both grammars are identical. Note that this is just the normal subset relation for each of the words in the lexicon G': $\text{lex}_G(a) \subset \text{lex}_{G'}(a)$ for every a in the lexicon of G', with $\text{lex}_G(a)$ non-empty. Keep in mind that in what follows we will also use the subset relation symbol to signify inclusion of the generated *languages*; the intended meaning should always be clear from the context.

size of a grammar The size of a grammar is simply the sum of the sizes of the occurrences of categories in the lexicon, where the size of a category is its number of occurrences of base categories (variables or s).

$G \sqsubseteq G'$ This reflexive relation between G and G' holds when there exists a substitution σ such that $\sigma(G) \subset G'$ which does not identify different categories of a given word, but this is always the case when the grammar is rigid.

$\text{sPF}(G)$ As said earlier, $\text{sPF}(G)$ is the the set of s proof structures generated by a Lambek categorial grammar G.

$GF(D)$ Given a set D of structured examples i.e. a set of s proof frames, the grammar $GF(D)$ is define as in the examples above: it is obtained by collecting the categories of each word in the various examples of D.

$RG(D)$ Given a set of sPF D, $RG(D)$ is, whenever it exists, the rigid grammar/lexicon obtained by applying the most general unifier to $GF(D)$.

Proposition 3. *Given a grammar G, the number of grammars H such that $H \sqsubseteq G$ is finite.*

Proof. There are only finitely many grammars which are included in G, since G is a finite set of assignments. Whenever $\sigma(H) = K$ for some substitution σ the size of H is smaller or equal to the size of K, and, up to renaming, there are only finitely many grammars smaller than a given grammar.

By definition, if $H \sqsubseteq G$ then there exist $K \subset G$ and a substitution σ such that $\sigma(H) = K$. Because there are only finitely many K such that $K \subset G$, and for every K there are only finitely many H for which there could exist a substitution σ with $\sigma(H) = K$ we conclude that there are only finitely many H such that $H \sqsubseteq G$. □

From the definition of $\sqsubseteq$ and from proposition 1 one immediately has:

Proposition 4. *If $G \sqsubseteq G'$ then $\text{sPF}(G) \subset \text{sPF}(G')$.*

Proposition 5. *If $GF(D) \sqsubseteq G$ then $D \subset \text{sPF}(G)$.*

Proof. By construction of $GF(D)$, we have $D \subset \text{sPF}(GF(D))$. In addition, because of proposition 4, we have $\text{sPF}(GF(D)) \subset \text{sPF}(G)$. □

Proposition 6. *If $RG(D)$ exists then $D \subset \text{sPF}(RG(D))$.*

Proof. By definition $RG(D) = \sigma_u(GF(D))$ where σ_u is the most general unifier of all the categories of each word. So we have $GF(D) \sqsubseteq RG(D)$, and applying proposition 5 with $G = RG(D)$ we obtain $D \subset \text{sPF}(RG(D))$. □

Proposition 7. *If $D \subset sPF(G)$ then $GF(D) \sqsubseteq G$.*

Proof. By construction of $GF(D)$, there is exactly one occurrence of a given category variable x in an sPF of D categorised as done in the example. Now, viewing the same sPF as an sPF of $sPF(G)$ at the place corresponding to x there is a category label, say T. Doing so for every category variable, we can define a substitution by $\sigma(x) = T$ for all category variables x: indeed because x occurs once, such a substitution is well defined. When this substitution is applied to $GF(D)$ it yields a grammar which only contains assignments from G — by applying the substitution to the whole sPF , it remains a well-categorised sPF , and in particular the categories on the conclusions corresponding to the words must coincide — if it is the linear formula F then the corresponding Lambek category is $F^{\bullet}$, see subsection 4.1. □

Proposition 8. *When $D \subset sPF(G)$ with G a rigid grammar, the grammar $RG(D)$ exists and $RG(D) \sqsubseteq G$.*

Proof. By proposition 7 we have $GF(D) \sqsubseteq G$, so there exists a substitution σ such that $\sigma(GF(D)) \subset G$.

As G is rigid, σ unifies all the categories of each word. Hence there exists a unifier of all the categories of each word, and $RG(D)$ exists.

$RG(D)$ is defined as the application of most general unifier σ_u to $GF(D)$. By the definition of a most general unifier, which works as usual even though we unify sets of categories, there exists a substitution τ such that $\sigma = \tau \circ \sigma_u$.

Hence $\tau(RG(D)) = \tau(\sigma_u(GF(D))) = \sigma(GF(D)) \subset G$;
thus $\tau(RG(D)) \subset G$, hence $RG(D) \sqsubseteq G$. □

Proposition 9. *If $D \subset D' \subset sPF(G)$ with G a rigid grammar then $RG(D) \sqsubseteq RG(D') \sqsubseteq G$.*

Proof. Because of proposition 8 both $RG(D)$ and $RG(D')$ exist. We have $D \subset D'$ and $D' \subset sPF(RG(D'))$, so $D \subset sPF(RG(D'))$; hence, by proposition 8 applied to D and $G = RG(D')$ (a rigid grammar) we have $RG(D) \sqsubseteq RG(D')$. □

Theorem 4. *The algorithm RG for learning rigid Lambek grammars converges in the sense of Gold.*

Proof. Take $D_i, i \in \omega$ an increasing sequence of sets of examples in $sPF(G)$ enumerating $sPF(G)$, in other words $\cup_{i \in \omega} D_i = sPF(G)$:

$$D_1 \subset D_2 \subset \cdots D_i \subset D_{i+1} \cdots \subset sPF(G)$$

Because of proposition 8 for every $i \in \omega$ $RG(D_i)$ exists and because of proposition 9 these grammars define an increasing sequence of grammars w.r.t. $\sqsubseteq$ which by proposition 8 is bounded by G:

$$RG(D_1) \sqsubseteq RG(D_2) \sqsubseteq \cdots RG(D_i) \sqsubseteq RG(D_{i+1}) \cdots \sqsubseteq G$$

As they are only finitely many grammars below G w.r.t. $\sqsubseteq$ (proposition 3) this sequence is stationary after a certain rank, say N, that is, for all $n \geq N$ $RG(D_n) = RG(D_N)$.

Let us show that the langue generated is the one to be learnt, let us prove that $\mathsf{sPF}(RG(D_N)) = \mathsf{sPF}(G)$ by proving the two inclusions:

1. Firstly, let us prove that $\mathsf{sPF}(RG(D_N)) \supset \mathsf{sPF}(G)$ Let π_f be an sPF in $\mathsf{sPF}(G)$. Since $\cup_{i \in \omega} D_i = \mathsf{sPF}(G)$ there exists a p such that $\pi_f \in \mathsf{sPF}(D_p)$.
   - If $p < N$, because $D_p \subset D_N$, $\pi_f \in D_N$, and by proposition 6 $\pi_f \in \mathsf{sPF}(RG(D_N))$.
   - If $p \geq N$, we have $RG(D_p) = RG(D_N)$ since the sequence of grammars is stationary after N. By proposition 6 we have $D_p \subset \mathsf{sPF}(RG(D_p))$ hence $\pi_f \in \mathsf{sPF}(RG(D_N)) = \mathsf{sPF}(RG(D_p))$.
   In all cases, $\pi_f \in \mathsf{sPF}(RG(D_N))$.
2. Let us finally prove that $\mathsf{sPF}(RG(D_N)) \subset \mathsf{sPF}(G)$: Since $RG(D_N) \sqsubseteq G$, by proposition 4 we have $\mathsf{sPF}(RG(D_N)) \subset \mathsf{sPF}(G)$ □

This exactly shows that the algorithm proposed in section 6 converges in the sense of Gold's definition (1).

8 Learning Product Free Lambek Grammars from Natural Deduction Frames

The reader may well find that the structure of the positive examples that we learn from, sorts of proofnets are rather sophisticated structures to learn from and he could think that our learning process is a drastic simplification w.r.t. standard work using functor argument structures.

Let us first see that normal natural deductions are quite a sensible structure to learn Lambek grammars from. Tiede [40] observed that natural deductions in the Lambek calculus (be they normal or not) are plain trees, defined by two unary operators (\ and / introduction rules) and two binary operators (\ and / elimination rules), from formulae as leaves (hypotheses, cancelled or free). As opposed to the intuitionistic case, there is no need to specify which hypothesis are cancelled by the introduction rules, as they may be inferred inductively: a \ (respectively /) introduction rule cancels the left-most (respectively right-most) free hypothesis. He also observed that *normal* natural deductions should be considered as the proper parse structures, since otherwise any possible syntactic structure (a binary tree) is possible. Therefore is is natural to learn Lambek grammars from normal natural deduction frames — natural deductions from which category names have been erased but the final s. Indeed, s natural deduction frames are to Lambek categorial grammars what the functor-argument (FA) structures are to AB categorial grammars — these FA structures are the standard structures used for learning AB grammars by Buszkowski, Penn and Kanazawa [12,22].

The purpose of this section is to exhibit a one to one correspondence between cut-free proof nets of the product free Lambek calculus and normal natural deductions, thus justifying the use of proof frames for learning Lambek grammars.

When there is no product, proof frames are the same as natural deduction frames that we initially used in [10]. They generalise the standard FA structures, and when the product is used, natural deduction become quite tricky [3,1] and there are the only structures one can think about.

The correspondence between on one hand natural deduction or the isomorphic λ-terms and on the other hand, proof nets, can be traced back to [36] (for second order lambda calculus) but the the closest result is the one for linear λ-calculus [13].

8.1   Proofnets and Natural Deduction: Climbing Principal Branches

As said in section 3, the formulae of product free Lambek calculus are defined by $\mathcal{C} ::= s \mid B \mid \mathcal{C} \backslash \mathcal{C} \mid \mathcal{C} / \mathcal{C}$ hence their linear counterpart are a strict subset of the polarised linear formulae of subsection 4.1:

$$\begin{cases} \mathsf{L}_h^\circ ::= \mathsf{P} & \mid (\mathsf{L}_h^\bullet \wp \, \mathsf{L}_h^\circ) \mid (\mathsf{L}_h^\circ \wp \, \mathsf{L}_h^\bullet) \\ \mathsf{L}_h^\bullet ::= \mathsf{P}^\perp & \mid (\mathsf{L}_h^\circ \otimes \mathsf{L}_h^\bullet) \mid (\mathsf{L}_h^\bullet \otimes \mathsf{L}_h^\circ) \end{cases}$$

Let us call these formulae the *heterogeneous* polarised positive or negative formulae. In these heterogeneous formulae the connectives $\wp$ and $\otimes$ only apply to a pair of formulae with opposite polarity. The translation from Lambek categories to linear formulae and vice versa from subsection 4.1 are the same.

One may think that a proof net corresponds to a sequent calculus proof which itself corresponds to a natural deduction: as shown in our book [26], this is correct, as far as one does not care about *cuts* — which are problematic in non commutative calculi, see e.g.[24]. As it is well known in the case of intuitionnistic logic, cut-free and normal are different notions [41], and proof net are closer to sequent calculus in some respects. If one translate inductively, rule by rule, a natural deduction into a sequent calculus or into a proof net, the elimination rule from A and $A \backslash B$ yields a cut on the $A \backslash B$ formula, written $A^\perp \wp B$ in linear logic. We shall see how this can be avoided. .

From Normal Natural Deductions to Cut-Free Proof Nets. Let us briefly recall some basic facts on natural deduction for the product free Lambek calculus, from our book [26, section 2.6 pages 33-39]. In particular we shall need the following notation a formula C, and a sequence of length p of pairs consisting of a letter ε_i (where $\varepsilon_i \in \{l, r\}$) and a formula G_i we denote by

$$C[(\varepsilon_1, G_1), \dots, (\varepsilon_p, G_p)]$$

the formula defined as follows:

if $p = 0$ $C[] = C$
if $\varepsilon_p = l$ $C[(\varepsilon_1, G_1), \dots, (\varepsilon_{p-1}, G_{p-1}), (\varepsilon_p, G_p)] =$
    $G_p \backslash C[(\varepsilon_1, G_1), \dots, (\varepsilon_{p-1}, G_{p-1})]$
if $\varepsilon_p = r$ $C[(\varepsilon_1, G_1), \dots, (\varepsilon_{p-1}, G_{p-1}), (\varepsilon_p, G_p)] =$
    $C[(\varepsilon_1, G_1), \dots, (\varepsilon_{p-1}, G_{p-1})] / G_p$

The rule below requires at least two free hyp.

A leftmost free hyp.
$$\dots[A]\dots\dots$$
$$\vdots$$
$$\frac{B}{A\setminus B}\; \setminus_i \text{ binding } A$$

$$\frac{\begin{array}{cc}\Delta & \Gamma\\ \vdots & \vdots\\ A & A\setminus B\end{array}}{B}\; \setminus_e$$

The rule below requires at least two free hyp.

A rightmost free hyp.
$$\dots\dots[A]\dots$$
$$\vdots$$
$$\frac{B}{B\,/\,A}\; /_i \text{ binding } A$$

$$\frac{\begin{array}{cc}\Gamma & \Delta\\ \vdots & \vdots\\ B\,/\,A & A\end{array}}{B}\; /_e$$

Fig. 5. Natural deduction rule for product free Lambek calculus

An important property of normal natural deductions is that whenever the last rule is an elimination rule, there is a principal branch leading from the conclusion to a free hypothesis [26, proposition 2.10 page 35] When a rule $\setminus_e$ (resp. $/_e$) is applied between a right premise $A\setminus X$ (resp. a left premise $X\,/\,A$) and a formula A as its left (resp. right) premise, the premise $A\setminus X$ (resp. a left premise $X\,/\,A$) is said to be the *principal* premise. In a proof ending with an elimination rule, a *principal branch* is a path from the root $C = X_0$ to a leaf $C[(\varepsilon_1,G_1),\dots,(\varepsilon_p,G_p)] = X_p$ such that one has $X_i = C[(\varepsilon_1,G_1),\dots,(\varepsilon_i,G_i)]$ and also $X_{i+1} = C[(\varepsilon_1,G_1),\dots,(\varepsilon_{i+1},G_{i+1})]$ and X_i is the conclusion of an elimination rule, $\setminus_e$ if $\varepsilon_{i+1} = l$ and $/_e$ if $\varepsilon_{i+1} = r$, with principal premise X_{i+1} and G_{i+1} as the other premise.

Let d be a normal natural deduction with conclusion C and hypotheses $H_1,\dots,H_n$. It is inductively turned into a proof net with conclusions $H_n-,\dots,$ $H_1-, C+$ as follows (we only consider $\setminus$ because $/$ is symmetrical).

- If d is just an hypothesis A which is at the same time its conclusion the corresponding proof net is the axiom $A, A^{\perp}$.
- If d ends with a $\setminus$ intro, from $A, H_1,\dots,H_n \vdash B$ to $H_1,\dots,H_n \vdash A\setminus B$, by induction hypothesis we have a proof net with conclusions $(H_n)-,\dots,(H_1)-,$ $A-, B+$. The heterogeneous $\wp$ rule applies since $B+$ is heterogeneous positive and $A-$ heterogeneous negative. A $\wp$ rule yields a proof net with conclusions $(H_n)-,\dots,(H_1)-, A-\wp B+$, and $A-\wp B+$ is precisely $(A\setminus B)+$
- The only interesting case is when d ends with an elimination rule, say $\setminus_e$. In this case there is a principal branch, say with hypothesis $C[(\varepsilon_1,G_1),\dots,$ $(\varepsilon_p,G_p)]$ which is applied to G_i's. Let us call $\Gamma_i = H_i^1,\dots,H_i^{k_i}$ the hypotheses of G_i, and let d_i be the proof of G_i from Γ_i. By induction hypothesis we have a proof net π_i with conclusions $(\Gamma_i)-,(G_i)+$. Let us define the proof net π^k of conclusion $C^k- = C[(\varepsilon_1,G_1),\dots,(\varepsilon_k,G_k)]-, \Gamma_i$ for $i \leq k$ and $C+$ by:

- if $k = 0$ then it is an axiom $C^\perp, C$ (consistent with the translation of an axiom)
- otherwise π^{k+1} is obtained by a times rule between the conclusions C^k- of π^k and $G_{k+1}+$ of π_{k+1} When $\varepsilon_i = r$ then the conclusion chose the conclusion of this link to $G_{k+1} + \otimes C^k-$ that is $C^k - /G_{k+1}+ = C^{k+1}-$ and when $\varepsilon_i = l$ the conclusion is $C^k - \otimes G_{k+1}+$ that is $G_{k+1} + \backslash C^k- = C^{k+1}-$. hence, in any case the conclusions of π^{k+1} are $C^{k+1}+$ $C+$ and the Γ_i for $i \leq k + 1$.

The translation of d is simply π^p, which has the proper conclusions.

As there is no cut in any of the translation steps, the result is a cut-free proof net.

From Cut-Free Proof Nets to Normal Natural Deductions. There is an algorithm that performs the reverse translation, presented for multiplicative linear logic and linear lambda terms in [13]. It strongly relies on the correctness criterion, which makes sure that everything happens as indicated during the algorithm and that it terminates. This algorithm always points at a sub formula of the proof net. Going *up* means going to an immediate sub formula, and going *down* means considering the immediate super formula.

1. Enter the proof net by its unique output conclusion.
2. Go up until you reach an axiom. Because of the polarities, during this upwards path, because of polarities, you only meet $\wp$-links, which correspond to $\backslash$ and $/$ introduction rules — λ_r and λ_l if one uses Lambek λ-terms. The hypotheses that are cancelled (the variables that are abstracted) are the ones on the non output premises — place a name on them.
3. Use the axiom link and go down with the input polarity. Hence you only meet $\otimes$ links (*) until you reach a conclusion or a $\wp$ link. In both cases, it is the head-variable of the λ-term. If it is the premise of a $\wp$-link, then it is necessarily a $\wp$ link on the path of step 2 (because of the correctness criterion) and the hypothesis of the principal branch is cancelled by $\backslash$ and $/$ introduction rules that we met during step 2 (the head variable bound by some of these λ_r or λ_l of the previous step). Otherwise it the hypothesis of the principal branch is free (the head variable is free).
4. The deductions (λ-terms) that are the arguments of the hypothesis of the principal branch (the head variable) are the ones on the output premises of the $\otimes$ links (*) that we met at step 3. They should be translated as we just did, going up from theses output formulae, starting again at step 2.

8.2 Learning Product Free Lambek Grammars from Natural Deduction

Because of the bijective correspondence between cut free product free proof nets and normal product free natural deduction we also have a correspondence between such structure without names but the S main conclusion. Hence if one wishes to it is possible to learn product free Lambek grammars from natural deduction without names but the final S, as we did in [10] Such structures are simply the generalisation to Lambek calculus of the FA structures that are commonly used for AB-grammars by [12,22].

9   Conclusion and Possible Extensions

A first criticism that can be addressed to our learning algorithm is that the rigid condition on Lambek grammars is too restrictive. One can say, as in [22] that k-valued grammars can be learned by doing all the possible unification that lead to less than k categories. Every successful unification of grammar with less than k categories should be kept, because it can thereafter work with other types, hence this approach is computationnally intractable. An alternative is to use a precise part of speech tagger and to consider word with different categories as distinct. This can be done and looks more sound and could be done partly with statistical techniques. [37,25]

The principal weakness of identification in the limit is that too much structure is required. Ideally, one would like to learn directly from strings, but in the case of Lambek grammars it has been shown to be impossible in [14]. One may think that it could be possible to try every possible structure on sentences as strings of words as done in [22] for basic categorial grammars. Unfortunately, in the case of Lambek grammars, with or without product, this cannot be done. Indeed, there can be infinitely many structures corresponding to a sentence, because a cancelled hypothesis does not have to be anchored in one the finitely many words of the sentence. Hence we ought to learn from structured sentences.

From the point of view of first language acquisition we know that some structure is available, but it is unlikely that the structured sentences are proof frames that are are partial categorial parse structure. The real input available to the learner is a mixture of prosodic and semantic information, and no one knows how to formalise these structures in order to simulate the natural data for language learning. From a computational linguistic perspective, our result is not as bad as one may think. Indeed, there exist tools that annotate corpora, and one may implement other tools that turn standard annotations into other more accurate annotations. These shallow processes may lead to structures from which one can infer the proper structure for algorithm like the one we presented in this paper. In the case of proof nets, as observed long ago, axioms express the consumption of valency. This the reason why, apart from the structure of the formulae, the structure of the proof frames is not so different from dependency annotations and can be used to infer categorial structures see e.g. [37,25]. However, the automatic acquisition of wide-coverage grammars for natural language processing applications, certainly requires a combination of machine learning techniques and of identification in the limit à la Gold, although up to now there are not so many such works.

Grammatical formalisms that can be represented in Lambek grammars can also be learnt like we did in this paper. For instance categorial version of Stabler's minimalist grammars [38] can be learnt that way as the attempts by Fulop or us show [15,10] This should be even better with the so-called Categorial Minimalist grammars of Lecomte, Amblard and us [1,2].

References

1. Amblard, M.: Calculs de représentations sémantiques et syntaxe générative: les grammaires minimalistes catégorielles. PhD thesis, Université Sciences et Technologies - Bordeaux I (September 2007)
2. Amblard, M., Lecomte, A., Retoré, C.: Categorial minimalist grammars: From generative grammar to logical form. Linguistic Analysis 36(1-4), 273–306 (2010)
3. Amblard, M., Retoré, C.: Natural deduction and normalisation for partially commutative linear logic and lambek calculus with product. In: Cooper, S.B., Kent, T.F., Löwe, B., Sorbi, A. (eds.) Computation and Logic in the Real World (Computing in Europe 2007). Quaderni del Dipartimento di Scienze Matematiche e Informatiche Roberto Magari, vol. ID487, pp. 28–35. Università degli Studi di Siena (September 2007)
4. Angluin, D.: Finding patterns common to a set of strings. Journal of Computer and Sytem Science 21(1), 46–62 (1980)
5. Angluin, D.: Inductive inference of formal languages from positive data. Information and Control 45, 117–135 (1980)
6. Bar-Hillel, Y.: A quasi arithmetical notation for syntactic description. Language 29, 47–58 (1953)
7. Bar-Hillel, Y., Gaifman, C., Shamir, E.: On categorial and phrase-structure grammars. Bulletin of the Research Council of Israel F(9), 1–16 (1963)
8. Berwick, R.C., Pietroski, P., Yankama, B., Chomsky, N.: Poverty of the stimulus revisited. Cognitive Science 35(5), 1207–1242 (2011)
9. Bonato, R.: Uno studio sull'apprendibilità delle grammatiche di Lambek rigide — A study on learnability for rigid Lambek grammars. Tesi di Laurea & Mémoire de D.E.A, Università di Verona & Université Rennes 1 (2000)
10. Bonato, R., Retoré, C.: Learning rigid Lambek grammars and minimalist grammars from structured sentences. In: Popelìnskỳ, L., Nepil, M. (eds.) Proceedings of the third workshop on Learning Language in Logic, LLL 2001. FI MU Report series, vol. FI-MU-RS-2001-08, pp. 23–34. Faculty of Informatics – Masaryk University, Strabourg (September 2001)
11. Buszkowski, W.: Discovery procedures for categorial grammars. In: van Benthem, J., Klein, E. (eds.) Categories, Polymorphism and Unification. Universiteit van Amsterdam (1987)
12. Buszkowski, W., Penn, G.: Categorial grammars determined from linguistic data by unification. Studia Logica 49, 431–454 (1990)
13. de Groote, P., Retoré, C.: Semantic readings of proof nets. In: Kruijff, G.J., Morrill, G., Oehrle, D. (eds.) Formal Grammar, pp. 57–70. FoLLI, Prague (1996), http://hal.archives-ouvertes.fr/hal-00823554
14. Foret, A., Le Nir, Y.: Lambek rigid grammars are not learnable from strings. In: COLING 2002, 19th International Conference on Computational Linguistics, Taipei, Taiwan, vol. 1, pp. 274–279 (August 2002)
15. Fulop, S.: The Logic and Learning of Language. Trafford on Demand Pub. (2004)
16. Gleitman, L., Liberman, M. (eds.): An invitation to cognitive sciences, vol. 1. Language. MIT Press (1995)

17. Gold, E.M.: Language identification in the limit. Information and control 10, 447–474 (1967)
18. Guerrini, S.: A linear algorithm for mll proof net correctness and sequentialization. Theoretical Computer Science 412(20), 1958–1978 (2011)
19. Johnson, K.: Gold's theorem and cognitive science. Philosophy of Science 71, 571–592 (2004)
20. Joshi, A., Vijay-Shanker, K., Weir, D.: The convergence of mildly context-sensitive grammar formalisms. In: Sells, P., Schieber, S., Wasow, T. (eds.) Fundational Issues in Natural Language Processing. MIT Press (1991)
21. Kanazawa, M.: Learnable classes of categorial grammars. PhD thesis, Universiteit van Amsterdam (1994)
22. Kanazawa, M.: Learnable classes of categorial grammars. Studies in Logic, Language and Information. FoLLI & CSLI distributed by Cambridge University Press (1998)
23. Lambek, J.: The mathematics of sentence structure. American Mathematical Monthly, 154–170 (1958)
24. Melliès, P.A.: A topological correctness criterion for multiplicative non commutative logic. In: Ehrhard, T., Girard, J.Y., Ruet, P., Scott, P. (eds.) Linear Logic in Computer Science. London Mathematical Society Lecture Notes, vol. 316, pp. 283–321. Cambridge University press (2004)
25. Moot, R.: Semi-automated extraction of a wide-coverage type-logical grammar for French. In: Proceedings of Traitement Automatique des Langues Naturelles (TALN), Montreal (2010)
26. Moot, R., Retoré, C.: The logic of categorial grammars: A deductive account of natural language syntax and semantics. LNCS, vol. 6850. Springer, Heidelberg (2012)
27. Morrill, G.: Categorial Grammar: Logical Syntax, Semantics, and Processing. OUP, Oxford (2011)
28. Morrill, G.: Incremental processing and acceptability. Computational Linguistics 26(3), 319–338 (2000); preliminary version: UPC Report de Recerca LSI-98-46-R (1998)
29. Nicolas, J.: Grammatical inference as unification. Rapport de Recherche RR-3632. INRIA (1999)
30. Pentus, M.: Lambek grammars are context-free. In: Logic in Computer Science. IEEE Computer Society Press (1993)
31. Piattelli-Palmarini, M. (ed.): Théories du langage, théories de l'apprentissage — le débat Chomsky Piaget. Editions du Seuil. Number 138 in Points (1975)
32. Pinker, S.: Language acquisition. In: [16], ch. 6, pp. 135–182
33. Pinker, S.: Why the child holded the baby rabbits. In: [16], ch. 5, pp. 107–133
34. Pullum, G.K., Scholz, B.C.: Empirical assessment of stimulus poverty arguments. The Linguistic Review 19, 9–50 (2002)
35. Reali, F., Christiansen, M.H.: Uncovering the richness of the stimulus: Structure dependence and indirect statistical evidence. Cognitive Science 29(6), 1007–1028 (2005)
36. Retoré, C.: Le système F en logique linéaire. Mémoire de D.E.A. (dir.: J.-Y. Girard), Université Paris 7 (1987)
37. Sandillon-Rezer, N.-F., Moot, R.: Using tree transducers for grammatical inference. In: Pogodalla, S., Prost, J.-P. (eds.) LACL 2011. LNCS (LNAI), vol. 6736, pp. 235–250. Springer, Heidelberg (2011)
38. Stabler, E.: Derivational minimalism. In: Retoré, C. (ed.) LACL 1996. LNCS (LNAI), vol. 1328, pp. 68–95. Springer, Heidelberg (1997)

39. Tellier, I.: How to split recursive automata. In: Clark, A., Coste, F., Miclet, L. (eds.) ICGI 2008. LNCS (LNAI), vol. 5278, pp. 200–212. Springer, Heidelberg (2008)
40. Tiede, H.J.: Deductive Systems and Grammars: Proofs as Grammatical Structures. PhD thesis, Illinois Wesleyan University (1999), http://www.iwu.edu/htiede/
41. Zucker, J.: The correspondence between cut-elimination and normalisation i, ii. Annals of Mathematical Logic 7, 1–156 (1974)

Multi-Sorted Residuation

Wojciech Buszkowski

The Adam Mickiewicz University in Poznań, Poland
buszko@amu.edu.pl

Abstract. Nonassociative Lambek Calculus (**NL**) is a pure logic of residuation, involving one binary operation (product) and its two residual operations defined on a poset [26]. Generalized Lambek Calculus **GL** involves a finite number of basic operations (with an arbitrary number of arguments) and their residual operations [7]. In this paper we study a further generalization of **GL** which admits operations whose arguments and values can be of different sorts. This logic is called *Multi-Sorted Lambek Calculus* **mL**. We also consider its variants with lattice and boolean operations. We discuss some basic properties of these logics (completeness, decidability, complexity and others) and the corresponding algebras.

1 Introduction

Nonassociative Lambek Calculus (**NL**) was introduced in [26] as a weaker variant of the Syntactic Calculus [25], the latter nowadays called (Associative) Lambek Calculus (**L**). Lambek's motivation for **NL** was linguistic: to block some over-generation, appearing when sentences are parsed by means of **L**. For example, *John likes poor Jane* and *John likes him* justify the following typing:

$$\text{John, Jane: } n, \text{ likes: } (n\backslash s)/n, \text{ poor: } n/n, \text{ him: } (s/n)\backslash s\,,$$

which yields type s of *John likes poor him* in **L**, but not in **NL**.

Besides linguistic interpretations, usually related to *type grammars*, these calculi became popular in some groups of logicians, as basic *substructural logics*. **L** admitting Exchange and sequents $\Rightarrow A$ (i.e. sequents with the empty antecedent) is equivalent to the $\{\otimes, \rightarrow\}$−fragment of Linear Logic of Girard, and without Exchange to an analogous fragment of Noncommutative Linear Logic of Abrusci. Full Lambek Calculus (**FL**), i.e. **L** with 1, 0 (optionally) and lattice connectives $\sqcup, \sqcap$, and its nonassociative version **FNL** are treated as basic substructural logics in the representative monograph [11] (**FNL** is denoted **GL** from 'groupoid logic', but we use the latter symbol in a different meaning). Recall that substructural logics are nonclassical logics whose Gentzen-style sequent systems omit some structural rules (Exchange, Weakening, Contraction). This class contains (among others) relevant logics (omit Weakening) and multi-valued logics (omit Contraction); they can be presented as axiomatic extensions of **FL**.

Studies in substructural logics typically focus on associative systems in which *product* $\otimes$ is associative. Nonassociative systems are less popular among logicians, although they are occasionally considered as a close companion of the

C. Casadio et al. (Eds.): Lambek Festschrift, LNCS 8222, pp. 136–155, 2014.

former. In the linguistic community, some work has been done in Nonassociative Lambek Calculus, treated as a natural framework for parsing structured expressions. This approach is dominating in Moortgat's studies on type grammars; besides nonassociative product and its residuals $\backslash, /$, Moortgat considers different unary modalities and their residuals which allow a controlled usage of certain structural rules [30]. Recently, Moortgat [31] also admits a dual residuation triple, which leads to some Grishin-style nonassociative systems. Nonassociative Lambek Calculus was shown context-free in [5] (the product-free fragment) and [20] (the full system). A different proof was given by Jäger [15], and its refinement yields the polynomial time complexity and the context-freeness of **NL** augmented with (finitely many) assumptions [6].

A straightforward generalization of **NL** admits an arbitrary number of generalized product operations of different arities together with their residuals. The resulting system, called *Generalized Lambek Calculus*, was studied in the author's book (*Logical Foundations of Ajdukiewicz-Lambek Categorial Grammars*, in Polish, 1989) and later papers [6,9,7] (also with lattice and boolean operations). In this setting the associative law is not assumed, as not meaningful for non-binary operations.

The present paper introduces a further generalization of this framework: different product operations are not required to act on the same universe. For instance, one may consider an operation $f : A \times B \mapsto C$ with residuals $f^{r,1} : C \times B \mapsto A$ and $f^{r,2} : A \times C \mapsto B$ and another operation $g : A' \times B' \mapsto C'$ with residuals $g^{r,1}, g^{r,2}$. Here A, B, C represent certain ordered algebras: posets, semi-lattices, lattices, boolean algebras etc., and one assumes the residuation law: $f(x,y) \leq_C z$ iff $x \leq_A f^{r,1}(z,y)$ iff $y \leq_B f^{r,2}(x,z)$.

This approach seems quite natural: in mathematics one often meets residuated operations acting between different universes, and such operations can also be used in linguistics (see section 2). The resulting multi-sorted residuation logic extends **NL**, and we show here that it inherits many essential proof-theoretic, model-theoretic and computational properties of **NL**. For instance, without lattice operations it determines a polynomial consequence relation; with distributive lattice or boolean operations the consequence relation remains decidable in opposition to the case of **L**.

The multi-sorted framework can further be generalized by considering categorical notions, but this generalization is not the same as cartesian-closed categories, studied by Lambek and others; see e.g. [28,27]. Instead of a single category with object-constructors $A \times B, A^B, ^B A$, corresponding to the algebraic $a \otimes b, a \backslash b, b / a$, one should consider a multicategory whose morphisms are residuated maps. We do not develop this approach here.

In section 2 we define basic notions, concerning residuated maps, and provide several illustrations. In particular, we show how multi-sorted residuated maps can be used in modal logics and linguistics.

In section 3 we consider multi-sorted (heterogeneous) residuation algebras: abstract algebraic models of multi-sorted residuation logics. We discuss canonical embeddings of such algebras into complex algebras of multi-sorted relational frames, which yield some completeness theorems for multi-sorted residuation logics. The multi-sorted perspective enables one to find more uniform proofs of embedding theorems even for the one-sort case.

The multi-sorted residuation logics are defined in section 4; the basic system is *Multi-Sorted Lambek Calculus* **mL**, but we also consider some extensions of it. In general, basic properties of one-sort residuation logics are preserved by multi-sorted logics. Therefore we omit most proofs. Some events, however, only appear in the multi-sorted world (e.g. classical paraconsistent theories).

Some ideas of this paper have been presented in the author's talk 'Many-sorted gaggles' at the conference *Algebra and Coalgebra Meet Proof Theory*, Prague, 2012 [8].

2   Residuated Maps

Let $(P_1, \leq_1)$, $(P_2, \leq_2)$ be posets. A map $f : P_1 \mapsto P_2$ is said to be *residuated*, if the co-image $f^{-1}[x^\downarrow]$ of any principal downset $x^\downarrow \subseteq P_2$ is a principal downset in P_1 [4]. Equivalently, there exists a *residual* map $f^r : P_2 \mapsto P_1$ such that

$$(\text{uRES}) \quad f(x) \leq_2 y \text{ iff } x \leq_1 f^r(y)$$

for all $x \in P_1, y \in P_2$.

NL is a logic of one binary operation $\otimes$ on a poset $(P, \leq)$ such that, for any $w \in P$, the maps $\lambda x.x \otimes w$ and $\lambda x.w \otimes x$ from P to P are residuated. Equivalently, the binary operation $\otimes$ admits two *residual* operations $\backslash, /$, satisfying:

$$(\text{bRES}) \quad x \otimes y \leq z \text{ iff } y \leq x \backslash z \text{ iff } x \leq z/y,$$

for all $x, y, z \in P$.

It is natural to consider a more general situation. A map $f : P_1 \times \cdots \times P_n \mapsto P$, where $(P_i, \leq_i)$, for $i = 1, \ldots, n$, and $(P, \leq)$ are posets, is said to be *residuated*, if, for any $i = 1, \ldots, n$, the unary maps $\lambda x.f(w_1, \ldots, x : i, \ldots, w_n)$ are residuated, for all $w_1 \in P_1, \ldots, w_n \in P_n$. (Here $x : i$ means that x is the i-th argument of f; clearly $w_i \in P_i$ is dropped from the latter list.) Equivalently, the map f admits n residual maps $f^{r,i}$, for $i = 1, \ldots, n$, satisfying:

$$(\text{RES}) \quad f(x_1, \ldots, x_n) \leq z \text{ iff } x_i \leq_i f^{r,i}(x_1, \ldots, z : i, \ldots, x_n),$$

for all $x_1 \in P_1, \ldots, x_n \in P_n, z \in P$, where:

$$f^{r,i} : P_1 \times \cdots \times P : i \times \cdots \times P_n \mapsto P_i.$$

Every *identity* map $I(x) = x$ from P to P is residuated, and its residual is the same map. We write $\bar{P}_{(n)}$ for $P_1 \times \cdots \times P_n$. If $f : \bar{P}_{(n)} \mapsto P$ and $g : \bar{Q}_{(m)} \mapsto P_i$ are residuated, then their composition $h : P_1 \times \cdots P_{i-1} \times \bar{Q}_{(m)} \times P_{i+1} \cdots \times P_n \mapsto P$ is residuated, where one sets:

$$h(\ldots, y_1, \ldots, y_m, \ldots) = f(\ldots, g(y_1, \ldots, y_m), \ldots).$$

WARNING. The residuated maps are not closed under a stronger composition operation which from $f, g_1, \ldots, g_k$ yields $h(\bar{x}) = f(g_1(\bar{x}), \ldots, g_k(\bar{x}))$, where $\bar{x}$ stands for $(x_1, \ldots, x_n)$. This composition is considered in recursion theory.

Consequently, posets and residuated maps form a multicategory; posets and unary residuated maps form a category. Notice that an $n-$ary residuated map from $\bar{P}_{(n)}$ to P need not be residuated, if considered as a unary map, defined on the product poset. This can easily be seen, if one notices that an $n-$ary residuated map must be completely additive in each argument, it means:

$$f(\ldots, \bigvee_t x_i^t, \ldots) = \bigvee_t f(\ldots, x_i^t, \ldots),$$

if $\bigvee_t x_i^t$ exists. (If $P_1, \ldots, P_n, P$ are complete lattices, then f is residuated iff it is completely additive in each argument.) Treated as a unary residuated map, it should satisfy a stronger condition: preserve bounds with respect to the product order:

$$f(\bigvee_t (x_1^t, \ldots, x_n^t)) = \bigvee_t f(x_1^t, \ldots, x_n^t).$$

A more concrete example is as follows. Let $(P, \leq)$ be a bounded poset, and let $\otimes$ be a binary residuated map from P^2 to P. We have $\bot \otimes \top = \bot$ and $\top \otimes \bot = \bot$. Then $\otimes^{-1}[\{\bot\}]$ contains the pairs $(\bot, \top), (\top, \bot)$ whose l.u.b. (in the product poset) is $(\top, \top)$. But, in general, $\top \otimes \top \neq \bot$, hence $\otimes^{-1}[\{\bot\}]$ need not be a principal downset. If all universes are complete lattices, then every unary residuated map from the product lattice is an $n-$ary residuated map in the above sense.

If f is a residuated map from $(P, \leq_P)$ to $(Q, \leq_Q)$, then f^r is a residuated map from $(Q, \geq_Q)$ to $(P, \geq_P)$, and f is the residual of f^r. For an $n-$ary residuated map $f : P_1 \times \cdots \times P_n \mapsto Q$, $f^{r,i}$ is a residuated map from $P_1 \times \cdots \times Q^{op} \times \cdots \times P_n$ to P_i^{op}, where P^{op} denotes the poset dual to P; the $i-$th residual of $f^{r,i}$ is f, and the $j-$th residual $(j \neq i)$ is $g(x_1, \ldots, x_n) = f^{r,j}(x_1, \ldots, x_j : i, \ldots, x_i : j, \ldots, x_n)$. Accordingly there is a symmetry between all maps $f, f^{r,1}, \ldots, f^{r,n}$, not explicit in the basic definition. These symmetries will be exploited in section 3.

Residuated maps appear in many areas of mathematics, often defined as Galois connections. A *Galois connection* between posets $(P_1, \leq_1), (P_2, \leq_2)$ is a pair $f : P_1 \mapsto P_2$, $g : P_2 \mapsto P_1$ such that, for all $x \in P_1, y \in P_2$, $x \leq_1 g(y)$ iff $y \leq_2 f(x)$. Clearly, f, g is a Galois connection iff g is the residual of f when $\leq_2$ is replaced by its reversal. In opposition to residuated maps, the first (second) components of Galois connections are not closed under composition (hence residuated maps lead to a more elegant framework [4]).

Residuated maps in mathematics usually act between different universes, like in the classical Galois example: between groups and fields. On the other hand, the logical theory of residuation focused, as a rule, on the one-universe case, and similarly for the algebraic theory. One considers different kinds of *residuation algebras*, e.g. residuated semigroups (groupoids), (nonassociative) residuated lattices, their expansions with unary operations, and so on, together with the corresponding logics; see e.g. [4,11]. Typically all operations are (unary or binary) operations in the algebra. The situation is similar in linguistic approaches, traditionally developed in connection with type grammars based on different variants of the Lambek calculus.

We provide some examples of residuated maps.

$\mathcal{P}(W)$ is the powerset of W. A residuated map from $\mathcal{P}(V_1) \times \cdots \times \mathcal{P}(V_n)$ to $\mathcal{P}(W)$ can be defined as follows. Let $R \subseteq W \times V_1 \times \cdots \times V_n$. For $(X_1, \ldots, X_n)$, where $X_j \subseteq V_j$, for $j = 1, \ldots, n$, one defines:

$$f_R(X_1, \ldots, X_n) = \{y \in W : (\exists x_1 \in X_1, \ldots, x_n \in X_n) R(y, x_1, \ldots, x_n)\}.$$

f_R is residuated, and its residual maps are:

$$f_R^{r,i}(X_1, \ldots, Y : i, \ldots, X_n) = \{x \in V_i : f_R(X_1, \ldots, \{x\} : i, \ldots, X_n) \subseteq Y\}.$$

For $n = 1$ and $V_1 = W$, f_R is the $\Diamond$−modality determined by the Kripke frame (W, R), $R \subseteq W^2$; see e.g. [3]. Precisely, it is the operation corresponding to $\Diamond$ in the complex algebra of the frame. Analogously, for $V_i = W$, $i = 1, \ldots, n$, f_R corresponds to the $\Diamond$ determined by the multi-modal frame (W, R), $R \subseteq W^{n+1}$. To get the correspondence, the truth definition should be: $y \models \Diamond \varphi$ iff, for some x, $R(y, x)$ and $x \models \varphi$, and similarly for the multi-modal case. If one defines: $\|\varphi\| = \{x \in W : x \models \varphi\}$, then $\Diamond(\|\varphi\|) = \|\Diamond(\varphi)\|$, where the first $\Diamond$ is the operation f_R, and the second one is the corresponding modal connective.

If R is not symmetric, then f_R^r does not equal the $\Box$−modality corresponding to $\Diamond$, namely $\Box(X) = -\Diamond(-X)$. One often writes $\Box^\downarrow$ for f_R^r. Modal logics are usually presented with the modal pair $\Diamond, \Box$, but without $\Box^\downarrow$. Some exceptions are temporal logics with their residual pairs F, H and P, G, and some substructural modal logics. Let us notice that every normal modal logic which is complete with respect to a class of Kripke frames can conservatively be expanded by adding $\Box^\downarrow$, the residual of $\Diamond$. Such expansions inherit basic properties of normal modal logics, and they can be studied by certain methods of substructural logics.

Dynamic logics make the connection between R and $\Diamond$ explicit; one writes $\langle R \rangle$ for the $\Diamond$ determined by R, and $[R]$ for its De Morgan dual; instead of R one writes a program term interpreted as R.

A greater flexibility can be attained by treating $\Diamond$ as a binary map from $(\mathcal{P}(W^2)) \times \mathcal{P}(W)$ to $\mathcal{P}(W)$: $\Diamond(R, X) = \{y \in W : (\exists x \in X) R(y, x)\}$. In this setting $\Diamond = f_S$, where $S \subseteq W \times W^2 \times W$ consists of all tuples $(y, (y, x), x)$ such that $x, y \in W$. Notice that S is a *logical* relation, since it is invariant under permutations of W.

Consequently the binary $\Diamond$ is residuated. We have:

$$\Diamond^{r,2}(R, X) = [R]^\downarrow(X) = [R^\smile](X) = \{x \in W : \Diamond(R, \{x\}) \subseteq X\}.$$

The other residual:

$$\Diamond^{r,1}(X, Y) = \{(x, y) \in W^2 : \Diamond(\{(x, y)\}, Y) \subseteq X\} =$$

$$= \{(x, y) \in W^2 : x \in X \sqcup y \notin Y\}$$

yields the greatest relation R such that $\Diamond(R, Y) \subseteq X$. It is not a standard operation in dynamic logics, but it may be quite useful. If φ, ψ are formulas,

$\Diamond^{r,1}(\|\varphi\|, \|\psi\|)$ is interpreted as the largest (nondeterministic) program R such that, for any input satisfying the pre-condition $\neg\varphi$, every outcome of R satisfies the post-condition $\neg\psi$. Besides known laws of dynamic logic, in the extended language one can express new laws, e.g.:

$$\Diamond^{r,1}(\|\varphi \wedge \psi\|, \|\chi\|) = \Diamond^{r,1}(\|\varphi\|, \|\chi\|) \cap \Diamond^{r,1}(\|\psi\|, \|\chi\|),$$

$$\Diamond^{r,1}(\|\varphi\|, \|\psi \vee \chi\|) = \Diamond^{r,1}(\|\varphi\|, \|\psi\|) \cap \Diamond^{r,1}(\|\varphi\|, \|\chi\|).$$

(In general, if f is residuated, then $f^{r,i}$ preserves all existing meets in the i−th argument, and sends the existing joins to the corresponding meets in any other argument.) Clearly the binary $\Diamond$ with its residuals is an example of a multi-sorted residuation triple. They are logical operations in the above sense.

Other examples of logical multi-sorted residuated maps are the relative product map $\circ : \mathcal{P}(U \times V) \times \mathcal{P}(V \times W) \mapsto \mathcal{P}(U \times W)$, the Cartesian product map $\times : \mathcal{P}(V) \times \mathcal{P}(W) \mapsto \mathcal{P}(V \times W)$, and the disjoint union map $\uplus : \mathcal{P}(V) \times \mathcal{P}(W) \mapsto \mathcal{P}(V \uplus W)$.

Given any map $g : V_1 \times \cdots \times V_n \mapsto W$, by $R(g)$ we denote the relation: $R(g)(y, x_1, \ldots, x_n)$ iff $y = g(x_1, \ldots, x_n)$ (the graph of g). The residuated map $f_{R(g)}$ will be denoted by p_g. This construction appears in numerous applications. We mention some examples connected with linguistics.

A standard interpretation of **NL** involves binary *skeletal* trees, i.e. trees whose leaves but no other nodes are labeled by certain symbols. Clearly skeletal trees can be represented as bracketed strings over some set of symbols. Let $\Sigma = \{a, b\}$. Then $[a, [b, a]]$ represents the tree on Figure 1.

Fig. 1. A binary skeletal tree

The formulas of **NL** are interpreted as sets of skeletal trees (over an alphabet Σ), and the product connective $\otimes$ is interpreted as p_*, where $*$ is the concatenation of skeletal trees: $t_1 * t_2 = [t_1, t_2]$.

If skeletal trees are replaced with labeled trees whose internal nodes are labeled by category symbols, then instead of one operation $*$ one must use a family of operations $*_A$, one for each category symbol A. One defines: $t_1 *_A t_2 = [t_1, t_2]_A$. Often binary operations are not sufficient; one needs n−ary operations for $n = 1, 2, 3, \ldots$. For instance, a ternary operation o_A sends (t_1, t_2, t_3) to $[t_1, t_2, t_3]_A$. This leads to the formalism of Generalized Lambek Calculus.

In the above setting we admit that an n−ary operation is defined on all possible n−tuples of trees. As a result, we generate a huge universe of trees, many of them being completely useless for syntactic analysis. This overgeneration can be eliminated, if one restricts the application of an operation to those tuples which satisfy additional constraints. To formalize this idea we might admit partial operations, which would essentially complicate the algebraic and logical details.

Here we describe another option, involving multi-sorted operations. Let G be a context-free grammar (CFG) in a normal form: every production rule of G is of the form $A \to B_1, \ldots, B_n$, where $n \geq 1$ and A, B_i are nonterminals, or $A \to a$, where A is a nonterminal, a is a terminal symbol from Σ. The rules of the first form are called *tree rules*, and those of the second form are called *lexical rules*.

Let T_A denote the set of all labeled trees whose root is labeled by A. With any tree rule r we associate an operation o_r; if r is $A \to B_1, \ldots, B_n$, then $o_r : T_{B_1} \times \cdots T_{B_n} \mapsto T_A$ is defined as follows: $o_r(t_1, \ldots, t_n) = [t_1, \ldots, t_n]_A$.

L_A denotes the set of all *lexical* trees $[a]_A$ such that $A \to a$ is a lexical rule. D_A denotes the set of all (complete) derivation trees of G whose root is labeled by A.

The sets T_A with the operations o_r form a multi-sorted algebra, and the sets $D_A \subseteq T_A$ with the same operations (naturally restricted) form a subalgebra of this algebra; it is the subalgebra generated by the lexical trees. Precise definitions of these notions will be given in section 3. Speaking less formally, if one starts from lexical trees and applies operations o_r, then the generated trees are precisely the derivation trees of G. For instance, let the rules of G be $r_1 : S \to S, B$; $r_2 : S \to A, B$; $A \to a$; $B \to b$. Figure 2 shows a tree in D_S.

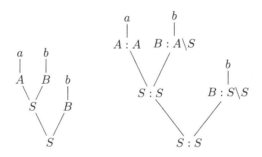

Fig. 2. The tree $o_{r_1}(o_{r_2}([a]_A, [b]_B), [b]_B)$ and its typed version

A type grammar G' equivalent to G assigns: $a : A$, $b : A\backslash S$, $S\backslash S$. To attain a full coincidence of derivation trees we assign types to lexical trees: $[a]_A : A$, $[b]_B : A\backslash S$, $S\backslash S$. Then, **NL** (actually the pure reduction calculus **AB**) yields essentially the derivation trees of G; see Figure 2. The label $A : \alpha$ means that the tree with root A is of type α.

The grammar G' should be modified to be fully compatible with the multi-sorted framework. One should take $[b]_B : A\backslash_2 S$, $S\backslash_1 S$. Then, in the algebra of sets of trees one interprets o_{r_i} as the operation $p_i = p_{o_{r_i}}$, and $\backslash_i$ is interpreted as the 2-nd residual of p_i. The typing of non-lexical subtrees of the above tree agrees with basic reduction laws $p_i(X, p_i^{r,2}(X, Y)) \subseteq Y$, which follow from (RES).

The above example illustrates one of many possible applications of multi-sorted operations in language description: a type grammar describes syntactic trees generated by a CFG. The CFG may provide a preliminary syntactic analysis, while the type grammar gives a more subtle account, or the grammars may focus on different features (like in applications of product pregroups [24,10]).

Another obvious option is a multi-level grammar, which handles both the syntactic and the semantic level; a two-sorted meaning map m sends syntactic trees into semantic descriptions (m need not be residuated, but the powerset map p_m certainly is). We can also imagine a joint description of strings (unstructured expressions) and trees (structured expressions) with a forgetting map from structures to strings; also expressions from two different languages with translation maps. Other examples will be mentioned in section 4.

3   Multi-Sorted Residuation Algebras

According to [7], a *residuated algebra* (RA) is a poset $(A, \leq)$ with a family F of residuated operations on A; each $n-$ary operation $f \in F$ admits n residual operations $f^{r,i}$, $1 \leq i \leq n$. (In [7], $o, o/i$ are used instead of $f, f^{r,i}$.) One also considers residuated algebras with lattice operations $\sqcup, \sqcap$ and Boolean negation or Heyting implication. The corresponding logics are Generalized Lambek Calculus and its extensions. The term 'residuated algebra' was coined after 'residuated lattice', used in the literature on substructural logics. Here we prefer 'residuation algebra', since the operations (not the algebra) are residuated; also 'residuated lattice' seems (even more) unlucky, since the residuals are not directly related to the lattice operations.

A *multi-sorted residuation algebra* (mRA) is a family $\{A_s\}_{s \in S}$ of ordered algebras with a family F of residuated maps; each map $f \in F$ is assigned a unique type $s_1, \ldots, s_n \to s$, where $s_i, s \in S$, and $f : A_{s_1} \times \cdots \times A_{s_n} \mapsto A_s$. S is the set of *sorts*. So a map f of type $s_1, \ldots, s_n \to s$ admits n residual maps:

$$f^{r,i} : A_{s_1} \times \cdots \times A_s : i \times \cdots \times A_{s_n} \mapsto A_{s_i}.$$

The ordered algebras A_s are always posets, but some of them can also admit semilattice, lattice, boolean or Heyting operations. A mRA is often denoted $\mathcal{A} = (\{A_s\}_{s \in S}, F)$ (we also write $F_\mathcal{A}$ for F).

A *subalgebra* of $\mathcal{A}$ is a family $\{B_s\}_{s \in S}$ such that $B_s \subseteq A_s$ and this family is closed under the operations from $F_\mathcal{A}$ and their residuals (dropping residuals, one obtains a standard notion of a subalgebra of a multi-sorted algebra). Clearly a subalgebra of a mRA is also a mRA with appropriately restricted operations.

Two mRAs $\mathcal{A}, \mathcal{B}$ are said to be *similar*, if they have the same set of sorts S, $F_\mathcal{A} = \{f_i\}_{i \in I}$, $F_\mathcal{B} = \{g_i\}_{i \in I}$, and f_i, g_i are of the same type, for any $i \in I$; we also assume that A_s, B_s are of the same type, for any $s \in S$ (it means: both are posets or lattices, semilattices, etc.). A *homomorphism* from $\mathcal{A}$ to $\mathcal{B}$, which are similar, is a family $\{h_s\}_{s \in S}$ such that $h_s : A_s \mapsto B_s$ is a homomorphism of ordered algebras, and the following equations hold, for any f_j of type $s_1, \ldots, s_n \to s$ and all $1 \le i \le n$:

(HOM1) $h_s(f_j(a_1, \ldots, a_n)) = g_j(h_{s_1}(a_1), \ldots, h_{s_n}(a_n))$,

(HOM2) $h_{s_i}(f_j^{r,i}(a_1, \ldots, a_n)) = g_j^{r,i}(h_{s_1}(a_1), \ldots, h_s(a_i) : i, \ldots, h_{s_n}(a_n))$.

We assume (HOM1) for all $a_1 \in A_{s_1}, \ldots, a_n \in A_{s_n}$, and (HOM2) for all $a_k \in A_{s_k}$, for $k \ne i$, and $a_i \in A_s$. An *embedding* is a homomorphism $\{h_s\}_{s \in S}$ such that every h_s is an embedding, it means: $a \le_{A_s} b$ iff $h_s(a) \le_{B_s} h_s(b)$, for all $a, b \in A_s$.

Standard examples of mRAs are *complex algebras* of multi-sorted frames $(\{V_s\}_{s \in S}, \mathcal{R})$ such that every V_s is a set, and $\mathcal{R}$ is a family of relations, each $R \in \mathcal{R}$ having a unique type $s_1, \ldots, s_n \to s$, and $R \subseteq V_s \times V_{s_1} \times \cdots \times V_{s_n}$. The given relation R determines a residuated map f_R, as defined in section 2. *The complex mRA associated with the frame is defined as* $(\{\mathcal{P}(V_s)\}_{s \in S}, \{f_R\}_{R \in \mathcal{R}})$. Clearly every $\mathcal{P}(V_s)$ is a boolean algebra of sets, and the ordering on $\mathcal{P}(V_s)$ is inclusion.

If all algebras A_s in $\mathcal{A}$ are of the same type, say posets or distributive lattices, admitting boolean algebras (we only consider these types; see the remarks at the end of this section), then $\mathcal{A}$ *can be embedded in the complex algebra of some multi-sorted frame*. This result generalizes known results on *canonical embeddings* of modal algebras, tracing back to [16,17]. Closely related results for *gaggles* (restricted to one sort) have been presented in [2]. Below we sketch a proof for many sorts, which seems more uniform than those in [2]: we make use of some order dualities and antitone operators to reduce the case of residual operations to that of basic (additive) operations.

Let $\mathcal{A} = (\{A_s\}_{s \in S}, F)$ be a mRA with all ordered algebras of the same type. We define the *canonical* frame $\mathcal{A}^c$ as follows. V_s is defined as the set of:

- all proper upsets of A_s, if A_s is a poset,
- all prime filters of A_s, if A_s is a distributive lattice (a boolean algebra).

A proper upset is a nonempty upset, different from A_s. A filter is an upset closed under meets, and a proper filter is a filter being a proper upset. A prime filter of A_s is a proper filter $X \subseteq A_s$ such that, for all $a, b \in A_s$, $a \sqcup b \in X$ entails $a \in X$ or $b \in X$. The prime filters of a boolean algebra are precisely its ultrafilters.

Let $g \in F$ be of type $s_1, \ldots, s_n \to s$. The relation $R[g] \subseteq V_s \times V_{s_1} \times \cdots \times V_{s_n}$ is defined as follows:

(CAN1) $R[g](Y, X_1, \ldots, X_n)$ iff $p_g(X_1, \ldots, X_n) \subseteq Y$,

where p_g is defined as in section 2. The complex mRA of $\mathcal{A}^c$ is defined as above. The canonical embedding $\{h_s\}_{s \in S}$ is defined as follows:

(CAN2) $h_s(a) = \{X \in V_s : a \in X\}$.

Clearly $h_s : A_s \mapsto \mathcal{P}(V_s)$. Also $a \le_{A_s} b$ iff $h_s(a) \subseteq h_s(b)$. The implication $(\Rightarrow)$ holds, since all elements of V_s are upsets. The implication $(\Leftarrow)$ is obvious for posets. If A_s is a distributive lattice and $a \leq_{A_s} b$ is not true, then there exists a prime filter $X \subseteq A_s$ such that $a \in X$, $b \notin X$.

h_s preserves lattice operations. If A_s is a distributive lattice and X is a prime filter of A_s, then $a \sqcap b \in X$ iff $a \in X$ and $b \in X$, and $a \sqcup b \in X$ iff $a \in X$ or $b \in X$, so $h_s(a \sqcap b) = h_s(a) \cap h_s(b)$ and $h_s(a \sqcup b) = h_s(a) \cup h_s(b)$. If A_s is a boolean algebra and $X \subseteq A_s$ is an ultrafilter, then $-a \in X$ iff $a \notin X$, so $h_s(-a) = -h_s(a)$.

We show that $\{h_s\}_{s \in S}$ preserves the operations in F and their residuals. Let $g \in F$ be of type $s_1, \ldots, s_n \to s$. We prove:

$$h_s(g(a_1, \ldots, a_n)) = f_{R[g]}(h_{s_1}(a_1), \ldots, h_{s_n}(a_n)). \tag{1}$$

The proof of (1) is correct for any g which in every argument preserves all finite joins, including the empty join, if it exists (this means: $g(a_1, \ldots, a_n) = \bot$ whenever $a_i = \bot$, for some i). For the case of posets, one only assumes that g is isotone in each argument and preserves the empty join.

We show $\subseteq$; the converse inclusion is easy. Let $X \in h_s(g(a_1, \ldots, a_n))$, hence $g(a_1, \ldots, a_n) \in X$. Since g is isotone in each argument, and X is an upset, then $p_g((a_1)^\uparrow, \ldots, (a_n)^\uparrow) \subseteq X$. One shows that for any $1 \leq i \leq n$: (EXT) there exist $X_1 \in h_{s_1}(a_1), \ldots, X_i \in h_{s_i}(a_i)$ such that $p_g(X_1, \ldots, X_i, (a_{i+1})^\uparrow, \ldots, (a_n)^\uparrow) \subseteq X$. Consequently, for $i = n$, one obtains $R[g](X, X_1, \ldots, X_n)$, for some $X_i \in h_{s_i}(a_i)$, $i = 1, \ldots, n$, which yields $X \in f_{R[g]}(h_{s_1}(a_1), \ldots, h_{s_n}(a_n))$.

(EXT) is proved by induction on i. Assume that it holds for all $j < i$. If A_{s_i} is a poset, we set $X_i = (a_i)^\uparrow$; it is proper, since $a_i \neq \bot$; otherwise $g(a_1, \ldots, a_n) = \bot$, hence $\bot \in X$, which is impossible. Let A_{s_i} be a distributive lattice. If $p_g(X_1, \ldots, Y : i, (a_{i+1})^\uparrow, \ldots, (a_n)^\uparrow) \subseteq X$ holds for $Y = A_{s_i}$, then X_i can be any prime filter containing a_i (it exists, since $a_i \neq \bot$). Otherwise one considers the family $\mathcal{F}$ of all proper filters $Y \subseteq A_{s_i}$ such that $p_g(X_1, \ldots, Y : i, (a_{i+1})^\uparrow, \ldots) \subseteq X$. $\mathcal{F}$ is nonempty, since $(a_i)^\uparrow \in \mathcal{F}$. By the maximality principle, $\mathcal{F}$ has a maximal element Z. One shows that Z is prime and sets $X_i = Z$.

Suppose that Z is not prime. Then there exist $a, b \notin Z$ such that $a \sqcup b \in Z$. One defines $Z_a = \{y \in A_{s_i} : (\exists x \in Z) a \sqcap x \leq y\}$, and similarly for Z_b. Z_a, Z_b are proper filters (we have $b \notin Z_a$ and $a \notin Z_b$) containing Z and different from Z (we have $a \in Z_a$ and $b \in Z_b$). Accordingly $Z_a, Z_b \notin \mathcal{F}$. Then, for some $x_1 \in X_1, \ldots, x_{i-1} \in X_{i-1}, z_1 \in Z$, $g(x_1, \ldots, x_{i-1}, a \sqcap z_1, a_{i+1}, \ldots, a_n) \notin X$ and, for some $y_1 \in X_1, \ldots, y_{i-1} \in X_{i-1}, z_2 \in Z$, $g(y_1, \ldots, y_{i-1}, b \sqcap z_2, a_{i+1}, \ldots, a_n) \notin X$. Define $u_j = x_j \sqcap y_j$, $z = z_1 \sqcap z_2$. We have $g(u_1, \ldots, u_{i-1}, a \sqcap z, a_{i+1}, \ldots, a_n) \notin X$ and $g(u_1, \ldots, u_{i-1}, b \sqcap z, a_{i+1}, \ldots, a_n) \notin X$. Since X is prime, the join of the latter elements does not belong to X, but it equals $g(u_1, \ldots, u_{i-1}, (a \sqcup b) \sqcap z, a_{i+1}, \ldots, a_n)$. This is impossible, since $Z \in \mathcal{F}$.

For $g^{r,i} : A_{s_1} \times \cdots \times A_s : i \times \cdots \times A_{s_n} \mapsto A_{s_i}$, we prove:

$$h_{s_i}(g^{r,i}(a_1, \ldots, a_n)) = f_{R[g]}^{r,i}(h_{s_1}(a_1), \ldots, h_s(a_i), \ldots, h_{s_n}(a_n)). \tag{2}$$

While the proof of (1) follows routine lines, tracing back to [16] (also see [2]), our proof of (2) is different. We reduce (2) to (1) by applying some dualities.

By A_s^{op}, $A_{s_i}^{op}$ we denote the algebras dual to A_s, A_{s_i}, respectively; the ordering in the dual algebra is the reversal of the ordering in the initial algebra. Thus, one interchanges $\bot$ with $\top$, and $\sqcup$ with $\sqcap$ in lattices.

By g' we denote the mapping from $A_{s_1} \times \cdots \times A_s^{op} : i \times \cdots \times A_{s_n}$ to $A_{s_i}^{op}$ which equals $g^{r,i}$ as a function. Since $g^{r,i}$ respects arbitrary meets in the i–th argument and turns joins into meets in the other arguments, then g' respects finite joins in each argument. So g' satisfies the requirements needed in the proof of (1). We, however, must replace $\mathcal{A}$ by $\mathcal{A}'$ in which A_s, A_{s_i} are replaced by $A_s^{op}, A_{s_i}^{op}$, respectively. Precisely, we assume that now $s_1, \ldots, s_n, s$ are different sorts, if even they are not different in $\mathcal{A}$. Actually our argument only depends on the fixed operations $g, g^{r,i}$, not on the whole frame $\mathcal{A}$, so we may modify it for the purposes of this argument.

In the canonical frame $(\mathcal{A}')^c$, $(V_s)'$ consists of all proper upsets of A_s^{op}, hence all proper downsets of A_s, if A_s is a poset, and all prime filters of A_s^{op}, hence all prime ideals of A_s, if A_s is a distributive lattice, and similarly for $(V_{s_i})'$. The homomorphism $\{k_s\}_{s \in S}$ is defined as $\{h_s\}_{s \in S}$ except that $\mathcal{A}$ is replaced by $\mathcal{A}'$, and similarly for the canonical frame. (1) yields:

$$k_{s_i}(g'(a_1, \ldots, a_n)) = f_{R[g']}(k_{s_1}(a_1), \ldots, k_s(a_i), \ldots, k_{s_n}(a_n)), \qquad (3)$$

where $f_{R[g']}$ is defined in the complex algebra of $(\mathcal{A}')^c$.

For any $t \in S$, $X \subseteq A_t$, we denote $-X = A_t - X$. For $U \subseteq V_t$, we denote $\sim_{V_t} U = V_t - U$, $U^\sim = \{-X : X \in U\}$. We define the auxiliary operations: $_t^*(-) : \mathcal{P}((V_t)') \mapsto \mathcal{P}(V_t)$ and $(-)_t^* : \mathcal{P}(V_t) \mapsto \mathcal{P}((V_t)')$, for $t = s$ and $t = s_i$:

$$_t^*(U) = \sim_{V_t} (U^\sim), \quad (V)_t^* = \sim_{(V_t)'} (V^\sim), \qquad (4)$$

for $U \subseteq (V_t)'$, $V \subseteq V_t$. We write $^*U$, V^* for $_t^*(U)$, $(V)_t^*$.

One easily shows $^*U = (\sim_{(V_t)'} U)^\sim$ and $V^* = (\sim_{V_t} V)^\sim$. The operations $^*(-)$ and $(-)^*$ are antitone and $(^*U)^* = U$, $^*(V^*) = V$. Also, for $t = s$ and $t = s_i$, we have $h_t(a) = {}^*(k_t(a))$, for any $a \in A_t$. For $t = s_j$, $j \neq i$, we have $k_t = h_t$. Since g and g' are equal as functions, then (3) yields:

$$h_{s_i}(g(a_1, \ldots, a_n)) = {}^*(f_{R[g']}(h_{s_1}(a_1), \ldots, (h_s(a_i))^*, \ldots, h_s(a_n))). \qquad (5)$$

To prove (2) it suffices to show:

$$^*(f_{R[g']}(V_1, \ldots, (V_i)^*, \ldots, V_n)) = f_{R[g]}^{r,i}(V_1, \ldots, V_n), \qquad (6)$$

for all $V_1 \subseteq V_{s_1}, \ldots, V_i \subseteq V_s, \ldots, V_n \subseteq V_{s_n}$.

One proves (6) by simple computations, using: $X \in {}^*U$ iff $(-X) \notin U$, for all $X \in V_t$, $U \subseteq (V_t)'$ and $X \in V^*$ iff $(-X) \notin V$, for all $X \in (V_t)'$, $V \subseteq V_t$. The following formulas are equivalent.

1. $X \in {}^*(f_{R[g']}(V_1, \ldots, (V_i)^*, \ldots, V_n))$,
2. $(-X) \notin f_{R[g']}(V_1, \ldots, (V_i)^*, \ldots, V_n)$,
3. $\neg R[g'](-X, X_1, \ldots, X_n)$, for all $X_j \in V_j$, $(j \neq i)$, and $X_i \in (V_i)^*$,
4. $\neg R[g'](-X, X_1, \ldots, X_n)$, for all $X_j \in V_j$, $(j \neq i)$, $(-X_i) \notin V_i$,

5. for all $X_j \in V_j$, $(j \neq i)$, $X_i \in (V_s)'$, if $-X_i \notin V_i$ then $\neg R[g'](-X, X_1, \ldots, X_n)$,

6. for all $X_j \in V_j$, $(j \neq i)$, $X_i \in (V_s)'$, if $R[g'](-X, X_1, \ldots, X_n)$ then $(-X_i) \in V_i$,

7. for all $X_j \in V_j$, $(j \neq i)$, $Y_i \in V_s$, if $R[g'](-X, X_1, \ldots, -Y_i, \ldots, X_n)$ then $Y_i \in V_i$,

8. $X \in f_{R[g]}^{r,i}(V_1, \ldots, V_n)$.

For the equivalence of formulas 7 and 8, we need further equivalences. The equivalences of formulas 2-3 and 5-6 below use the fact that, if Y is an upset of a poset $(A, \leq)$ and $a \in A$, then $a \in Y$ iff, for all $b \in A$, if $a \leq b$ then $b \in Y$.

1. $R[g'](-X, X_1, \ldots, -Y_i, \ldots, X_n)$,

2. $p_{g'}(X_1, \ldots, -Y_i, \ldots, X_n) \subseteq -X$,

3. for all $a_j \in X_j$, $(j \neq i)$, $a_i \in A_s, b \in A_{s_i}$, if $a_i \notin Y_i$ and $g^{r,i}(a_1, \ldots, a_n) \leq_{A_{s_i}^{op}} b$ then $b \notin X$,

4. for all $a_j \in X_j$, $(j \neq i)$, $a_i \in A_s, b \in A_{s_i}$, if $a_i \notin Y_i$ and $b \leq_{A_{s_i}} g^{r,i}(a_1, \ldots, a_n)$ then $b \notin X$,

5. for all $a_j \in X_j$, $(j \neq i)$, $a_i \in A_s, b \in A_{s_i}$, if $b \in X$ and $g(a_1, \ldots, b : i, \ldots, a_n) \leq_{A_s} a_i$ then $a_i \in Y_i$,

6. $p_g(X_1, \ldots, X : i, \ldots, X_n) \subseteq Y_i$,

7. $R[g](Y_i, X_1, \ldots, X : i, \ldots, X_n)$.

The proof is finished. As we point out in 4, the embedding results imply some basic completeness theorems and conservation results for multi-sorted substructural logics. Even for the one-sort case, the above proof brings something new. Even for a basic map $g : A^n \mapsto A$, hence also $g^{r,i} : A^n \mapsto A$, the second part of the proof introduces $g' : A \times \cdots \times A^{op} : i \times \cdots A \mapsto A^{op}$, which is a multi-sorted map. This shows that multi-sorted algebras can be useful for studying standard algebras.

The canonical embedding h preserves $\bot, \top$; we have $h_s(\bot) = \emptyset$ and $h_s(\top) = V_s$. As shown in [21], it also preserves units for binary operations and some non-classical negations; the complex algebra inherits such properties of basic operations as associativity and commutativity (but not idempotence) and preserves the equations of linear logics. These results have been adapted for symmetric residuation algebras (with one sort, but the proof also works for many sorts) in [22], using the * operators on $\mathcal{A}^c$, after [8].

At this moment, the author does not know whether the embedding theorem can be obtained for mRAs in which different A_s can have different types, e.g. some of them are posets, and some others are distributive lattices. The proof of (EXT) (see the proof of (1)) uses the fact that A_s is a lattice whenever A_{s_i} is a lattice (so the converse is needed in the proof of (2)). Obviously, the distributive law cannot be easily avoided; non-distributive lattices cannot be embedded in the complete lattices of sets.

4   Multi-Sorted Residuation Logics

Generalized Lambek Calculus (**GL**) is a logic of RAs. Formulas are formed out of variables by means of operation symbols (connectives) $o, o^{r,i}$ ($1 \leq i \leq n$, if

o is $n-$ary). The formal language contains a finite number of operation symbols. These operation symbols are *multiplicative* (or: intensional, according to a different tradition). One can also admit *additive* (or: extensional) symbols $\sqcup, \sqcap$, interpreted as lattice operations, and additive constants $\bot, \top$.

The algebraic form of the multiplicative **GL** admits sequents of the form $A \Rightarrow B$ such that A, B are formulas. The only axioms are

$$(\text{Id}) \ A \Rightarrow A \,,$$

and the inference rules strictly correspond to the residuation laws (RES): (R-RES) from $o(A_1, \ldots, A_n) \Rightarrow B$ infer $A_i \Rightarrow o^{r,i}(A_1, \ldots, B : i, \ldots, A_n)$, and conversely, (1-CUT) from $A \Rightarrow B$ and $B \Rightarrow C$ infer $A \Rightarrow C$.

An equivalent Gentzen-style system admits sequents of the form $X \Rightarrow A$ such that A is a formula, and X is a formula structure (tree). A formula structure is a formula or an expression of the form $(X_1, \ldots, X_n)_o$ such that each X_i is a formula structure. Here $(-)_o$ is the structural operation symbol corresponding to the $n-$ary multiplicative symbol o.

The axioms are (Id) and (optionally):

$$(\bot \Rightarrow) \ X[\bot] \Rightarrow A \ (\Rightarrow \top) \ X \Rightarrow \top$$

and the inference rules are:

$$(o \Rightarrow) \ \frac{X[(A_1, \ldots, A_n)_o] \Rightarrow A}{X[o(A_1, \ldots, A_n)] \Rightarrow A} \ (\Rightarrow o) \ \frac{X_1 \Rightarrow A_1; \ldots; X_n \Rightarrow A_n}{(X_1, \ldots, X_n)_o \Rightarrow o(A_1, \ldots, A_n)}$$

$$(o^{r,i} \Rightarrow) \ \frac{X[A_i] \Rightarrow B; \ (Y_j \Rightarrow A_j)_{j \neq i}}{X[(Y_1, \ldots, o^{r,i}(A_1, \ldots, A_n), \ldots, Y_n)_o] \Rightarrow B}$$

$$(\Rightarrow o^{r,i}) \ \frac{(A_1, \ldots, X : i, \ldots, A_n)_o \Rightarrow A_i}{X \Rightarrow o^{r,i}(A_1, \ldots, A_n)}$$

$$(\sqcup \Rightarrow) \ \frac{X[A] \Rightarrow C; \ X[B] \Rightarrow C}{X[A \sqcup B] \Rightarrow C} \ (\Rightarrow \sqcup) \ \frac{X \Rightarrow A_i}{X \Rightarrow A_1 \sqcup A_2}$$

$$(\sqcap \Rightarrow) \ \frac{X[A_i] \Rightarrow B}{X[A_1 \sqcap A_2] \Rightarrow B} \ (\Rightarrow \sqcap) \ \frac{X \Rightarrow A; \ X \Rightarrow B}{X \Rightarrow A \sqcap B}$$

$$(\text{CUT}) \ \frac{X[A] \Rightarrow B; \ Y \Rightarrow A}{X[Y] \Rightarrow B}$$

One can also admit constants, treated as nullary operation symbols; they do not possess residuals. For a constant o, one admits rules $(o \Rightarrow)$, $(\Rightarrow o)$ for $n = 0$ (the second one is an axiom):

$$(o \Rightarrow_0) \ \frac{X[()_o] \Rightarrow B}{X[o] \Rightarrow B} \ (\Rightarrow_0 o) \ ()_o \Rightarrow o \,.$$

If a constant has to play a special role, then one needs additional axioms or rules. That 1 is the unit of o (binary) can be axiomatized by means of the following structural rules and their reversals:

$$(1') \ \frac{X[Y] \Rightarrow A}{X[((\ )_1, Y)_o] \Rightarrow A} \ (1") \ \frac{X[Y] \Rightarrow A}{X[(Y, (\ )_1)_o] \Rightarrow A}.$$

The above system with additives has been studied in [7] and called there Full Generalized Lambek Calculus (**FGL**). Here we consider its multi-sorted version, called Multi-Sorted Full Generalized Lambek Calculus or, simply, Multi-Sorted Full Lambek Calculus (**mFL**). Its multiplicative fragment is referred to as Multi-Sorted Lambek Calculus (**mL**).

We fix a nonempty set S whose elements are called sorts. Each variable is assigned a unique sort; we write $p : s$. One admits a nonempty set $\mathcal{O}$ whose elements are called operation symbols. Each symbol $o \in \mathcal{O}$ is assigned a unique type of the form $s_1, \ldots, s_n \to s$, where $s_1, \ldots, s_n, s \in S, n \geq 1$. If $o : s_1, \ldots, s_n \to s$, then the language also contains operation symbols $o^{r,i}$ $(1 \leq i \leq n)$ such that $o^{r,i} : s_1, \ldots, s : i, \ldots, s_n \to s_i$. One also admits a (possibly empty) set $\mathcal{C}$ whose elements are called constants. Each constant o is assigned a unique sort.

One recursively defines sets F_s, for $s \in S$; the elements of F_s are called formulas of sort s. All variables and constants of sort s belong to F_s; if f is an operation symbol (basic o or residual $o^{r,i}$) of type $s_1, \ldots, s_n \to s$, $(n \geq 0)$, and A_i is a formula of sort s_i, for any $i = 1, \ldots, n$, then $f(A_1, \ldots, A_n)$ is a formula of sort s. In the presence of additives, if $A, B \in F_s$, then $A \sqcup B, A \sqcap B \in F_s$; optionally, also $\perp_s, \top_s \in F_s$. We write $A : s$ for $A \in F_s$.

Each formula of sort s is a formula structure of sort s; if $X_i : s_i$ for $i = 1, \ldots, n$, $(n \geq 0)$, and $o \in \mathcal{O}$ is of type $s_1, \ldots, s_n \to s$, then $(X_1, \ldots, X_n)_o$ is a formula structure of sort s. FS_s denotes the set of formula structures of sort s. We write $X : s$ for $X \in \mathrm{FS}_s$. An expression $X \Rightarrow A$ such that $X \in \mathrm{FS}_s, A \in F_s$ is called a sequent of sort s.

The axioms and rules of **mFL** are the same as for **FGL**, but we require that all formulas and sequents must have some sort. Clearly **mFL** is not a single system; we have defined a class of systems, each determined by the particular choice of S and $\mathcal{O}$. Every system from this class admits *cut elimination*, which was first shown for **NL** by Lambek [26].

As an example, we consider a system with one basic binary operation $\otimes$; we write $/$ and $\backslash$ for $\otimes^{r,1}$ and $\otimes^{r,2}$, respectively. We assume $\otimes : s, t \to u$, where s, t, u are different sorts. Hence $/ : u, t \to s$ and $\backslash : s, u \to t$. The following laws of **NL** are provable in **mL** (we use the infix notation).

(NL1) $(A/B) \otimes B \Rightarrow A, A \otimes (A\backslash B) \Rightarrow B$,
(NL2) $A \Rightarrow (A \otimes B)/B, A \Rightarrow B\backslash(B \otimes A)$,
(NL3) $A \Rightarrow B/(A\backslash B), A \Rightarrow (B/A)\backslash B$,
(NL4) $A/B \Leftrightarrow A/((A/B)\backslash A), A\backslash B \Leftrightarrow (B/(A\backslash B))\backslash B$,
(NL5) $A/B \Leftrightarrow ((A/B) \otimes B)/B, A\backslash B \Leftrightarrow A\backslash(A \otimes (A\backslash B))$.

We cannot build formulas of the form $(A \otimes B) \otimes C, (A/B)/C$ due to sort restrictions. As a consequence, not all laws of **NL** are provable; e.g. $(((A/B)/C)\otimes C) \otimes B \Rightarrow A$ is not. With new operations one can prove a variant of this law $(((A/B)/'C) \otimes' C) \otimes B \Rightarrow A$ under an appropriate sort assignment. We have

$A : u$, $B : t$, $A/B : s$. Assuming $C : v$, $(A/B)/'C : x$, we get $\otimes' : x, v \to s$, hence $/' : s, v \to x$. Notice that both the type of $\otimes$ and that of $\otimes'$ consists of three different sorts.

Applying cut elimination, one proves a general theorem: *every sequent provable in* **GL** (hence every sequent provable in **NL**) *results from some sequent provable in* **mL** *in which the type of each operation symbol consists of different sorts (in $s_1, \ldots, s_n \to s$ all sorts are different), after one has identified all sorts and some operation symbols and variables*. This can be shown by a transformation of a cut-free proof of $X \Rightarrow A$ with all axioms (Id) of the form $p \Rightarrow p$. In the new proof different axioms contain different variables of different sorts; then different premises of any rule have no common variable and no common sort. Every instance of $(\Rightarrow o)$ and $(o^{r,i} \Rightarrow)$ introduces a new operation symbol together with its structural companion and one new sort. Each sequent in the new proof satisfies the above condition. Furthermore, in each sequent, *every residuation family is represented by 0 or 2 symbols (counting structural symbols)*. Consequently, every sequent $A \Rightarrow B$ provable in **mL** contains an even number of operation symbols (this also holds for **L**).

Let us look at (NL5). $A \Leftrightarrow B$ means that both $A \Rightarrow B$ and $B \Rightarrow A$ are provable. The $(\Rightarrow)$ part of (NL5) is $A/B \Rightarrow ((A/B) \otimes B)/B$. In **mL** one proves $A/B \Rightarrow ((A/B) \otimes' B)/'B$ (the reader can find appropriate sorts); the symbol $/$ appears twice in the latter sequent, and the second residuation family is represented by $\otimes', /'$. For the $(\Leftarrow)$ part, the appropriate sequent is $((A/'B) \otimes' B)/B \Rightarrow A/B$. This transformation is impossible for **FGL**; e.g $(A \sqcap B)/C \Rightarrow (A/C) \sqcap (B/C)$ contains 3 occurrences of $/$.

S may consist of one sort only, so **GL** is a limit system from the **mL**-class. The above observations show that the apparently opposite case: each operation symbol has a type consisting of different sorts, leads to essentially the same (pure) logic provided that one admits infinite sets $S, \mathcal{O}$.

Some possible applications of **mL** in linguistics have been mentioned in section 2. Another one is subtyping. A 'large' type S (sentence) can be divided in several subtypes, sensitive to Tense, Number, Mode etc.; these subtypes can be represented by different variables (or: constants) of sort S. In **NL** this goal can be accomplished by additional assumptions: $S_i \Rightarrow S$, for any subtype S_i. With additives one can define $S = S_1 \sqcup \cdots \sqcup S_k$ and apply types dependent on features, e.g. 'John' is assigned 'np $\sqcap$ sing, 'boys' type 'np $\sqcap$ pl [19].

By routine methods, one can show that **mL** *is (strongly) complete with respect to mRAs based on posets*, and **mFL** *is (strongly) complete with respect to mRAs based on (optionally: bounded) lattices*. The strong completeness means that, for any set of sequents Φ (treated as nonlogical assumptions), the sequents derivable from Φ in the system are precisely the sequents valid in all models satisfying all sequents from Φ (a model is an algebra with a valuation of variables). In other words, the strong completeness of a system (with respect to a class of algebras) is equivalent to the completeness of the consequence relation of this system (with respect to the class of algebras).

To attain the completeness with respect to mRAs based on distributive lattices, we add the distributive law as a new axiom:

$$\text{(D)} \ A \sqcap (B \sqcup C) \Rightarrow (A \sqcap B) \sqcup (A \sqcap C)$$

for any formulas A, B, C of the same sort. The resulting system is denoted by **mDFL**. (D) expresses one half of one distributive law; the other half is provable (it holds in every lattice), and the second distributive law is derivable from the first one and basic lattice laws.

This version of **mDFL** does not admit cut elimination. Another version, admitting cut elimination, can be axiomatized like **DFL** in [23] with a structural operation symbol for $\sqcap$ and the corresponding structural rules (an idea originated by J.M. Dunn and G. Mints). We omit somewhat sophisticated details of this approach.

mDFL *is (strongly) complete with respect to mRAs based on distributive lattices.* Soundness is easy, and completeness can be proved, using the Lindenbaum-Tarski algebra (its multi-sorted version). The results from section 3 imply that **mDFL** is strongly complete with respect to the complex mRAs of multi-sorted frames. Soundness is obvious. For completeness, assume that $X \Rightarrow A$ is not derivable from Φ. By the above, there exist a model $(\mathcal{A}, \alpha)$ such that $X \Rightarrow A$ is not true in $(\mathcal{A}, \alpha)$ (it means: $\alpha(X) \leq \alpha(A)$ is not true), but all sequents from Φ are true in $(\mathcal{A}, \alpha)$. Let $\{h_s\}_{s \in S}$ be the canonical embedding of $\mathcal{A}$ in the complex algebra of the frame $\mathcal{A}^c$. The valuation α can be presented as $\{\alpha_s\}_{s \in S}$, where α_s is the restriction of α to variables of sort s (the values of α_s belong to A_s). Then, $\{h_s \circ \alpha_s\}_{s \in S}$ is a valuation in the complex algebra, and the resulting model satisfies all sequents from Φ, but $X \Rightarrow A$ is not true in this model. Ignoring additives, one can prove the same for **mL**. Consequently, the consequence relation of **mDFL** is a conservative extension of the consequence relation of **mL**.

The same holds for Multi-Sorted Boolean Lambek Calculus (**mBL**), which adds to **mDFL** a unary negation (complement) '$-$' and axioms:

$$\text{(N1)} \ A \sqcap -A \Rightarrow \bot \ \ \text{(N2)} \ \top \Rightarrow A \sqcup -A.$$

mBL *is (strongly) complete with respect to boolean mRAs* (all A_s are boolean algebras) as well as *the complex algebras of multi-sorted frames.* (One can also assume that $-A$ can be formed for A of some sorts only.) These results obviously entail the strong completeness of **mBL** and **mDFL** with respect to Kripke frames with standard (classical) clauses for boolean (lattice) operations: $x \models -A$ iff $x \not\models A$, $x \models A \sqcap B$ iff $x \models A$ and $x \models B$, and so on.

In **mBL**, for any operation o, one can define its De Morgan dual. This turns any residuation family to a dual residuation family, which satisfies (RES) with respect to dual orderings; in particular, it yields a faithful interpretation of Moortgat's Symmetric **NL** (without Grishin axioms; see [31]) in **BNL**, i.e. **NL** with boolean operations.

The consequence relation for **L** is undecidable; see [6]. The consequence relation for **mBL** (hence for **mDFL**, **mL**) is decidable (so the pure logics are decidable, too). The proof is similar to that for **DFGL**, **GL** in [9,7]. One shows

Strong Finite Model Property (SFMP): for any finite Φ, if $\Phi \not\vdash X \Rightarrow A$, then there exists a finite multi-sorted model $(\mathcal{A}, \alpha)$ such that all sequents from Φ are but $X \Rightarrow A$ is not true in $(\mathcal{A}, \alpha)$.

The proof of SFMP in [9,7] uses some interpolation property of sequent systems and a construction of algebras by means of nuclear completions. Different proofs are due to [18] for **BNL** (presented as a Hilbert-style system), by the method of filtration of Kripke frames, and [13] where FEP (see below) has been proved directly for some classes of algebras. Each of them can be adapted for multi-sorted logics.

SFMP yields the decidability of stronger logics: the universal theories of the corresponding classes algebras. Here we refer to a standard translation of substructural logics in first-order language: formulas of these logics correspond to terms and sequents to atomic formulas $t \leq u$. Multi-sorted logics require a multi-sorted first-order language; in particular, $A \Rightarrow B$, where A, B are of sort s, is translated into $t_A \leq_s t_B$, where t_A, t_B are terms of sort s which correspond to A, B.

A Horn formula is a first-order formula of the form $\varphi_1 \wedge \cdots \wedge \varphi_n \rightarrow \varphi_{n+1}$, where $n \geq 0$, such that each φ_i is an atomic formula $t \leq_s u$. An open formula is a propositional (boolean) combination of atomic formulas (so Horn formulas are open formulas). A universal sentence results from an open formula by the universal quantification of all variables.

Let $\mathcal{K}$ be a class of algebras. The universal theory of $\mathcal{K}$ is the set of all universal sentences valid in $\mathcal{K}$. The Horn theory of $\mathcal{K}$ is the set of all universally quantified Horn formulas valid in $\mathcal{K}$.

Let a logic $\mathcal{L}$ be strongly complete with respect to $\mathcal{K}$. Then the rules derivable in $\mathcal{L}$ correspond to the Horn formulas belonging to the universal theory of $\mathcal{K}$. Hence the decidability of the universal theory of some class of mRAs (say, boolean residuated groupoids) entails that the problem of derivability of rules in the corresponding logic (here **BNL**) is decidable.

A general, model-theoretic theorem states: *if $\mathcal{K}$ is closed under finite products (admitting the empty product, which yields the trivial algebra), then FMP of the Horn theory of $\mathcal{K}$ entails FMP of the universal theory of $\mathcal{K}$.* For finite languages, FMP of the universal theory of $\mathcal{K}$ is equivalent to Finite Embeddability Property (FEP) of $\mathcal{K}$: every finite, partial subalgebra of an algebra from $\mathcal{K}$ can be embedded in a finite algebra from $\mathcal{K}$. In the literature (see e.g. [11]), the above theorem is formulated for quasi-varieties (which are closed under arbitrary products) in the following form: SFMP for the Horn theory of a quasi-variety $\mathcal{K}$ entails FEP of $\mathcal{K}$, and the proof provides the embedding. Below we sketch another proof, which yields the general result, with arbitrary relation symbols in the language. Also, the usual one-sort algebras can be replaced by multi-sorted algebras. If $\{\mathcal{A}^i\}_{i \in I}$ is a class of similar mRAs, then $\prod_{i \in I} \mathcal{A}^i$ is defined in a natural way: its algebra of sort s equals $\prod_{i \in I} A_s^i$ with point-wise defined relations and lattice (boolean) operations; also the operations in F are defined point-wise. The basic classes of mRAs are closed under arbitrary products (they are multi-sorted quasi-varieties), so this theorem can be applied to them.

Let us sketch the proof. Let $\psi = \forall x_1 \ldots x_n \varphi$ be a universal sentence (φ is open). φ is logically equivalent to a CNF- formula $\varphi_1 \wedge \cdots \wedge \varphi_m$, each φ_i being a disjunction of finitely many atomic formulas and negated atomic formulas. So ψ is logically equivalent to the conjunction of ψ_i, $i = 1, \ldots, m$, where ψ_i is the universally quantified φ_i. Clearly ψ is valid in an algebra $\mathcal{A}$ iff each ψ_i is valid in $\mathcal{A}$, and the same holds for the validity in $\mathcal{K}$.

Assume that ψ is not valid in $\mathcal{K}$. Then, some sentence ψ_i is not valid. Assuming FMP of the Horn theory, we show that there is a finite algebra in $\mathcal{K}$ such that ψ_i is not true in this algebra. If φ_i consists of negated atomic formulas only, then ψ_i is not true in the trivial algebra, which is finite (an mRA is trivial iff all its algebras A_s are one-element algebras). So assume that φ_i is of the form:

$$\neg \chi_1 \vee \cdots \vee \neg \chi_k \vee \sigma_1 \vee \cdots \vee \sigma_p$$

where $k \geq 0$, $p \geq 1$, and all χ_j, σ_l are atomic. It is logically equivalent to:

$$\chi_1 \wedge \cdots \wedge \chi_k \to \sigma_1 \vee \cdots \vee \sigma_p \,.$$

Denote $\delta_j = \chi_1 \wedge \cdots \wedge \chi_k \to \sigma_j$. Since δ_j logically entails φ_i, then δ_j is not valid in $\mathcal{K}$, for $j = 1, \ldots, p$. By FMP of the Horn theory, there exists a finite model $(\mathcal{A}^j, \alpha_j)$ over $\mathcal{K}$ which falsifies δ_j. One easily shows that the product model (i.e. the product of all $\mathcal{A}^j$ with the product valuation) falsifies φ_i. Therefore ψ is not true in this product algebra, which finishes the proof.

Since SFMP of our logics is equivalent to FMP of the Horn theories of the corresponding classes of mRAs, then we obtain FMP of the universal theories, which yields their decidability.

The above proof yields: ψ_i is valid in $\mathcal{K}$ iff some δ_j is valid in $\mathcal{K}$. Accordingly, a decision method for the universal theory of $\mathcal{K}$ can be reduced to a decision method for the Horn theory of $\mathcal{K}$ (equivalently: for the consequence relation of the corresponding logic). Some proof-theoretic decision methods for the latter can be designed like for **DFGL** [9,7], but we skip all details here. We note that a Kripke frame falsifying $\Phi \vdash X \Rightarrow A$ (if it exists) can be found of size at most 2^n, where n is the number of subformulas occurring in this pattern (this was essentially shown in the three proofs of SFMP, mentioned above).

Although **mFL** is decidable, since it admits cut elimination (also FMP holds), the decidability of its consequence relation remains an open problem (even for **FNL**).

The consequence relation of **GL** is polynomial [6]; for the pure **NL** it was earlier shown in [12]. Associative systems **FL**, **DFL** and their various extensions are PSPACE-complete [14]; the proof of PSPACE-hardness (by a reduction of the validity of QBFs to the provability of sequents) essentially relies upon the associative law. Without associativity, by a modification of this proof we can prove the PSPACE-hardness of the consequence relation of **FNL**, **FGL**, **DFGL**, **mFL**, **mDFL** (with at least one binary operation), but the precise complexity of the pure logics is not known. **BNL**, **BGL**, **mBL** are PSPACE-hard, like the modal logic **K**; see e.g. [3].

In [6,9] it has been shown that the type grammars based on the multiplicative systems and the systems with additives and distribution, also enriched with finitely many assumptions, are equivalent to CFGs.

BGL (i.e. **GL** with boolean operations) is a conservative extension of **K**; it follows from the fact that both **K** and **BGL** are complete with respect to all Kripke frames. (This is obvious, if F contains a unary operation; otherwise, one can reduce an $n-$ary operation to a unary one by fixing some arguments.) A provable formula A of **K** is represented as the provable sequent $\top \Rightarrow A$ of **BGL**; a provable sequent sequent $A \Rightarrow B$ of **BGL** is represented as the provable formula $A \to B$ of **K**. **mBL** can be treated as a multi-sorted classical modal logic.

Interestingly, some theories based on multi-sorted classical modal logics are paraconsistent: the inconsistency in one sort need not cause the total inconsistency. In algebraic terms, it means that there exist mRAs $\mathcal{A}$ in which some, but not all, algebras A_s are trivial (one-element). Let $A_s = \{a\}$, and let A_t be non-trivial with $\perp_t \in A_t$. Then $f(a) = \perp_t$ is the only residuated map $f : A_s \mapsto A_t$ (notice $a = \perp_s$), and f^r is the constant map: $f^r(x) = a$, for all $x \in A_t$.

There are many natural connections between substructural logics, studied here, and (multi-)modal logics; an early discussion can be found in [1]. Some results, discussed above, have been adapted for one-sort systems admitting special modal axioms (e.g. T, 4, 5) in [29] (FEP, polynomial complexity). This research program seems promising.

References

1. van Benthem, J.: Language in Action. Categories, Lambdas and Dynamic Logic. North Holland, Amsterdam (1991)
2. Bimbó, K., Dunn, J.M.: Generalized Galois Logics. Relational Semantics of Nonclassical Logical Calculi. CSLI Lecture Notes, vol. 188 (2008)
3. Blackburn, P., de Rijke, M., Venema, Y.: Modal Logic. Cambridge University Press, Cambridge (2001)
4. Blyth, T.S.: Lattices and Ordered Algebraic Structures. Springer, London (2010)
5. Buszkowski, W.: Generative Capacity of Nonassociative Lambek Calculus. Bull. Pol. Acad. Scie. Math. 34, 507–516 (1986)
6. Buszkowski, W.: Lambek Calculus with Nonlogical Axioms. In: Casadio, C., Scott, P.J., Seely, R.A.G. (eds.) Language and Grammar. Studies in Mathematical Linguistics and Natural Language. CSLI Lecture Notes, vol. 168, pp. 77–93 (2005)
7. Buszkowski, W.: Interpolation and FEP for logics of residuated algebras. Logic Journal of the IGPL 19(3), 437–454 (2011)
8. Buszkowski, W.: Many-sorted gaggles. A Talk at the Conference Algebra and Coalgebra Meet Proof Theory (ALCOP 2012). Czech Academy of Sciences, Prague (2012), http://www2.cs.cas.cz/~horcik/alcop2012/slides/buszkowski.pdf
9. Buszkowski, W., Farulewski, M.: Nonassociative Lambek Calculus with Additives and Context-Free Languages. In: Grumberg, O., Kaminski, M., Katz, S., Wintner, S. (eds.) Francez Festschrift. LNCS, vol. 5533, pp. 45–58. Springer, Heidelberg (2009)

10. Casadio, C.: Agreement and Cliticization in Italian: A Pregroup Analysis. In: Dediu, A.-H., Fernau, H., Martín-Vide, C. (eds.) LATA 2010. LNCS, vol. 6031, pp. 166–177. Springer, Heidelberg (2010)
11. Galatos, N., Jipsen, P., Kowalski, T., Ono, H.: Residuated Lattices: An Algebraic Glimpse at Substructural Logics. Elsevier (2007)
12. de Groote, P., Lamarche, F.: Classical Nonassociative Lambek Calculus. Studia Logica 71(2), 355–388 (2002)
13. Haniková, Z., Horčik, R.: Finite Embeddability Property for Residuated Groupoids (submitted)
14. Horčik, R., Terui, K.: Disjunction property and complexity of substructural logics. Theoretical Computer Science 412, 3992–4006 (2011)
15. Jäger, G.: Residuation, structural rules and context-freeness. Journal of Logic, Language and Information 13, 47–59 (2004)
16. Jónsson, B., Tarski, A.: Boolean algebras with operators. Part I. American Journal of Mathematics 73, 891–939 (1952)
17. Jónsson, B., Tarski, A.: Boolean algebras with operators. Part II. American Journal of Mathematics 74, 127–162 (1952)
18. Kaminski, M., Francez, N.: Relational semantics of the Lambek calculus extended with classical propositional logic. Studia Logica (to appear)
19. Kanazawa, M.: The Lambek Calculus Enriched with Additional Connectives. Journal of Logic, Language and Information 1(2), 141–171 (2002)
20. Kandulski, M.: The equivalence of nonassociative Lambek categorial grammars and context-free grammars. Zeitschrift f. Math. Logik und Grundlagen der Mathematik 34, 41–52 (1988)
21. Kołowska-Gawiejnowicz, M.: On Canonical Embeddings of Residuated Groupoids. In: Casadio, C., et al. (eds.) Lambek Festschrift. LNCS, vol. 8222, pp. 253–267. Springer, Heidelberg (2014)
22. Kołowska-Gawiejnowicz, M.: Powerset Residuated Algebras. Logic and Logical Philosophy (to appear)
23. Kozak, M.: Distributive Full Lambek Calculus has the Finite Model Property. Studia Logica 91(2), 201–216 (2009)
24. Kusalik, T.: Product pregroups as an alternative to inflectors. In: Casadio, C., Lambek, J. (eds.) Computational Algebraic Approaches to Natural Language, p. 173. Polimetrica, Monza (2002)
25. Lambek, J.: The mathematics of sentence structure. American Mathematical Monthly 65, 154–170 (1958)
26. Lambek, J.: On the calculus of syntactic types. In: Jakobson, R. (ed.) Structure of Language and its Mathematical Aspects, pp. 166–178. AMS, Providence (1961)
27. Lambek, J.: From Categorial Grammar to Bilinear Logic. In: Schroeder-Heister, P., Došen, K. (eds.) Substructural Logics, pp. 207–237. Clarendon Press, Oxford (1993)
28. Lambek, J., Scott, P.J.: Introduction to higher order categorical logic. Cambridge University Press, Cambridge (1986)
29. Lin, Z.: Nonassociative Lambek Calculus with Modalities: Interpolation, Complexity and FEP (submitted)
30. Moortgat, M.: Categorial Type Logic. In: van Benthem, J., ter Meulen, A. (eds.) Handbook of Logic and Language, pp. 93–177. Elsevier, Amsterdam (1997)
31. Moortgat, M.: Symmetric Categorial Grammar. Journal of Philosophical Logic 38(6), 681–710 (2009)

Italian Clitic Patterns in Pregroup Grammar: State of the Art

Claudia Casadio[1] and Aleksandra Kiślak-Malinowska[2]

[1] University G. D'Annunzio Chieti IT
[2] Faculty of Mathematics and Computer Science
University of Warmia and Mazury, Olsztyn Poland
casadio@unich.it, akis@uwm.edu.pl

Abstract. Pregroup calculus and pregroup grammars are introduced by Lambek as an algebraic tool for the grammatical analysis of natural languages and the computation of strings of words and sentences. Some interesting aspects of natural languages have been profitably handled by means of pregroups. In the present paper we focus on a chosen aspect of Italian grammar - clitic pronouns - and show how to tackle it by means of different types of pregroup grammars. We start with classical pregroup grammars, proceed to product pregroup grammars and then introduce tupled pregroup grammars. Advantages and disadvantages of the different approaches are discussed and compared.

Keywords: Pregroup, pregroup grammar, product pregroup grammars, tupled pregroup grammars, Italian clitics, natural language computation.

To Jim Lambek whose work and insights are always inspiring

1 Introduction

In this paper we present a summary of the results obtained in the treatment of Italian clitic patterns in terms of pregroup grammars (PG): classical PG, product PG and tupled PG, respectively. Casadio and Lambek [6] have worked with the first type of grammar obtaining the analysis of a rather large pattern of clitic constructions; the second approach has been developed in some details in [8,9] and [13], following ideas of Kusalik [14] and Lambek [20]; Kiślak–Malinowska [12] has recently developed an approach in terms of tupled PG, then Casadio and Kiślak–Malinowska [10] have extended it to Italian clitics showing the advantages both in terms of computability and of correct linguistic description and generation. In the present paper we intend to briefly summarize these three different PG approaches, and proceed at discussing some differences and some interesting applications of tupled PG.

2 The Classical Approach

The calculus of pregroups is developed in [16,19] as an alternative to the Syntactic Calculus [15], a well known model of categorial grammar [22]. Pregroups are a

C. Casadio et al. (Eds.): Lambek Festschrift, LNCS 8222, pp. 156–171, 2014.

particular kind of substructural logic that is compact and non-commutative [3,4]. In fact the calculus is a non-conservative extension of classical non-commutative multiplicative linear logic [1]: the *left* and *right* 'iterable' adjoints of pregroups have as their counterparts the *left* and *right* 'iterable' negations of non-commutative multiplicative linear logic; in principle formulas can occur with n (left or right) adjoints (n $\geq$ 2), although 2 appears as being sufficient for linguistic applications; see [20,7,5,11].

2.1 Basic Properties of Classical Pregroups

A *pregroup* $\{G, \,.\,, 1, {}^{\ell}, {}^{r}, \rightarrow\}$ is a partially ordered monoid in which each element a has a *left adjoint* a^{ℓ}, and a *right adjoint* a^{r} such that

$$a^{\ell} a \rightarrow 1 \rightarrow a\,a^{\ell}$$
$$a\,a^{r} \rightarrow 1 \rightarrow a^{r} a$$

where the dot ".", that is usually omitted, stands for the pregroup operation, the compact conjunction or multiplication with unit 1, the arrow denotes the partial order, the rules $a^{\ell} a \rightarrow 1$, $a\,a^{r} \rightarrow 1$ are called *contractions*, and the opposite rules $1 \rightarrow a\,a^{\ell}$, $1 \rightarrow a^{r} a$ are called *expansions*. From the point of view of linguistics, the constant 1 represents the empty string of types, and the operation "." is interpreted as concatenation. The following principles state that the constant 1 is self-dual, adjoints are unique and contravariant, and iterated adjoints admit a kind of De Morgan rule allowing a left (right) adjoint to distribute over a formula, inverting the order of its constituents

$$1^{\ell} = 1 = 1^{r},$$

$$\frac{a \rightarrow b}{b^{\ell} \rightarrow a^{\ell}} \quad , \quad \frac{a \rightarrow b}{b^{r} \rightarrow a^{r}} \quad , \quad \frac{b^{\ell} \rightarrow a^{\ell}}{a^{\ell\ell} \rightarrow b^{\ell\ell}} \quad , \quad \frac{b^{r} \rightarrow a^{r}}{a^{rr} \rightarrow b^{rr}}$$

$$(a \cdot b)^{\ell} = b^{\ell} \cdot a^{\ell} \quad , \quad (a \cdot b)^{r} = b^{r} \cdot a^{r} ,$$

In the pregroup calculus the following equalities and rules are provable

$$a^{r\ell} = a = a^{\ell r},$$

$$a^{\ell\ell} a^{\ell} \rightarrow 1 \rightarrow a^{\ell} a^{\ell\ell} \quad , \quad a^{r} a^{rr} \rightarrow 1 \rightarrow a^{rr} a^{r} ,$$

the former expresses the property of cancellation of double opposite adjoints, the latter the contraction and expansion of identical *left* and *right* double adjoints respectively. Just *contractions* $a^{\ell} a \rightarrow 1$ and $a\,a^{r} \rightarrow 1$ are needed to determine constituent analysis and grammaticality of linguistic expressions, and to prove that a string of words is a sentence; on the other hand, *expansions* $1 \rightarrow a\,a^{\ell}$, $1 \rightarrow a^{r} a$ are useful for representing general structural properties of a given language (see e.g. [7,19]).

At the syntactic level, a pregroup is *freely generated* by a partially ordered set of *basic* types. From each basic type a we form *simple* types by taking single or repeated adjoints: ... $a^{\ell\ell}, a^{\ell}, a, a^{r}, a^{rr}$... . A compound type or just a *type* is a string of simple types: $a_1\,a_2 \ldots a_n$. A basic type is a type (for n = 1).

2.2   Pregroup Grammar for a Fragment of Italian

Assuming the rules and definitions given above we obtain what we shall call a classical PG grammar, in which only type assignments to the elements of the lexicon, adjoints cancellations and contractions are needed to determine the grammaticality (or well formation) of strings of words and select, on this basis, the set of strings that are (well formed) sentences of the given language.

In the classical approach, the free pregroup for a fragment of Italian is generated by the following partially ordered set of basic types (see [6]):

s	declarative sentences
$i, \tilde{\imath}, \bar{\imath}, i^*, j, \bar{\jmath}, \bar{\bar{\imath}}$	infinitive clauses
π	subject
o	direct object
ω	indirect object
λ	locative phrase

The type π is assigned to the subject, in *nominative* case, the types o, ω, λ, to the arguments of the verb, in *accusative, dative* and *locative* case respectively; the type $\bar{\bar{\imath}}$ is the maximal element of a set of types $i, \tilde{\imath}, \bar{\imath}, i^*, j, \bar{\jmath}$, assigned to verbal morphems and expressions introducing a variety of infinitival clauses.

From these basic types, appropriate types can be formed for different kinds of verbal expressions. For example, the types given in the list below are assigned to intransitive verbs like *correre* [to run], *arrivare* [to arrive] and transitive or ditransitive verbs taking different kinds of complements like *vedere* [to see], *obbedire* [to obey], *dare* [to give], *mettere* [to put]

$$(1) \quad vedere: \quad i \ , \ i \ o^\ell$$
$$(2) \quad obbedire: i \ , \quad i \ \omega^\ell$$
$$(3) \quad dare: \qquad i \ \omega^\ell o^\ell \ , \quad i \ o^\ell \omega^\ell$$
$$(4) \quad mettere: i \ \lambda^\ell o^\ell \ , \quad i \ o^\ell \lambda^\ell$$
$$(5) \quad correre: \quad i \ , \ i \ \lambda^\ell$$
$$(6) \quad arrivare: i^* \ , \quad i^* \lambda^\ell \ .$$

The star on i^* is a reminder that the perfect tense of verbs like *arrivare* is to be formed with the auxiliary *essere* [to be] rather than *avere* [to have], producing infinitival phrases of type i^*, rather than of type i, like e.g. *Io sono arrivato* [I have arrived] vs. *Io ho corso* [I have run]. The verbs in (3) and (4) receive two types since their arguments in Italian can occurr in both orders.

The following examples show how the verb types combine via contraction with the types of their arguments to give the expected infinitives of type i: simple types are assigned to verbal complements such as the direct object phrases *un quadro, un libro*, of type o, the indirect object phrase *a Carla*, of type ω, the locative phrases *sul tavolo, a Roma*: of type λ.

(1) *vedere* $\underbrace{un\ quadro}$ [to see a picture]
     $(i\ o^\ell)$ $o\ \rightarrow\ i$

(2) *obbedire* $\underbrace{a\ Mario}$ [to obey to Mario]
     $(i\ \omega^\ell)$ $\omega\ \rightarrow\ i$

(3) *dare* $\underbrace{un\ libro}$ $\underbrace{a\ Carla}$ [to give a book to Carla]
     $(i\ \omega^\ell\ o^\ell)$ o $\omega\ \rightarrow\ i$

(4) *mettere* $\underbrace{un\ libro}$ $\underbrace{sul\ tavolo}$ [to put a book on the table]
     $(i\ \lambda^\ell\ o^\ell)$ o $\lambda\ \rightarrow\ i$

(5) *arrivare* $\underbrace{a\ Roma}$ [to arrive to Rome]
     $(i^*\ \lambda^\ell)$ $\lambda\ \rightarrow\ i^*$

Italian, like Spanish and Portuguese, has both preverbal and postverbal clitic pronouns. We list below the types assigned by the classical PG grammar to preverbal clitics in the accusative and dative cases

Accusative	$mi,\ ti,\ ci,\ vi\ :\ \bar{\bar{j}}\,o^{\ell\ell}\,i^\ell$
	$lo,\ la,\ li,\ le\ :\ j\,o^{\ell\ell}\,i^\ell$
Dative	$mi,\ ti,\ ci,\ vi,\ gli,\ le\ :\ \bar{\bar{j}}\,\omega^{\ell\ell}\,i^\ell\ ,\ \bar{\bar{j}}^*\,\omega^{\ell\ell}\,i^{*\ell}$
	$me,\ te,\ ce,\ ve,\ se,\ glie\ :\ \bar{\bar{j}}\,\omega^{\ell\ell}\,j^\ell$
	$se\ :\ \bar{\bar{j}}^*\,\omega^{\ell\ell}\,j^\ell$

To type preverbal clitics we introduce four new basic types for infinitives j , j^*, $\bar{\bar{j}}$, $\bar{\bar{j}}^*$ and postulate their order conditions: $j \rightarrow \bar{\bar{j}}$, $j^* \rightarrow \bar{\bar{j}}^*$, but $i \nrightarrow j \nrightarrow \bar{i}$. It follows that infinitives of type $\bar{\bar{j}}$ cannot be preceded by any clitics and infinitives of type j only by clitics such as *me* and *ce*. We also obtain clitic clusters such as

$$\underset{(\bar{\bar{j}}\ \omega^{\ell\ell}\ j^\ell)}{me}\ .\ \underset{(\ j\ o^{\ell\ell}\ i^\ell)}{lo}\ ,\ \rightarrow\ \bar{\bar{j}}\ \omega^{\ell\ell}\ o^{\ell\ell}\ i^\ell\ .$$

Here are some illustrations of preverbal clitics where the under-links show how contractions apply to produce the calculation of the resulting type

$$\underset{(\bar{\bar{j}}\ \omega^{\ell\ell}\ o^{\ell\ell}\ i^\ell)}{me}\ .\ \underset{}{lo}\ \underset{(i\ o^\ell\omega^\ell)}{dare}\ ,\ \rightarrow\ \bar{\bar{j}} \qquad\qquad \underset{(\bar{\bar{j}}\ \lambda^{\ell\ell}\ o^{\ell\ell}\ i^\ell)}{ce}\ .\ \underset{}{lo}\ \underset{(i\ o^\ell\lambda^\ell)}{mettere}\ ,\ \rightarrow\ \bar{\bar{j}}$$

$$\begin{array}{ll} lo \quad vedere \\ (j \; o^{\ell\ell} \; i^{\ell}) \, (i \; o^{\ell}) \;\; \to \;\; j \end{array} \quad , \quad \begin{array}{ll} ci \quad arrivare \; . \\ (\bar{j}* \; \lambda^{\ell\ell} \; i*^{\ell}) \, (i*\lambda^{\ell}) \;\; \to \;\; \bar{j}* \end{array}$$

Partial cliticization can be obtained with double complements verbs

$$\begin{array}{lll} mi \quad dare \; un \; libro \\ (\bar{j} \; w^{\ell\ell} \; i^{\ell}) \, (i \; w^{\ell}o^{\ell}) \quad\; o \;\; \to \;\; \bar{j} \end{array} \quad , \quad \begin{array}{lll} lo \quad dare \quad a \; Mario \\ (j \; o^{\ell\ell} \; i^{\ell}) \, (i \; o^{\ell}w^{\ell}) \;\; \omega \;\; \to \;\; j \end{array} \; ,$$

We conclude this section with two examples of declarative sentences in the present tense involving pre-verbal cliticization, where C_{11} is a verb conjugation matrix as explained below (for more details see [6])

$$
\begin{array}{llll}
(io) & te \;\; . \;\; lo \quad do & \text{(I give it to you)} \\
= \; io & C_{11} (\quad te \;\; . \quad lo \quad dare \quad) \\
& \pi_1 \; (\pi_1^r \; s_1 \; \bar{\imath}^{\ell}) \, (\bar{j} \; w^{\ell\ell}o^{\ell\ell} \; i^{\ell}) \, (i \; o^{\ell}w^{\ell}) \to s_1 & (\; \bar{j} \to \bar{\imath} \;)
\end{array}
$$

$$
\begin{array}{llll}
Dario & lo \;\; vuole \;\; vedere & \text{(Dario wants to see him)} \\
= \; Dario & C_{13} (\;\; lo \quad volere \;) \;\; vedere \\
& \pi_3 \; (\pi_3^r \; s_1 \; \bar{\imath}^{\ell}) \, (j \; o^{\ell\ell} \; i^{\ell}) \, (i \bar{\imath}^{\ell})(i \; o^{\ell}) \to s_1 & (\; j \to \bar{j} \to \bar{\imath} \;)
\end{array}
$$

3   Product Pregroup Grammars

To account for the rich morphological system of Romance languages such as French or Italian, Lambek [17,19,20] proposed an analysis of inflected verbs which consisted in two parts: the first one, called the *inflector*, the second one, called the *infinitive*. In doing so, the verb and its inflectional morphology are described by the communication of those two parts. The inflector is responsible for the conjugated form of the verb, whereas the infinitive contains additional information concerning specific verb features, like transitivity of the verb, possible complements of the verb etc. Essentially, the meaning of this decomposition is that the inflector modifies the infinitive and as a result one gets a suitable type of the verb in a certain grammatical form including tense and person.

Therefore, according to Lambek (e.g. [16,2]) to each verb V we associate a matrix $C_{jk}(V)$, with j referring to the tense and k referring to the person. For example, the matrix for the English verb *be* is given as follows

$$C_{jk} \, (be) \to \begin{pmatrix} am & are & is \\ was & were & was \end{pmatrix}$$

Taking single elements of the matrix we obtain the following forms, assuming from the lexicon: $i \leq j$, where j (as well as i, $\bar{i}$, $\bar{j}$ etc.) stand for infinitive clauses, the subscripts 1, 2, 3 in π stand for the first, second, third person, the subscripts 1 and 2 in s stand for the present tense and the past tense respectively

$C_{13}(go)$ = (he) goes
$(\pi_3^r s_1 j^l) \, i \rightarrow \pi_3^r s_1$

$C_{23}(go)$ = (he) went
$(\pi_3^r s_2 j^l) \, i \rightarrow \pi_3^r s_2$

$C_{11}(like)$ = (I) like   and $C_{11}(like)$ the boy = (I) like the boy
$(\pi_1^r s_1 j^l)(io^l) \rightarrow \pi_1^r s_1 o^l$ $(\pi_1^r s_1 j^l)(io^l) \, o \rightarrow \pi_1^r s_1$

An interesting approach is elaborated by Kusalik [14], who presented an alternative analysis which replaced Lambek's free pregroup with a system called *product pregroups*, allowing for a separation between those two aspects of verbs. This seems to work better for the cases which appeared to be problematic for Lambek's former approach. The product pregroup grammar is understood as an intersection of two (or more) pregroup grammars. In the lexicon this time one can find types belonging to the first and the second pregroup grammar (or possibly more). The first can be seen as a usual pregroup grammar checking sentencehood at the syntactic level. The second is introduced to account for feature checking and similar grammatical operations applying in parallel with the syntactic ones. The calculations must end successfully on both (or more) levels in order to accept the string of words as a sentence.

It has been shown, that if G_1 and G_2 are pregroup grammars, the language defined by the pregroup $G_1 \times G_2$ is an intersection of two context-free languages, and there is an algorithm of polynomial complexity for determining whether a given string of types in $G_1 \times G_2$ reduces to $\mathbf{1}$. ($\mathbf{1}$ is understood as a vector of coordinates such that each coordinate belongs to the given pregroup grammar. It is usually $s \in P_1$ for the first grammar and $1 \in P_i$, $2 \leq i \leq k$ for the additional grammars.) Given a product of k free pregroups, the language can be expressed as the intersection of k context-free languages. The fact that the finite products of free pregroups are computationally no more complex than a free pregroup itself means that they can be used as a model of the grammatical structure of a given natural language.

Product pregroup grammars are used by Lambek [20] while trying to analyze feature agreement in French. He makes use of two pregroup grammars, one for syntactic types and the other for the feature types. Kusalik [14] has used the product of three pregroup grammars for analyzing English sentences. Let's now look at this approach on the basis of some Italian sentences, where the types assigned to words come from Casadio [9]. Consider some examples first

$$vedere\ il\ ragazzo\ \text{[to see the boy]}$$
$$vedere\ la\ ragazza\ \text{[to see the girl]}$$
$$vedere\ i\ ragazzi\ \text{[to see the boys]}$$
$$vedere\ le\ ragazze\ \text{[to see the girls]}$$
$$io^l \qquad o \qquad \qquad \to i$$

where the introduction of an inflector will justify for the Italian sentence *Mario vede la ragazza* [Mario sees the girl], namely:

$$Mario\ vede \qquad \qquad la\ ragazza.$$
$$Mario\ C_{13}(vedere)\ \ la\ ragazza$$
$$\pi_3 \qquad (\pi_3^r s_1 \bar{\bar{i}}^l)(io^l)\ o \qquad \qquad \to s_1$$

When the accusative objects *il ragazzo, la ragazza, i ragazzi, le ragazze* are changed into personal pronouns in accusative case they become *lo, la, li, le*, respectively. Then, in Italian, their position in the sentence changes, and it needs to be preverbal. Thus one gets

$$lo \qquad vedere\ \text{[to see him]}$$
$$la \qquad vedere\ \text{[to see her]}$$
$$li \qquad vedere\ \text{[to see them (masculine)]}$$
$$le \qquad vedere\ \text{[to see them (feminine)]}$$
$$(jo^{ll}i^l)\ (io^l) \quad \to j$$

Here j (as well as $i,\ \bar{i},\ \bar{j}$ etc.) stand for infinitive clauses. All constraints and partial order concerning them are at the moment irrelevant and we will not bother the reader with too many details. Taking into consideration infinitive clauses in present perfect tense, we get

$$avere\ visto \ \ il\ ragazzo\ \text{[to have seen the boy]}$$
$$avere\ visto \ \ la\ ragazza\ \text{[to have seen the girl]}$$
$$avere\ visto \ \ i\ ragazzi\ \ \ \text{[to have seen the boys]}$$
$$avere\ visto \ \ le\ ragazze\ \text{[to have seen the girls]}$$
$$(ip_2^l)\ (p_2o^l)\ o \qquad \qquad \to i$$

Here p_2 stands for the past participle of the verb *vedere* [to see]. Now the nouns in accusative case *il ragazzo, la ragazza, i ragazzi, le ragazze* are changed again into personal pronouns in accusative case *lo, la, li, le*: and it causes problems. This is due to some grammatical peculiarity of Italian: while using personal pronouns in accusative case together with present perfect tense, the past participle endings should be changed according to the pronoun's gender and number. It looks as below

$$lo \qquad avere\ visto\ \text{[to have seen him]}$$
$$la \qquad avere\ vista\ \text{[to have seen her]}$$
$$li \qquad avere\ visti\ \text{[to have seen them (masculine)]}$$
$$le \qquad avere\ viste\ \text{[to have seen them (feminine)]}$$
$$(jo^{ll}i^l)\ (ip_2^l)\ (p_2o^l) \to j$$

From now on we face an over-generation. On the syntactic ground the wrong word order can be blocked. For example *avere visto lo* cannot be accepted in Italian. If the personal pronouns are got rid of and replaced by nouns as verbal arguments no changes should be made, and *avere visto* will be proper in all cases, irrespective of the gender and the number of the noun (e. g. *avere visto una ragazza* [to have seen a girl]). Changing the past participle ending is only needed while using a personal pronoun instead of the noun in present perfect tense. One can say *lo avere visto*, but *lo avere visti* would not be correct, because there is a lack of feature agreement. A similar problem arises while considering intransitive verbs which form the present perfect tense with the verb *essere* [to be]. In that case the past participle ending must agree with the personal pronoun or the noun that is the sentential subject: *Maria/Piero deve essere arrivata/o* [Maria/Piero has to be arrived].

Product pregroup grammars seem to be appropriate in similar cases. A new pregroup for feature agreement can be introduced for that purpose and four new types can be defined into the lexicon of this second PG: π_{ms}, π_{fs}, π_{mp}, π_{fp}, where m, f stand for masculine and feminine, whereas s, p stand for singular and plural. Then each string of words from Italian can be typed on both levels (using the product of the two PGs), and computations can be performed in parallel. If both computations are successful, the sentence is accepted, otherwise it is rejected. The first type assignment (syntactic level) should end with a single i or j type (infinitive clause), while the second one should end with 1, corresponding to feature matching. For example, making use of two free pregroups the following result can be obtained

$$
\begin{array}{llll}
\textbf{\textit{lo}} & \textit{avere} & \textit{visto} & \\
jo^{ll}i^l & ip_2^l & p_2o^l & \to j \\
\pi_{ms} & 1 & \pi_{ms}^r & \to 1
\end{array}
\qquad
\begin{array}{llll}
\textbf{\textit{lo}} & \textit{avere} & \textit{visti} & \\
jo^{ll}i^l & ip_2^l & p_2o^l & \to j \\
\pi_{ms} & 1 & \pi_{mp}^r & \not\to 1
\end{array}
$$

As it can be seen above, the second sentence is rejected at the level of feature cheking and the string of words *lo avere visti* cannot be accepted.

4   Clitics in terms of Tupled Pregroup Grammars

Differently from classical and product PGs, tupled pregroup grammars are based on a lexicon consisting of tuples of ordered pairs whose first element is a type and the second one is a symbol from the alphabet. Elements of the lexicon are also called lexical entries. The idea of taking certain elements of an alphabet together in one tuple can be motivated and explained with the fact that in natural languages certain words tend to occur together in the sentence, as for example *prepositions* with *nouns*, *pronouns* with *verbs*, *clitic pronouns* with certain *verbs*, etc. In formal languages one may wish to copy each item of an alphabet (as in the copying language $\{xx \mid x \in \{a, b\}^*\}$), to have the same number of occurrences of certain elements, etc.

Let s_i be elements of a finite alphabet $\Sigma \cup \epsilon$ (we can think of them as the words of a natural language, ϵ being an empty string), and let $\mathbb{P}$ be a set of simple types, partially ordered by $\leq$. Types ($\mathbb{T}_{\mathbb{P}}$) are of the form $p_1^r\, p_2^r...p_k^r\, v\, w_1^l\, w_2^l...w_m^l$, for $p_1, p_2, ...p_k, v, w_1, w_2, ..., w_m \in \mathbb{P}$, $k, m \geq 0$, where p^r and p^l are called *right* and *left* adjoints of p, respectively, for any $p \in \mathbb{P}^1$. The latter fulfil the following inequations[2]: $p^l p \leq 1$ and $pp^r \leq 1$.

The lexical entries take the following form

$$\begin{pmatrix} t_1\ t_2\ ...\ t_k \\ s_1\ s_2\ ...\ s_k \end{pmatrix}$$

Here $s_1, ..., s_k$ are elements of the alphabet and $t_1, ..., t_k$ are types. (The reader can consult the dictionary of a given fragment of Italian language, given in [10], in order to see how they look like).

A merge operation applying to any pair of tuples is defined as follows

$$\begin{pmatrix} t_1\ ...\ t_i \\ s_1\ ...\ s_i \end{pmatrix} \bullet \begin{pmatrix} t_{i+1}\ ...\ t_k \\ s_{i+1}\ ...\ s_k \end{pmatrix} = \begin{pmatrix} t_1\ ...\ t_k \\ s_1\ ...\ s_k \end{pmatrix}$$

An operation of deleting i-th coordinate, for any k-tuple $k > 0$ and any $1 \leq i \leq k$ is defined as follows

$$\begin{pmatrix} t_1\ ...\ t_{i-1}\ t_i\ t_{i+1}\ ...\ t_k \\ s_1\ ...\ s_{i-1}\ s_i\ s_{i+1}\ ...\ s_k \end{pmatrix}_{-i} = \begin{pmatrix} t_1\ ...\ t_{i-1}\ t_{i+1}\ ...\ t_k \\ s_1\ ...\ s_{i-1}\ s_{i+1}\ ...\ s_k \end{pmatrix}$$

Let us define a binary relation on tupled pregroup expressions, denoted by $\Rightarrow$ that holds in the following cases, for any tuples e_1, e_2 and sequence of tuples α, β

$$(Mrg) \qquad \alpha\ e_1\ e_2\ \beta \Rightarrow\ \alpha\ e_1 \bullet e_2\ \beta$$

$$(Move)\ \alpha \begin{pmatrix} t_1\ ...\ t_k \\ s_1\ ...\ s_k \end{pmatrix} \beta \Rightarrow \alpha \begin{pmatrix} t_i t_j \\ s_i s_j \end{pmatrix} \bullet \begin{pmatrix} t_1\ ...\ t_k \\ s_1\ ...\ s_k \end{pmatrix}_{-i-j} \beta$$

(Move) applies to any k-tuple ($k > 1$), for any $1 \leq i \leq k$ and $1 \leq j \leq k$.

The type in any coordinate can be contracted, for any a, b such that $a \leq b$

$$(GCon) \qquad \alpha \begin{pmatrix} ...\ \overset{xb^l a y}{s}\ ... \end{pmatrix} \beta \Rightarrow\ \alpha \begin{pmatrix} ...\ \overset{xy}{s}\ ... \end{pmatrix} \beta$$

$$(GCon) \qquad \alpha \begin{pmatrix} ...\ \overset{xab^r y}{s}\ ... \end{pmatrix} \beta \Rightarrow\ \alpha \begin{pmatrix} ...\ \overset{xy}{s}\ ... \end{pmatrix} \beta$$

---

[1] Every type must contain exactly one simple type v (without the superscript r or l) and may contain an arbitrary number (possibly equal to zero) of right and left adjoints.

[2] Presenting types in that form is a great simplification, but it is done on purpose for our linguistic applications; for more details concerning tupled pregroup grammars see [23].

Additionally, (Mrg) can be applied to a pair of tuples only when in one tuple all types are of the form v (without left and right adjoints) and $(Move)$ takes two items of a tuple if one of the types is of the form v.

A tupled pregroup grammar is $G = (\Sigma, \mathbb{P}, \leq, \mathbb{I}, S)$, where Σ is a finite alphabet, $\mathbb{P}$ is a set partially ordered by $\leq$, $\mathbb{I}$ is a relation $\mathbb{I} \subset (\mathbb{T}_{\mathbb{P}} \times (\Sigma \cup \{\epsilon\})^*$, S is a designated type (in our applications the type of a sentence). The language of the grammar is $L_G = \{s | \mathbb{I}^* \overset{*}{\Rightarrow} \binom{S}{s}\}$

In our analysis we will make use of the following types

s_1 type of a sentence in a present tense

π_3 third person subject

$\overset{\approx}{i}$ infinitive of bitransitive verb without the ending e,
   for example *dar* instead of *dare*

$\tilde{i}$ infinitive of a transitive verb without the ending e,
   for example *veder* instead of *vedere*

   We make use of the following partial order: $\overset{\approx}{i} \leq \tilde{i}$

i infinitive of the verb, for example *dare*

$\bar{i}$ infinitive form of the verb phrase with a clitic,
   for example *darla, darmi, darmela, poterla avere vista*

$\overset{=}{i}$ infinitive form of the verb phrase required by modal verb without e,
   for example *la avere vista*

   We make use of the following partial order: $i \leq \bar{i}$.
   It can be explained by the fact that a bitransitive verb may act as
   a transitive one, as in *darmi, darla*, but it cannot be the other way around
   - we cannot say *vedermela*.

   We do not compare $\overset{=}{i}$ with $\bar{i}$.

o_4 an accusative (direct) object (as i.e. *una ragazza, una bella ragazza*, etc.)

o_3 a dative (indirect) object (as *a me, a una ragazza*, etc.)

$\hat{o}_4$ an accusative (direct) personal pronoun (*mi, ti, lo, la, ci, vi, li, le, ne*),
   occurring in the sentence independently (as a single lexical entry)

$\hat{o}_3$ a dative (direct) personal pronoun (*mi, ti, gli, le, ci, vi*),
   occurring in the sentence independently (as a single lexical entry)

$\check{o}_3$ a dative (direct) personal pronoun (*me, te, gli, ce, ve*),
   occurring in the sentence independently (as a single lexical entry)
   together with an accusative object (for example *te la, me li*)

$\hat{o}_4$ an accusative (direct) personal pronoun occurring within a tuple,
   accompanied with the verb as in $\begin{pmatrix} i & \hat{o}_4 \\ dare & la \end{pmatrix}$

$\hat{o}_3$ a dative(direct) personal pronoun occurring within a tuple,
   accompanied with the verb as in $\begin{pmatrix} i & \hat{o}_3 \\ dare & mi \end{pmatrix}$

$\check{o}_3$ a dative (direct) personal pronoun occurring within a tuple, accompanied
   with the verb and an accusative personal pronoun as in $\begin{pmatrix} i & \check{o}_3 & \hat{o}_4 \\ dare & me & la \end{pmatrix}$

We make use of the following partial order: $\hat{o}_4 \leq \hat{\hat{o}}_4$, $\hat{o}_3 \leq \hat{\hat{o}}_3$, $\check{o}_3 \leq \check{\check{o}}_3$

Note: we do not compare either o_4 with $\hat{o}_4$ or o_3 with $\hat{o}_3$.

$\bar{p}_2$ the plain form of past participle (*visto, dato*) of the verb which takes an auxiliary verb *avere* to form past participle

p_2 the full form of past participle (*visto una ragazza, dato una mela a me*) of the verb with an auxiliary verb *avere*

Let the dictionary (showing only tuples used in our examples) be as follows, for more see [10]

$$I = \left\{ \begin{pmatrix} \pi_3 \\ Mario \end{pmatrix} \begin{pmatrix} \hat{o}_4^r \pi_3^r s_1 i^l \\ vuole \end{pmatrix} \begin{pmatrix} i & \hat{o}_4 \\ vedere & la \end{pmatrix} \begin{pmatrix} \tilde{i} & \hat{o}_4 \\ veder & la \end{pmatrix} \begin{pmatrix} \bar{\tilde{i}} i^l \\ poter \end{pmatrix} \begin{pmatrix} i \\ avere \end{pmatrix} \right.$$

$$\left. \begin{pmatrix} \tilde{\bar{i}} \\ aver \end{pmatrix} \begin{pmatrix} \bar{p}_2 & \hat{o}_4 \\ vista & la \end{pmatrix} \begin{pmatrix} \bar{p}_2 & \check{o}_3 & \hat{o}_4 \\ data & me & la \end{pmatrix} \begin{pmatrix} \bar{p}_2^r \hat{o}_4^r \hat{o}_3^r \tilde{i}^r \bar{\tilde{i}} \\ \epsilon \end{pmatrix} \begin{pmatrix} \hat{o}_4^r \tilde{i}^r \bar{\tilde{i}} \\ \epsilon \end{pmatrix} \begin{pmatrix} \bar{p}_2^r i^r \hat{o}_4^r \bar{\tilde{i}} \\ \epsilon \end{pmatrix} \ldots \right\}$$

Single tuples are put together with those containing two or three coordinates - the first word in the tuple being usually a verb in its inflected form (we wish to remind the reader that the order of an ordered pair - type, word - within a tuple is not important and it can be switched without any consequences). Empty strings play the role of ordering words or strings of words in the sentence. The idea behind the tuples is that some words are indispensable in the sentence and have to occur together in order to fulfill certain features and peculiarities of the grammar. In comparison with former approaches the types assigned to words are less complicated and not so numerous, the proposed tupled PG 'catches' all acceptable clitic patterns and excludes those considered wrong by Italian grammarians.

Italian clitics exhibit two basic patterns: clitic pronouns can occur both in preverbal and post-verbal position, keeping the same relative order: locative/indirect object/direct object. Subjects do not allow clitic counterparts and concerning the other verbal arguments, clitic unstressed elements can be attached both to the main verb or to auxiliaries and modal verbs. Therefore the set of clitic types will include types for the direct object (accusative), types for the indirect object (dative), and types for the locative object, for details see [21]. In our work we consider transitive and ditransitive verbs. The general patterns for ditransitive verbs with accusative and dative objects as well as their preverbal and postverbal cliticization in Italian are as shown below. We consider the transitive verb *vedere* [to see] and ditransitive verb *dare* [to give] in their inflected form or accompanied by modal verbs in different tenses. Underlined particles are direct objects (noun phrases or clitic pronouns) and those with double underline are indirect objects. Clitics are in bold. The meaning of the sentences in English are *Mario vede una ragazza - Mario sees a girl* and *Mario da una mela a me - Mario gives an apple to me*, then we consider all their possible variations in different tenses and word orders (also with clitic pronouns).

1. Mario vede una ragazza.

2. Mario **la** vede.

3. Mario vuole vedere una ragazza.

4. Mario **la** vuole vedere.

5. Mario ha visto una ragazza.

6. Mario **la** ha vist**a***. (Mario l'ha vista).

7. Mario ha voluto vedere una ragazza.

8. Mario **la** ha voluto vedere.

9. Mario ha voluto veder**la**.

10. Mario vuole potere vedere una ragazza.

11. Mario vuole potere aver**la** vist**a***.

12. Mario vuole poter**la** avere vist**a***

13. Mario **la** ha voluto avere vist**a***.

14. Mario ha voluto aver**la** vist**a***.

15. Mario **la** ha voluto potere avere vist**a***.

16. Mario ha voluto potere aver**la** vist**a***.

....

17. Mario da una mela a me.

18. Mario da a me una mela.

19. Mario **la** da a me.

20. Mario **mi** da una mela.

21. Mario **me la** da.

22. Mario vuole dare una mela a me.

23. Mario vuole dare a me una mela.

24. Mario **la** vuole dare a me.

25. Mario **mi** vuole dare una mela.

26. Mario **me la** vuole dare.

27. Mario vuole dar**la** a me.

28. Mario vuole dar**mi** una mela.

29. Mario vuole dar**mela**.

30. Mario ha dato una mela a me.

31. Mario ha dato a me una mela.

32. Mario **mi** ha dato una mela.

33. Mario **la** ha dat**a*** a me.

34. Mario **me la** ha dat**a***.

35. Mario ha voluto dare una mela a me.

36. Mario **la** ha voluto dare a me.

...

37. Mario vuole potere aver**la** data a me.

38. Mario vuole poter**la** avere data a me.

39. Mario vuole aver**mi** dato una mela.

40. Mario vuole aver**la** dat**a*** a me.

41. Mario vuole aver**mela** dat**a***.

...

* It is a peculiarity of Italian (also of other Romance languages) - that when using the accusative clitic (femminine, singular) in Present Perfect tense we need also to change the ending of the Past Participle (feature agreement). That posed a problem in former approaches and led to overgeneration. For that purpose Lambek [20] decided to use product pregroups for French.

On the basis of the types in the dictionary and of the rules transforming the lexical entries, we can exclude a variety of non-sentences (marked with ¬) like

¬ Mario la ha voluto vederla.

¬ Mario la ha visto.

¬ Mario vuole la avere vista.

¬ Mario ha la voluto avere vista.

¬ La Mario ha voluto avere vista.

¬ Mario ha voluto la potere avere vista.

¬ Mario mi la da.

¬ Mario la me da.

¬ Mario la da me.

¬ Mario mi da la.

¬ Mario me da una mela.

¬ Mario mi ha voluto darla.

¬ Mario la ha voluto darmi.

¬ Mario ha voluto me la dare.

Making use of lexical entries in the dictionary and transforming them according to the rules of (Mrg), $(Move)$ and $(GCon)$ we can justify the correctness of the sentence **Mario la vuole vedere** and proceed as follows

$$\begin{pmatrix} \pi_3 \\ Mario \end{pmatrix} \begin{pmatrix} \hat{o}_4^r \pi_3^r s_1 i^l \\ vuole \end{pmatrix} \begin{pmatrix} i & \hat{o}_4 \\ vedere \; la \end{pmatrix} \Rightarrow$$

$$\begin{pmatrix} \pi_3 \\ Mario \end{pmatrix} \begin{pmatrix} \hat{o}_4^r \pi_3^r s_1 i^l & i & \hat{o}_4 \\ vuole & vedere \; la \end{pmatrix} \Rightarrow \begin{pmatrix} \pi_3 \\ Mario \end{pmatrix} \begin{pmatrix} \hat{o}_4 \hat{o}_4^r \pi_3^r s_1 i^l & i \\ la \; vuole & vedere \end{pmatrix} \Rightarrow$$

$$\begin{pmatrix} \pi_3 \\ Mario \end{pmatrix} \begin{pmatrix} \pi_3^r s_1 i^l & i \\ la \; vuole \; vedere \end{pmatrix} \Rightarrow \begin{pmatrix} \pi_3 \\ Mario \end{pmatrix} \begin{pmatrix} \pi_3^r s_1 i^l i \\ la \; vuole \; vedere \end{pmatrix} \Rightarrow$$

$$\begin{pmatrix} \pi_3 \\ Mario \end{pmatrix} \begin{pmatrix} \pi_3^r s_1 \\ la \; vuole \; vedere \end{pmatrix} \Rightarrow \begin{pmatrix} \pi_3 & \pi_3^r s_1 \\ Mario \; la \; vuole \; vedere \end{pmatrix} \Rightarrow$$

$$\begin{pmatrix} \pi_3 \pi_3^r s_1 \\ Mario \; la \; vuole \; vedere \end{pmatrix} \Rightarrow \begin{pmatrix} s_1 \\ Mario \; la \; vuole \; vedere \end{pmatrix}$$

In the sentence **Mario vuole vederla** beside the tuple with two coordinates we also need a tuple with an empty string, and proceed in the following way

$$\begin{pmatrix} \pi_3 \\ Mario \end{pmatrix} \begin{pmatrix} \pi_3^r s_1 \tilde{i}^l \\ vuole \end{pmatrix} \begin{pmatrix} \tilde{i} & \hat{o}_4 \\ veder \; la \end{pmatrix} \begin{pmatrix} \hat{o}_4^r \tilde{i}^r \tilde{i} \\ \epsilon \end{pmatrix} \Rightarrow$$

$$\begin{pmatrix} \pi_3 \\ Mario \end{pmatrix} \begin{pmatrix} \pi_3^r s_1 \tilde{i}^l \\ vuole \end{pmatrix} \begin{pmatrix} \tilde{i} & \hat{o}_4 & \hat{o}_4^r \tilde{i}^r \tilde{i} \\ veder \; la & \epsilon \end{pmatrix} \Rightarrow$$

$$\begin{pmatrix} \pi_3 \\ Mario \end{pmatrix} \begin{pmatrix} \pi_3^r s_1 \tilde{i}^l \\ vuole \end{pmatrix} \begin{pmatrix} \hat{o}_4 \hat{o}_4^r \tilde{i}^r \tilde{i} & \tilde{i} \\ la & veder \end{pmatrix} \Rightarrow$$

$$\begin{pmatrix} \pi_3 \\ Mario \end{pmatrix} \begin{pmatrix} \pi_3^r s_1 \tilde{i}^l \\ vuole \end{pmatrix} \begin{pmatrix} \tilde{i}^r \tilde{i} & \tilde{i} \\ la \; veder \end{pmatrix} \Rightarrow$$

$$\begin{pmatrix} \pi_3 \\ Mario \end{pmatrix} \begin{pmatrix} \pi_3^r s_1 \tilde{i}^l \\ vuole \end{pmatrix} \begin{pmatrix} \tilde{i} \tilde{i}^r \tilde{i} \\ vederla \end{pmatrix} \Rightarrow \begin{pmatrix} \pi_3 \\ Mario \end{pmatrix} \begin{pmatrix} \pi_3^r s_1 \tilde{i}^l \\ vuole \end{pmatrix} \begin{pmatrix} \tilde{i} \\ vederla \end{pmatrix} \Rightarrow$$

Note that we can proceed with combining together the first and the second tuple or the second and the third one (the first and the third contain only simple types). The same situation might have been observed at the beginning, instead of starting with (Mrg) on the third and the fourth tuple we could have chosen the first and the second.

$$\begin{pmatrix} \pi_3 \\ Mario \end{pmatrix} \begin{pmatrix} \pi_3^r s_1 \tilde{i}^l & \tilde{i} \\ vuole \; vederla \end{pmatrix} \Rightarrow \begin{pmatrix} \pi_3 \\ Mario \end{pmatrix} \begin{pmatrix} \pi_3^r s_1 \tilde{i}^l \tilde{i} \\ vuole \; vederla \end{pmatrix} \Rightarrow$$

$$\begin{pmatrix} \pi_3 \\ Mario \end{pmatrix} \begin{pmatrix} \pi_3^r s_1 \\ vuole \; vederla \end{pmatrix} \Rightarrow \begin{pmatrix} \pi_3 & \pi_3^r s_1 \\ Mario \; vuole \; vederla \end{pmatrix} \Rightarrow$$

$$\begin{pmatrix} \pi_3 \pi_3^r s_1 \\ Mario \; vuole \; vederla \end{pmatrix} \Rightarrow \begin{pmatrix} s_1 \\ Mario \; vuole \; vederla \end{pmatrix}$$

A more complicated example could be **Mario vuole poterla avere vista**. Here one should proceed as follows:

$$\begin{pmatrix} \pi_3 \\ Mario \end{pmatrix} \begin{pmatrix} \pi_3^r s_1 \bar{i}^l \\ vuole \end{pmatrix} \begin{pmatrix} \bar{i}\,\bar{\bar{i}}^l \\ poter \end{pmatrix} \begin{pmatrix} i \\ avere \end{pmatrix} \begin{pmatrix} \bar{p}_2 & \hat{o}_4 \\ vista & la \end{pmatrix} \begin{pmatrix} \bar{p}_2^r i^r \hat{o}_4^{r\bar{\bar{i}}} \\ \epsilon \end{pmatrix} \Rightarrow$$

$$\begin{pmatrix} \pi_3 \\ Mario \end{pmatrix} \begin{pmatrix} \pi_3^r s_1 \bar{i}^l \\ vuole \end{pmatrix} \begin{pmatrix} \bar{i}\,\bar{\bar{i}}^l \\ poter \end{pmatrix} \begin{pmatrix} i \\ avere \end{pmatrix} \begin{pmatrix} \bar{p}_2 & \hat{o}_4 & \bar{p}_2^r i^r \hat{o}_4^{r\bar{\bar{i}}} \\ vista & la & \epsilon \end{pmatrix} \Rightarrow$$

$$\begin{pmatrix} \pi_3 \\ Mario \end{pmatrix} \begin{pmatrix} \pi_3^r s_1 \bar{i}^l \\ vuole \end{pmatrix} \begin{pmatrix} \bar{i}\,\bar{\bar{i}}^l \\ poter \end{pmatrix} \begin{pmatrix} i \\ avere \end{pmatrix} \begin{pmatrix} \bar{p}_2 \bar{p}_2^r i^r \hat{o}_4^{r\bar{\bar{i}}} & \hat{o}_4 \\ vista & la \end{pmatrix} \Rightarrow$$

$$\begin{pmatrix} \pi_3 \\ Mario \end{pmatrix} \begin{pmatrix} \pi_3^r s_1 \bar{i}^l \\ vuole \end{pmatrix} \begin{pmatrix} \bar{i}\,\bar{\bar{i}}^l \\ poter \end{pmatrix} \begin{pmatrix} i \\ avere \end{pmatrix} \begin{pmatrix} i^r \hat{o}_4^{r\bar{\bar{i}}} & \hat{o}_4 \\ vista & la \end{pmatrix} \Rightarrow$$

$$\begin{pmatrix} \pi_3 \\ Mario \end{pmatrix} \begin{pmatrix} \pi_3^r s_1 \bar{i}^l \\ vuole \end{pmatrix} \begin{pmatrix} \bar{i}\,\bar{\bar{i}}^l \\ poter \end{pmatrix} \begin{pmatrix} i & i^r \hat{o}_4^{r\bar{\bar{i}}} & \hat{o}_4 \\ avere & vista & la \end{pmatrix} \Rightarrow$$

$$\begin{pmatrix} \pi_3 \\ Mario \end{pmatrix} \begin{pmatrix} \pi_3^r s_1 \bar{i}^l \\ vuole \end{pmatrix} \begin{pmatrix} \bar{i}\,\bar{\bar{i}}^l \\ poter \end{pmatrix} \begin{pmatrix} i i^r \hat{o}_4^{r\bar{\bar{i}}} & \hat{o}_4 \\ avere\ vista & la \end{pmatrix} \Rightarrow$$

$$\begin{pmatrix} \pi_3 \\ Mario \end{pmatrix} \begin{pmatrix} \pi_3^r s_1 \bar{i}^l \\ vuole \end{pmatrix} \begin{pmatrix} \bar{i}\,\bar{\bar{i}}^l \\ poter \end{pmatrix} \begin{pmatrix} \hat{o}_4^{r\bar{\bar{i}}} & \hat{o}_4 \\ avere\ vista & la \end{pmatrix} \Rightarrow$$

$$\begin{pmatrix} \pi_3 \\ Mario \end{pmatrix} \begin{pmatrix} \pi_3^r s_1 \bar{i}^l \\ vuole \end{pmatrix} \begin{pmatrix} \bar{i}\,\bar{\bar{i}}^l \\ poter \end{pmatrix} \begin{pmatrix} \hat{o}_4 \hat{o}_4^{r\bar{\bar{i}}} \\ la\ avere\ vista \end{pmatrix} \Rightarrow$$

$$\begin{pmatrix} \pi_3 \\ Mario \end{pmatrix} \begin{pmatrix} \pi_3^r s_1 \bar{i}^l \\ vuole \end{pmatrix} \begin{pmatrix} \bar{i}\,\bar{\bar{i}}^l \\ poter \end{pmatrix} \begin{pmatrix} \bar{\bar{i}} \\ la\ avere\ vista \end{pmatrix} \Rightarrow$$

$$\begin{pmatrix} \pi_3 \\ Mario \end{pmatrix} \begin{pmatrix} \pi_3^r s_1 \bar{i}^l \\ vuole \end{pmatrix} \begin{pmatrix} \bar{i}\,\bar{\bar{i}}^l & \bar{\bar{i}} \\ poter\ la\ avere\ vista \end{pmatrix} \Rightarrow$$

$$\begin{pmatrix} \pi_3 \\ Mario \end{pmatrix} \begin{pmatrix} \pi_3^r s_1 \bar{i}^l \\ vuole \end{pmatrix} \begin{pmatrix} \bar{i}\,\bar{\bar{i}}^l\bar{\bar{i}} \\ poterla\ avere\ vista \end{pmatrix} \Rightarrow$$

$$\begin{pmatrix} \pi_3 \\ Mario \end{pmatrix} \begin{pmatrix} \pi_3^r s_1 \bar{i}^l \\ vuole \end{pmatrix} \begin{pmatrix} \bar{i} \\ poterla\ avere\ vista \end{pmatrix} \Rightarrow \ \ldots\ldots \Rightarrow$$

$$\begin{pmatrix} s_1 \\ Mario\ vuole\ poterla\ avere\ vista \end{pmatrix}$$

Note that during the derivation process the types occurring within tuples are consistent with our assumptions. We obtain the tuple $\begin{pmatrix} \bar{\bar{i}} \\ la\ avere\ vista \end{pmatrix}$ with the type $\bar{\bar{i}}$ for *la avere vista* acting as a complement of a short form of the modal verb (for example *poter*) and building together the expression of the form $\bar{i}$ - infinitive form of the verb phrase with a clitic - *poterla avere vista*.

To complete the examples we present one with the ditransitive verb **Mario vuole avermela data**.

$$\begin{pmatrix} \pi_3 \\ Mario \end{pmatrix} \begin{pmatrix} \pi_3^r s_1 \bar{i}^l \\ vuole \end{pmatrix} \begin{pmatrix} \tilde{i} \\ aver \end{pmatrix} \begin{pmatrix} \bar{p}_2 & \check{o}_3 & \hat{o}_4 \\ data & me & la \end{pmatrix} \begin{pmatrix} \bar{p}_2^r \hat{o}_4^r \hat{o}_3^r \tilde{i}^r \bar{\bar{i}} \\ \epsilon \end{pmatrix} \Rightarrow$$

$$\begin{pmatrix} \pi_3 \\ Mario \end{pmatrix} \begin{pmatrix} \pi_3^r s_1 \bar{i}^l \\ vuole \end{pmatrix} \begin{pmatrix} \tilde{i} \\ aver \end{pmatrix} \begin{pmatrix} \bar{p}_2 & \check{o}_3 & \hat{o}_4 & \bar{p}_2^r \hat{o}_4^r \hat{o}_3^r \tilde{i}^r \bar{i} \\ data & me & la & \epsilon \end{pmatrix} \Rightarrow$$

$$\begin{pmatrix} \pi_3 \\ Mario \end{pmatrix} \begin{pmatrix} \pi_3^r s_1 \bar{i}^l \\ vuole \end{pmatrix} \begin{pmatrix} \tilde{i} \\ aver \end{pmatrix} \begin{pmatrix} \bar{p}_2 \bar{p}_2^r \hat{o}_4^r \hat{o}_3^r \tilde{i}^r \bar{i} & \check{o}_3 & \hat{o}_4 \\ data & me & la \end{pmatrix} \Rightarrow$$

$$\begin{pmatrix} \pi_3 \\ Mario \end{pmatrix} \begin{pmatrix} \pi_3^r s_1 \bar{i}^l \\ vuole \end{pmatrix} \begin{pmatrix} \tilde{i} \\ aver \end{pmatrix} \begin{pmatrix} \hat{o}_4^r \hat{o}_3^r \tilde{i}^r \bar{i} \;\; \check{o}_3 \;\; \hat{o}_4 \\ data \quad me \;\; la \end{pmatrix} \Rightarrow$$

$$\begin{pmatrix} \pi_3 \\ Mario \end{pmatrix} \begin{pmatrix} \pi_3^r s_1 \bar{i}^l \\ vuole \end{pmatrix} \begin{pmatrix} \tilde{i} \\ aver \end{pmatrix} \begin{pmatrix} \hat{o}_4 \hat{o}_4^r \hat{o}_3^r \tilde{i}^r \bar{i} \;\; \check{o}_3 \\ la \; data \;\; me \end{pmatrix} \Rightarrow$$

$$\begin{pmatrix} \pi_3 \\ Mario \end{pmatrix} \begin{pmatrix} \pi_3^r s_1 \bar{i}^l \\ vuole \end{pmatrix} \begin{pmatrix} \tilde{i} \\ aver \end{pmatrix} \begin{pmatrix} \hat{o}_3^r \tilde{i}^r \bar{i} \;\; \check{o}_3 \\ la \; data \; me \end{pmatrix} \Rightarrow$$

$$\begin{pmatrix} \pi_3 \\ Mario \end{pmatrix} \begin{pmatrix} \pi_3^r s_1 \bar{i}^l \\ vuole \end{pmatrix} \begin{pmatrix} \tilde{i} \\ aver \end{pmatrix} \begin{pmatrix} \check{o}_3 \hat{o}_3^r \tilde{i}^r \bar{i} \\ me \; la \; data \end{pmatrix} \Rightarrow$$

$$\begin{pmatrix} \pi_3 \\ Mario \end{pmatrix} \begin{pmatrix} \pi_3^r s_1 \bar{i}^l \\ vuole \end{pmatrix} \begin{pmatrix} \tilde{i} \\ aver \end{pmatrix} \begin{pmatrix} \tilde{i}^r \bar{i} \\ me \; la \; data \end{pmatrix} \Rightarrow$$

$$\begin{pmatrix} \pi_3 \\ Mario \end{pmatrix} \begin{pmatrix} \pi_3^r s_1 \bar{i}^l \\ vuole \end{pmatrix} \begin{pmatrix} \tilde{i} \qquad \tilde{i}^r \bar{i} \\ aver \; me \; la \; data \end{pmatrix} \Rightarrow$$

$$\begin{pmatrix} \pi_3 \\ Mario \end{pmatrix} \begin{pmatrix} \pi_3^r s_1 \bar{i}^l \\ vuole \end{pmatrix} \begin{pmatrix} \tilde{\tilde{i}} \tilde{i}^r \bar{i} \\ avermela \; data \end{pmatrix} \Rightarrow$$

$$\begin{pmatrix} \pi_3 \\ Mario \end{pmatrix} \begin{pmatrix} \pi_3^r s_1 \bar{i}^l \\ vuole \end{pmatrix} \begin{pmatrix} \bar{i} \\ avermela \; data \end{pmatrix} \Rightarrow \; .... \Rightarrow$$

$$\begin{pmatrix} s_1 \\ Mario \; vuole \; avermela \; data \end{pmatrix}$$

5   Conclusions

In this paper we have compared three different appoaches to pregroup grammar: classical PG, product PG and tupled PG. By analysing a number of representative examples taken from Italian, with particular reference to the contexts in which clitic pronouns occur, we have shown the undeniable advantages of tupled PG with respect to the other two approaches. The analysis presented here for Italian clitics can be extended without much difficulties to other Romance languages such as French, Spanish and Portuguese. It will be interesting, on this respect, to compare the different sets of tuples projected from the different dictionaries defined by means of a tupled pregroup grammar.

References

1. Abrusci, M.: Classical Conservative Extensions of Lambek Calculus. Studia Logica 71, 277–314 (2002)
2. Bargelli, D., Lambek, J.: An algebraic approach to French sentence structure. In: de Groote, P., Morrill, G., Retoré, C. (eds.) LACL 2001. LNCS (LNAI), vol. 2099, pp. 62–78. Springer, Heidelberg (2001)
3. Buszkowski, W.: Lambek Grammars Based on Pregroups. In: de Groote, P., Morrill, G., Retoré, C. (eds.) LACL 2001. LNCS (LNAI), vol. 2099, pp. 95–109. Springer, Heidelberg (2001)
4. Buszkowski, W.: Pregroups: Models and Grammars. In: de Swart, H. (ed.) RelMiCS 2001. LNCS, vol. 2561, pp. 35–49. Springer, Heidelberg (2002)
5. Buszkowski, W.: Type Logics and Pregroups. Studia Logica 87(2-3), 145–169 (2007)

6. Casadio, C., Lambek, J.: An Algebraic Analysis of Clitic Pronouns in Italian. In: de Groote, P., Morrill, G., Retoré, C. (eds.) LACL 2001. LNCS (LNAI), vol. 2099, pp. 110–124. Springer, Heidelberg (2001)
7. Casadio, C., Lambek, J.: A Tale of Four Grammars. Studia Logica 71(2), 315–329 (2002)
8. Casadio, C.: Cliticization and Agreement in a Two-Pregroups Grammar. Linguistic Analysis 36(1-4), 73–92 (2010)
9. Casadio, C.: Agreement and Cliticization in Italian: A Pregroup Analysis. In: Dediu, A.-H., Fernau, H., Martín-Vide, C. (eds.) LATA 2010. LNCS, vol. 6031, pp. 166–177. Springer, Heidelberg (2010)
10. Casadio, C., Kiślak-Malinowska, A.: Tupled Pregroups. A Study of Italian Clitic Patterns. In: Proceedings LENLS 9, (JSAI) (2012) ISBN 978-4-915905-51-3
11. Kiślak-Malinowska, A.: Polish Language in Terms of Pregroups. In: Casadio, C., Lambek, J. (eds.) Computational Algebraic Approaches to Natural Language, pp. 145–172. Polimetrica (2008)
12. Kiślak-Malinowska, A.: Some Aspects of Polish Grammar in Terms of Tupled Pregroups. Linguistic Analysis, 93–119 (2010)
13. Kiślak-Malinowska, A.: Extended Pregroup Grammars Applied to Natural Languages. In: Logic and Logical Philosophy, vol. 21, pp. 229–252 (2012)
14. Kusalik, T.: Product Pregroups as an Alternative to Inflectors. In: Casadio, C., Lambek, J. (eds.) Computational Algebraic Approaches to Natural Language, pp. 173–189. Polimetrica (2008)
15. Lambek, J.: The Mathematics of Sentence Structure. American Mathematics Monthly 65, 154–169 (1958)
16. Lambek, J.: Type Grammars Revisited. In: Lecomte, A., Perrier, G., Lamarche, F. (eds.) LACL 1997. LNCS (LNAI), vol. 1582, pp. 1–27. Springer, Heidelberg (1999)
17. Lambek, J.: Type Grammars as Pregroups. Grammars 4(1), 21–39 (2001)
18. Lambek, J.: A computational Algebraic Approach to English Grammar. Syntax 7(2), 128–147 (2004)
19. Lambek, J.: From Word to Sentence. Polimetrica (2008)
20. Lambek, J.: Exploring Feature Agreement in French with Parallel Pregroup Computations. Journal of Logic, Language and Information, 75–88 (2010)
21. Monachesi, P.: A lexical Approach to Italian Cliticization. Lecture Notes 84. CSLI, Stanford (1999)
22. Moortgat, M.: Categorical Type Logics. In: van Benthem, J., ter Meulen, A. (eds.) Handbook of Logic and Language, pp. 93–177. Elsevier, Amsterdam (1997)
23. Stabler, E.: Tupled pregroup grammars. In: Casadio, C., Lambek, J. (eds.) Computational Algebraic Approaches to Natural Language, pp. 23–52. Polimetrica (2008)

On Associative Lambek Calculus
Extended with Basic Proper Axioms

Annie Foret

IRISA, University of Rennes 1
Campus de Beaulieu, 35042 Rennes cedex, France
foret@irisa.fr

Abstract. The purpose of this article is to show that the associative
Lambek calculus extended with basic proper axioms can be simulated by
the usual associative Lambek calculus, with the same number of types per
word in a grammar. An analogue result had been shown for pregroups
grammars [1]. We consider Lambek calculus with product, as well as the
product-free version.

1 Introduction

The associative Lambek calculus (L) has been introduced in [6], we refer to [3,8]
for details on (L) and its non-associative variant (NL). The pregroup formalism
(PG) has been later introduced [7] as a simplification of Lambek calculus. These
formalisms are considered for the syntax modeling and parsing of various natural
languages. In contrast to (L), pregroups allow some kind of postulates ; we
discuss this point below.

Postulates in Pregroups. The order on primitive types has been introduced in
Pregroups (PG) to simplify the calculus for simple types. The consequence is
that PG is not fully lexicalized. From the results in [1], this restriction is not so
important because a PG using an order on primitive types can be transformed
into a PG based on a simple free pregroup using a pregroup morphism, s.t. : its
size is bound by the size of the initial PG times the number of primitive types
(times a constant which is approximatively 4), moreover, this transformation
does not change the number of types that are assigned to a word (a k-valued
PG is transformed into a k-valued PG).

Postulates in the Lambek Calculus. Postulates (non-logical axioms) in Lambek
calculus have also been considered. We know from [2,5], that :

(i) the associative version (L) with nonlogical axioms generate ϵ-free r.e. languages (the result also holds for L without product). The proof in the case
with product is based on *binary grammars* whose production are of the form :

$$p \rightarrow q \ , \ p \rightarrow q\,r \ , \ p\,q \rightarrow r$$

C. Casadio et al. (Eds.): Lambek Festschrift, LNCS 8222, pp. 172–187, 2014.

for which is constructed a language-equivalent categorial grammar $L(\Phi(G))$ where $\Phi(G)$ is a finite set of non-logical axioms.

(ii) the non-associative version (NL) with nonlogical axioms generate context-free languages [5].

This article adresses the associative version (L). It is organized as follows : section 2 gives a short background on categorial grammars and on L extended with proper axioms $L(\Phi)$; section 3 gives some preliminary facts on $L(\Phi)$, when Φ corresponds to a preorder $\leq$ on primitive types (written $\Phi_\leq$) ; section 4 defines the simulation (written h) ; section 5 gives the main results on the h simulation ; section 6 gives the lemmas and proof details. Section 7 concludes.

Such a result also aims at clarifying the properties of classes of rigid and k-valued type logical grammars (TLG).

2   Categorial Grammars, their Languages and Systems

2.1   Categorial Grammars and their Languages

A *categorial grammar* is a structure $G = (\Sigma, I, S)$ where: Σ is a finite alphabet (the words in the sentences); given a set of types $Tp(Pr)$, where Pr denotes a set of primitive types, $I : \Sigma \mapsto \mathcal{P}^f(Tp(Pr))$ is a function that maps a finite set of types from each element of Σ (the possible categories of each word); $S \in Tp(Pr)$ is the *main type* associated to correct sentences.

Language. Given a relation on $Tp(Pr)^*$ called the derivation relation on types : a sentence $v_1 \ldots v_n$ then belongs to the *language of* G, written $\mathcal{L}(G)$, provided its words v_i can be assigned types X_i whose sequence $X_1 \ldots X_n$ derives S according to the derivation relation on types.

An *AB-grammar* is a categorial grammar $G = (\Sigma, I, S)$, such that its set of types $Tp(Pr)$ is constructed from Pr (primitive), using two binary connectives $/$, $\backslash$, and its language is defined using two deduction rules:

A , $A \backslash B \vdash B$ (Backward elimination, written $\backslash_e$)

B / A , $A \vdash B$ (Forward elimination, written $/_e$)

For example, using $\backslash_e$*, the string of types* $(N, N \backslash S)$ *associated to "John swims" entails* S*, the type of sentences. Another typical example is* $(N, ((N \backslash S) / N), N))$ *associated to "John likes Mary", where the right part is associated to "likes Mary".*

Lambek-grammars AB-grammars are the basis of a hierarchy of type-logical grammars (TLG). The associative Lambek calculus (L) has been introduced in [6], we refer to [3] for details on (L) and its non-associative variant (NL). A sequent-style presentation of (L) is detailed after.

The above examples illustrating AB-grammars also hold for (L) and (NL).

The pregroup formalism has been introduced in [7] as a simplification of Lambek calculus [6]. See [7] for a definition.

2.2 Type Calculus for (L)

By a sequent we mean a pair written $\Gamma \vdash A$, where Γ is a sequence of types of $Tp(Pr)$ and A is a type in $Tp(Pr)$. We give a "Gentzen style" sequent presentation, by means of introduction rules on the left or on the right of a sequent :

Lambek Calculus (associative)	(Gentzen style)

$$\frac{\Gamma, A, \Gamma' \vdash C \quad \Delta \vdash A}{\Gamma, \Delta, \Gamma' \vdash C}\ \mathbf{Cut} \qquad\qquad A \vdash A$$

$$\frac{\Gamma \vdash A \quad \Delta, B, \Delta' \vdash C}{\Delta, B\,/\,A, \Gamma, \Delta' \vdash C}\,/L \qquad \frac{\Gamma, A \vdash B}{\Gamma \vdash B\,/\,A}\,/R$$

$$\frac{\Gamma \vdash A \quad \Delta, B, \Delta' \vdash C}{\Delta, \Gamma, A \setminus B, \Delta' \vdash C}\,\backslash L \qquad \frac{A, \Gamma \vdash B}{\Gamma \vdash A \setminus B}\,\backslash R$$

$$\frac{\Delta, A, B, \Delta' \vdash C}{\Delta, A \bullet B, \Delta' \vdash C}\,\bullet L \qquad \frac{\Gamma \vdash A \quad \Gamma' \vdash B}{\Gamma, \Gamma' \vdash A \bullet B}\,\bullet R$$

The calculus denoted by L consists in this set of rules and has the extra requirement when applying a rule : the left-handside of a sequent cannot be empty. We may consider the system restricted to $/$ and $\setminus$ or its full version, where the set of types has a product type constructor $\bullet$ (non-commutative). The Cut rule can be eliminated from the type system (proving the same sequents). This property with the subformula property entail the decidability of the system.

2.3 Type Calculus for (L) Enriched with Postulates

$L(\Phi)$. In the general setting (as in [5]) nonlogical axioms are of the form :
   $A \vdash B$, where $A, B \in Tp(Pr)$
   and $L(\Phi)$ denotes the system L with all $A \vdash B$ from Φ as new axioms.
   The calculus corresponds to adding a new rule of the form : $\dfrac{A \vdash B \in \Phi}{A \vdash B}\ Ax_\Phi$

$L(\Phi_\le)$. In the following of the paper, we shall restrict to axioms of the form :
   $p \vdash q$, where p, q are primitive (elements of Pr). Moreover, to keep the parallel with pregroups, we consider a preorder $\le$ on a *finite* set of primitive types Pr and consider : $L(\Phi_\le)$ where $\Phi_\le$ is the set of axioms $p \vdash q$ whenever $p \le q$, for $p, q \in$ Pr.

   The calculus corresponds to adding a new rule of the form : $\dfrac{p \le q}{p \vdash q}\ Ax_\le$

Some Remarks and Known Facts

On Axioms. As in L, we get an equivalent version of $L(\Phi)$, where axioms $A \vdash A$ in the type calculus are supposed basic (A primitive).

A Remark on Substitutions. In general $L(\Phi)$ is not substitution closed, see [4].

Facts on Models. [4] discusses several completeness results, in particular, L is *strongly complete* with respect to *residuated semigroups* (RSG in short)[1] : the sequents provable in $L(\Phi)$ are those which are true in all RSG where all sequents from Φ are true.

3   Some Preliminary Facts with Basic Postulates

3.1   Cut Elimination and the Subformula Property

Proposition 1. *Let $\leq$ denote a preorder on the set of primitive types, and $\Phi_\leq$ denote the corresponding set of axioms. The type calculus $L(\Phi_\leq)$ admits cut elimination and the subformula property : every derivation of $\Gamma \vdash A$ in $L(\Phi_\leq)$ can be transformed into a cut-free derivation in $L(\Phi_\leq)$ of the same sequent, such that all formulas occurring in it are subformulas of this sequent.*

Proof Sketch. The proof is standard (see [8]), on derivations, by induction on (d, r) where r is the number of rules above the cut rule (to be eliminated) and d is the depth (as a subformula tree) of the cut formula (that disappears by the cut rule). The proof shows how to remove one cut having the smallest number of rules above it, by a case analysis considering the subproof $\mathcal{D}_l$ which ends at the left premise of the cut rule and the subproof $\mathcal{D}_r$ which ends at the right of the cut rule.

The only new specific case is when $\mathcal{D}_l$ and $\mathcal{D}_r$ are both axioms :

Original derivation	New derivation
$\dfrac{\dfrac{p_i \leq p_j}{p_i \vdash p_j} \quad \dfrac{p_j \leq p_k}{p_j \vdash p_k}}{p_i \vdash p_k} cut$	$\dfrac{p_i \leq p_k}{p_i \vdash p_k}$

Observe that the transitivity of $\leq$ on Pr is crucial here.

Corollary 1. *Let $\leq$ denote a preorder on the set of primitive types, and $\Phi_\leq$ denote the corresponding set of axioms. The type calculus $L(\Phi_\leq)$ is decidable.*

These above propositions apply for full L and product-free L.

---

[1]   A *residuated semigroup* (RSG) is a structure $(M, \leq, ., \backslash, /)$ such that $(M, \leq)$ is a nonempty poset, $\forall a, b, c \in M : a.b \leq c$ iff $b \leq a \backslash c$ iff $a \leq c / b$ (residuation), . is associative ; $\Gamma \vdash B$ is said true in a model (M, μ), where M is a RSG and μ from Pr into M iff $\mu(\Gamma) \leq \mu(B)$, where μ from Pr into M is extended as usual by $\mu(A \backslash B) = \mu(A) \backslash \mu(B), \mu(A / B) = \mu(A) / \mu(B), \mu(A \bullet B) = \mu(A).\mu(B), \mu(\Gamma, \Delta) = \mu(\Gamma).\mu(\Delta)$.

3.2 Rule Reversibility

Proposition 2 (Reversibility of $/R$ and $\backslash R$).

In the type calculus $L(\Phi_\leq)$: $\begin{cases} \Gamma \vdash B \,/\, A \text{ iff } \Gamma, A \vdash B \\ \Gamma \vdash A \,\backslash\, B \text{ iff } A, \Gamma \vdash B \end{cases}$

Proof ($\rightarrow$) : easy by induction on the derivation, according to the last rule.

This proposition holds for the full calculus and its product-free version. In the full calculus, the reversibility of rule $\bullet L$ also holds.

Main Type in the Product-Free Calculus. For a type-formula built over $Pr \,/\, , \backslash \,,$ its main type is :

- the formula if it is primitive ;
- the main type of B if it is of the form $B \,/\, A$ or the form $A \,\backslash\, B$.

In the product-free case, any type A can thus be written (ommitting parenthesis) as $X_1 \backslash \ldots \backslash X_n \backslash p_A / Y_m / \ldots / Y_1$ where p_A is the main type of A. Reversibility then gives : $\Gamma \vdash A$ in $L(\Phi_\leq)$ iff $X_1, ..., X_n, \Gamma, Y_m, ..., Y_1 \vdash p_A$ in $L(\Phi_\leq)$.

3.3 Count Checks

This notion will be useful for proofs on the simulation defined in section 4.

Polarity. We first recall the notion of *polarity* of an occurrence of $p \in Pr$ in a formula : p is positive in p ; if p is positive in A, then p is positive in $B \backslash A$, $A \,/\, B$, $A \bullet B$, $B \bullet A$, and p is negative in $A \backslash B$, $B \,/\, A$; if p is negative in A, then p is negative in $B \backslash A$, $A \,/\, B$, $A \bullet B$, $B \bullet A$, and p is positive in $A \backslash B$, $B \,/\, A$.

For a sequent $\Gamma \vdash B$, the *polarity* of an occurrence of $p \in Pr$ in B is the same as its polarity in B, but the *polarity* of an occurrence of p in Γ is the opposite of its polarity in the formula of Γ.

In the presence of non-logical axioms Φ on primitive types, a *count check property* can be given as follows :

Proposition 3 (Count check in $L(\Phi)$, on primitive types). *If $\Gamma \vdash B$ is provable in $L(\Phi)$, then for each primitive type p that is not involved in any axiom $p \vdash q$ in $L(\Phi)$ where $p \neq q$: the number of positive occurrences of p in $\Gamma \vdash B$ equals the number of negative occurrences of p in $\Gamma \vdash B$.*

The proof is easy by induction on derivations.

3.4 A Duplication Method

As is the case for pregroups [1], we may propose to duplicate assignments for each primitive type occurring in a basic postulate $p_i \leq p_j$. We give more details below.

Definition 1 (polarized duplication sets).

1. *We write* $(q)^{\uparrow}_{\leq} = \{p_j \mid q \leq p_j\}$ *and* $(q)^{\downarrow}_{\leq} = \{p_j \mid p_j \leq q\}$ *for primitive types.*
2. *We use the following operations on sets of types, that extend* $/\,,\backslash\,,\bullet$:

$$\mathbb{T}_1 \mathbin{/\!/} \mathbb{T}_2 = \{X_1 / X_2 \mid X_1 \in \mathbb{T}_1 \text{ and } X_2 \in \mathbb{T}_2\}$$
$$\mathbb{T}_1 \mathbin{\backslash\!\backslash} \mathbb{T}_2 = \{X_1 \backslash X_2 \mid X_1 \in \mathbb{T}_1 \text{ and } X_2 \in \mathbb{T}_2\}$$
$$\mathbb{T}_1 \odot \mathbb{T}_2 = \{X_1 \bullet X_2 \mid t_1 \in \mathbb{T}_1 \text{ and } X_2 \in \mathbb{T}_2\}$$
$$\mathbb{T}_1 \circ \mathbb{T}_2 \circ ... \circ \mathbb{T}_n = \{X_1, X_2, ... X_n \mid X_1 \in \mathbb{T}_1 \ X_2 \in \mathbb{T}_2 ... X_n \in \mathbb{T}_n\} \text{ for sequences}$$

3. *We define the upper-duplication* $Dupl^{\uparrow}_{\leq}(.)$ *and lower-duplication* $Dupl^{\downarrow}_{\leq}(.)$ *inductively on types, for* $\delta \in \{\uparrow,\downarrow\}$, *where we write* $op(\uparrow) =\downarrow$, $op(\downarrow) =\uparrow$:

$$Dupl^{\uparrow}_{\leq}(q) = (q)^{\uparrow}_{\leq} \text{ and } Dupl^{\downarrow}_{\leq}(q) = (q)^{\downarrow}_{\leq} \text{ for primitive types.}$$
$$Dupl^{\delta}_{\leq}(X_1 / X_2) = Dupl^{\delta}_{\leq}(X_1) \mathbin{/\!/} Dupl^{op(\delta)}_{\leq}(X_2)$$
$$Dupl^{\delta}_{\leq}(X_1 \backslash X_2) = Dupl^{op(\delta)}_{\leq}(X_1) \mathbin{\backslash\!\backslash} Dupl^{\delta}_{\leq}(X_2)$$
$$Dupl^{\delta}_{\leq}(X_1 \bullet X_2) = Dupl^{\delta}_{\leq}(X_1) \odot Dupl^{\delta}_{\leq}(X_2)$$
$$\text{and } Dupl^{\delta}_{\leq}(X_1, X_2, \ldots, X_n) = Dupl^{\delta}_{\leq}(X_1) \circ Dupl^{\delta}_{\leq}(X_2) \circ ... \circ Dupl^{\delta}_{\leq}(X_n)$$

This amounts to consider all replacements, according to $\leq$ *and the two polarities.*

Proposition 4 (Simulation 1). *For* $p \in Pr$ *(primitive) :*

if $\boxed{X_1, \ldots X_n \vdash p \text{ in } L(\Phi_{\leq})}$ *then* $\exists X'_1 \in Dupl^{\uparrow}_{\leq}(X_1) \ldots \exists X'_n \in Dupl^{\uparrow}_{\leq}(X_n)$ *such that* $X'_1, \ldots X'_n \vdash p$ *in* L *(without postulates).*

Proof Sketch. See annex.

Drawbacks. However this transformation does not preserve the size of the lexicon in general, nor the k-valued class of grammars to which the original lexicon belongs.

4   Simulation over k-valued Classes

4.1   Basic Definitions

Using morphisms-based encodings will enable to stay in a k-valued class and to keep a strong parse similarity (through the simulation).

Definition 2 (preorder-preserving mapping).
Let $(P, \leq)$ *and* $(P', \leq')$ *denote two sets of primitive types with a preorder. Let* h *denote a mapping from types of* $Tp(P)$ *(with* $\leq$ *on* P*) to types of* $Tp(P')$ *(with* $\leq'$ *on* P'*)*

- *h is a type-homomorphism iff*
  1. $\forall X, Y \in Tp(P) : h(X / Y) = h(X) / h(Y)$
  2. $\forall X, Y \in Tp(P) : h(X \backslash Y) = h(X) \backslash h(Y)$
  3. $\forall X, Y \in Tp(P) : h(X \bullet Y) = h(X) \bullet h(Y)$

- h is said monotonic iff
   4a. $\forall X, Y \in Tp(P)$:
      if $X \vdash Y$ *in* $L(\Phi_\leq)$ *then* $h(X) \vdash h(Y)$ *in* $L(\Phi_{\leq'})$ *[Monotonicity]*
- h is said preorder-preserving iff
   4b. $\forall p_i, p_j \in P$: *if* $p_i \leq p_j$ *then* $h(p_i) \vdash h(p_j)$ *in* $L(\Phi_{\leq'})$.

Condition (4b) ensures (4a) for a type-homomorphism. This can be shown by induction on derivations. Next sections define and study a type-homomorphism that fullfills all these conditions.

4.2   Construction on One Component

We consider the type calculus without empty sequents on the left, and with product. The result also holds for the product-free calculus, because the constructed simulation does not add any product.

In this presentation, we allow to simulate either a fragment (represented as Pr below) or the whole set of primitive types ; for example, we may want not to transform isolated primitive types, or to proceed incrementally.

Primitive Types. Let $P = \{p_1, \ldots, p_n\}$ and $P = Pr \cup Pr'$, denote the set of primitive types, in which Pr a connex component, where no element of Pr is related by $\leq$ to an element of Pr', and each element of Pr is related by $\leq$ to another element of Pr.
We introduce new letters q_0, q_1 and β_k for each p_k of Pr (no new postulate) [2].
We take as preordered set $P' = Pr' \cup \{q_0, q_1\} \cup \{\beta_k \mid p_k \in Pr\}$,
$\leq'$ denotes the restriction of $\leq$ on Pr' (Pr' may be empty).

Notation. We write $X \vdash_{\leq'} Y$ for a sequent provable in the type calculus $L(\Phi_{\leq'})$ and we write $X \vdash_\leq Y$ for a sequent provable in the type calculus $L(\Phi_\leq)$

We now define the simulation-morphism h for Pr as follows:

Definition 3 (Simulation-morphism h for Pr).

$h(X\ /\ Y) = h(X)\ /\ h(Y)$	*for* $p_i \in Pr$	*for* $p_i \in Pr'$
$h(X \setminus Y) = h(X) \setminus h(Y)$	*let* $Num^\uparrow(p_i) = \{k \mid p_i \leq p_k\} = \{i_1 \ldots i_k\}$	$h(p_i) = p_i$
$h(X \bullet Y) = h(X) \bullet h(Y)$	*s. t.* $i_1 < \ldots < i_k$ $\quad h(p_i) = q_0\ /\ exp(q_1, \beta_{i_1}.\ \ldots\ .\beta_{i_k})$	

where
   $exp(X, \beta) = \beta\ /\ (X \setminus \beta)$
and the notation is extended to sequences on the right by :
   $exp(X, \epsilon) = X$
   $exp(X, \beta_{i_1}.\ \ldots\ .\beta_{i_{k-1}}.\beta_{i_k}) = \beta_{i_k}\ /\ (exp(X, \beta_{i_1}.\ \ldots\ .\beta_{i_{k-1}}) \setminus \beta_{i_k})$
   $= exp(exp(X, \beta_{i_1}.\ \ldots\ .\beta_{i_{k-1}}), \beta_{i_k})$

---

[2] q_0, q_1 can also be written q_{0Pr}, q_{1Pr} if necessary w.r.t. Pr.

Notation. In the following, in expressions of the form $exp(X, \Pi)$, Π is assumed to denote a sequence (possibly empty) $\beta_{k_1} \ldots \beta_{k_n}$ (where β_k is the new letter for p_k of Pr) ; we will then write $Num(\Pi) = Num(\beta_{k_1} \ldots \beta_{k_n}) = \{k_1, \ldots, k_n\}$.

Fact. The h mapping of definition 3 is a type-homomorphism by construction.

Next sections will show that it is monotonic and a simulation (verifying the converse of monotonicity).

5   Main Results

Proposition 5 (Preorder-preserving property). *The homomorphism h of definition 3 satisfies :   (4b.) $\forall p_i, p_j \in P$: if $p_i \leq p_j$ then $h(p_i) \vdash h(p_j)$ in $L(\Phi_{\leq'})$.*

Proof. This is a corollary of this type-raise property : $A \vdash B / (A \setminus B)$; we have $A \vdash exp(A, \Pi)$ and more generally : if $\{k \mid \beta_k \in \Pi\} \subseteq \{k \mid \beta_k \in \Pi'\}$ then $exp(A, \Pi) \vdash exp(A, \Pi')$; by construction, if $p_i \leq p_j$ then $Num^\uparrow(p_j) \subseteq Num^\uparrow(p_i)$, hence the result.

Proposition 6 (Equivalence property). *The homomorphism h of definition 3 satisfies :*
$\forall X, Y \in Tp(P)$: $h(X) \vdash h(Y)$ holds in $L(\Phi_{\leq'})$ iff $X \vdash Y$ holds in $L(\Phi_{\leq})$

Proof. For the $\leftarrow$ part, this is a corollary of the preorder-preserving property, that entails monotonicity, for a type-homomorphism. For the $\rightarrow$ part, see lemmas in the next section.

Proposition 7 (Grammar Simulation). *Given a grammar $G = (\Sigma, I, S)$ and a preorder $\leq$ on the primitive types P, we define h from types on $(P, \leq)$ to types on $(P', \leq')$ such that $P = Pr \cup Pr'$, where Pr is a connex component, as in definition 3.   We construct a grammar on $(P', \leq')$ and $L(\phi_{\leq'})$ as follows :*

$G' = (\Sigma, h(I), h(S))$
where $h(I)$ is the assignment of $h(X_i)$ to a_i for $X_i \in I(a_i)$,

as a result we have : $\mathcal{L}(G) = \mathcal{L}(G')$

Note. This corresponds to the standard case of grammar, when $h(S)$ is primitive.

This proposition can apply the transformation to *the whole set* of primitive types, thus providing a fully lexicalized grammar G' (no order postulate). A similar result holds to *a fragment Pr of $P = Pr \cup Pr'$.*

A Remark on Constructions to Avoid. For other constructions based on the same idea of chains of type-raise, we draw the attention on the fact that a simplication such as h' below would not be correct. Suppose $\Phi_\leq$ consists in $p_0 \leq p_1$ as postulate, define h' a type-morphism such that

$h'(p_0) = exp(q, \beta_0)$ and $h'(p_1) = exp(q, \beta_0.\beta_1)$,

this is preorder-preserving : we have $h'(p_0) \vdash h'(p_1)$,

but this is not a correct simulation, because

$h'(p_1), h'(p_0 \setminus p_0) \vdash h'(p_1)$ whereas $p_1 (p_0 \setminus p_0) \not\vdash p_1$ (in $L(\Phi_\leq)$).

In more details, the sequent on the left is proved by :

$h'(p_0), h'(p_0) \setminus h'(p_0), h'(p_0) \setminus \beta_1 \vdash \beta_1$,

then by $\setminus R$: $h'(p_0) \setminus h'(p_0), h'(p_0) \setminus \beta_1 \vdash h'(p_0) \setminus \beta_1$,

then by $/L$: $\beta_1 / h'(p_0) \setminus \beta_1, h'(p_0) \setminus h'(p_0), h'(p_0) \setminus \beta_1 \vdash \beta_1$, then apply $/R$.

6   Lemmas

Fact (1) [count checks for new letters]

for $X \in Tp^+(P)$: if $Y_1, h(X), Y_2 \vDash_{\leq'} Z$ and X is not empty, then :

(a) the number of positive occurrences of q_0 or q_1 in $Y_1, Y_2 \vdash Z$ equals the number of negative occurrences of q_0 or q_1 in $Y_1, Y_2 \vdash Z$

(b) the number of positive occurrences of $\alpha \in \{\beta_k \mid p_k \in Pr\}$ in $Y_1, Y_2 \vdash Z$ equals the number of negative occurrences of α in $Y_1, Y_2 \vdash Z$

Proof. (a) is a consequence of the *count check property* for q_0 and for q_1, and of the following fact : by construction, in $h(X)$ the number of positive occurrences of q_0 equals the number of negative occurrences of q_1, and the number of negative occurrences of q_0 equals the number of positive occurrences of q_1.

(b) is a consequence of the *count check property* for α, and of the following fact : by construction, $h(X)$ has the same number of positive occurrences of $\alpha \in \{\beta_k \mid p_k \in Pr\}$ as its number of negative occurrences.

Note. Thus by (a), the presence of a formula $h(X)$ in a sequent imposes some equality constraints on the counts of q_0 and q_1.

Fact (2) [interactions with new letters]

for $X \in Tp^*(P)$ and $\alpha, \alpha' \in \{q_0, q_1\} \cup \{\beta_k \mid p_k \in Pr\}$:

(a) $h(X), \alpha \vDash_{\leq'} \alpha'$ is impossible when X is not empty, unless $(\alpha, \alpha') = (q_1, q_0)$

(b) $h(X), \alpha \vDash_{\leq'} exp(q_1, \Pi)$ where $\Pi \neq \epsilon$ implies X is empty and $\alpha = q_1$

(c) $h(X), \alpha, exp(q_1, \Pi") \setminus \beta \vDash_{\leq'} \beta$, where $\beta \in \{\beta_k \mid p_k \in Pr\}$ implies X is empty and $\alpha = q_1$

Proof. The proof is technical, see Annex.

Fact (3) [chains of type-raise]

if $exp(q_1, \Pi') \mathrel{\vDash_{\leq'}} exp(q_1, \Pi'')$ then $Num(\Pi') \subseteq Num(\Pi'')$

Proof. We show a simpler version when $\Pi' = \beta_{k_1}$ (the general case follows from type-raise properties $A \mathrel{\vDash_{\leq'}} exp(A, \Pi)$; also if Π' is empty, the assertion is obvious).

We proceed by induction on the length of Π'' and consider $exp(q_1, \beta_{k_1}) \mathrel{\vDash_{\leq'}} exp(q_1, \Pi'')$, that is $\beta_{k_1} / (q_1 \setminus \beta_{k_1}) \mathrel{\vDash_{\leq'}} exp(q_1, \Pi'')$. The case Π'' empty is impossible ; we write $\Pi'' = \Pi_2.\beta_{k_2}$; the sequent is $\beta_{k_1} / (q_1 \setminus \beta_{k_1})$, $exp(q_1, \Pi_2) \setminus \beta_{k_2} \mathrel{\vDash_{\leq'}} \beta_{k_2}$; the end of the derivation has two possibilities:

$$\frac{\boxed{\beta_{k_1} / (q_1 \setminus \beta_{k_1}) \mathrel{\vDash_{\leq'}} exp(q_1, \Pi_2)} \quad \beta_{k_2} \mathrel{\vDash_{\leq'}} \beta_{k_2}}{\beta_{k_1} / (q_1 \setminus \beta_{k_1}), \, exp(q_1, \Pi_2) \setminus \beta_{k_2} \mathrel{\vDash_{\leq'}} \beta_{k_2}}$$

we get in this case the assertion by rec. : $k_1 \in Num(\Pi_2)$ ($\subseteq Num(\Pi'')$)

or

$$\frac{exp(q_1, \Pi_2) \setminus \beta_{k_2} \mathrel{\vDash_{\leq'}} (q_1 \setminus \beta_{k_1}) \quad \boxed{\beta_{k_1} \mathrel{\vDash_{\leq'}} \beta_{k_2}}}{\beta_{k_1} / (q_1 \setminus \beta_{k_1}), \, exp(q_1, \Pi_2) \setminus \beta_{k_2} \mathrel{\vDash_{\leq'}} \beta_{k_2}}$$

From which we get the assertion : $k_1 = k_2$ ($\in Num(\Pi'')$).

Main Lemma

(main) if $h(X) \mathrel{\vDash_{\leq'}} h(Y)$ then $X \mathrel{\vDash_{\leq}} Y$ \hfill (where X and Y in $Tp^+(P)$).

Sketch of Proof. We distinguish several cases, depending on the form of Y and of $h(Y)$, and proceed by (joined) induction on the total number of connectives in X, Y :

- for cases where Y is primitive, we recall that $P = Pr \cup Pr'$, where Pr is a connex component and $\leq'$ has no postulate on Pr ; there are two subcases (detailed later) depending on $p_i \in Pr$ or $p_i \in Pr'$:
  (o) for $p_i \in Pr'$ and $X \in Tp^+(P)$: $h(X) \mathrel{\vDash_{\leq'}} p_i$ implies $X \mathrel{\vDash_{\leq}} p_i$
  (i) if $h(X) \mathrel{\vDash_{\leq'}} q_0 / exp(q_1, \Pi')$ then $\forall k \in Num(\Pi') : X \mathrel{\vDash_{\leq}} p_k$
  where Π' is a sequence of β_{k_j} (this corresponds to $Y = p_i \in Pr$)
  we will show (ii) an equivalent version of (i) as follows :
  (ii) if $h(X)$, $exp(q_1, \Pi') \mathrel{\vDash_{\leq'}} q_0$ then $\forall k \in Num(\Pi') : X \vdash p_k$
  (see proof details after for (o) (i) (ii))
- (iii) if $h(Y)$ is of the form $h(D / C)$ and $Y = D / C$
  $h(X) \mathrel{\vDash_{\leq'}} h(Y)$ iff $h(X), h(C) \mathrel{\vDash_{\leq'}} h(D)$
  by induction $X, C \mathrel{\vDash_{\leq}} D$ \hfill hence $X \mathrel{\vDash_{\leq}} D / C$ by the $/R$ right rule
- (iv) if $h(Y)$ of the form $h(C \setminus D)$, $Y = C \setminus D$, the case is similar to (iii)
- (v) if $h(Y)$ of the form $h(C \bullet D)$ *(see proof details after, partly similar to (o))*

Main Lemma Part (o)
(o) for $p_i \in Pr'$ and $X \in Tp^+(P)$: $h(X) \mathrel{\vert\approx'} p_i$ implies $X \mathrel{\vert\approx} p_i$

Proof Details: we discuss on the derivation ending for $h(X) \mathrel{\vert\approx'} p_i$:

- if this is an axiom $h(X) = p_i = h(p_i) = X$
- if this is inferred from a postulate on Pr', $p_j \leq p_i$ then also $X = p_j \mathrel{\vert\approx} p_i$
- if $/L$ is the last rule, there are two cases
  - if the rule introduces $h(B) / h(A)$, s. t. X has the form
    $X = \Delta, B / A, \Gamma, \Delta'$
    $$\frac{h(\Gamma)\mathrel{\vert\approx'}h(A) \qquad h(\Delta)\,h(B)\,h(\Delta')\mathrel{\vert\approx'}p_i}{h(\Delta), h(B) / h(A), h(\Gamma), h(\Delta')\mathrel{\vert\approx'}p_i} \quad \boxed{\text{by rec. } (main+(o)):} \quad \frac{\Gamma\mathrel{\vert\approx}A \qquad \Delta, B, \Delta'\mathrel{\vert\approx}p_i}{\text{by rule } /L : \Delta, B / A, \Gamma, \Delta'\mathrel{\vert\approx}p_i}$$
  - if the rule introduces $h(p_i) = q_0 / exp(q_1, \Pi')$,
    the end is of the form $\dfrac{\boxed{h(\Gamma) \mathrel{\vert\approx'} exp(q_1, \Pi')} \qquad h(\Delta),\, q_0,\, h(\Delta')\vdash' p_i}{h(\Delta),\, q_0 / exp(q_1, \Pi'),\, h(\Gamma),\, h(\Delta')\mathrel{\vert\approx'}p_i}$
    which is impossible according to $\boxed{\text{Fact (1)}}$
- if $\backslash L$ is the last rule, the case is similar to the first subcase for $/L$ above
- if the last rule is $\bullet L$ introducing $h(A)\bullet h(B)$, we apply rec. (o) to the antecedent, then $\bullet L$
- the right rules are impossible   ∎

Main Lemma Part (v) for $X \in Tp^+(P)$: $h(X) \mathrel{\vert\approx'} h(C_1 \bullet C_2)$ implies $X \mathrel{\vert\approx} C_1 \bullet C_2$

Proof Details: we discuss on the derivation ending for $h(X) \mathrel{\vert\approx'} h(Y)$ where $Y = C_1 \bullet C_2$:

- this cannot be an axiom, a postulate, $/R$, or $\backslash R$
- if $/L$ is the last rule, there are two cases
  - if the rule introduces $h(B) / h(A)$, s. t. X has the form
    $X = \Delta, B / A, \Gamma, \Delta'$
    $$\frac{h(\Gamma)\mathrel{\vert\approx'}h(A) \qquad h(\Delta)\,h(B)\,h(\Delta')\mathrel{\vert\approx'}h(Y)}{h(\Delta), h(B) / h(A), h(\Gamma), h(\Delta')\mathrel{\vert\approx'}h(Y)} \quad \boxed{\text{by rec. } (main+(v)):} \quad \frac{\Gamma\mathrel{\vert\approx}A \qquad \Delta\,B\,\Delta'\mathrel{\vert\approx}Y}{\text{by rule } /L : \Delta, B / A, \Gamma, \Delta'\mathrel{\vert\approx}Y}$$
  - if the rule introduces $h(p_i) = q_0 / exp(q_1, \Pi')$,
    the end is of the form $\dfrac{\boxed{h(\Gamma) \mathrel{\vert\approx'} exp(q_1, \Pi')} \qquad h(\Delta),\, q_0,\, h(\Delta')\vdash' h(Y)}{h(\Delta),\, q_0 / exp(q_1, \Pi'),\, h(\Gamma),\, h(\Delta')\mathrel{\vert\approx'}h(Y)}$
    which is impossible according to $\boxed{\text{Fact (1)}}$
- if $\backslash L$ is the last rule, the case is similar to the first subcase for $/L$ above
- if the last rule is $\bullet L$ introducing $h(A)\bullet h(B)$, we apply rec. (v) to the antecedent, then $\bullet L$
- if the last rule is $\bullet R$ introducing $h(C_1)\bullet h(C_2)$ then X has the form Δ, Δ', such that :

$$\frac{h(\Delta)\mathrel{\vert\approx'}h(C_1) \qquad h(\Delta')\mathrel{\vert\approx'}h(C_2)}{h(\Delta), h(\Delta')\mathrel{\vert\approx'}h(Y)} \qquad \boxed{\text{by rec. } (main):} \quad \frac{\Delta\mathrel{\vert\approx}C_1 \qquad \Delta'\mathrel{\vert\approx}C_2}{\text{by rule } \bullet R : \Delta, \Delta'\mathrel{\vert\approx}Y}$$

∎

Main Lemma Part (ii) if $h(X)$, $exp(q_1, \Pi') \models_{\leq'} q_0$ then $\forall k \in Num(\Pi') : X \vdash p_k$

Proof Details : we show a simpler version when $\Pi' = \beta_{k_1}$ (the general case follows from type-raise properties $A \models_{\leq'} exp(A, \Pi)$, and if Π' is empty, the assertion is obvious). The sequent is $h(X)$, $\beta_{k_1} / (q_1 \setminus \beta_{k_1}) \models_{\leq'} q_0$:

- if $/ L$ is the last rule, there are two cases (it cannot introduce $exp(q_1, \Pi')$ being rightmost)
  - if the rule introduces $\boxed{h(B) / h(A)}$, s. t. X has the form $X = \Delta, B / A, \Gamma, \Delta'$ there are two subcases :

$$\frac{h(\Gamma)\models_{\leq'}h(A) \quad h(\Delta), h(B), h(\Delta'), exp(q_1,\Pi')\models_{\leq'}q_0}{h(\Delta), h(B) / h(A), h(\Gamma), h(\Delta'), exp(q_1,\Pi')\models_{\leq'}q_0}$$

$\qquad\qquad\qquad\qquad$ by $\boxed{\text{by global rec + rec (ii)}}$:

$$\frac{\Gamma\vdash A \quad \Delta, B, \Delta'\vdash p_{k_1}}{\text{by rule } /L : \Delta, B / A, \Gamma, \Delta'\vdash p_{k_1}}$$

$\qquad$ or

$$\frac{h(\Gamma),h(\Delta'), exp(q_1,\Pi')\models_{\leq}h(A) \quad \boxed{h(\Delta), h(B)\models_{\leq'}q_0}}{h(\Delta), h(B) / h(A), h(\Gamma), h(\Delta'), exp(q_1,\Pi')\models_{\leq'}q_0}$$

$\qquad\qquad$ impossible, see $\boxed{\text{Fact (1)}}$:

  - if the rule introduces $\boxed{h(p_i) = q_0 / exp(q_1, \Pi'')}$, in $h(X)$, s. t. X has the form $X = \Delta, p_i, \Gamma, \Delta'$

- if $\dfrac{\boxed{h(\Gamma)\models_{\leq'} exp(q_1,\Pi'')} \quad h(\Delta), q_0, h(\Delta'), exp(q_1,\Pi')\vdash q_0}{h(\Delta), q_0 / exp(q_1,\Pi''), h(\Gamma), h(\Delta'), exp(q_1,\Pi')\vdash q_0}$ impossible, see $\boxed{\text{Fact (1)}}$

- if $\dfrac{\boxed{h(\Gamma),h(\Delta'),exp(q_1,\Pi')\models_{\leq'}exp(q_1,\Pi'')} \quad h(\Delta),q_0\models_{\leq'} q_0}{h(\Delta),q_0 / exp(q_1,\Pi''),h(\Gamma),h(\Delta'),exp(q_1,\Pi')\models_{\leq'}q_0}$ Γ, Δ', Δ are empty by Fact (2)

$\qquad\qquad$ and $Num(\Pi') = \{\beta_{k_1}\} \subseteq Num(\Pi'')$ by Fact (3),

$\qquad\qquad\qquad$ we get $X = p_i \leq p_{k_1}$

- if $\setminus L$ is the last rule, it introduces $h(A) \setminus h(B)$, similar to the first subcase for $/L$ above

$$\frac{h(\Gamma)\models_{\leq'}h(A) \quad h(\Delta), h(B), h(\Delta'), exp(q_1,\Pi')\models_{\leq'}q_0}{h(\Delta), h(\Gamma), h(A) \setminus h(B), h(\Delta'), exp(q_1,\Pi')\models_{\leq'}q_0}$$

$\qquad\qquad\qquad\qquad$ $\boxed{\text{by global rec + rec (ii)}}$:

$$\frac{\Gamma\vdash A \quad \Delta, B, \Delta'\vdash p_{k_1}}{\text{by rule } \setminus L : \Delta, \Gamma, A \setminus B, \Delta'\vdash p_{k_1}}$$

- if the last rule is $\bullet L$ introducing $h(A)\bullet h(B)$, we apply rec. (ii) to the antecedent, then $\bullet L$
- the right rules and the axiom rule are impossible $\qquad\qquad\qquad$ ∎

7 Conclusion and Discussion

Former Work in Pregroups. The order on primitive types has been introduced in PG to simplify the calculus for simple types. The consequence is that PG is not fully lexicalized. We had proven in [1] that this restriction is not so important because a PG using an order on primitive types can be transformed into a PG based on a simple free pregroup using a pregroup morphism, s.t. :

- *its size* is bound by the size of the initial PG times the number of primitive types (times a constant which is approximatively 4),
- moreover, this transformation does not change the number of types that are assigned to a word (a k-valued PG is transformed into a k-valued PG).

The AB case. In constrast to pregroups (and L) rigid AB-grammars with basic postulates are more expressive than rigid AB-grammars as shown by the following language ; let $L = \{a, ab\}$, and $G = \{a \mapsto x, b \mapsto y\}$ where $x, y \in Tp(Pr)$, suppose T_1, T_2 are parse trees using G, for a and ab respectively

- in the absence of postulates, we have from T_1 and T_2 : $y = x \setminus x$ in which case abb should also belong to the language, contradiction;
- if basic postulates are allowed, we can take $x = S_1$ and then $y = S_1 \setminus S$, with $S_1 \leq S$, generating $L = \{a, ab\}$.

$L = \{a, ab\}$ cannot be handled by a rigid AB-grammar without postulate, whereas it is with postulates.

A similar situation might hold for extensions based on AB, such as Categorial Dependency Grammars (CDG).

In L and Related Formalisms. The work in this paper shows a result similar to [1], for L extended with an order on primitive types. The result holds for both versions with or without product. *A similar result should hold for NL and some other related calculi*, but it does not hold for AB as shown above.

Such a simulation result aims at clarifying properties of the extended calculus, in particular in terms of generative capacity and hierarchies of grammars. Another interest of the extended calculus is to allow some parallels in grammar design (type assignments, acquisition methods) between both frameworks (pregroups and (L)).

References

1. Béchet, D., Foret, A.: Fully lexicalized pregroup grammars. In: Leivant, D., de Queiroz, R. (eds.) WoLLIC 2007. LNCS, vol. 4576, pp. 12–25. Springer, Heidelberg (2007)
2. Buszkowski, W.: Some decision problems in the theory of syntactic categories. Zeitschrift f. Math. Logik u. Grundlagen der Mathematik 28, 539–548 (1982)
3. Buszkowski, W.: Mathematical linguistics and proof theory. In: van Benthem, J., ter Meulen, A. (eds.) Handbook of Logic and Language, ch. 12, pp. 683–736. North-Holland Elsevier, Amsterdam (1997)
4. Buszkowski, W.: Type Logics in Grammar. Trends in logic. Studia Logica Library, vol. 21, pp. 321–366. Springer (2003)
5. Buszkowski, W.: Lambek calculus with nonlogical axioms. In: Language and Grammar, Studies in Mathematical Linguistics and Natural Language, pp. 77–93. CSLI Publications (2005)
6. Lambek, J.: The mathematics of sentence structure. American Mathematical Monthly 65 (1958)
7. Lambek, J.: Type grammars revisited. In: Lecomte, A., Perrier, G., Lamarche, F. (eds.) LACL 1997. LNCS (LNAI), vol. 1582, pp. 1–27. Springer, Heidelberg (1999)
8. Moot, R., Retoré, C.: The Logic of Categorial Grammars. LNCS, vol. 6850, pp. 23–63. Springer, Heidelberg (2012)

Annex : Details of Proofs

Proof sketch for the Simulation based on Duplication. We write $\vdash_{\preceq}$ for $\vdash$ in $L(\Phi_{\leq})$, and we consider a version where axioms $A \vdash A$ are such that A is primitive.

In full L. We show by induction on the length of derivation of $\Gamma \vdash_{\preceq} Z$ in $L(\Phi_{\leq})$ without cut, that more generally : if $\Gamma \vdash_{\preceq} Z$ and $\Gamma = X_1, \ldots X_n$ then
(a) if Z is primitive then $\exists X_1' \in Dupl_{\leq}^{\uparrow}(X_1) \ldots \exists X_n' \in Dupl_{\leq}^{\uparrow}(X_n)$ $X_1', \ldots X_n' \vdash Z$ in L (without postulates)
(b) $\exists X_1' \in Dupl_{\leq}^{\uparrow}(X_1) \ldots \exists X_n' \in Dupl_{\leq}^{\uparrow}(X_n)$ and $\exists Z^- \in Dupl_{\leq}^{\downarrow}(Z)$ such that : $X_1', \ldots X_n' \vdash Z^-$ in L (without postulates)
We first show (b) separately :
- in the axiom case $p \vdash p$ in $L(\Phi_{\leq})$: we take $p \vdash p$ in L
- in the non-logical axiom case $p \vdash q$, where $p \leq q$, in $L(\Phi_{\leq})$: we take $q \vdash q$ in L
- for a rule introducing $/$, $\backslash$ or $\bullet$ on the right (b) is shown easily by rec. on the antecedent then the same rule in L, because for $A' \in Dupl_{\leq}^{\uparrow}(A)$ and $B^- \in Dupl_{\leq}^{\downarrow}(B)$, we get $A' \backslash B^- \in Dupl_{\leq}^{\downarrow}(A \backslash B)$ and $B^- / A' \in Dupl_{\leq}^{\downarrow}(B / A)$ and for $A^- \in Dupl_{\leq}^{\downarrow}(A)$ and $B^- \in Dupl_{\leq}^{\downarrow}(B)$, we get $A^- \bullet B^- \in Dupl_{\leq}^{\downarrow}(A \bullet B)$
- we detail the $/L$ case :

$$\cfrac{\Gamma \vdash_{\preceq} A \quad \Delta_1, B, \Delta_2 \vdash_{\preceq} Z}{\Delta_1, B / A, \Gamma, \Delta_2 \vdash_{\preceq} Z} /L \quad \cfrac{\begin{array}{c}\text{by rec. } \exists \Gamma' \in Dupl_{\leq}^{\uparrow}(\Gamma),\\ \exists A^- \in Dupl_{\leq}^{\downarrow}(A), \exists Z^- \in Dupl_{\leq}^{\downarrow}(Z)\\ \exists \Delta_i' \in Dupl_{\leq}^{\uparrow}(\Delta_i), \exists B' \in Dupl_{\leq}^{\uparrow}(B)\\ \Gamma' \vdash_{\preceq} A^- \quad \bar{\Delta}_1', B', \Delta_2' \vdash_{\preceq} Z^-\end{array}}{\Delta_1', B' / A^-, \Gamma, \Delta_2' \vdash_{\preceq} Z^-} /L$$
$$\text{where } B' / A^- \in Dupl_{\leq}^{\uparrow}(B / A)$$

- the other cases follow similarly the rule and structure without difficulty.
We now show (a) using (b), we suppose Z is primitive :
- the axiom cases are similar to (b)
- a right rule is not possible
- we detail the $/L$ case :

$$\cfrac{\Gamma \vdash_{\preceq} A \quad \Delta_1, B, \Delta_2 \vdash_{\preceq} Z}{\Delta_1, B / A, \Gamma, \Delta_2 \vdash_{\preceq} Z} /L \quad \cfrac{\begin{array}{c}\text{by rec. (b)} \exists \Gamma' \in Dupl_{\leq}^{\uparrow}(\Gamma), \exists A^- \in Dupl_{\leq}^{\downarrow}(A)\\ \text{by rec. (a)} \exists \Delta_i' \in Dupl_{\leq}^{\uparrow}(\Delta_i), \exists B' \in Dupl_{\leq}^{\uparrow}(B)\\ \Gamma' \vdash_{\preceq} A^- \quad \Delta_1', B', \Delta_2' \vdash_{\preceq} Z\end{array}}{\Delta_1', B' / A^-, \Gamma, \Delta_2' \vdash_{\preceq} Z} /L$$
$$\text{where } B' / A^- \in Dupl_{\leq}^{\uparrow}(B / A)$$

- the other cases follow similarly the rule and structure without difficulty. ∎

Proof of Fact (2) by joined induction for (abc) on the derivation. We consider (for (bc)) a version of the calculus where axioms are on primitives such that for a deduction of $\Gamma \vdash Y \,/\, Z$, there is a deduction of not greater length for $\Gamma, Z \vdash Y$.
Part (2)(a) : we first consider $h(X)\,\alpha \underset{\leq}{\vdash}_{'} \alpha'$ and the last rule in a derivation :

- if this is an axiom, then X is empty
- if $/L$ is the last rule, there are two cases (with subcases)

  - if the rule introduces $\boxed{h(B)\,/\,h(A)}$, s. t. X is $\Delta,\ B\,/\,A,\ \Gamma,\ \Delta'$

$$\frac{h(\Gamma)\underset{\leq}{\vdash}_{'}h(A) \qquad h(\Delta),\,h(B),\,h(\Delta'),\,\alpha\underset{\leq}{\vdash}_{'}\alpha'}{h(\Delta),\,h(B)\,/\,h(A),\,h(\Gamma),\,h(\Delta'),\,\alpha\underset{\leq}{\vdash}_{'}\alpha'} \qquad \boxed{\text{by rec. (a), } h(B) \text{ being not empty}}$$

  or

$$\frac{h(\Gamma,\Delta'),\,\alpha\underset{\leq}{\vdash}_{'}h(A) \qquad h(\Delta),\,h(B)\underset{\leq}{\vdash}_{'}\alpha'}{h(\Delta),\,h(B)\,/\,h(A),\,h(\Gamma),\,h(\Delta'),\,\alpha\underset{\leq}{\vdash}_{'}\alpha'} \qquad \text{impossible, see } \boxed{\text{Fact}(1)}$$

  - if the rule introduces $\boxed{h(p_i) = q_0\,/\,exp(q_1,\Pi'')}$, in $h(X)$ s. t. X is $\Delta,\ p_i,\ \Gamma,\ \Delta'$

$$\frac{\boxed{h(\Gamma)\underset{\leq}{\vdash}_{'}exp(q_1,\Pi'')} \qquad h(\Delta),\,q_0,\,h(\Delta'),\,\alpha\underset{\leq}{\vdash}_{'}\alpha'}{h(\Delta),\,q_0\,/\,exp(q_1,\Pi''),\,h(\Gamma),\,h(\Delta'),\,\alpha\underset{\leq}{\vdash}_{'}\alpha'} \qquad \text{impossible, see } \boxed{\text{Fact (1)}}$$

    or

$$\frac{h(\Gamma,\,\Delta'),\,\alpha\underset{\leq}{\vdash}_{'}exp(q_1,\Pi'') \qquad \boxed{h(\Delta),q_0\underset{\leq}{\vdash}_{'}\alpha'}}{h(\Delta),\,q_0\,/\,exp(q_1,\Pi''),\,h(\Gamma\,\Delta'),\,\alpha\underset{\leq}{\vdash}_{'}\alpha'} \qquad \text{we get} : \alpha \in \{q_0,q_1\} \text{ by } \boxed{\text{Fact (1)}}$$

    but then $\boxed{\text{by rec. (2)a}}$ Δ is empty and $\alpha' = q_0$

    also $\boxed{\text{by (b) and rec. (2)a}}$ Γ, Δ' is empty and $\alpha = q_1$, thus (a).

- if $\setminus L$ is the last rule, it introduces $\boxed{h(A)\setminus h(B)}$, similar to the first subcase for $/\,L$ above

$$\frac{h(\Gamma)\underset{\leq}{\vdash}_{'}h(A) \qquad h(\Delta),\,h(B),\,h(\Delta'),\,\alpha\underset{\leq}{\vdash}_{'}\alpha'}{h(\Delta),\,h(\Gamma),\,h(A)\setminus h(B),\,h(\Delta')\,\alpha\underset{\leq}{\vdash}_{'}\alpha'} \qquad \boxed{\text{by rec. (a), } h(B) \text{ being not empty}}$$

- if $\bullet L$ is the last rule, it introduces $\boxed{h(A)\bullet h(B)}$, we apply rec. (a) to the antecedent

Part (2)(bc) : we then consider (b) $h(X), \alpha \vdash_{\leq'} exp(q_1, \Pi)$, suppose $\Pi = \beta.\Pi''$ and its equivalent form (c) if $h(X), \alpha, exp(q_1, \Pi'') \backslash \beta \vdash_{\leq'} \beta$ (where Π'' may be ϵ) then X is empty and $\alpha = q_1$. We discuss the last rule in a derivation for (c).

- if this is an axiom, this is impossible. The right rules are also not possible for (c).
- if $/L$ is the last rule, there are two cases (with subcases)
  - if the rule introduces $\boxed{h(B) \,/\, h(A)}$, s. t. X is $\Delta, B \,/\, A, \Gamma, \Delta'$

$$\frac{h(\Gamma) \vdash_{\leq'} h(A) \qquad h(\Delta), h(B), h(\Delta'), \alpha, exp(q_1, \Pi'') \backslash \beta \vdash_{\leq'} \beta}{h(\Delta), h(B) \,/\, h(A), h(\Gamma), h(\Delta'), \alpha, exp(q_1, \Pi'') \backslash \beta \vdash_{\leq'} \beta}$$

$\boxed{\text{impossible by rec. (c), h(B) being not empty}}$

or

$$\frac{h(\Gamma), h(\Delta'), \alpha \vdash_{\leq'} h(A) \qquad h(\Delta), h(B), exp(q_1, \Pi'') \backslash \beta \vdash_{\leq'} \beta}{h(\Delta), h(B) \,/\, h(A), h(\Gamma), h(\Delta'), \alpha, exp(q_1, \Pi'') \backslash \beta \vdash_{\leq'} \beta}$$

impossible, see $\boxed{\text{Fact}(1)}$

or

$$\frac{h(\Gamma), h(\Delta'), \alpha, exp(q_1, \Pi'') \backslash \beta \vdash_{\leq'} h(A) \qquad h(\Delta), h(B) \vdash_{\leq'} \beta}{h(\Delta), h(B) \,/\, h(A), h(\Gamma), h(\Delta'), \alpha, exp(q_1, \Pi'') \backslash \beta \vdash_{\leq'} \beta}$$

impossible, see $\boxed{\text{Fact}(1)}$

  - if the rule introduces $\boxed{h(p_i) = q_0 \,/\, exp(q_1, \Pi_i)}$, in $h(X)$, where $X = \Delta, p_i, \Gamma, \Delta'$

$$\frac{h(\Gamma) \vdash_{\leq'} exp(q_1, \Pi_i) \qquad h(\Delta), q_0, h(\Delta'), \alpha, exp(q_1, \Pi'') \backslash \beta \vdash_{\leq'} \beta}{h(\Delta), q_0 \,/\, exp(q_1, \Pi_i), h(\Gamma), h(\Delta'), \alpha, exp(q_1, \Pi'') \backslash \beta \vdash_{\leq'} \beta}$$

impossible, see $\boxed{\text{Fact (1)}}$

or

$$\frac{h(\Gamma, \Delta'), \alpha \vdash_{\leq'} exp(q_1, \Pi_i) \qquad h(\Delta), q_0, exp(q_1, \Pi'') \backslash \beta \vdash_{\leq'} \beta}{h(\Delta), q_0 \,/\, exp(q_1, \Pi_i), h(\Gamma, \Delta'), \alpha, exp(q_1, \Pi'') \backslash \beta \vdash_{\leq'} \beta}$$

impossible, by rec. (c)

or

$$\frac{h(\Gamma, \Delta'), \alpha, exp(q_1, \Pi'') \backslash \beta \vdash_{\leq'} exp(q_1, \Pi_i) \qquad h(\Delta), q_0 \vdash_{\leq'} \beta}{h(\Delta), q_0 \,/\, exp(q_1, \Pi_i), h(\Gamma, \Delta'), \alpha, exp(q_1, \Pi'') \backslash \beta \vdash_{\leq'} \beta}$$

impossible, see $\boxed{\text{Fact (1)}}$

- if $\backslash L$ is the last rule,
  - if it introduces $\boxed{h(A) \backslash h(B)}$ in $h(X)$: similar to the first subcase for $/L$

$$\frac{h(\Gamma) \vdash_{\leq'} h(A) \qquad h(\Delta), h(B), h(\Delta'), \alpha, exp(q_1, \Pi'') \backslash \beta \vdash_{\leq'} \beta}{h(\Delta), h(\Gamma), h(A) \backslash h(B), h(\Delta'), \alpha, exp(q_1, \Pi'') \backslash \beta \vdash_{\leq'} \beta}$$

$\boxed{\text{by rec. (c)}}$, $h(B)$ being not empty

  - if the rule introduces $\boxed{exp(q_1, \Pi'') \backslash \beta}$ with $X = \Delta, \Gamma$

$$\frac{h(\Gamma), \alpha \vdash_{\leq'} exp(q_1, \Pi'') \qquad h(\Delta), \beta \vdash_{\leq'} \beta}{h(\Delta), h(\Gamma), \alpha, exp(q_1, \Pi'') \backslash \beta \vdash_{\leq'} \beta}$$

$\boxed{\text{by rec.(2bc)}}$ Γ is empty, and $\alpha = q_1$
$\boxed{\text{by rec (2a)}}$, Δ is also empty

- if $\bullet L$ is the last rule, it introduces $\boxed{h(A) \bullet h(B)}$, we apply rec. (c) to the antecedent, the case is impossible

■

Classical Structures Based on Unitaries

Peter Hines

University of York, UK
`peter.hines@york.ac.uk`

Abstract. Starting from the observation that distinct notions of copying have arisen in different categorical fields (logic and computation, contrasted with quantum mechanics) this paper addresses the question of when, or whether, they may coincide.

Provided all definitions are strict in the categorical sense, we show that this can never be the case. However, allowing for the defining axioms to be taken up to canonical isomorphism, a close connection between the *classical structures* of categorical quantum mechanics, and the categorical property of *self-similarity* familiar from logical and computational models becomes apparent.

The required canonical isomorphisms are non-trivial, and mix both typed (multi-object) and untyped (single-object) tensors and structural isomorphisms; we give coherence results that justify this approach.

We then give a class of examples where distinct self-similar structures at an object determine distinct matrix representations of arrows, in the same way as classical structures determine matrix representations in Hilbert space. We also give analogues of familiar notions from linear algebra in this setting such as changes of basis, and diagonalisation.

Keywords: Category theory, self-similarity, categorical quantum mechanics, classical structures, untyped systems.

Dedicated to J. Lambek, on the occasion of his 90th birthday.
I hope to be as productive as he currently is when I am half his age.

1   Introduction

1.1   Background

Analogies are often drawn between quantum mechanics and linear logic [8], simply based on shared structural properties. In both cases, the structural operation of *copying* or *cloning* is forbidden [29], as is the structural operation of *contraction* or *deletion* [26] (this is deletion *against a copy*, and should be strongly distinguished from *erasure* or the deletion of a single copy).

A deeper investigation reveals that the no-cloning and no-deleting principles of quantum mechanics are about *arbitrary* quantum states, or states for which no information is known. Indeed, when a quantum state is known to be a member of some given orthonormal basis (usually taken to be the *computation basis* of

C. Casadio et al. (Eds.): Lambek Festschrift, LNCS 8222, pp. 188–210, 2014.
© Springer-Verlag Berlin Heidelberg 2014

quantum computing), limited forms of copying and deleting are possible — this is via the *fanout* primitive that plays an important role in quantum algorithms and protocols [30].

Similarly, in linear logic, the structural copying and contraction rules are not completely discarded (leading to 'substructural logics'), but are severely restricted via a typing system based on two modalities, !( ) and ?( ), commonly called 'of course' and 'why not'. An interesting approach to concrete models of these modalities was given in the Geometry of Interaction series of representations of linear logic [9,10], where they were derived, in an essentially untyped setting, by iterating bijections exhibiting *self-similarity* (the categorical identity $S \cong S \otimes S$). We refer to [14,18] for details of this construction from a categorical viewpoint.

In [4], the restricted notion of copying available in quantum information was used to give a new abstract characterisation of the notion of orthonormal basis in quantum mechanics, via a special form of Frobenius algebra within a category (see Section 3 for details). These are based on a paired monoid-comonoid structures in categories, satisfying additional conditions.

In [19], an apparent structural similarity between the *classical structures* of quantum mechanics, and the *self-similar structures* familiar from the Geometry of Interaction (and indeed, untyped computing and logical systems generally) was noted. Although the emphasis of [19] was primarily on models of meaning in linguistics and natural language processing, it also raised the interesting question of whether this correspondence is precise, or merely an analogy – this is the question addressed in this paper.

1.2   The Results of This Paper

This paper addresses the question of whether the *classical structures* used in categorical quantum mechanics (based on monoid co-monoid pairs with additional structure) can ever be built from unitary maps — can the monoid and co-monoid arrows be mutually inverse unitaries? From a simplistic perspective, the answer is negative (Corollary 2 and Corollary 3); however when we allow the defining conditions of a classical structure to be taken *up to canonical isomorphism*, not only is this possible, but the required conditions (at least, using the redefinition of [2]) may be satisfied by any pair of mutually inverse unitaries (Theorem 2) with the correct typing (i.e. exhibiting self-similarity) in a † monoidal category.

However, the required canonical isomorphisms are non-trivial, and mix 'typed' and 'untyped' (i.e. multi-object and single-object) monoidal tensors and canonical isomorphisms. We study these, and refer to [17] for general techniques that will reduce questions of coherence in this setting to the well-established coherence results found in [23].

We illustrate this connection with a concrete example, and show how in this setting, self-similar structures play an identical role to that played by classical structures in finite-dimensional Hilbert space — that of specifying and manipulating matrix representations. We also give analogues of notions such as 'changes of basis' and 'diagonalisation' in this setting.

2   Categorical Preliminaries

The general area of this paper is firmly within the field of † monoidal categories. However, due to the extremal settings we consider, we will frequently require monoidal categories without a unit object. We axiomatise these as follows:

Definition 1. *Let C be a category. We say that C is **semi-monoidal** when there exists a **tensor** $(\_\otimes\_) : C \times C \to C$ together with a natural indexed family of **associativity isomorphisms** $\{ \tau_{A,B,C} : A \otimes (B \otimes C) \to (A \otimes B) \otimes C\}_{A,B,C \in Ob(C)}$ satisfying MacLane's **pentagon condition** $(\tau_{A,B,C} \otimes 1_D)\tau_{A,B\otimes C,D}(1_A \otimes \tau_{B,C,D}) = \tau_{A\otimes B,C,D}\tau_{A,B,C\otimes D}$.*

*When there also exists a natural object-indexed natural family of **symmetry isomorphisms** $\{\sigma_{X,Y} : X \otimes Y \to Y \otimes X\}_{X,Y \in Ob(C)}$ satisfying MacLane's **hexagon condition** $\tau_{A,B,C}\sigma_{A\otimes B,C}\tau_{A,B,C} = (\sigma_{A,C} \otimes 1_B)\tau_{A,C,B}(1_A \otimes \sigma_{B,C})$ we say that $(C, \otimes, \tau, \sigma)$ is a **symmetric semi-monoidal category**. A semi-monoidal category $(C, \otimes, \tau_{\_,\_,\_})$ is called **strictly associative** when $\tau_{A,B,C}$ is an identity arrow[1], for all $A, B, C \in Ob(C)$. A functor $\Gamma : C \to D$ between two semi-monoidal categories $(C, \otimes_C)$ and $(D, \otimes_D)$ is called (strictly) **semi-monoidal** when $\Gamma(f \otimes_C g) = \Gamma(f) \otimes_D \Gamma(g)$. A semi-monoidal category $(C, \otimes)$ is called **monoidal** when there exists a **unit object** $I \in Ob(C)$, together with, for all objects $A \in Ob(C)$, distinguished isomorphisms $\lambda_A : I \otimes A \to A$ and $\rho_A : A \otimes I \to A$ satisfying MacLane's **triangle condition** $1_U \otimes \lambda_V = (\rho_U \otimes 1_V)\tau_{U,I,V}$ for all $U, V \in Ob(C)$.*

*A **dagger** on a category C is simply a duality that is the identity on objects; that is, a contravariant endofunctor $(\ )^\dagger : C \to C$ satisfying $(1_A)^\dagger = 1_A$ and $\left((f)^\dagger\right)^\dagger = f$, for all $A \in Ob(C)$ and $f \in C(A, B)$. An arrow $U \in C(X, Y)$ is called **unitary** when it is an isomorphism with inverse given by $U^{-1} = U^\dagger \in C(Y, X)$.*

When C has a (semi-) monoidal tensor $\_ \otimes \_ : C \times C \to C$, we say that $(C, \otimes)$ is † (semi-) monoidal when $(\ )^\dagger$ is a (semi-) monoidal functor, and all canonical isomorphisms are unitary.

Remark 1. **Coherence for semi-monoidal categories** A close reading of [23] will demonstrate that MacLane's coherence theorems for associativity and commutativity are equally applicable in the presence or absence of a unit object. The theory of Saavedra units [20] also demonstrates that the properties of the unit object are independent of other categorical properties (including associativity). Motivated by this, we give a simple method of adjoining a strict unit object to a semi-monoidal category that is left-inverse to the obvious forgetful functor.

Definition 2. *Let $(C, \otimes)$ be a semi-monoidal category. We define its **unit augmentation** to be the monoidal category given by the following procedure: We*

---

[1] This is not implied by equality of objects $A \otimes (B \otimes C) = (A \otimes B) \otimes C$, for all $A, B, C \in Ob(C)$. Although MacLane's pentagon condition is trivially satisfied by identity arrows, naturality with respect to the tensor may fail. Examples we present later in this paper illustrate this phenomenon.

first take the coproduct of C with the trivial group $\{1_I\}$, considered as a single-object dagger category. We then extend the tensor of C to the whole of $C \coprod I$ by taking $\_ \otimes I = Id_{C \coprod I} = I \otimes \_$.

It is straightforward that the unit augmentation of a semi-monoidal category is a monoidal category; a full proof, should one be needed, is given as an appendix to [17]. Similarly, it is a triviality that if $(C, \otimes)$ is † semi-monoidal, then its unit augmentation is dagger monoidal.

The connection of the above procedure with MacLane's coherence theorems for associativity and commutativity should then be clear; any diagram that commutes in C also commutes in the unit augmentation; conversely any diagram (not containing the unit object) that commutes in the unit augmentation also commutes in C. Thus MacLane's coherence theorems (with the obvious exclusion of the unit object) also hold in the semi-monoidal and unitless cases.

3   Classical Structures and Their Interpretation

Classical structures were introduced in [4] as an abstract categorical interpretation of *orthonormal bases* in Hilbert spaces and the special role that these play in quantum mechanics (i.e. as sets of compatible disjoint measurement outcomes). This intuition was validated in [5], where it is proved that in the category of finite-dimensional Hilbert spaces, there is a bijective correspondence between orthonormal bases and classical structures. Mathematically, classical structures are symmetric † Frobenius algebras in † monoidal categories satisfying a simple additional condition.

Definition 3. *Let $(C, \otimes, I, (\ )^\dagger)$ be a strictly associative monoidal category. A **Frobenius algebra** consists of a co-monoid structure $(\Delta : S \to S \otimes S, \top : S \to I)$ and a monoid structure $(\nabla : S \otimes S \to S, \bot : I \to S)$ at the same object, where the monoid / comonoid pair satisfy the **Frobenius condition***

$$(1_S \otimes \nabla)(\Delta \otimes 1_S) = \Delta\nabla = (\nabla \otimes 1_S)(1_S \otimes \Delta)$$

Expanding out the definitions of a monoid and a comonoid structure gives:

- **(associativity)** $\nabla(1_S \otimes \nabla) = \nabla(\nabla \otimes 1_S) \in C(S \otimes S \otimes S, S)$.
- **(co-associativity)** $(\Delta \otimes 1_S)\Delta = (1_S \otimes \Delta)\Delta \in C(S, S \otimes S \otimes S)$.
- **(unit)** $\nabla(\bot \otimes 1_S) = \nabla(1_S \otimes \bot)$.
- **(co-unit)** $(\top \otimes_S)\Delta = 1_X \otimes \top)\Delta$.

*A Frobenius algebra $(S, \Delta, \nabla, \top, \bot)$ in a † monoidal category is called a **dagger Frobenius algebra** when it satisfies $\Delta^\dagger = \nabla$ and $\top^\dagger = \bot$.*

*Let $(C, \otimes)$ be a symmetric † monoidal category, with symmetry isomorphisms $\sigma_{X,Y} \in C(X \otimes Y, Y \otimes X)$. A † Frobenius algebra is called **commutative** when $\sigma_{S,S}\Delta = \Delta$, and hence $\nabla = \nabla\sigma_{S,S}$. A **classical structure** is then a commutative † Frobenius algebra satisfying the following additional condition:*

- *(**The classical structure condition**) $\Delta^\dagger$ is left-inverse to Δ, so $\nabla\Delta = 1_S$.*

Remark 2. The intuition behind a classical structure is that it describes related notions of copying and deleting (the comonoid and monoid structures). The underlying intuition is that, although arbitrary quantum states are subject to the no-cloning and no-deleting theorems [29,26], quantum states that are 'classical' (i.e. members of some fixed orthonormal basis – the 'computational basis' of quantum computation) can indeed be both copied and deleted (against a copy) using the fan-out maps and their inverses [30].

An aim of this paper is to compare such a notion of copying with a distinct notion of copying that arose independently in models of resource-sensitive logical and computational systems [9,10], and to demonstrate connections, via the theory of untyped categorical coherence, between these notions.

3.1   Classical Structures without Units

As noted in [2], when considering the theory of classical structures in arbitrary separable Hilbert spaces, is often necessary to generalise Definition 3 to the setting where unit objects are not considered – i.e. to lose the *unit* and *co-unit* axioms. We refer to [2] for a study of how much of the theory of [4] carries over to this more general setting, and give the following formal definition, which is a key definition of [2] in the strict, semi-monoidal setting:

Definition 4. *Let* $(\mathcal{C}, \otimes)$ *be a strictly associative semi-monoidal category. An* **Abramsky-Heunen (A.-H.) dagger Frobenius algebra** *consists of a triple* $(S \in Ob(\mathcal{C}), \Delta : S \to S \otimes S, \nabla = \Delta^{\dagger} : S \otimes S \to S)$ *satisfying*

1. **(associativity)** $\nabla(1_S \otimes \nabla) = (1 \otimes \nabla)\nabla \in \mathcal{C}(S \otimes S \otimes S, S).$
2. **(Frobenius condition)** $\Delta\nabla = (1_S \otimes \nabla)(\Delta \otimes 1_S) \in \mathcal{C}(S \otimes S, S \otimes S)$

An A-H † Frobenius algebra is an **A-H classical structure** *when* $(\mathcal{C}, \otimes)$ *is symmetric, and the following two conditions are satisfied:*

3. **(Classical structure condition)** $\nabla\Delta = 1_S,$
4. **(Commutativity)** $\sigma_{S,S}\Delta = \Delta.$

3.2   Classical Structures, and Identities Up to Isomorphism

It is notable that the definitions of the previous sections are based on strictly associative tensors. Consider the definition presented of a monoid within a category, $(1_A \otimes \nabla)\nabla = (\nabla \otimes 1_A)\nabla$. Drawing this as a commutative diagram

$$
\begin{array}{ccc}
A \otimes A & \xleftarrow{\ \nabla\ } A \xrightarrow{\ \nabla\ } & A \otimes A \\
{\scriptstyle \nabla \otimes 1_A}\downarrow & & \downarrow{\scriptstyle 1_A \otimes \nabla} \\
(A \otimes A) \otimes A & \xrightarrow[\ Id.\ ]{} & A \otimes (A \otimes A)
\end{array}
$$

demonstrates that this definition relies on the identity of objects $A \otimes (A \otimes A) = (A \otimes A) \otimes A$ required for strict associativity[2] in an essential way. The definition of a co-monoid requires the same identification of objects.

Similarly, the Frobenius condition $(1_A \otimes \nabla)(\Delta \otimes 1_A) = \Delta \nabla$ may be drawn as

$$
\begin{array}{ccc}
A \otimes A & \xrightarrow{\ \nabla\ } A \xrightarrow{\ \Delta\ } & A \otimes A \\
{\scriptstyle \Delta \otimes 1_A} \big\downarrow & & \big\uparrow {\scriptstyle 1_A \otimes \nabla} \\
A \otimes A \otimes A & \xrightarrow[\quad Id \quad]{} & A \otimes A \otimes A
\end{array}
$$

Remark 3. A significant feature of this paper is the relaxation of these strict identities, to allow the above definitions to be satisfied up to canonical isomorphisms. When making this generalisation, the choice of canonical isomorphisms seems to be straightforward enough; however, there are other possibilities. We take a more general view and allow the axioms above to be satisfied up to any canonical isomorphisms for which there exists a suitable theory of coherence.

4   Self-similarity, and † Self-similar Structures

By contrast with the strongly physical intuition behind classical structures, self-similar structures were introduced to study infinitary and type-free behaviour in logical and computational systems. Their definition is deceptively simple – they are simply a two-sided form of the 'classical structure' condition of Definition 3. The following definition is based on [14,15]:

Definition 5. *Let* $(\mathcal{C}, \otimes)$ *be a semi-monoidal category. A* **self-similar structure** $(S, \lhd, \rhd)$ *is an object* $S \in Ob(\mathcal{C})$, *together with two mutually inverse arrows*

- **(code)** $\lhd \in \mathcal{C}(S \otimes S, S)$.
- **(decode)** $\rhd \in \mathcal{C}(S, S \otimes S)$.

satisfying $\rhd \lhd = 1_{S \otimes S}$ *and* $\lhd \rhd = 1_S$. *A* **dagger self-similar structure** *is a self-similar structure in a* † *monoidal category with unitary code / decode arrows.*

Remark 4. Recall from Remark 2 the intuition of the classical structures of categorical quantum mechanics as a (restricted form of) copying and deleting that is applicable to computational basis states only. The very simple definition of a self-similar structure above is also clearly describing a notion of copying, albeit at the level of objects rather than arrows; simply, there are canonical arrows that provide isomorphisms between one copy of an object, and two copies of an object. A key theme of this paper is the relationship between these two notions of copying: whether the monoid / comonoid structure of an A.H. classical structure can also be a † self-similar structure, and whether a classical structure can also define a monoid / comonoid satisfying the Frobenius condition, &c.

---

[2] We emphasise that such identities of objects are a necessary, but not sufficient, condition for strict associativity of a tensor; see the footnote to Definition 1.

Instead of a simple yes/no answer, we will observe a close connection with the theory of categorical coherence and strictification. In the strict case, requiring unitarity of the monoid / comonoid arrows implies a collapse to the unit object (Corollaries 2 and 3), whereas, up to a certain set of (non-trivial) canonical isomorphisms, † self-similar structures do indeed satisfy the conditions for an A.-H. classical structure (Theorems 2 and 3).

We will first require many preliminary results on self-similar structures and their relationship with the theory of monoidal categories; we start by demonstrating that † self-similar structures are unique up to unique unitary:

Proposition 1. *Let* $(S, \lhd, \rhd)$ *be a* † *self-similar structure of a* † *semi-monoidal category* $(\mathcal{C}, \otimes, (\ )^\dagger)$. *Then*

1. *Given an arbitrary unitary* $U \in \mathcal{C}(S, S)$, *then* $(S, U\lhd, \rhd U^\dagger)$ *is also a* † *self-similar structure.*
2. *Given* † *self-similar structures* $(S, \lhd, \rhd)$ *and* $(S, \lhd', \rhd')$, *there exists a unique unitary* $U \in \mathcal{C}(S, S)$ *such that* $\lhd' = U\lhd \in \mathcal{C}(S \otimes S, S)$ *and* $\rhd' = \rhd U^\dagger \in \mathcal{C}(S, S \otimes S)$.

Proof.

1. Since U is unitary, $U \lhd \rhd U^\dagger = 1_S$ and $\rhd U^\dagger U \lhd = 1_{S \otimes S}$. Thus, as the composite of unitaries is itself unitary, $(S, U\lhd, \rhd U^\dagger)$ is a † self-similar structure.
2. We define $U = \lhd'\rhd \in \mathcal{C}(S, S)$, giving its inverse as $U^{-1} = \lhd\rhd' = U^\dagger$. The following diagrams then commute:

and $U = \lhd'\rhd$ is the unique unitary satisfying this condition.

4.1   The 'Internal' Monoidal Tensor of a Self-Similar Structure

We now demonstrate a close connection between self-similar structures and untyped (i.e. single-object) categorical properties:

Theorem 1. *Let* $(S, \lhd, \rhd)$ *be a self-similar structure of a semi-monoidal category* $(\mathcal{C}, \otimes, \tau_{\_,\_,\_})$. *Then the code / decode arrows determine a semi-monoidal tensor*

$$\_ \otimes_{\lhd\rhd} \_ : \mathcal{C}(S, S) \times \mathcal{C}(S, S) \to \mathcal{C}(S, S)$$

on the endomorphism monoid of S *given by, for all* $a, b \in \mathcal{C}(S, S)$,

$$a \otimes_{\lhd\rhd} b = \lhd(a \otimes b)\rhd \in \mathcal{C}(S, S)$$

The associativity isomorphism for this semi-monoidal structure is given by

$$\tau_{\lhd\rhd} = \lhd(\lhd \otimes 1_S)\tau_{S,S,S}(1_S \otimes \rhd)\rhd$$

When $(S, \otimes)$ is symmetric, with symmetry isomorphisms $\sigma_{X,Y} \in \mathcal{C}(X \otimes Y, Y \otimes X)$ then $\_ \otimes_{\lhd\rhd} \_ : \mathcal{C}(S,S) \times \mathcal{C}(S,S) \to \mathcal{C}(S,S)$ is a symmetric semi-monoidal tensor, with symmetry isomorphism $\sigma_{\lhd\rhd} \in \mathcal{C}(S,S)$ given by $\sigma_{\lhd\rhd} = \lhd\sigma_{S,S}\rhd$.

Proof. This is a standard result of the categorical theory of self-similarity; see [14,15,19] for the general construction, and [14,21] for numerous examples based on inverse monoids.

Definition 6. *Let $(S, \lhd, \rhd)$ be a self-similar structure of a semi-monoidal category $(\mathcal{C}, \otimes, \tau_{\_,\_,\_})$. We refer to the semi-monoidal tensor*

$$\_ \otimes_{\lhd\rhd} \_ : \mathcal{C}(S,S) \times \mathcal{C}(S,S) \to \mathcal{C}(S,S)$$

*given in Theorem 1 above as the **internalisation of** $(\otimes)$ by $(S, \lhd, \rhd)$. We similarly refer to the canonical associativity isomorphism $\tau_{\lhd\rhd} \in \mathcal{C}(S,S)$ (resp. symmetry isomorphism $\sigma_{\lhd\rhd} \in \mathcal{C}(S,S)$ as the **associativity isomorphism (resp. symmetry isomorphism) induced by** $(S, \lhd, \rhd)$.*

Remark 5. It is proved in [17] (See also Appendix B of [19]) that strict associativity for single-object semi-monoidal categories is equivalent to degeneracy (i.e. the single object being a unit object for the tensor). Thus, even when $(\mathcal{C}, \otimes)$ is strictly associative, the associativity isomorphism induced by $(S, \lhd, \rhd)$ given by $\tau_{\lhd\rhd} = \lhd(\lhd \otimes 1_S)(1_S \otimes \rhd)\rhd$ is not the identity (at least, provided S is not the unit object for $\_ \otimes_{\lhd\rhd} \_$).

The following simple corollary of Theorem 1 above is taken from [19].

Corollary 1. *Let $(S, \lhd, \rhd)$ be a $\dagger$ self-similar structure of a $\dagger$ semi-monoidal category $(\mathcal{C}, \otimes, \tau_{\_,\_,\_})$. Then $\_ \otimes_{\lhd\rhd} \_ : \mathcal{C}(S,S) \times \mathcal{C}(S,S) \to \mathcal{C}(S,S)$, the internalisation of $\_ \otimes \_$ by $(S, \lhd, \rhd)$, is a $\dagger$ semi-monoidal tensor.*

Proof. This is immediate from the property that $\lhd^{\dagger} = \rhd$, and the definition of $\_ \otimes_{\lhd\rhd} \_$ and the canonical isomorphism $\tau_{\lhd\rhd} \in \mathcal{C}(S,S)$ in terms of unitaries.

5 $\dagger$ Self-similar Structures as Lax A-H Classical Structures

We now demonstrate that, up to certain canonical coherence isomorphisms a $\dagger$ self-similar structure $(S, \lhd, \rhd)$ of a symmetric $\dagger$ semi-monoidal category $(\mathcal{C}, \otimes, \tau_{\_,\_,\_}, \sigma_{\_,\_})$ satisfies the axioms for an A-H classical structure. The precise coherence isomorphisms required are those generated by

- The semi-monoidal coherence isomorphisms $\{\tau_{\_,\_,\_}, \sigma_{\_,\_}\}$ of $(\mathcal{C}, \otimes)$
- The induced coherence isomorphisms $\{\tau_{\lhd\rhd}, \sigma_{\lhd\rhd}\}$ of $(\mathcal{C}(S,S), \otimes_{\lhd\rhd})$
- The semi-monoidal tensors $\_ \otimes \_$ and $\_ \otimes_{\lhd\rhd} \_$

Theorem 2. *Let $(S, \triangleleft, \triangleright)$ be a † self-similar structure of a symmetric † semi-monoidal category $(\mathcal{C}, \otimes, \tau_{\_,\_,\_}, \sigma_{\_,\_})$. Then the following conditions hold:*

- **(Lax associativity)** $\triangleleft(\triangleleft \otimes 1_S)\tau_{S,S,S} = \tau_{\triangleleft\triangleright} \triangleleft (1_S \otimes \triangleleft)$
- **(Lax Frobenius condition)** $\triangleright\tau_{\triangleleft\triangleright}^{-1}\triangleleft = (1_S \otimes \triangleleft)\tau_{S,S,S}^{-1}(\triangleright \otimes 1_S)$
- **(Classical structure condition)** $\triangleleft\triangleright = 1_S$
- **(Lax symmetry)** $\sigma_{S,S}\triangleright = \triangleright\sigma_{\triangleleft\triangleright}$

Proof. The following proof is based on results of [19].

Conditions 1. and 2. above follow from the commutativity of the following diagram

$$
\begin{array}{ccc}
S \xrightarrow{\triangleright} S \otimes S \xrightarrow{1_S \otimes \triangleright} S \otimes (S \otimes S) \\
\tau_{\triangleleft\triangleright} \downarrow \qquad\qquad\qquad\qquad\qquad \downarrow \tau_{S,S,S} \\
S \xleftarrow{\triangleleft} S \otimes S \xleftarrow{\triangleleft \otimes 1_S} (S \otimes S) \otimes S
\end{array}
$$

which is simply the definition of the induced associativity isomorphism. Condition 3. follows immediately from the definition of a † self-similar structure, and condition 4. is simply the definition of the induced symmetry isomorphism.

Remark 6. For the above properties to be taken seriously as lax versions of the axioms for an A-H classical structure, there needs to be some notion of coherence relating the semi-monoidal tensor $\_ \otimes \_ : \mathcal{C} \times \mathcal{C} \to \mathcal{C}$ and its canonical isomorphisms, to the semi-monoidal tensor $\_\otimes_{\triangleleft\triangleright} \_ : \mathcal{C}(S,S) \times \mathcal{C}(S,S) \to \mathcal{C}(S,S)$ and its canonical isomorphisms. A general theory of coherence for self-similarity and associativity is given in [17]; a simple case of this is also applicable in the † symmetric case.

It may be wondered whether the induced isomorphisms are necessary in theorem 2 above – can we not have a † self-similar structure satisfying analogous conditions solely based on the canonical isomorphisms of $(\mathcal{C}, \otimes)$? The following corollary demonstrates that this can only be the case when S is degenerate — i.e. the unit object for some monoidal category.

Corollary 2. *Let $(S, \triangleleft, \triangleright)$ be a self-similar structure of a semi-monoidal category $(\mathcal{C}, \otimes, \tau_{\_,\_,\_})$. Then the following condition*

- **(Overly restrictive Frobenius condition)** $\triangleright\triangleleft = (1_S \otimes \triangleleft)\tau_{S,S,S}^{-1}(\triangleright \otimes 1_S)$

implies that S is degenerate – i.e. the unit object for some monoidal category.

Proof. By definition, the associativity isomorphism for the internalisation of $(\otimes)$ is given by

$$\tau_{\triangleleft\triangleright} = \triangleleft(1_S \otimes \triangleleft)\tau_{S,S,S}^{-1}(\triangleright \otimes 1_S)\triangleright$$

Thus as $\triangleleft$ and $\triangleright$ are mutually inverse unitaries, the overly restrictive Frobenius condition implies that $\tau_{\triangleleft\triangleright} = 1_S$. However, as proved in [17] (see also Appendix B of [19]), single-object semi-monoidal categories are strictly associative exactly when their unique object is a unit object of some monoidal category.

An alternative perspective of Corollary 2 is the following:

Corollary 3. *Let (S, Δ, ∇) be an A-H classical structure satisfying the precise axioms[3] of Definition 4. Unitarity of Δ implies that S is the unit object of a monoidal category.*

Despite Corollaries 2 and 3 above, it is certainly possible for a self-similar structure to satisfy all the axioms for a Frobenius algebra up to a single associativity isomorphism; however, this must be the induced associativity isomorphism of Definition 6, as we now demonstrate:

Theorem 3. *Let $(S, \lhd, \rhd)$ be a † self-similar structure of a strictly associative † semi-monoidal category $(\mathcal{C}, \otimes, (\ )^{\dagger})$. Then the defining conditions of an A.-H. † Frobenius algebra are satisfied up to a single associativity isomorphism as follows:*

$$- \lhd(\lhd \otimes 1_S) = \tau_{\lhd\rhd} \lhd (1_s \otimes \lhd)$$
$$- (\rhd \otimes 1_S)(1_S \otimes \rhd) = \rhd\tau_{\lhd\rhd}^{-1}\lhd$$

Proof. This is simply the result of Theorem 2 in the special case where the monoidal tensor $\_ \otimes \_ : \mathcal{C} \times \mathcal{C} \to \mathcal{C}$ is strictly associative. Note that even though $\_ \otimes \_ : \mathcal{C} \times \mathcal{C} \to \mathcal{C}$ is strictly associative, its internalisation $\otimes_{\lhd\rhd} : \mathcal{C}(S,S) \times \mathcal{C}(S,S) \to \mathcal{C}(S,S)$ cannot be strictly associative; rather, from Theorem 1 the required associativity isomorphism is given by $\tau_{\lhd\rhd} = \lhd(\lhd \otimes 1_S)(1_S \otimes \rhd)\rhd \neq 1_S$.

6 An Illustrative Example

In the following sections, we will present the theory behind an example of a † self-similar structure that determines matrix representations of arrows in a similar manner to how classical structures in finite-dimensional Hilbert space determine matrix representations of linear maps. Our example is deliberately chosen to be as 'non-quantum' as possible, in order to explore the limits of the interpretations of pure category theory: it comes from a setting where all isomorphisms are unitary, all idempotents commute, and the lattice of idempotents satisfies distributivity rather than some orthomodularity condition. Many of these properties are determined by the particular form of dagger operation used; we will work with *inverse categories*.

7 Inverse Categories as † Categories

Inverse categories arose from the algebraic theory of semigroups, but the extension to a categorical definition is straightforward and well-established. We also refer to [3] for the more general *restriction categories* that generalise inverse categories in the same way that restriction monoids generalise inverse monoids.

---

[3] We strongly emphasise that this corollary does not hold if we allow the axioms of Definition 4 to hold up to canonical isomorphism, as demonstrated in Theorems 2 and 3.

Definition 7. Inverse categories
An **inverse category** *is a category $\mathcal{C}$ where every arrow $f \in \mathcal{C}(X,Y)$ has a unique* **generalised inverse** *$f^{\ddagger} \in \mathcal{C}(Y,X)$ satisfying $ff^{\ddagger}f = f$ and $f^{\ddagger}ff^{\ddagger} = f^{\ddagger}$. A single-object inverse category is called an* **inverse monoid.**

Remark 7. **Uniqueness of generalised inverse operations** Inverse monoids and semigroups were defined and studied long before inverse categories; the definition of an inverse category is thus rather 'algebraic' in nature, given by requiring the existence of unique arrows satisfying certain properties – this is in contrast to a more functorial definition. However, uniqueness implies that there can be at most one (object-indexed) operation $(\ )_{XY} : \mathcal{C}(X,Y) \to \mathcal{C}(Y,X)$ that takes each arrow to some generalised inverse satisfying the above axioms. We will therefore treat $(\ )^{\ddagger}$ as an (indexed) bijection of hom-sets, and ultimately (as we demonstrate in Theorem 7 below) a contravariant functor.

The following result is standard, and relates generalised inverses and idempotent structures of inverse monoids (see, for example [21]).

Lemma 1. *Let $M, (\ )^{\ddagger}$ be an inverse monoid. Then for all $a \in M$, the element $a^{\ddagger}a$ is idempotent, and the set of idempotents E_M of M is a commutative submonoid of M where every element is its own generalised inverse.*

Proof. These are standard results of inverse semigroup theory, relying heavily on the *uniqueness* of generalised inverses. The key equivalence between commutativity of idempotents and uniqueness of generalised inverses is due to [24].

Based on the above, the following is folklore:

Theorem 4. *Let $\mathcal{C}, (\ )^{\ddagger}$ be an inverse category. Then the operation $(\ )^{\ddagger}$ is a dagger operation, and all isomorphisms of $\mathcal{C}$ are unitary.*

Proof. The technical results we require are straightforward generalisations of well-established inverse semigroup theory, so are simply given in outline.
First observe that it is implicit from the definition that, on objects $X^{\ddagger} = X \in Ob(\mathcal{C})$. We now prove that $(\ )^{\ddagger}$, with this straightforward extension to objects, is a contravariant involution.
To demonstrate contravariant functoriality, observe that $(gf)(gf)^{\ddagger}gf = gf$ for all $f \in \mathcal{C}(X,Y)$ and $g \in \mathcal{C}(Y,Z)$. However, $ff^{\ddagger}$ and $g^{\ddagger}g$ are both idempotents of Y, and thus commute. Hence $(gf)f^{\ddagger}g^{\ddagger}(gf) = gg^{\ddagger}gff^{\ddagger}f = gf$ and so $(gf)^{\ddagger} = f^{\ddagger}g^{\ddagger}$ as required.
To see that $(\ )^{\ddagger}$ is involutive, note that by definition $f^{\ddagger}\left(f^{\ddagger}\right)^{\ddagger}f^{\ddagger} = f^{\ddagger}$, for all $f \in \mathcal{C}(X,Y)$. However, also from the definition, $f^{\ddagger}ff^{\ddagger} = f^{\ddagger}$ and again by uniqueness, $\left(f^{\ddagger}\right)^{\ddagger} = f$.
Thus $(\ )^{\ddagger}$ is a contravariant involution that acts trivially on objects. To see that all isomorphisms are unitary, consider an arbitrary isomorphism $u \in \mathcal{C}(X,Y)$. Then trivially, $uu^{-1}u = u \in \mathcal{C}(X,Y)$ and $u^{-1}uu^{-1} = u^{-1} \in \mathcal{C}(Y,X)$. Uniqueness of generalised inverses then implies that $u^{-1} = u^{\ddagger}$, and hence u is unitary.

Corollary 4. *Let C be an inverse category with a semi-monoidal tensor ($\otimes$). Then $(C, \otimes, (\ )^{\ddagger})$ is a dagger semi-monoidal category.*

Proof. Given arbitrary $f \in C(A, B)$ and $g \in C(X, Y)$, then by functoriality

$$(f \otimes g)(f^{\ddagger} \otimes g^{\ddagger})(f \otimes g) = (ff^{\ddagger}f \otimes gg^{\ddagger}g) = (f \otimes g)$$

However, by definition

$$(f \otimes g)(f \otimes g)^{\ddagger}(f \otimes g) = (ff^{\ddagger}f \otimes gg^{\ddagger}g) = (f \otimes g)$$

and by uniqueness, $(f \otimes g)^{\ddagger} = f^{\ddagger} \otimes g^{\ddagger}$. Also, since all isomorphisms are unitary, all canonical isomorphisms are unitary.

7.1   The Natural Partial Order on Hom-Sets

All inverse categories have a naturally defined partial order on their hom-sets:

Definition 8. *Let $C, (\ )^{\ddagger}$ be an inverse category. For all $A, B \in Ob(C)$, the relation $\trianglelefteq_{A,B}$ is defined on $C(A, B)$, as follows:*

$$f \trianglelefteq_{A,B} g \quad \text{iff} \quad \exists \ e^2 = e \in C(A, A) \ \text{s.t.} \ f = ge$$

*It is immediate that, for all $A, B \in Ob(C)$, the relation $\trianglelefteq_{A,B}$ is a partial order on $C(A, B)$, called the **natural partial order**.*

Convention: *When it is clear from the context, we omit the subscript on $\trianglelefteq$.*

We may rewrite the above non-constructive definition more concretely:

Lemma 2. *Given $f \trianglelefteq g \in C(X, Y)$, in some inverse category, then $f = gf^{\ddagger}f$.*

Proof. By definition, $f = ge$, for some $e^2 = e \in C(X, X)$. Thus $fe = f$, since e is idempotent. From the defining equation for generalised inverses, $f = ff^{\ddagger}f = gef^{\ddagger}f$. As $f^{\ddagger}f$ is idempotent, and idempotents commute, $f = gf^{\ddagger}fe$. However, we have already seen that $fe = f$, and hence $f = gf^{\ddagger}f$.

A very useful tool in dealing with the natural partial order is the following lemma, which is again a classic result of inverse semigroup theory rewritten in a categorical setting (see also [11] where it is rediscovered under the name 'passing a message through a channel').

Lemma 3. Pushing an idempotent through an arrow *Let $C, (\ )^{\ddagger})$ be an inverse category. Then for all $f \in C(X, Y)$, and $e^2 = e \in C(X, X)$, there exists an idempotent $e'^2 = e' \in C(Y, Y)$ satisfying $e'f = fe$.*

Proof. We define $e' = fef^{\ddagger} \in C(Y, Y)$. By Lemma 1, $e'^2 = fef^{\ddagger}fef^{\ddagger} = ff^{\ddagger}feef^{\ddagger} = fef^{\ddagger} = e'$. Further, $e'f = fef^{\ddagger}f = ff^{\ddagger}fe = fe$, as required.

Proposition 2. *The natural partial order is a congruence — that is, given $f \unlhd$ $h \in \mathcal{C}(X, Y)$ and $g \unlhd k \in \mathcal{C}(Y, Z)$, then $gf \unlhd kh \in \mathcal{C}(X, Z)$.*

Proof. By definition, there exists idempotents $p^2 = p \in \mathcal{C}(X, X)$ and $q^2 = q \in \mathcal{C}(Y, Y)$ such that $f = hp \in \mathcal{C}(X, Y)$ and $g = kq \in \mathcal{C}(Y, Z)$, and hence $gf = hpkq \in \mathcal{C}(X, Z)$. We now use the 'passing an idempotent through an arrow' technique of Lemma 3 to deduce the existence of an idempotent $p' \in \mathcal{C}(X, X)$ such that $pk = kp' \in \mathcal{C}(X, Y)$. Hence $gf = khp'q$. However, by Part 3. of Lemma 3, $p'q$ is idempotent, and hence $gf \unlhd kh$, as required.

Corollary 5. *Every locally small inverse category $(\mathcal{C}, (\ )^{\ddagger})$ is enriched over the category* **Poset** *of partially ordered sets.*

Proof. Expanding out the definition of categorical enrichment will demonstrate that the crucial condition is that proved in Proposition 2 above.

7.2   A Representation Theorem for Inverse Categories

A classic result of inverse semigroup theory is the Wagner-Preston representation theorem [27,28] which states that every inverse semigroup S is isomorphic to some semigroup of partial isomorphisms on some set. This implies the usual representation theorem for groups as subgroups of isomorphisms on sets. There exists a natural generalisation of this theorem to inverse categories:

Definition 9. *The inverse category* **pIso** *is defined as follows:*

- **(Objects)** *All sets.*
- **(Arrows)** **pIso**(X, Y) *is the set of all partial isomorphisms from X to Y. In terms of diagonal representations, it is the set of all subsets $f \subseteq Y \times X$ satisfying*

$$b = y \ \Leftrightarrow \ y = a \ \ \forall \ (b, a), (y, x) \in f$$

- **(Composition)** *This is inherited from the category* **Rel** *of relations on sets in the obvious way.*
- **(Generalised inverse)** *This is given by $f^{\ddagger} = \{(x, y) : (y, x) \in f$; the obvious restriction of the relational converse.*

The category **pIso** *has zero arrows, given by $0_{XY} = \emptyset \subseteq Y \times X$. This is commonly used to define a notion of* **orthogonality** *by*

$$f \perp g \in \mathcal{C}(X, Y) \ \ \Leftrightarrow \ \ g^{\ddagger} f = 0_X \ and \ gf^{\ddagger} = 0_Y$$

Remark 8. The category $(\mathbf{pIso}, \uplus)$ is well-equipped with self-similar structures; one of the most heavily-studied [14,21,15] is the natural numbers $\mathbb{N}$, although any countably infinite set will suffice. As demonstrated in an Appendix to [16], there is a 1:1 correspondence between self-similar structures at $\mathbb{N}$ and points of the Cantor set (excluding a subset of measure zero). Other examples include the Cantor set itself [14,15] and other fractals [22].

Theorem 5. *Every locally small inverse category* $(\mathcal{C}, (\ )^{\ddagger})$ *is isomorphic to some subcategory of* $(\mathbf{pIso}, (\ )^{\ddagger})$.

Proof. This is proved in [13], and significantly prefigured (for small categories) in [3].

The idempotent structure and natural partial ordering on **pIso** is particularly well-behaved, as the following standard results demonstrate:

Proposition 3.

1. *The natural partial order of* **pIso** *may be characterised in terms of diagonal representations by* $f \trianglelefteq g \in \mathbf{pIso}(X, Y)$ *iff* $f \subseteq g \in Y \times X$.
2. *All idempotents* $e^2 = e \in \mathbf{pIso}(X, X)$ *are simply partial identities* $1_{X'}$ *for some* $X' \subseteq X$, *and thus* **pIso** *is isomorphic to its own Karoubi envelope.*
3. *The meet and join w.r.t. the natural partial order are given by, for all* $f, g \in \mathbf{pIso}(X, Y)$

$$f \vee g = f \cup g \quad and \quad f \wedge g = f \cap g$$

when these exist. Therefore, set of idempotents at an object is a distributive lattice.
4. *Given an arbitrarily indexed set* $\{f_j \in \mathbf{pIso}(X, Y)\}_{j \in J}$ *of pairwise-orthogonal elements, together with arbitrary* $a \in \mathbf{pIso}(\mathbf{W}, \mathbf{X})$ *and* $b \in \mathbf{pIso}(\mathbf{Y}, \mathbf{Z})$, *then* $\bigvee_{j \in J} f_j \in \mathbf{pIso}(X, Y)$ *exists, as does* $\bigvee_{j \in J} b f_j a \in \mathbf{pIso}(W, Z)$, *and*

$$b \left(\bigvee_{j \in J} f_j \right) a = \bigvee_{j \in J} (b f_j a)$$

Proof. These are all standard results for the theory of inverse categories; 1. is a straightforward consequence of the definition of the natural partial order, and 2.-4. follow as simple corollaries.

8   Monoidal Tensors and Self-similarity in pIso

We have seen that **pIso** is a † category; it is also a † monoidal category with respect to two distinct monoidal tensors - the Cartesian product $\_ \times \_$ and the disjoint union $\_ \uplus \_$. For the purposes of this paper, we will study the disjoint union. We make the following formal definition:

Definition 10. *We define the* **disjoint union** $\uplus : \mathbf{pIso} \times \mathbf{pIso} \to \mathbf{pIso}$ *to be the following monoidal tensor:*

- **(Objects)** $A \uplus B = A \times \{0\} \cup B \times \{1\}$, *for all* $A, B \in Ob(\mathbf{pIso})$.
- **Arrows)** *Given* $f \in \mathbf{pIso}(A, B)$ *and* $g \in \mathbf{pIso}(X, Y)$, *we define* $f \uplus g = inc_{00}(f) \cup inc_{11}(g) \subseteq (B \uplus Y) \times (A \uplus X)$ *where* inc_{00} *is the canonical (for the Cartesian product) isomorphism* $B \times A \cong B \times \{0\} \times A \times \{0\}$, *and similarly,* $inc_{11} : Y \times X \cong Y \times \{1\} \times X \times \{1\}$.

It is immediate that (**pIso**, ⊎) *is a †-monoidal tensor since, as a simple consequence of the definition of generalised inverses, all isomorphisms are unitary.*

By contrast with the behaviour of disjoint union in (for example) the category of relations, it is neither a product nor a coproduct on **pIso**. Despite this, it has analogues of projection & inclusion maps:

Definition 11. *Given* $X, Y \in Ob(\mathbf{pIso})$, *the arrows* $\iota_l \in \mathbf{pIso}(X, X \uplus Y)$ *and* $\iota_r \in \mathbf{pIso}(Y, X \uplus Y)$ *are defined by* $\iota_l(x) = (x, 0) \in X \uplus Y$ *and* $\iota_r(y) = (y, 1) \in X \uplus Y$. *By convention, we denote their generalised inverses by*

$$\pi_l :\in \mathbf{pIso}(\mathbf{X} \uplus \mathbf{Y}, \mathbf{X}) \text{ and } \pi_r \in \mathbf{pIso}(\mathbf{X} \uplus \mathbf{Y}, \mathbf{Y})$$

respectively, giving

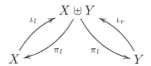

Following [14] we refer to these arrows as the **projection** *and* **inclusion** *arrows; they are sometimes [12,1] called* **quasi-projections / injections**, *in order to emphasise that they are not derived from categorical products / coproducts. By construction, the projections / inclusions satisfy the following four identities:*

$$\pi_r \iota_l = 0_{XY} \quad, \quad \pi_l \iota_r = 0_{YX} \quad, \quad \pi_l \iota_l = 1_X \quad, \quad \pi_r \iota_r = 1_Y$$

As noted in [14,15], the above arrows can be 'internalised' by a self-similar structure $(S, \triangleleft, \triangleright)$, in a similar way to canonical isomorphisms (see Theorem 6). Doing so will give an embedding of a well-studied inverse monoid into **pIso**$(\mathbf{S}, \mathbf{S})$.

Definition 12. Polycyclic monoids
The 2 generator **polycyclic monoid** P_2 *is defined in [25] to be the inverse monoid given by the generating set* $\{p, q\}$, *together with the relations*

$$pp^{-1} = 1 = qq^{-1} \quad, \quad pq^{-1} = 0 = qp^{-1}$$

Remark 9. This inverse monoid is also familiar to logicians as the (multiplicative part of) the *dynamical algebra* of [9,6]. It is also familiar from the theory of state machines as the syntactic monoid of a pushdown automaton with a binary stack [7], and to pure mathematicians as the monoid of partial homeomorphisms of the Cantor set [14].

The following result on polycyclic monoids will prove useful:

Lemma 4. P_2 *is congruence-free; i.e. the only composition-preserving equivalence relations on* P_2 *are either the universal congruence* $r \sim s$ *for all* $r, s \in P_2$, *or the identity congruence* $r \sim s \Leftrightarrow r = s$ *for all* $r, s \in P_2$.

Proof. This is a special case of a general result of [25]. Congruence-freeness is an example of Hilbert-Post completeness; categorically, it is closely related to the 'no simultaneous strictification' theorem of [17].

The following result, generalising a preliminary result of [14,15], makes the connection between embeddings of polycyclic monoids and internalisations of projection/ injection arrows of **pIso** precise:

Theorem 6. *Let S be a self-similar object (and hence a † self-similar object) of* **pIso**. *We say that an inverse monoid homomorphism $\phi : P_2 \to$ **pIso**(S, S) is a* **strong embedding** *when it satisfies the condition*

$$\phi(p^\dagger p) \vee \phi(q^\dagger q) = 1_S$$

*Then every strong embedding $\phi : P_2 \to$ **pIso**(S, S) uniquely determines, and is uniquely determined by, a † self-similar structure at S.*

Proof. Let $\pi_l, \pi_r \in$ **pIso**$(S \uplus S, S$ and $\iota_l, \iota_r \in$ **pIso**$(S, S \uplus S)$ be the projections / inclusions of Definition 11, and let $(S, \lhd, \rhd)$ be a self-similar structure. We define $\phi_{\lhd\rhd} : P_2 \to$ **pIso**(S, S) by its action on the generators of P_2, giving

$$\phi_{\lhd\rhd}(p) = \pi_l \rhd \quad \text{and} \quad \phi_{\lhd\rhd}(q) = \pi_r \rhd$$

Their generalised inverses are then $\phi_{\lhd\rhd}(p^\dagger) = \lhd\iota_l$ and $\phi_{\lhd\rhd}(q^\dagger) = \lhd\iota_r$. Thus

$$\phi_{\lhd\rhd}(p)\phi_{\lhd\rhd}(p^\dagger) = \pi_l \rhd \lhd\iota_l = 1_S = \pi_r \rhd \lhd\iota_r = \phi_{\lhd\rhd}(q)\phi_{\lhd\rhd}(q^\dagger)$$

Similarly, $\phi_{\lhd\rhd}(p)\phi_{\lhd\rhd}(q^\dagger) = \pi_l \rhd \lhd\iota_r = 0_S = \pi_r \rhd \lhd\iota_l = \phi_{\lhd\rhd}(q)\phi_{\lhd\rhd}(p^\dagger)$ and so $\phi_{\lhd\rhd}$ is a homomorphism. Since P_2 is congruence-free it is also an embedding. To demonstrate that it is also a strong embedding,

$$1_S = \lhd 1_{S \uplus S} \rhd = \lhd(\iota_l\pi_l \vee \iota_r\pi_r)\rhd$$

$$= \lhd\iota_l\pi_l \rhd \vee \lhd \iota_r\pi_r\rhd = \phi_{\lhd\rhd}(p^\dagger p) \vee \phi_{\lhd\rhd}(q^\dagger q)$$

as required. Further, given another self-similar structure (S, c, d) satisfying $\phi_{dc} = \phi_{\lhd\rhd}$, then $\lhd = c \in$ **pIso**$(S \uplus S, S)$ and $\rhd = d \in$ **pIso**$(S, S \uplus S)$.

Conversely, let $\phi : P_2 \to$ **pIso**(S, S) be a strong embedding, and consider the two arrows $\iota_l\phi(p) \in$ **pIso**$(S, S \uplus S)$ and $\iota_r\phi(q) \in$ **pIso**$(S, S \uplus S)$. it is straightforward that these are orthogonal; we thus define

$$\rhd_\phi = \iota_l\phi(p) \vee \iota_r\phi(q) \in \mathbf{pIso}(S, S \uplus S)$$

and take $\lhd_\phi = \rhd_{\lhd\rhd}^\dagger$. The strong embedding condition implies that $\lhd_\phi\rhd_\phi = 1_S$ and $\rhd_\phi\lhd_\phi = 1_{S \uplus S}$; thus we have a self-similar structure, as required. Further, given another strong embedding $\psi : P_2 \to$ **pIso**(S, S), then $\lhd_\phi = \lhd_\psi$ iff $\phi = \psi$.

8.1 Matrix Representations from Self-similar Structures

We are now in a position to demonstrate how self-similar structures in **pIso** determine matrix representations of arrows.

Theorem 7. *Let $S \in Ob(\mathbf{pIso})$ be a self-similar object. Then every self-similar structure $(S, \lhd, \rhd)$ determines matrix representations of arrows of* **pIso**(S, S).

Proof. We use the correspondence between self-similar structures and strong embeddings of polycyclic monoids given in Theorem 6. Given arbitrary $f \in$ **pIso(S, S)**, we define $[f]_{\diamond}$, the **matrix representation of f determined by** $(S, \lhd, \rhd)$ to be the following matrix:

$$[f]_{\diamond} = \begin{pmatrix} \phi_{\diamond}(p) f \phi_{\diamond}(p^{\ddagger}) & \phi_{\diamond}(p) f \phi_{\diamond}(q^{\ddagger}) \\ \phi_{\diamond}(q) f \phi_{\diamond}(p^{\ddagger}) & \phi_{\diamond}(q) f \phi_{\diamond}(q^{\ddagger}) \end{pmatrix}$$

Given two such matrices of this form, we interpret their **matrix composition** as follows:

$$\begin{pmatrix} g_{00} & g_{01} \\ g_{10} & g_{11} \end{pmatrix} \begin{pmatrix} f_{00} & f_{01} \\ f_{10} & f_{11} \end{pmatrix} = \begin{pmatrix} g_{00} f_{00} \vee g_{01} f_{10} & g_{00} f_{01} \vee g_{01} f_{11} \\ g_{10} f_{00} \vee g_{11} f_{10} & g_{10} f_{01} \vee g_{11} f_{11} \end{pmatrix}$$

that is, the usual formula for matrix composition, with summation interpreted by join in the natural partial order – provided that the required joins exist. We prove that this composition is defined for matrix representations determined by a fixed self-similar structure.

In what follows, we abuse notation, for clarity, and refer to $p, q, p^{\ddagger}, q^{\ddagger} \in$ **pIso(S, S)** *instead of $\phi_{\diamond}(p), \phi_{\diamond}(q), \phi_{\diamond}(p^{\ddagger}), \phi_{\diamond}(q^{\ddagger}) \in$ **pIso(S,S)**. As this proof is based on a single fixed self-similar structure at S, we may do this without ambiguity.*

Consider the entry in the top left hand corner of $[g]_{\diamond}[f]_{\diamond}$. Expanding out the definition will give this as $pgp^{\ddagger}pfp^{\ddagger} \vee pgq^{\ddagger}qfp^{\ddagger}$. To demonstrate that these two terms are orthogonal, $\left(pgp^{\ddagger}pfp^{\ddagger}\right)^{\ddagger}\left(pgq^{\ddagger}qfp^{\ddagger}\right) = pf^{\ddagger}p^{\ddagger}pg^{\ddagger}p^{\ddagger}pgq^{\ddagger}qfp^{\ddagger}$. Appealing to the 'pushing an idempotent through an arrow' technique of Proposition 3 gives the existence of some idempotent $e^2 = e$ such that

$$\left(pgp^{\ddagger}pfp^{\ddagger}\right)^{\ddagger}\left(pgq^{\ddagger}qfp^{\ddagger}\right) = pf^{\ddagger}p^{\ddagger}peg^{\ddagger}gq^{\ddagger}qfp^{\ddagger}$$

Again appealing to this technique gives the existence of some idempotent $E^2 = E$ such that $\left(pgp^{\ddagger}pfp^{\ddagger}\right)^{\ddagger}\left(pgq^{\ddagger}qfp^{\ddagger}\right) = pf^{\ddagger}p^{\ddagger}Epq^{\ddagger}qfp^{\ddagger}$. However, $pq^{\ddagger} = 0$ and hence $\left(pgp^{\ddagger}pfp^{\ddagger}\right)^{\ddagger}\left(pgq^{\ddagger}qfp^{\ddagger}\right) = 0$. as required. An almost identical calculation will give that $\left(pgq^{\ddagger}qfp^{\ddagger}\right)^{\ddagger}\left(pgp^{\ddagger}pfp^{\ddagger}\right) = 0$ and thus these two terms are orthogonal, so the required join exists.

The proof of orthogonality for the other three matrix entries is almost identical; alternatively, it may be derived using the obvious isomorphism of P_2 that interchanges the roles of p and q.

It remains to show that composition of matrix repesentations of elements coincides with composition of these elements; we now prove that $[g]_{\diamond}[f]_{\diamond} = [gf]_{\diamond}$. By definition,

$$[gf]_{\diamond} = \begin{pmatrix} pgfp^{\ddagger} & pgfq^{\ddagger} \\ qgfp^{\ddagger} & qgfq^{\ddagger} \end{pmatrix}$$

As the (implicit) embedding of P_2 is strong, $1_S = p^\ddagger p \vee q^\ddagger q$. We may then substitute $g(p^\ddagger p \vee q^\ddagger q)f$ for gf in the above to get

$$[gf]_{\triangleleft\triangleright} = \begin{pmatrix} pg(p^\ddagger p \vee q^\ddagger q)fp^\ddagger & pg(p^\ddagger p \vee q^\ddagger q)fq^\ddagger \\ qg(p^\ddagger p \vee q^\ddagger q)fp^\ddagger & qg(p^\ddagger p \vee q^\ddagger q)fq^\ddagger \end{pmatrix}$$

Expanding this out using the distributivity of composition over joins gives the definition of $[g]_{\triangleleft\triangleright}[f]_{\triangleleft\triangleright}$, and hence $[gf]_{\triangleleft\triangleright} = [g]_{\triangleleft\triangleright}[f]_{\triangleleft\triangleright}$, as required.

Finally, we need to prove that the representation of arrows as matrices determined by the self-similar structure $(S, \triangleleft, \triangleright)$ is faithful — that is, $a = b \in \mathbf{pIso}(\mathbf{S}, \mathbf{S})$ iff $[b]_{\triangleleft\triangleright} = [a]_{\triangleleft\triangleright}$ (where equality of matrices is taken as component-wise equality).

The ($\Rightarrow$) implication is immediate from the definition. For the other direction, $[b]_{\triangleleft\triangleright} = [a]_{\triangleleft\triangleright}$ when the following four identities are satisfied:

$$pap^\ddagger = pbp^\ddagger \quad paq^\ddagger = pbq^\ddagger$$

$$qap^\ddagger = qbp^\ddagger \quad qaq^\ddagger = qbq^\ddagger$$

Prefixing/ suffixing each of these identities with the appropriate choice selection taken from $\{p, q, p^\ddagger, q^\ddagger\}$ will give the following identities:

$$p^\ddagger pap^\ddagger p = p^\ddagger pbp^\ddagger p \quad p^\ddagger paq^\ddagger q = p^\ddagger pbq^\ddagger q$$

$$q^\ddagger qap^\ddagger p = q^\ddagger qbp^\ddagger p \quad q^\ddagger qaq^\ddagger q = q^\ddagger qbq^\ddagger q$$

Now observe that these four elements are pairwise-orthogonal. We may take their join, and appeal to distributivity of composition over join to get

$$(p^\ddagger p \vee q^\ddagger q)a(p^\ddagger p \vee q^\ddagger q) = (p^\ddagger p \vee q^\ddagger q)a(p^\ddagger p \vee q^\ddagger q)$$

However, as the implicit embedding of P_2 is strong, $(p^\ddagger p \vee q^\ddagger q) = 1_S$ and thus $a = b$, as required.

Remark 10. It may seem somewhat disappointing that a self-similar structure $(S, \triangleleft, \triangleright)$ simply determines (2×2) matrix representations of arrows of $\mathbf{pIso}(\mathbf{S}, \mathbf{S})$, rather than matrix representations of arbitrary orders. This is not quite the case, but there is a subtlety to do with the behaviour of the internalisation of the tensor $\_ \uplus \_ : \mathbf{pIso} \times \mathbf{pIso} \to \mathbf{pIso}$. It is immediate from the definition that the internalisation of this tensor by a self-similar structure has the obvious matrix representation: $[f \uplus_{\triangleleft\triangleright} g]_{\triangleleft\triangleright} = \begin{pmatrix} f & 0_S \\ 0_S & g \end{pmatrix}$. However, recall from Remark 5 that the internalisation $\_ \otimes_{\triangleleft\triangleright} \_$ of an arbitrary tensor $\_ \otimes \_$ can never be strictly associative, even when $\_ \otimes \_$ itself is associative. Thus, in our example in $(\mathbf{pIso}, \uplus)$, arbitrary $(n \times n)$ matrices, in the absence of additional bracketing information, cannot *ambiguously* represent arrows. It is of course possible to have unambiguous $n \times n$ matrix representations that are determined by binary treeS whose leaves are labelled with a single formal symbol, and whose nodes are labelled by self-similar structures at S – however, this is beyond the scope of this paper!

8.2 Isomorphisms of Self-similar Structures as 'Changes of Matrix Representation'

We have seen in Proposition 1 that † self-similar structures are unique up to unique unitary. We now relate this to the correspondence in $(\mathbf{pIso}, \uplus)$ between † self-similar structures, strong embeddings of P_2, and matrix representations.

Lemma 5. *Let $(S, \lhd, \rhd)$ and (S, c, d) be two † self-similar structures at the same object of $(\mathbf{pIso}, \uplus)$, and let U be the unique isomorphism (following Proposition 1) making the following diagram commute:*

The two strong embeddings $\phi_{\lhd\rhd}, \phi_{(c,d)} : P_2 \to \mathbf{pIso}(S, S)$ determined by these self-similar structures (as in Theorem 6) are mutually determined by the following identities:

$$\phi_{(c,d)}(p) = \phi_{\lhd\rhd}(p)U^{-1} \qquad \phi_{(c,d)}(q) = \phi_{\lhd\rhd}(q)U^{-1}$$

$$\phi_{(c,d)}(p^{\ddagger}) = U\phi_{\lhd\rhd}(p^{\ddagger}) \qquad \phi_{(c,d)}(q^{\ddagger}) = U\phi_{\lhd\rhd}(q^{\ddagger})$$

Proof. By construction, $c = U\lhd$ and $d = \rhd U^{-1}$. Thus $\phi_{(c,d)}(p) = \pi_l d = \pi_l \rhd U^{-1} = \phi_{\lhd\rhd}(p)U^{-1}$. Taking duals (generalised inverses) gives $\phi_{(c,d)}(p^{\ddagger}) = U \lhd u_l = U\phi_{\lhd\rhd}(p^{\ddagger})$. The other two identities follow similarly.

The above connection between the embeddings of P_2 given by two self-similar structures allows us to give the transformation between matrix representations of arrows given by two self-similar structures:

Theorem 8. *Let $(S, \lhd, \rhd)$ and (S, c, d) be two self-similar structures at the same object of $(\mathbf{pIso}, \uplus)$, and let the matrix representations of some arrow $f \in \mathbf{pIso}(S, S)$ given by $(S, \lhd, \rhd)$ and (S, c, d) respectively be*

$$[f]_{\lhd\rhd} = \begin{pmatrix} \alpha & \beta \\ \gamma & \delta \end{pmatrix} \quad and \quad [f]_{(c,d)} = \begin{pmatrix} \alpha' & \beta' \\ \gamma' & \delta' \end{pmatrix}$$

Then $[f]_{(c,d)}$ is given in terms of $[f]_{\lhd\rhd}$ by the following matrix composition:

$$\begin{pmatrix} \alpha' & \beta' \\ \gamma' & \delta' \end{pmatrix} = \begin{pmatrix} u_{00}^{\ddagger} & u_{10}^{\ddagger} \\ u_{01}^{\ddagger} & u_{11}^{\ddagger} \end{pmatrix} \begin{pmatrix} \alpha & \beta \\ \gamma & \delta \end{pmatrix} \begin{pmatrix} u_{00} & u_{01} \\ u_{10} & u_{11} \end{pmatrix}$$

where

$$\begin{pmatrix} u_{00} & u_{01} \\ u_{10} & u_{11} \end{pmatrix} = \begin{pmatrix} \phi_{\lhd\rhd}(p)\phi_{(c,d)}(p^{\ddagger}) & \phi_{\lhd\rhd}(p)\phi_{(c,d)}(q^{\ddagger}) \\ \phi_{\lhd\rhd}(q)\phi_{(c,d)}(p^{\ddagger}) & \phi_{\lhd\rhd}(q)\phi_{(c,d)}(q^{\ddagger}) \end{pmatrix}$$

Proof. Long direct calculation, expanding out the definition of the above matrix representations, will demonstrate that

$$\begin{pmatrix} \alpha' & \beta' \\ \gamma' & \delta' \end{pmatrix} = \begin{pmatrix} \phi_{(c,d)}(p)f\phi_{(c,d)}(p^\ddagger) & \phi_{(c,d)}(p)f\phi_{(c,d)}(q^\ddagger) \\ \phi_{(c,d)}(q)f\phi_{(c,d)}(p^\ddagger) & \phi_{(c,d)}(q)f\phi_{(c,d)}(q^\ddagger) \end{pmatrix}$$

as a consequence of the identities

$$\phi_{\diamondsuit}(p^\ddagger p) \vee \phi_{\diamondsuit}(q^\ddagger q) = 1_S = \phi_{(c,d)}(p^\ddagger p) \vee \phi_{(c,d)}(q^\ddagger q)$$

8.3 Diagonalisations of Matrices via Isomorphisms of Self-similar Structures

A useful application of basis changes in linear algebra is to construct diagonalisations of matrices. For a matrix $M = \begin{pmatrix} A & B \\ C & D \end{pmatrix}$ over a vector space $V = V_1 \oplus V_2$, a *diagonalisation* is a linear isomorphism D satisfying $D^{-1}MD = \begin{pmatrix} A' & 0 \\ 0 & B' \end{pmatrix}$, for some elements A', B'. We demonstrate how this notion of diagonalisation has a direct analogue at self-similar objects of $(\mathbf{pIso}, \uplus)$, and provide a necessary and sufficient condition (and related construction) for an arrow to be diagonalised by an isomorphism of self-similar structures.

Definition 13. Diagonalisation at self-similar objects of $(\mathbf{pIso}, \uplus)$
*Let $(S, \triangleleft, \triangleright)$ be a self-similar structure of $(\mathbf{pIso}, \uplus)$ and let $\in \mathbf{pIso}(S, S)$ be an arrow with matrix representation $[f]_{\diamondsuit} = \begin{pmatrix} \alpha & \beta \\ \gamma & \delta \end{pmatrix}$. We define a **diagonalisation** of this matrix representation to be a self-similar structure (S, c, d) such that $[f]_{(c,d)} = \begin{pmatrix} \lambda & 0 \\ 0 & \mu \end{pmatrix}$, so the matrix conjugation given in Theorem 8 satisfies*

$$\begin{pmatrix} \lambda & 0 \\ 0 & \mu \end{pmatrix} = \begin{pmatrix} u_{00}^\ddagger & u_{10}^\ddagger \\ u_{01}^\ddagger & u_{11}^\ddagger \end{pmatrix} \begin{pmatrix} \alpha & \beta \\ \gamma & \delta \end{pmatrix} \begin{pmatrix} u_{00} & u_{01} \\ u_{10} & u_{11} \end{pmatrix}$$

We now characterise when the matrix representation of an arrow (w.r.t. a certain self-similar structure) may be diagonalised by another self-similar structure:

Theorem 9. *Let $(S, \triangleleft, \triangleright)$ and (S, c, d) be self-similar structures of $(\mathbf{pIso}, \uplus)$ at the same object, giving rise to strong embeddings $\phi_{\diamondsuit}, \phi_{(c,d)} : P_2 \to \mathbf{pIso}(S, S)$ and (equivalently) internalisations of the disjoint union*

$$\_ \uplus_{\diamondsuit} \_ \text{-,} \_ \uplus_{(c,d)} \_ \text{-} : \mathbf{pIso}(S, S) \times \mathbf{pIso}(S, S) \to \mathbf{pIso}(S, S)$$

The matrices representations that may be diagonalised by the unique isomorphism between $(S, \triangleleft, \triangleright)$ and (S, c, d) are exactly those of the form

$$\begin{pmatrix} \phi_{\diamondsuit}(p)(X \uplus_{(c,d)} Y)\phi_{\diamondsuit}(p^\ddagger) & \phi_{\diamondsuit}(p)(X \uplus_{(c,d)} Y)\phi_{\diamondsuit}(q^\ddagger) \\ \phi_{\diamondsuit}(q)(X \uplus_{(c,d)} Y)\phi_{\diamondsuit}(p^\ddagger) & \phi_{\diamondsuit}(q)(X \uplus_{(c,d)} Y)\phi_{\diamondsuit}(q^\ddagger) \end{pmatrix}$$

Proof. In the following proof, we abuse notation slightly for purposes of clarity. We will denote

$$\phi_{\lhd\rhd}(p), \phi_{\lhd\rhd}(q), \phi_{\lhd\rhd}(p^{\ddagger}), \phi_{\lhd\rhd}(q^{\ddagger}) \in \mathbf{pIso}(\mathbf{S}, \mathbf{S})$$

by $p, q, p^{\ddagger}, q^{\ddagger} \in \mathbf{pIso}(\mathbf{S}, \mathbf{S})$, and similarly, denote

$$\phi_{(c,d)}(p), \phi_{(c,d)}(q), \phi_{(c,d)}(p^{\ddagger}), \phi_{(c,d)}(q^{\ddagger}) \in \mathbf{pIso}(\mathbf{S}, \mathbf{S})$$

by $r, s, r^{\ddagger}, s^{\ddagger} \in \mathbf{pIso}(\mathbf{S}, \mathbf{S})$.

Given arbitrary arrows $X, Y \in \mathbf{pIso}(\mathbf{S}, \mathbf{S})$, then $[X \uplus_{(c,d)} Y]_{(c,d)} = \begin{pmatrix} X & 0 \\ 0 & Y \end{pmatrix}$, and all diagonal matrix representations (w.r.t. (S, c, d)) are of this form. Let us now conjugate such a diagonal matrix by inverse of the matrix U derived from Theorem 8; this gives $U^{-1}[X \uplus_{(c,d)} Y]_{(c,d)} U =$

$$\begin{pmatrix} pr^{-1}Xrp^{-1} \vee qr^{-1}Yrq^{-1} & pr^{-1}Xrq^{-1} \vee ps^{-1}Ysq^{-1} \\ qr^{-1}Xrp^{-1} \vee qs^{-1}Ysp^{-1} & qr^{-1}Xrq^{-1} \vee qs^{-1}Ysq^{-1} \end{pmatrix}$$

Comparing this with the explicit form of the internalisation of the disjoint union by the self-similar structure (S, c, d) gives

$$\begin{pmatrix} p(X \uplus_{(c,d)} Y)p^{-1} & p(X \uplus_{(c,d)} Y)q^{-1} \\ q(X \uplus_{(c,d)} Y)p^{-1} & q(X \uplus_{(c,d)} Y)q^{-1} \end{pmatrix}$$

Therefore all matrices of this form are diagonalised by the unique isomorphism from $(S, \lhd, \rhd)$ to (S, c, d). Conversely, as $X, Y \in \mathbf{pIso}(\mathbf{S}, \mathbf{S})$ were chosen arbitrarily, all matrices diagonalised by this unique isomorphism are of this form.

Remark 11. It is worth emphasising that the above theorem characterises those matrix representations that may be diagonalised by a particular self-similar structure; it does not address the question of whether there exists a self-similar structure that diagonalises a particular matrix representation. For the particular example of $\mathbb{N}$ as a self-similar object, an arrow $f \in \mathbf{pIso}(\mathbb{N}, \mathbb{N})$ is diagonalisable iff there exists a partition of $\mathbb{N}$ into disjoint infinite subsets $A \cup B = \mathbb{N}$ such that $f(A) \subseteq A$ and $f(B) \subseteq B$. Simple cardinality arguments will demonstrate that this question is undecidable in general.

9   Conclusions

If nothing else, this paper has hopefully demonstrated that, although superficially dissimilar, the notions of copying derived from quantum mechanics and from logic (and categorical linguistics) are nevertheless closely connected. However, these connections are not apparent unless we allow for the definitions in both cases to be taken up to canonical isomorphisms.

Acknowledgements. The author wishes to acknowledge useful discussions, key ideas, and constructive criticism from a variety of sources, with particular thanks to S. Abramsky, B. Coecke, C. Heunen, M. Lawson, P. Panangaden, and P. Scott.

References

1. Abramsky, S., Haghverdi, E., Scott, P.: Geometry of interaction and linear combinatory algebras. Mathematical Structures in Computer Science 12(5) (2002)
2. Abramsky, S., Heunen, C.: H*-algebras and nonunital frobenius algebras: First steps in infinite-dimensional categorical quantum mechanics. In: Clifford Lectures, AMS Proceedings of Symposia in Applied Mathematics, vol. 71, pp. 1–24 (2012)
3. Cockett, J.R.B., Lack, S.: Restriction categories i: categories of partial maps. Theoretical Computer Science 270, 223–259 (2002)
4. Coecke, B., Pavlovic, D.: Quantum measurements without sums. In: Chen, G., Kauffman, L., Lamonaco, S. (eds.) Mathematics of Quantum Computing and Technology. Chapman & Hall (2007) arxiv.org/quant-ph/0608035
5. Coecke, B., Pavlovic, D., Vicary, J.: A new description of orthogonal bases. Mathematical Structures in Computer Science 23, 555–567 (2013)
6. Danos, V., Regnier, L.: Local and asynchronous beta reduction. In: Proceedings of the Eighth Annual IEEE Symp. on Logic in Computer Science (1993)
7. Gilman, R.: Formal languages and infinite groups. In: Baumslag, G., Epstein, D., Gilman, R., Short, H., Sims, C. (eds.) Geometric and Computational Perspectives on Infinite Groups. Discrete Mathematics and Theoretical Computer Science, vol. 25, pp. 27–51. American Mathematical Society (1996)
8. Girard, J.-Y.: Linear logic. Theoretical Computer Science 50, 1–102 (1987)
9. Girard, J.-Y.: Geometry of interaction 1. In: Proceedings Logic Colloquium 1988, pp. 221–260. North-Holland (1988)
10. Girard, J.-Y.: Geometry of interaction 2: Deadlock-free algorithms. In: Martin-Löf, P., Mints, G. (eds.) COLOG 1988. LNCS, vol. 417, pp. 76–93. Springer, Heidelberg (1988)
11. Girard, J.-Y.: Geometry of interaction 3: Accommodating the additives. Advances in Linear Logic 222, 329 (1995)
12. Haghverdi, E.: A categorical approach to linear logic, geometry of proofs and full completeness. PhD thesis, University of Ottawa (2000)
13. Heunen, C.: On the functor l2. In: Coecke, B., Ong, L., Panangaden, P. (eds.) Computation, Logic, Games and Quantum Foundations. LNCS, vol. 7860, pp. 107–121. Springer, Heidelberg (2013)
14. Hines, P.: The algebra of self-similarity and its applications. PhD thesis, University of Wales, Bangor (1997)
15. Hines, P.: The categorical theory of self-similarity. Theory and Applications of Categories 6, 33–46 (1999)
16. Hines, P.: A categorical analogue of the monoid semi-ring construction. Mathematical Structures in Computer Science 23(1), 55–94 (2013)
17. Hines, P.: Coherence in hilbert's hotel (2013) arXiv:1304.5954
18. Hines, P.: Girard's!() as a reversible fixed-point operator (2013) arXiv:1309.0361 [math.CT]
19. Hines, P.: Types and forgetfulness in categorical linguistics and quantum mechanics. In: Grefenstette, E., Heunen, C., Sadrzadeh, M. (eds.) Categorical Information Flow in Physics and Linguistics, pp. 215–248. Oxford University Press (2013)

20. Kock, J.: Elementary remarks on units in monoidal categories. Math. Proc. Cambridge Phil. Soc. 144, 53–76 (2008)

21. Lawson, M.V.: Inverse semigroups: the theory of partial symmetries. World Scientific, Singapore (1998)

22. Leinster, T.: A general theory of self-similarity. Advances in Mathematics 226, 2935–3017 (2011)

23. MacLane, S.: Categories for the working mathematician, 2nd edn. Springer, New York (1998)

24. Munn, W.D., Penrose, R.: A note on inverse semigroups. Mathematical Proceedings of the Cambridge Philosophical Society 51, 396–399 (1955)

25. Nivat, M., Perrot, J.: Une généralisation du monöide bicyclique. Comptes Rendus de l'Académie des Sciences de Paris 27, 824–827 (1970)

26. Pati, A.K., Braunstein, S.L.: Impossibility of deleting an unknown quantum state. Nature 404, 104 (2000)

27. Preston, G.B.: Representation of inverse semi-groups. J. London Math. Soc. 29, 411–419 (1954)

28. Wagner, V.V.: Generalised groups. Proceedings of the USSR Academy of Sciences 84, 1119–1122 (1952)

29. Wooters, W., Zurek, W.: A single quantum cannot be cloned. Nature 299, 802–803 (1982)

30. Høyer, P., Špalek, R.: Quantum fan-out is powerful. Theory of Computing 1, 81–103 (2005)

Initial Algebras of Terms
with Binding and Algebraic Structure

Bart Jacobs and Alexandra Silva

Institute for Computing and Information Sciences,
Radboud University Nijmegen
{bart,alexandra}@cs.ru.nl

Abstract. One of the many results which makes Joachim Lambek famous is: an initial algebra of an endofunctor is an isomorphism. This fixed point result is often referred to as "Lambek's Lemma". In this paper, we illustrate the power of initiality by exploiting it in categories of algebra-valued presheaves $\mathcal{EM}(T)^{\mathbb{N}}$, for a monad T on **Sets**. The use of presheaves to obtain certain calculi of expressions (with variable binding) was introduced by Fiore, Plotkin, and Turi. They used set-valued presheaves, whereas here the presheaves take values in a category $\mathcal{EM}(T)$ of Eilenberg-Moore algebras. This generalisation allows us to develop a theory where more structured calculi can be obtained. The use of algebras means also that we work in a linear context and need a separate operation ! for replication, for instance to describe strength for an endofunctor on $\mathcal{EM}(T)$. We apply the resulting theory to give systematic descriptions of non-trivial calculi: we introduce non-deterministic and weighted lambda terms and expressions for automata as initial algebras, and we formalise relevant equations diagrammatically.

Dedicated to Joachim Lambek on the occasion of his 90th birthday.

1 Introduction

In [22] Joachim Lambek proved a basic result that is now known as "Lambek's Lemma". It says: an initial algebra $F(A) \to A$ of an endofunctor $F \colon \mathbf{C} \to \mathbf{C}$ is an isomorphism. The proof is an elegant, elementary exercise in diagrammatic reasoning. The functor F can be seen as an abstraction of a signature, describing the arities of operations. The initial algebra, if it exists, is then the algebra whose carrier is constructed inductively, using these operations for the formation of terms. This "initial algebra semantics" forms the basis of the modern perspective on expressions (terms, formulas) defined by operations, in logic and in computer science. An early reference is [11]. A map going out of an initial algebra, obtained by initiality, provides denotational semantics of expressions. By construction this semantics is "compositional", which is a way of saying that it is a homomorphism of algebras.

A more recent development is to use the dual approach, involving coalgebras $X \to G(X)$, see [28,17]; they capture the operational semantics of languages,

C. Casadio et al. (Eds.): Lambek Festschrift, LNCS 8222, pp. 211–234, 2014.

describing the behaviour in terms of elementary steps. By duality, Lambek's Lemma also applies in this setting: a final coalgebra is an isomorphism. One of the highlights of the categorical approach to language semantics is the combined description of both denotational and operational semantics in terms of algebras and coalgebras of a functors (typically connected via distributive laws, see [19] for an overview).

This paper contains the first part of a study elaborating such a combined approach, using algebra-valued presheaves. It concentrates on obtaining structured terms as initial algebras. Historically, the first examples of initial algebras appeared in the category **Sets**, of sets and functions. But soon it became clear that initiality (or finality) in more complicated categories gives rise to a richer theory, involving additional language structure or stronger initiality principles.

- Induction with parameters (also called recursion) can be captured via initiality in "simple" slice categories, see *e.g.* [14]; similarly, initiality in (ordinary) slice categories gives a dependent form of induction, see [1].
- Expressions with variable binding operations like $\lambda x.\, xx$ in the lambda calculus can be described via initiality in presheaf categories, as shown in [8].
- This approach is extended to more complicated expressions, like in the π-calculus process language, see [10,9].
- The basics of Zermelo-Fraenkel set theory can be developed in a category with a suitably rich collection of map, see [18].
- Final coalgebras in Kleisli categories of a monad give rise to so-called trace semantics, see [12].

This paper contains an extension of the second point: the theory developed in [8] uses set-valued presheaves $\mathbb{N} \to$ **Sets**. In contrast, here we use algebra-valued presheaves $\mathbb{N} \to \mathcal{EM}(T)$, where $\mathcal{EM}(T)$ is the category of Eilenberg-Moore algebras of a monad T on **Sets**. The concrete examples elaborated at the end of the paper involve (1) the finite power set monad $\mathcal{P}_{\mathsf{fin}}$, with algebras $\mathcal{EM}(\mathcal{P}_{\mathsf{fin}}) = \mathbf{JSL}$, the category of join semilattices, and (2) the multiset monad $\mathcal{M}_S$ for a semiring S, with semimodules over S as algebras. The initial algebras in the category of presheaves over these Eilenberg-Moore categories have non-determinism and resource-sensitivity built into the languages, see Section 5. Moreover, in the current framework we can diagrammatically express relevant identifications, like the (β)-rule for the lambda calculus and the "trace equation" of Rabinovich [27].

The use of a category of algebras $\mathcal{EM}(T)$ instead of **Sets** requires a bit more care. The first part of the paper is devoted to extending the theory developed in [8] to this algebraic setting. The most important innovation that we need is the replication operation !, known from linear logic. Since Eilenberg-Moore categories are typically linear — with tensors $\otimes$ instead of cartesian products $\times$ — we have to use replication ! explicitly if we need non-linear features. For instance, substitution $t[s/v]$, that is, replacing all occurrences of the variable v in t by a term s, may require that we use s multiple times, namely if the variable v occurs multiple times in t. Hence the type of s must be $!\,Terms$, involving explicit replication !.

Since substitution is defined by initiality (induction) with parameters, like in [8], we need to adapt this approach. Specifically, we need to use replication ! in the so-called "strength" operation of a functor. It leads to a strength map of the form:

$$F(A) \otimes {!}B \xrightarrow{\quad strength \quad} F(A \otimes {!}B).$$

It turns out that such a strength map always exists, see Proposition 1.

In the end, in Section 5, we show how an initial algebra — of an endofunctor on such a rich universe as given by algebra-valued presheaves — gives rise to a very rich language in which many features are built-in and provided implicitly by the categorical infrastructure. This forms an illustration of the desirable situation where the formalism does the work for you. It shows the strength of the concepts that Joachim Lambek already worked on, almost fifty years ago.

2   Monads and Their Eilenberg-Moore Categories

This section recalls the basics of the theory of monads, as needed here. For more information, see *e.g.* [24,2,23,4]. A monad is a functor $T: \mathbf{C} \to \mathbf{C}$ together with two natural transformations: a unit $\eta: \mathrm{id}_{\mathbf{C}} \Rightarrow T$ and multiplication $\mu: T^2 \Rightarrow T$. These are required to make the following diagrams commute, for $X \in \mathbf{C}$.

A comonad on $\mathbf{C}$ is a monad on the opposite category $\mathbf{C}^{\mathrm{op}}$.

We briefly describe the examples of monads on **Sets** that we use in this paper.

- The lift monad $1 + (-)$ maps a set X to the set $1 + X$ containing an extra point, and a function $f: X \to Y$ to $1 + f: 1 + X \to 1 + Y$ given by $(1 + f)(\kappa_1(*)) = \kappa_1(*)$ and $(1 + f)(\kappa_2(x)) = \kappa_2(f(x))$. Here we write the coprojections of the coproduct as maps $\kappa_i: X_i \to X_1 + X_2$. The unit of the lift monad is the second injection $\eta(x) = \kappa_2(x)$ and multiplication is given by $\mu = [\kappa_1, \mathrm{id}]$.
- The powerset monad $\mathcal{P}$ maps a set X to the set $\mathcal{P}(X)$ of subsets of X, and a function $f: X \to Y$ to $\mathcal{P}(f): \mathcal{P}(X) \to \mathcal{P}(Y)$ given by direct image. Its unit is given by singleton $\eta(x) = \{x\}$ and multiplication by union $\mu(\{X_i \in \mathcal{P}X \mid i \in I\}) = \bigcup_{i \in I} X_i$. We also use the *finite* powerset monad $\mathcal{P}_{\mathrm{fin}}$ which sends a set X to the set of finite subsets of X.
- For a semiring S, the multiset monad $\mathcal{M} = \mathcal{M}_S$ is defined on a set X as:

$$\mathcal{M}(X) = \{\varphi: X \to S \mid \mathrm{supp}(\varphi) \text{ is finite }\}.$$

This monad captures multisets $\varphi \in \mathcal{M}(X)$, where the value $\varphi(x) \in S$ gives the multiplicity of the element $x \in X$. When $S = \mathbb{N}$, this is sometimes called the bag monad. On functions $f \colon X \to Y$, $\mathcal{M}$ is defined as

$$\mathcal{M}(f)(\varphi)(y) = \sum_{x \in f^{-1}(y)} \varphi(x).$$

The support set of a multiset $\varphi \in \mathcal{M}(X)$ is defined as $\mathrm{supp}(\varphi) = \{x \in X \mid \varphi(x) \neq 0\}$. The finite support requirement is necessary for $\mathcal{M}$ to be a monad. The unit of $\mathcal{M}$ is given by $\eta(x) = (x \mapsto 1)$ for $x \in X$ and the multiplication by

$$\mu(\Phi)(x) = \sum_{\varphi \in \mathrm{supp}(\Phi)} \Phi(\varphi) \cdot \varphi(x), \quad \text{for } \Phi \in \mathcal{M}\mathcal{M}(X).$$

Here, and later in the examples, we use the notation $(x \mapsto s)$ to denote a function $\varphi \colon X \to S$ assigning s to x and 0 to all other $x' \in X$.

For an arbitrary monad $T = (T, \eta, \mu)$ we write $\mathcal{EM}(T)$ for the category of Eilenberg-Moore algebras. Its objects are maps of the form $\alpha \colon T(X) \to X$ satisfying $\alpha \circ \eta = \mathrm{id}$ and $\alpha \circ \mu = \alpha \circ T(\alpha)$. A homomorphism $(\alpha \colon T(X) \to X) \longrightarrow (\beta \colon T(Y) \to Y)$ in $\mathcal{EM}(T)$ is a map $f \colon X \to Y$ between the underlying objects satisfying $f \circ \alpha = \beta \circ T(f)$. This yields a category, with obvious forgetful functor $\mathcal{U} \colon \mathcal{EM}(T) \to \mathbf{Sets}$. This functor $\mathcal{U}$ has a left adjoint $\mathcal{F}$, mapping an object X to the free algebra $\mathcal{F}(X) = (\mu \colon T^2(X) \to T(X))$ on X.

A category of algebras inherits all limits from the underlying category. In the special case of a monad T on $\mathbf{Sets}$, all colimits also exist in $\mathcal{EM}(T)$, by a result of Linton, see [2, §9.3, Prop. 4]. This category $\mathcal{EM}(T)$ is also symmetric monoidal closed, provided the monad T is commutative (*i.e.* monoidal), see [21,20]. We write the tensors as $\otimes$, with tensor unit I, and exponential $\multimap$. The free functor $\mathcal{F} \colon \mathbf{Sets} \to \mathcal{EM}(T)$ preserves the monoidal structure: $\mathcal{F}(X \times Y) \cong \mathcal{F}(X) \otimes \mathcal{F}(Y)$ and $\mathcal{F}(1) \cong I$, where 1 is a singleton set. As a result the set $A \multimap B$, for $A, B \in \mathcal{EM}(T)$, contains the homomorphisms $A \to B$ in $\mathcal{EM}(T)$.

The free algebra adjunction $\mathcal{F} \dashv \mathcal{U}$ induces a comonad $\mathcal{F}\mathcal{U} \colon \mathcal{EM}(T) \to \mathcal{EM}(T)$ that we write as $!_T$, or simply as $!$. This comonad $! = \mathcal{F}\mathcal{U}$ is relevant in the context of linear logic, see [13], whence the notation $!$; its Eilenberg-Moore coalgebras can be understood as bases [16]. This comonad $!$ on $\mathcal{EM}(T)$ also preserves the monoidal structure: $!(A \times B) \cong !A \otimes !B$ and $!1 \cong I$. Via these isomorphisms one sees that objects of the form $!A$ carry a comonoid structure (ε, Δ) given by:

$$
\begin{array}{ccc}
!A \xrightarrow{\;!(\langle\rangle)\;} !1 & \qquad & !A \xrightarrow{\;!(\langle\mathrm{id},\mathrm{id}\rangle)\;} !(A \times A) \\
\underset{\varepsilon}{\searrow} \;\; \downarrow{\scriptstyle\cong} & & \underset{\Delta}{\searrow} \;\; \downarrow{\scriptstyle\cong} \\
I & & !A \otimes !A.
\end{array}
\tag{1}
$$

These maps yield projection and diagonal operations for weakening and contraction. Also, they turn the functor $(-) \otimes !A \colon \mathcal{EM}(T) \to \mathcal{EM}(T)$ into a comonad.

Typically this ! is used to introduce "classical" computation in the "linear" setting $\mathcal{EM}(T)$, since (linear) algebra maps $!A \to B$ in $\mathcal{EM}(T)$ are in bijective correspondence with (ordinary) functions $\mathcal{U}(A) \to \mathcal{U}(B)$. In a linear setting a map $A \to B$ consumes/uses the input/resource A exactly once. But maps of the form $!A \to B$ can use A an arbitrary number of times.

In the sequel, we shall be using initial algebras of functors in Eilenberg-Moore categories $\mathcal{EM}(T)$. Initiality corresponds to induction. For "stronger" forms of initiality, with parameters, one needs strong functors. We recall the basic fact that each functor $H \colon \mathbf{Sets} \to \mathbf{Sets}$ is strong, via a strength map $\mathrm{st} \colon H(X) \times Y \to H(X \times Y)$, given by: $\mathrm{st}(u, y) = H(\lambda x. \langle x, y \rangle)(u)$. For instance, for the powerset functor $\mathcal{P}$ this yields: $\mathrm{st}(u, y) = \{\langle x, y \rangle \mid x \in u\}$. In this set the element y is used multiple times. In the next result, giving a strength in a linear setting, such multiple use leads to an occurrence of the comonad !. This new result extends the uniform set-theoretic definition of strength to an algebraic context.

Proposition 1. *Let T be a commutative monad on $\mathbf{Sets}$, and $H \colon \mathcal{EM}(T) \to \mathcal{EM}(T)$ be an arbitrary functor. For algebras A, B there is a "non-linear" strength map:*

$$H(A) \otimes !B \xrightarrow{\quad \mathrm{st} \quad} H(A \otimes !B). \tag{2}$$

This strength map is natural in A and B, and makes the following unit and associativity diagrams commute.

$$
\begin{array}{ccc}
H(A) \otimes !1 & \xrightarrow{\ \mathrm{st}\ } & H(A \otimes !1) \\
 & \searrow{\scriptstyle \cong} & \downarrow{\scriptstyle \cong} \\
 & & H(A)
\end{array}
$$

$$
\begin{array}{ccc}
\bigl(H(A) \otimes !B\bigr) \otimes !C & \xrightarrow{\ \mathrm{st}\otimes\mathrm{id}\ } H(A \otimes !B) \otimes !C \xrightarrow{\ \mathrm{st}\ } & H\bigl((A \otimes !B) \otimes !C\bigr) \\
{\scriptstyle \cong}\downarrow & & \downarrow{\scriptstyle \cong} \\
H(A) \otimes (!B \otimes !C) & & H\bigl(A \otimes (!B \otimes !C)\bigr) \\
{\scriptstyle \cong}\downarrow & & \downarrow{\scriptstyle \cong} \\
H(A) \otimes !(B \times C) & \xrightarrow{\hspace{4cm}\mathrm{st}\hspace{4cm}} & H\bigl(A \otimes !(B \times C)\bigr)
\end{array}
$$

Proof. We recall from [21,20] that the tensor $\otimes$ of algebras comes with a special map $\otimes \colon A \times B \to A \otimes B$ which is bilinear (a homomorphism in each variable separately), and also universal, in the following manner: for each bilinear map $f \colon A \times B \to C$ there is a unique algebra map $\overline{f} \colon A \otimes B \to C$ with $\overline{f} \circ \otimes = f$. Thus, there is a bijective correspondence:

$$
\frac{\text{bilinear maps } \ A \times B \longrightarrow C}{\text{algebra maps } \ A \otimes B \longrightarrow C}
$$

For the definition of strength we use the following correspondences.

$$
\frac{\dfrac{H(A) \otimes !B \xrightarrow{\;\;st\;\;} H(A \otimes !B)}{!B \longrightarrow H(A) \multimap H(A \otimes !B)}}{\mathcal{U}(B) \longrightarrow \mathcal{U}\Big(H(A) \multimap H(A \otimes !B) \Big)}
$$

We shall construct this last map, and then obtain strength st via these correspondences. For an element $b \in \mathcal{U}(B)$, applying the unit $\eta = \eta_{\mathcal{U}(B)}$ of the monad yields an element $\eta(b) \in \mathcal{U}(\mathcal{F}(\mathcal{U}(B))) = \mathcal{U}(!B)$. Since $\otimes$ is bilinear, the map $(-) \otimes \eta(b) \colon A \to A \otimes !B$ is a homomorphism of algebras. Applying the functor H yields another homomorphism:

$$
H(A) \xrightarrow{\;\;H\big((-)\otimes\eta(b)\big)\;\;} H(A \otimes !B)
$$

This algebra map is an element of the set $\mathcal{U}\Big(H(A) \multimap H(A \otimes !B) \Big)$. Hence we are done.

We prove naturality of strength, and leave the unit and associativity diagrams to the interested reader. For algebra maps $f \colon A \to C$ and $g \colon B \to D$ we wish to show that the following diagram commutes.

$$
\begin{array}{ccc}
H(A) \otimes !B & \xrightarrow{\;\;st\;\;} & H(A \otimes !B) \\
{\scriptstyle H(f)\otimes !g}\Big\downarrow & & \Big\downarrow{\scriptstyle H(f\otimes !g)} \\
H(C) \otimes !D & \xrightarrow[\;\;st\;\;]{} & H(C \otimes !D)
\end{array}
$$

Thus, we need to show for each $b \in \mathcal{U}(B)$ that the two algebra maps $H(A) \to H(C \otimes !D)$, obtained as:

$$
H\big(f \otimes !g\big) \circ H\big((-) \otimes \eta(b)\big) \qquad \text{and} \qquad H\big((-) \otimes \eta(g(b))\big) \circ H(f)
$$

are the same. But this is obvious by naturality of η.   □

It is not hard to see that an algebra of the form $A \otimes !B$ is isomorphic to a copower $\mathcal{U}(B) \cdot A = \coprod_{b \in B} A$, since for every algebra C,

$$
\frac{\dfrac{A \otimes !B \longrightarrow C}{\mathcal{U}(B) \longrightarrow (A \multimap C)}}{\dfrac{(A \longrightarrow C)_{b \in B}}{\coprod_{b \in B} A \longrightarrow C}}
$$

Such tensors/copowers $A \otimes !B$ are used in [25] to model state-based computation, where the state A can be used only linearly.

Example 2. As illustration, and for future reference, we shall describe the strength map (2) explicitly for several functors $H\colon \mathcal{EM}(T) \to \mathcal{EM}(T)$.

1. If H is the identity functor, then obviously the strength map $A \otimes !B \to A \otimes !B$ is the identity.
2. If H is the constant functor K_C, sending everything to the algebra C, then the strength map is defined via the projection ε obtained via (1):

$$K_C(A) \otimes !B = C \otimes !B \xrightarrow{\mathrm{id} \otimes \varepsilon} C \otimes I \cong C = K_C(A \otimes !B).$$

3. If $H = H_1 + H_2$, with strength maps st^i for H_i we get a new strength map $\mathrm{st}\colon (H_1(A) + H_2(A)) \otimes !B \to H_1(A \otimes !B) + H_2(A \otimes !B)$ determined by:

$$\mathrm{st}(\kappa_1 s \otimes t) = \kappa_1 \mathrm{st}^1(s \otimes t) \qquad \mathrm{st}(\kappa_2 s \otimes t) = \kappa_2 \mathrm{st}^2(s \otimes t).$$

This follows since the map st is defined via Currying in:

$$!B \xrightarrow{\hspace{2cm}} \Big(H_1(A) + H_2(A)\Big) \multimap \Big(H_1(A \otimes !B) + H_2(A \otimes !B)\Big)$$

The linear map on the right is clearly obtained via a sum $+$ of the existing strength maps st^i.

4. Similarly, if $H = H_1 \times H_2$, then:

$$\mathrm{st}(\langle s_1, s_2 \rangle \otimes t) = \langle \mathrm{st}^1(s_1 \otimes t), \mathrm{st}^2(s_2 \otimes t) \rangle.$$

5. For a tensor $H = H_1 \otimes H_2$ we need the duplication map Δ from (1) in:

$$\mathrm{st}((s_1 \otimes s_2) \otimes t) = \mathrm{st}^1(s_1 \otimes t) \otimes \mathrm{st}^2(s_2 \otimes t).$$

Diagrammatically this is:

$$(H_1(A) \otimes H_2(A)) \otimes !B$$
$$\downarrow {\scriptstyle \mathrm{id} \otimes \Delta}$$
$$(H_1(A) \otimes H_2(A)) \otimes (!B \otimes !B) \xrightarrow{\cong} (H_1(A) \otimes !B) \otimes (H_2(A) \otimes !B)$$
$$\qquad\qquad\qquad\qquad\qquad {\scriptstyle \mathrm{st}^1 \otimes \mathrm{st}^2} \downarrow$$
$$H_1(A \otimes !B) \otimes H_2(A \otimes !B).$$

6. Finally we look at functors of the form $H = !H_1$. Then:

$$\mathrm{st}(s \otimes t) = !\big(\mathrm{st}^1((-) \otimes t)\big)(s),$$

where ! is applied to the map $\mathrm{st}^1((-) \otimes t)\colon H_1(A) \to H_1(A \otimes !B)$.

With this strength map (2) we can now use the following strengthened formulation of initiality, involving additional parameters. This version of initiality is standard, see *e.g.* [14], except that it is formulated here in a linear setting with !. In principle, such strengthened versions of induction can be formulated more generally in terms of comonads and distributive laws, like in [32], but such generality is not needed here.

Lemma 3. *Let* $H\colon \mathcal{EM}(T) \to \mathcal{EM}(T)$ *be an endofunctor on the category of Eilenberg-Moore algebras of a commutative monad T on* **Sets**. *If this functor H has an initial algebra $a\colon H(A) \xrightarrow{\cong} A$, then: for each $c\colon H(C) \otimes !B \to C$ there is a unique map $h\colon A \otimes !B \to C$ in* $\mathcal{EM}(T)$ *in a commuting diagram:*

$$
\begin{array}{ccc}
H(A) \otimes !B \xrightarrow{\ \mathrm{id}\otimes\Delta\ } H(A) \otimes !B \otimes !B \xrightarrow{\ \mathrm{st}\otimes\mathrm{id}\ } H(A \otimes !B) \otimes !B \xrightarrow{H(h)\otimes\mathrm{id}} H(C) \otimes !B \\
{\scriptstyle a\otimes\mathrm{id}}\Big\downarrow{\scriptstyle\cong} \hspace{8cm} \Big\downarrow{\scriptstyle c} \\
A \otimes !B \xrightarrow{\hspace{9cm}} C \\
\hspace{4.5cm}{\scriptstyle h}
\end{array}
$$

$$(3)$$

Proof. Use initiality of a wrt. the algebra $H(!B \multimap C) \to (!B \multimap C)$ obtained by abstraction $\Lambda(-)$ in:

$$
c' = \Lambda\Big(H(!B \multimap C) \otimes !B \xrightarrow{\ \mathrm{id}\otimes\Delta\ } H(!B \multimap C) \otimes !B \otimes !B
$$
$$
\Big\downarrow{\scriptstyle \mathrm{st}\otimes\mathrm{id}}
$$
$$
H\big((!B \multimap C) \otimes !B\big) \otimes !B \xrightarrow{H(\mathrm{ev})\otimes\mathrm{id}} H(C) \otimes !B \xrightarrow{\ c\ } C \Big).
$$

$\square$

For future use we mention the following basic result. The proof is left as exercise to the reader.

Lemma 4. *Consider a situation:*

$$
H \,\circlearrowleft\, \mathbf{A} \underset{G}{\overset{F}{\rightleftarrows}} \mathbf{A} \qquad \text{where } F \dashv G.
$$

In the presence of an isomorphism $HF \cong FH$, if $U_X \in \mathbf{A}$ is (carries) the free H-algebra on $X \in \mathbf{A}$, then $F(U_X)$ is the free H-algebra on $F(X)$. $\square$

2.1 Intermezzo: Strength for Endofunctors on Hilbert Spaces

In Proposition 1 we have seen that strength maps exist in an algebraic context if we restrict ourselves to objects of the form $!B$, which, as we have seen in (1), come equipped with a comonoid structure for weakening and contraction. In a quantum context such comonoid structure is described in terms of Frobenius algebras. In [5] it shown that on a finite-dimensional Hilbert space such a Frobenius algebra structure corresponds to an orthonormal basis. The next result shows that in this situation one also define strength maps, much like in the proof of Proposition 1. (This will not be used in the sequel.)

Proposition 5. *Let $H\colon$ **FdHilb** $\to$ **FdHilb** be an arbitrary endofunctor on the category of finite-dimensional Hilbert spaces and (continuous) linear maps. For a* *space $W \in$ **FdHilb** with orthonormal basis $B = \{b_1, \ldots, b_n\}$ there is a collection of strength maps:*

$$H(V) \otimes W \xrightarrow{\quad \mathrm{st}^B \quad} H(V \otimes W)$$

natural in V. The unit and associativity diagrams commute for these strength maps.

Proof. Each base vector $b_i \in W$ yields a (continuous) linear map $(-) \otimes b_i \colon V \to V \otimes W$, and thus by applying the functor H we get $H((-) \otimes b_i) \colon H(V) \to H(V \otimes W)$. For an arbitrary vector $w \in W$ we write $w = \sum_i w_i b_i$ and define $f(u, w) = \sum_i w_i H((-) \otimes b_i)(u) \in H(V \otimes W)$. This yields a bilinear map $H(V) \times W \to H(V \otimes W)$, and thus a unique linear map $\mathrm{st}^B \colon H(V) \otimes W \to H(V \otimes W)$, with $\mathrm{st}^B(u \otimes w) = f(u, w)$. □

As already suggested above, this strength map in **FdHilb** really depends on the basis. This can be illustrated in a simple example (using Hilbert spaces over $\mathbb{C}$). Take as endofunctor $H(X) = X \otimes X$, with $V = \mathbb{C}$ and $W = \mathbb{C}^2 = \mathbb{C} \oplus \mathbb{C}$ in **FdHilb**. The strength map $H(\mathbb{C}) \otimes \mathbb{C}^2 \to H(\mathbb{C} \otimes \mathbb{C}^2)$ then amounts to a map $\mathbb{C}^2 \to \mathbb{C}^4$, using that $\mathbb{C}$ is the tensor unit. But we shall not drop this $\mathbb{C}$.

First we take the standard basis $B = \{(1, 0), (0, 1)\}$ on $\mathbb{C}^2$. The resulting strength map $\mathrm{st}^B \colon H(\mathbb{C}) \otimes \mathbb{C}^2 \to H(\mathbb{C} \otimes \mathbb{C}^2)$ is given by:

$$\begin{aligned}
\mathrm{st}^B &\big((u \otimes v) \otimes (w_1, w_2)\big) \\
&= w_1 H\big((-) \otimes (1, 0)\big)(u \otimes v) + w_2 H\big((-) \otimes (0, 1)\big)(u \otimes v) \\
&= w_1\big((u \otimes (1, 0)) \otimes (v \otimes (1, 0))\big) + w_2\big((u \otimes (0, 1)) \otimes (v \otimes (0, 1))\big) \\
&= w_1(uv, 0, 0, 0) + w_2(0, 0, 0, uv) \\
&= uv(w_1, 0, 0, w_2).
\end{aligned}$$

But if we take the basis $C = \{(1, 1), (1, -1)\}$ we get a different strength map:

$$\begin{aligned}
\mathrm{st}^C &\big((u \otimes v) \otimes (w_1, w_2)\big) \\
&= \mathrm{st}^C\big((u \otimes v) \otimes \big(\tfrac{w_1 + w_2}{2}(1, 1) + \tfrac{w_1 - w_2}{2}(1, -1)\big)\big) \\
&= \tfrac{w_1 + w_2}{2} H\big((-) \otimes (1, 1)\big)(u \otimes v) + \tfrac{w_1 - w_2}{2} H\big((-) \otimes (1, -1)\big)(u \otimes v) \\
&= \tfrac{w_1 + w_2}{2}\big((u \otimes (1, 1)) \otimes (v \otimes (1, 1))\big) + \tfrac{w_1 - w_2}{2}\big((u \otimes (1, -1)) \otimes (v \otimes (1, -1))\big) \\
&= \tfrac{w_1 + w_2}{2}\big(uv, uv, uv, uv\big) + \tfrac{w_1 - w_2}{2}\big(uv, -uv, -uv, uv\big) \\
&= uv(w_1, w_2, w_2, w_1).
\end{aligned}$$

3   Functor Categories and Presheaves

Later on in this paper we will be using presheaf categories of the form $\mathcal{EM}(T)^{\mathbb{N}}$, where T is a monad on **Sets** and $\mathbb{N} \hookrightarrow$ **Sets** is the full subcategory with the natural numbers $n \in \mathbb{N}$ as objects, considered as n-element set $\{0, 1, \ldots, n - 1\}$.

In this section we collect some basic facts about such functor categories. Some of these results apply more generally.

For instance, if $\mathbf{A}$ is a (co)complete category, then so is the functor category $\mathbf{A}^{\mathbf{C}}$. Limits and colimits in this category are constructed elementwise. We write $\Delta\colon \mathbf{A} \to \mathbf{A}^{\mathbf{C}}$ for the functor that sends an object $X \in \mathbf{A}$ to the constant functor $\Delta(X)\colon \mathbf{C} \to \mathbf{A}$ that sends everything to X. This functor Δ should not be confused with the natural transformation Δ from (1). Each functor $F\colon \mathbf{A} \to \mathbf{B}$ yields a functor $F^{\mathbf{C}}\colon \mathbf{A}^{\mathbf{C}} \to \mathbf{B}^{\mathbf{C}}$, also by an elementwise construction: on objects $F^{\mathbf{C}}(P)(Y) = F(P(Y))$, and on morphisms (natural transformations): $F^{\mathbf{C}}(\sigma)_Y = F(\sigma_Y)$.

Lemma 6. *An adjunction $F \dashv G$ as on the left below yields an adjunction $F^{\mathbf{C}} \dashv G^{\mathbf{C}}$ as on the right.*

$$
\mathbf{A} \underset{G}{\overset{F}{\rightleftarrows}} \mathbf{B}
\qquad\qquad
\mathbf{A}^{\mathbf{C}} \underset{G^{\mathbf{C}}}{\overset{F^{\mathbf{C}}}{\rightleftarrows}} \mathbf{B}^{\mathbf{C}}
\qquad\qquad \square
$$

In the sequel we extend the notion $(-)^{\mathbf{C}}$ to bifunctors $F\colon \mathbf{A} \times \mathbf{A} \to \mathbf{A}$, yielding a new bifunctors $F^{\mathbf{C}}\colon \mathbf{A}^{\mathbf{C}} \times \mathbf{A}^{\mathbf{C}} \to \mathbf{A}^{\mathbf{C}}$, given by $F^{\mathbf{C}}(P, Q)(X) = F(P(X), Q(X))$. In particular, this yields a tensor product $\otimes^{\mathbf{C}}$ on $\mathbf{A}^{\mathbf{C}}$, assuming a tensor product $\otimes$ on $\mathbf{A}$. The tensor unit $I \in \mathbf{A}$ then yields a unit $\Delta(I) \in \mathbf{A}^{\mathbf{C}}$.

For an endofunctor like F we write $\mathbf{Alg}(F)$ and $\mathbf{CoAlg}(F)$ for the categories of (functor) algebras and coalgebras of F.

Lemma 7. *For an endofunctor $F\colon \mathbf{A} \to \mathbf{A}$ there is an obvious functor $\Delta\colon \mathbf{Alg}(F) \to \mathbf{Alg}(F^{\mathbf{C}})$, sending an F-algebra $a\colon F(X) \to X$ to a "constant" $F^{\mathbf{C}}$-algebra on $\Delta(X)$. We write this algebra as a^{Δ}, which is a natural transformation $F^{\mathbf{C}}(\Delta(X)) \Rightarrow \Delta(X)$, with components:*

$$
F^{\mathbf{C}}(\Delta(X))(Y) = F(X) \xrightarrow{\;(a^{\Delta})_Y = a\;} X = \Delta(X)(Y).
$$

This functor $\Delta\colon \mathbf{Alg}(F) \to \mathbf{Alg}(F^{\mathbf{C}})$ has a left adjoint if $\mathbf{C}$ has a final object 1, and a right adjoint if there is an initial object $0 \in \mathbf{C}$.

Similarly, there is a functor $\Delta\colon \mathbf{CoAlg}(F) \to \mathbf{CoAlg}(F^{\mathbf{C}})$ which has a left (resp. right) adjoint in presence of a final (resp. initial) object in $\mathbf{C}$.

Proof. Assume a final object $1 \in \mathbf{C}$. The "evaluate at 1" functor $(-)(1)\colon \mathbf{A}^{\mathbf{C}} \to \mathbf{A}$ yields a functor $\mathbf{Alg}(F^{\mathbf{C}}) \to \mathbf{Alg}(F)$. This gives a left adjoint to $\Delta\colon \mathbf{Alg}(F) \to \mathbf{Alg}(F^{\mathbf{C}})$ since there is a bijective correspondence:

$$
\frac{
\left(\begin{matrix} F^{\mathbf{C}}(P) \\ \beta \Downarrow \\ P \end{matrix} \right)
\;\xrightarrow{\;\varphi\;}\;
\left(\begin{matrix} F^{\mathbf{C}}(\Delta(X)) \\ \Downarrow a^{\Delta} \\ \Delta(X) \end{matrix} \right)
}{
\left(\begin{matrix} F(P(1)) \\ \beta_1 \downarrow \\ P(1) \end{matrix} \right)
\;\xrightarrow{\;f\;}\;
\left(\begin{matrix} F(X) \\ \downarrow a \\ X \end{matrix} \right)
}
$$

In this situation φ consists of a collection of maps $\varphi_Y\colon P(Y) \to X$, forming a map of algebras from β_Y to a, natural in Y—so that $\varphi_Z \circ P(g) = \varphi_Y$ for each $g\colon Y \to Z$ in $\mathbf{C}$. The correspondence can be described as follows.

- Given φ as above we take $\overline{\varphi} = \varphi_1\colon P(1) \to X$. This is clearly an algebra map $\beta_1 \to a$.
- Given $f\colon P(1) \to X$ we define a natural transformation $\overline{f}$ with components $\overline{f}_Y = f \circ P(!_Y)$, where $!_Y\colon Y \to 1$ is the unique map. It is not hard to see that $\overline{f}$ is natural and a map of algebras, as required.

One has $\overline{\overline{\varphi}} = \varphi$ and $\overline{\overline{f}} = f$. The rest of the lemma is obtained in the same manner. $\square$

This result is useful to obtain preservation of initial algebra or final coalgebras by the functor Δ.

In a similar manner an endofunctor $F\colon \mathbf{C} \to \mathbf{C}$ gives rise to a functor $\mathbf{A}^F\colon \mathbf{A}^{\mathbf{C}} \to \mathbf{A}^{\mathbf{C}}$, via $\mathbf{A}^F(P)(Y) = P(F(Y))$. In this situation a natural transformation $\sigma\colon F \Rightarrow G$ yields another natural transformation $\mathbf{A}^\sigma\colon \mathbf{A}^F \Rightarrow \mathbf{A}^G$ with components $(\mathbf{A}^\sigma)_{P,Y} = P(\sigma_Y)\colon P(F(Y)) \to P(G(Y))$. This is of interest when F is a monad or comonad.

Lemma 8. *Let $T\colon \mathbf{C} \to \mathbf{C}$ be a (co)monad. Then so is $\mathbf{A}^T\colon \mathbf{A}^{\mathbf{C}} \to \mathbf{A}^{\mathbf{C}}$.*

Proof. In the obvious way: for instance if $T = (T, \eta, \mu)$ is a monad, then $\mathbf{A}^T$ is also a monad, where $\eta_P^T = \mathbf{A}_P^\eta\colon P \Rightarrow \mathbf{A}^T(P)$ and $\mu_P^T = \mathbf{A}_P^\mu\colon \left(\mathbf{A}^T\right)^2(P) = \mathbf{A}^{T^2}(P) \Rightarrow \mathbf{A}^T(P)$ have components:

$$P(Y) \xrightarrow{\;P(\eta_Y)\;} P(T(Y)) \xleftarrow{\;P(\mu_Y)\;} P(T^2(Y)). \qquad \square$$

We now restrict to functor categories of the form $\mathbf{A}^{\mathbb{N}}$, where $\mathbb{N} \hookrightarrow \mathbf{Sets}$ is the full subcategory with natural numbers $n = \{0, 1, \ldots, n-1\}$ as objects. We shall introduce a "weakening" monad $\mathcal{W}$ on this category, via the previous lemma. This $\mathcal{W}$ comes from [8] where it is written as δ and where it is called context extension. But it is called differentiation in [10] and dynamic allocation in [9]. Here we prefer to call it weakening to emphasise this aspect of context extension. We write it as $\mathcal{W}$ and not as δ to avoid a conflict with the comultiplication operation of the comonad !. Before we proceed we recall that in the category $\mathbb{N}$ the number 0 is initial, and the number $n + m$ is the coproduct of $n, m \in \mathbb{N}$. In any category with coproducts, the functor $(-) + X$ is a monad. We apply this to the category $\mathbb{N}$ and get the lift monad $(-) + 1$. The previous lemma now gives the following result.

Lemma 9. *For an arbitrary category $\mathbf{A}$ define a monad $\mathcal{W} = \mathbf{A}^{(-)+1}\colon \mathbf{A}^{\mathbb{N}} \to \mathbf{A}^{\mathbb{N}}$, so that:*

$$\mathcal{W}(P)(n) \;=\; P(n+1) \qquad \mathcal{W}(P)(f) \;=\; P(f + \mathrm{id}_1) \qquad \mathcal{W}(\sigma)_n \;=\; \sigma_{n+1}.$$

The unit up: id $\Rightarrow \mathcal{W}$ *and multiplication* ctt: $\mathcal{W}^2 \Rightarrow \mathcal{W}$ *of this monad have components* $\text{up}_P \colon P \Rightarrow \mathcal{W}(P)$ *and* $\text{ctt}_P \colon \mathcal{W}(\mathcal{W}(P)) \to \mathcal{W}(P)$ *with:*

$$P(n) \xrightarrow{\text{up}_{P,n}=P(\kappa_1)} P(n+1) \xleftarrow{\text{ctt}_{P,n}=P([\text{id},\kappa_2])} P(n+2). \qquad \square$$

The map up sends an item in $P(n)$ to $P(n+1)$, where the context $n+1$ has an additional variable v_{n+1}. The abbreviation ctt stands for "contract"; it can be understood as substitution $[v_{n+1}/v_{n+2}]$, removing the last variable. There is also a "swap" map swp: $\mathcal{W}^2 \Rightarrow \mathcal{W}^2$ defined, following [8], as:

$$\mathcal{W}^2(P)(n) = P(n+2) \xrightarrow{\text{swp}_{P,n}=P(\text{id}+[\kappa_2,\kappa_1])} P(n+2) = \mathcal{W}^2(P)(n). \qquad (4)$$

This swap map can be understood as simultaneous substitution $[v_{n+1}/v_{n+2}, v_{n+2}/v_{n+1}]$.

The following trivial observation will be useful later on (combined with Lemma 4). It includes commutation with (co)limits, which are given in a pointwise manner in functor categories.

Lemma 10. *The weakening functor* $\mathcal{W} \colon \mathbf{A}^{\mathbb{N}} \to \mathbf{A}^{\mathbb{N}}$ *commutes with functors* $F^{\mathbb{N}}$ *defined pointwise:* $\mathcal{W}F^{\mathbb{N}} = F^{\mathbb{N}}\mathcal{W}$. *This also holds when* F *is a bifunctor, covering products* $\times$, *tensors* $\otimes$ *and coproducts* $+$. $\qquad \square$

In [8] it is shown, for the special case where $\mathbf{A} = \mathbf{Sets}$, that this functor $\mathcal{W}$ has both a left and a right adjoint. In our more general situation basically the same constructions can be used, but they require more care. We will describe left and right adjoints separately.

Lemma 11. *Assuming the category* $\mathbf{A}$ *has finite copowers* $n \cdot X = X + \cdots + X$ *(n times), the weakening functor* $\mathcal{W} \colon \mathbf{A}^{\mathbb{N}} \to \mathbf{A}^{\mathbb{N}}$ *has a left adjoint, given by* $Q \mapsto (-) \cdot Q(-)$; *this right-hand-side is the functor* $\mathbb{N} \to \mathbf{A}$ *given by* $n \mapsto n \cdot Q(n)$.

Proof. There is a bijective correspondence between components of natural transformations:

$$\frac{Q(n) \overset{\sigma_n}{=\!=\!\Rightarrow} \mathcal{W}(P)(n) = P(n+1)}{n \cdot Q(n) \underset{\tau_n}{=\!=\!\Rightarrow} P(n)}$$

Given σ_n we take the cotuple:

$$\overline{\sigma}_n = \left(n \cdot Q(n) \xrightarrow{\left[P([\text{id},i]) \circ \sigma_n \right]_{i \in n}} P(n) \right).$$

In the reverse direction, given τ, we take:

$$\overline{\tau}_n = \left(Q(n) \xrightarrow{Q(\kappa_1)} Q(n+1) \xrightarrow{\kappa_2} (n+1) \cdot Q(n+1) \xrightarrow{\tau_{n+1}} P(n+1) \right).$$

It is not hard see that these $\overline{(-)}$ constructions yield natural transformations and are each other's inverses. $\qquad \square$

Remark 12. In the sequel we typically use $\mathbf{A} = \mathcal{EM}(T)$, the Eilenberg-Moore category of a monad T on **Sets**. If we assume that the monad T is commutative, then, the category $\mathcal{EM}(T)$ is symmetric monodial closed (see [21,20]), with tensor unit $I = \mathcal{F}(1)$, where $\mathcal{F}$ is the free monad functor and 1 is the terminal object in the underlying category. In that case we can reorganise the left adjoint from the previous lemma, since:

$$
\begin{aligned}
n \cdot Q(n) \;&\cong\; n \cdot (I \otimes Q(n)) \\
&\cong\; (n \cdot I) \otimes Q(n) \qquad \text{using that } (-) \otimes X \text{ is a left adjoint} \\
&\cong\; (n \cdot \mathcal{F}(1)) \otimes Q(n) \\
&\cong\; \mathcal{F}(n \cdot 1) \otimes Q(n) \qquad \text{since free functors preserve coproducts} \\
&\cong\; \mathcal{F}(n) \otimes Q(n).
\end{aligned}
$$

In this Eilenberg-Moore situation we can thus describe the adjunction from the previous lemma as $\mathcal{F} \otimes^{\mathbb{N}} (-) \dashv \mathcal{W}$, where $\otimes^{\mathbb{N}}$ is the pointwise tensor on $\mathcal{EM}(T)^{\mathbb{N}}$, and where $\mathcal{F} \in \mathcal{EM}(T)^{\mathbb{N}}$ is the restriction of the free functor $\mathcal{F}\colon \mathbf{Sets} \to \mathcal{EM}(T)$ to $\mathbb{N} \hookrightarrow \mathbf{Sets}$. It corresponds to the free variable functor V from [8] — and will be used as such later on.

The right adjoint to weakening is more complicated. In the set-theoretic setting of [8] it is formulated in terms of sets of natural transformations. In the current more general context this construction has to be done internally, via a suitable equaliser.

Lemma 13. *If a category* $\mathbf{A}$ *is complete, then the weakening functor* $\mathcal{W}\colon \mathbf{A}^{\mathbb{N}} \to \mathbf{A}^{\mathbb{N}}$ *has a right adjoint.*

Proof. For a presheaf $Q \in \mathbf{A}^{\mathbb{N}}$ we define a new presheaf $\mathcal{R}(Q) \in \mathbf{A}^{\mathbb{N}}$, where $\mathcal{R}(Q)(n)$ is obtained as equaliser in $\mathbf{A}$:

$$
\mathcal{R}(Q)(n) \;\xrightarrow{\;e_n\;}\; \prod_{m \in \mathbb{N}} Q(m)^{((m+1)^n)} \;\underset{\psi}{\overset{\varphi}{\rightrightarrows}}\; \prod_{g\colon m \to m'} Q(m')^{((m+1)^n)}
$$

where the parallel maps φ, ψ are defined by:

$$
\begin{aligned}
\varphi &= \langle\!\langle\, Q(g + \mathrm{id}) \circ \pi_f \circ \pi_m \,\rangle_{f\colon n \to m+1}\rangle_{g\colon m \to m'} \\
\psi &= \langle\!\langle\, \pi_{(g+\mathrm{id}) \circ f} \circ \pi_{m'} \,\rangle_{f\colon n \to m+1}\rangle_{g\colon m \to m'}.
\end{aligned}
$$

We sketch the adjunction $\mathcal{W} \dashv \mathcal{R}$. For a natural transformation $\sigma\colon \mathcal{W}(P) \Rightarrow Q$ we have maps $\sigma_n\colon P(n+1) \to Q(n)$ in $\mathbf{A}$, and define:

$$
\sigma'_n = \langle\!\langle\, \sigma_m \circ P(f) \,\rangle_{f\colon n \to m+1}\rangle_{m \in \mathbb{N}} \;\colon\; P(n) \longrightarrow \prod_{m \in \mathbb{N}} Q(m)^{((m+1)^n)}
$$

We leave it to the reader to verify that $\varphi \circ \sigma'_n = \psi \circ \sigma'_n$. This equation yields a unique map $\overline{\sigma}_n\colon P(n) \to \mathcal{R}(Q)(n)$.

Conversely, given $\tau\colon P \Rightarrow \mathcal{R}(Q)$ we get $e_n \circ \tau_n\colon P(n) \to \prod_{m\in\mathbb{N}} Q(m)^{((m+1)^n)}$. Hence we take:

$$\bar{\tau}_n \;=\; \pi_{\mathrm{id}_{n+1}} \circ \pi_n \circ e_{n+1} \circ \tau_{n+1} \;\colon\; P(n+1) \longrightarrow Q(n).$$

Remaining details are left to the reader.   □

Along the same lines of the previous result one obtains the following standard result.

Lemma 14. *Let $(I, \otimes, \multimap)$ be the monoidal closed structure of a complete category $\mathbf{A}$. The pointwise monoidal structure $(\Delta(I), \otimes^{\mathbb{N}})$ on the functor category $\mathbf{A}^{\mathbb{N}}$ is then also closed.*   □

When $\mathbf{A} = \mathcal{EM}(T)$ like in Remark 12, we can thus write weakening as exponent $\mathcal{W} \cong [\mathcal{F}, -]$, using the adjunction $\mathcal{F} \otimes^{\mathbb{N}} (-) \dashv \mathcal{W}$.

What we still need is the following.

Lemma 15. *For a commutative monad T on $\mathbf{Sets}$, the weakening monad $\mathcal{W}\colon \mathcal{EM}(T)^{\mathbb{N}} \to \mathcal{EM}(T)^{\mathbb{N}}$ is a strong monad wrt. the pointwise monoidal structure $(\Delta(I), \otimes^{N})$. The strength map is given by:*

$$\mathcal{W}(P) \otimes^{\mathbb{N}} Q \xrightarrow{\;\mathrm{st}=\mathrm{id}\otimes\mathrm{up}\;} \mathcal{W}(P \otimes^{\mathbb{N}} Q).$$ □

4   Substitution

By now we have collected enough material to transfer the definition of substitution used in [8] for set-valued presheaves to the algebra-valued presheaves used in the present setting. Here we rely on Lemma 4, where in [8] a uniformity principle of fixed points is used. Alternatively, free constructions can be used. But we prefer the rather concrete approach followed below in order to have a good handle on substitution.

Proposition 16. *Let T be a commutative monad on $\mathbf{Sets}$, and $H\colon \mathcal{EM}(T)^{\mathbb{N}} \to \mathcal{EM}(T)^{\mathbb{N}}$ be a strong functor with:*

- *an isomorphism $\phi\colon HW \overset{\cong}{\Rightarrow} WH$;*
- *for each $P \in \mathcal{EM}(T)^{\mathbb{N}}$ a free H-algebra $H^*(P) \in \mathcal{EM}(T)^{\mathbb{N}}$ on P. This $H^*(P)$ can equivalently be described as initial algebra of the functor $P + H(-)\colon \mathcal{EM}(T)^{\mathbb{N}} \to \mathcal{EM}(T)^{\mathbb{N}}$ in:*

$$P + H\big(H^*(P)\big) \xrightarrow[\cong]{\;[\eta_P,\theta_P]\;} H^*(P) \tag{5}$$

In this situation one can define a substitution map:

$$\mathcal{W}H^*(\mathcal{F}) \otimes \,!H^*(\mathcal{F}) \xrightarrow{\;\mathrm{sbs}\;} H^*(\mathcal{F}),$$

where $\mathcal{F} \in \mathcal{EM}(T)^{\mathbb{N}}$ is the "free variable" presheaf given by restricting the free functor $\mathcal{F}\colon \mathbf{Sets} \to \mathcal{EM}(T)$.

One may read $\mathrm{sbs}_n(s \otimes U) = s[U/v_{n+1}]$ where U is of !-type. The type $\mathcal{W}H^*(\mathcal{F}) \otimes \,!H^*(\mathcal{F}) \to H^*(\mathcal{F})$ of the substitution map sbs is very rich and informative:

- The first argument s of type $\mathcal{W}H^*(\mathcal{F})$ describes the term s in an augmented context; hence there is a variable v_{n+1} in which substitution is going to happen;
- The second argument U of replication type $!H^*(\mathcal{F})$ is going to be substituted for the variable v_{n+1}. Since this variable may occur multiple times (zero or more) in s, we need to be able to use U multiple times. Hence the replication comonad ! is needed in its type. It does not occur in the set-theoretic (cartesian) setting used in [8].

Proof. By Lemma 4 we know that $\mathcal{W}H^*(\mathcal{F})$ is the free H-algebra on $\mathcal{W}(\mathcal{F})$, and thus an initial algebra of the functor $\mathcal{W}(\mathcal{F}) + H(-)\colon \mathcal{EM}(T)^{\mathbb{N}} \to \mathcal{EM}(T)^{\mathbb{N}}$, via:

$$\mathcal{W}(\mathcal{F}) + H\big(\mathcal{W}H^*(\mathcal{F})\big) \xrightarrow[\cong]{\mathrm{id}+\phi} \mathcal{W}(\mathcal{F}) + \mathcal{W}H\big(H^*(\mathcal{F})\big) =$$

$$\mathcal{W}\Big(\mathcal{F} + H\big(H^*(\mathcal{F})\big)\Big) \xrightarrow[\cong]{\mathcal{W}([\eta_{\mathcal{F}},\theta_{\mathcal{F}}])} \mathcal{W}H^*(\mathcal{F}).$$

The strong version of induction described in Lemma 3 implies that it suffices to produce a map of the form:

$$\Big(\mathcal{W}(\mathcal{F}) + H\big(H^*(\mathcal{F})\big)\Big) \otimes \,!H^*(\mathcal{F}) \longrightarrow H^*(\mathcal{F}).$$

so that the substitution map sbs: $\mathcal{W}H^*(\mathcal{F}) \otimes \,!H^*(\mathcal{F}) \to H^*(\mathcal{F})$ arises like in Diagram (3). Tensors $\otimes$ distribute over coproducts $+$, since $\mathcal{EM}(T)^{\mathbb{N}}$ is monoidal closed by Lemma 14, and so it suffices to produce two maps:

$$\mathcal{W}(\mathcal{F}) \otimes \,!H^*(\mathcal{F}) \longrightarrow H^*(\mathcal{F}) \qquad H\big(H^*(\mathcal{F})\big) \otimes \,!H^*(\mathcal{F}) \longrightarrow H^*(\mathcal{F}). \quad (6)$$

We get the second of these maps simply via projection and the initial algebra in (5):

$$H\big(H^*(\mathcal{F})\big) \otimes \,!H^*(\mathcal{F}) \longrightarrow H\big(H^*(\mathcal{F})\big) \xrightarrow{\ \theta_{\mathcal{F}}\ } H^*(\mathcal{F}).$$

For the first map in (6) we start with the isomorphisms:

$$\mathcal{W}(\mathcal{F})(n) = \mathcal{F}(n+1) \cong \mathcal{F}(n) + \mathcal{F}(1) \cong \mathcal{F}(n) + I = (\mathcal{F} + \Delta(I))(n).$$

Again using that $\otimes$ distributes over $+$, and that $\Delta(I)$ is the tensor unit in $\mathcal{EM}(T)^{\mathbb{N}}$ we see that the first map in (6) amounts to two maps:

$$\mathcal{F} \otimes \,!H^*(\mathcal{F}) \longrightarrow H^*(\mathcal{F}) \qquad !H^*(\mathcal{F}) \longrightarrow H^*(\mathcal{F}).$$

In the second case we recall that ! is a comonad, to that there is a counit $\varepsilon \colon !H^*(\mathcal{F}) \to H^*(\mathcal{F})$, and in the first case we project and use the unit η from (5):

$$\mathcal{F} \otimes !H^*(\mathcal{F}) \longrightarrow \mathcal{F} \xrightarrow{\ \eta_{\mathcal{F}}\ } H^*(\mathcal{F}). \qquad \square$$

In [8] it is shown that the substitution map satisfies certain equations. They are not of immediate relevance here.

5   Examples

We apply the framework developed in the previous sections to describe some term calculi as initial algebras in a category of algebra-valued presheaves $\mathcal{EM}(T)^{\mathbb{N}}$. For convenience, we no longer use explicit notation for the pointwise functors like $\otimes^{\mathbb{N}}$ or $!^{\mathbb{N}}$ on $\mathcal{EM}(T)^{\mathbb{N}}$ and simply write them as $\otimes$ and $!$. Hopefully this does not lead to (too much) confusion.

5.1   Non-deterministic Lambda Calculus

We start with the finite powerset monad $T = \mathcal{P}_{\mathrm{fin}}$, so that $\mathcal{EM}(\mathcal{P}_{\mathrm{fin}})$ is the category **JSL** of join semilattices and maps preserving $(\bot, \vee)$. Let $\Lambda \in \mathbf{JSL}^{\mathbb{N}}$ be the initial algebra of the endofunctor on $\mathcal{EM}(T)^{\mathbb{N}} = \mathbf{JSL}^{\mathbb{N}}$, given by:

$$P \longmapsto \mathcal{F} + \mathcal{W}(P) + (P \otimes !P),$$

where $\mathcal{F}(n) = \mathcal{P}_{\mathrm{fin}}(n)$ and $!P(n) = \mathcal{P}_{\mathrm{fin}}(P(n))$. Thus, Λ is the free algebra on $\mathcal{F}$ — written as $H^*(\mathcal{F})$ in Proposition 16 — for the functor $H(P) = \mathcal{W}(P) + (P \otimes !P)$.

We describe this initial algebra as a co-triple of (join-preserving) maps:

$$\mathcal{F} + \mathcal{W}(\Lambda) + (\Lambda \otimes !\Lambda) \xrightarrow[\cong]{\ [\mathrm{var,lam,app}]\ } \Lambda$$

Elements of the set of terms $\Lambda(n) \in \mathbf{JSL}$ with variables from $\{v_1, \ldots, v_n\}$ are inductively given by:

- $\mathrm{var}_n(V)$, where $V \in \mathcal{F}(n) = \mathcal{P}_{\mathrm{fin}}(n) = \mathcal{P}_{\mathrm{fin}}(\{v_1, v_2, \ldots, v_n\})$;
- $\mathrm{lam}_n(N) = \lambda v_{n+1}. N$, where $N \in \mathcal{W}(\Lambda)(n) = \Lambda(n+1)$;
- $\mathrm{app}(M, \{N_1, \ldots, N_k\}) = M \cdot \{N_1, \ldots, N_k\}$, where $M, N_1, \ldots, N_k \in \Lambda(n)$;
- $\bot \in \Lambda(n)$, and $M \vee N \in \Lambda(n)$, for $M, N \in \Lambda(n)$.

The join-preservation property that holds by construction for these maps yields:

$$
\begin{aligned}
\mathrm{var}_n(\emptyset) &= \bot & \mathrm{var}_n(V \cup V') &= \mathrm{var}_n(V) \vee \mathrm{var}_n(V') \\
\lambda v_{n+1}. \bot &= \bot & \lambda v_{n+1}. (M \vee N) &= (\lambda v_{n+1}. M) \vee (\lambda v_{n+1}. N) \\
\bot \cdot U &= \bot & (M \vee M') \cdot U &= (M \cdot U) \vee (M' \cdot U) \\
M \cdot \emptyset &= \bot & M \cdot (U \cup U') &= (M \cdot U) \vee (M \cdot U').
\end{aligned}
$$

The non-standard features of this term calculus are the occurrences of *sets* of variables V in $\mathrm{var}_n(V)$ and of *sets* of terms as second argument in application $M \cdot \{N_1, \ldots, N_k\}$. But using these join preservation properties they can be described also in terms of single variables/terms:

$$\mathrm{var}_n(\{v_1, \ldots, v_k\}) = \mathrm{var}_n(\{v_1\}) \vee \cdots \vee \mathrm{var}_n(\{v_k\})$$
$$M \cdot \{N_1, \ldots, N_k\} = M \cdot \{N_1\} \vee \cdots \vee M \cdot \{N_k\}.$$

Following Proposition 16 — recall that $\Lambda = H^*(\mathcal{F})$ — there is a substitution map sbs in $\mathbf{JSL}^{\mathbb{N}}$:

$$\mathcal{W}(\Lambda) \otimes \,!\Lambda \xrightarrow{\quad \mathrm{sbs} \quad} \Lambda.$$

The diagram that defines sbs, according to Lemma 3, can be split up in three separate diagrams, for variables, abstraction and application.

$$
\begin{array}{ccc}
\mathcal{W}(\mathcal{F}) \otimes \,!\Lambda & \xrightarrow{\;\cong\;} & \mathcal{F} \otimes \,!\Lambda + \,!\Lambda \\
{\scriptstyle \mathcal{W}(\mathrm{var}) \otimes \mathrm{id}} \downarrow {\scriptstyle \cong} & & \downarrow {\scriptstyle [\mathrm{var} \circ \pi_1, \mathsf{V}]} \\
\mathcal{W}(\Lambda) \otimes \,!\Lambda & \xrightarrow{\quad \mathrm{sbs} \quad} & \Lambda
\end{array}
\qquad (7)
$$

$$
\begin{array}{ccccccc}
\mathcal{W}^2(\Lambda) \otimes \,!\Lambda & \xrightarrow{\mathrm{id} \otimes \mathrm{up}} & \mathcal{W}^2(\Lambda) \otimes \mathcal{W}(!\Lambda) & = & \mathcal{W}(\mathcal{W}(\Lambda) \otimes \,!\Lambda) & \xrightarrow{\mathcal{W}(\mathrm{sbs})} & \mathcal{W}(\Lambda) \\
{\scriptstyle \mathrm{swp} \otimes \mathrm{id}} \downarrow {\scriptstyle \cong} & & & & & & \downarrow \\
\mathcal{W}^2(\Lambda) \otimes \,!\Lambda & & & & & & \\
{\scriptstyle \mathcal{W}(\mathrm{lam}) \otimes \mathrm{id}} \downarrow & & & & & & {\scriptstyle \mathrm{lam}} \\
\mathcal{W}(\Lambda) \otimes \,!\Lambda & & \xrightarrow{\qquad\qquad\qquad \mathrm{sbs} \qquad\qquad\qquad} & & & & \Lambda
\end{array}
\qquad (8)
$$

$$
\begin{array}{ccccc}
(\mathcal{W}(\Lambda) \otimes \,!\mathcal{W}(\Lambda)) \otimes \,!\Lambda & \xrightarrow{\;\mathrm{st}\;} & (\mathcal{W}(\Lambda) \otimes \,!\Lambda) \otimes \,!(\mathcal{W}(\Lambda) \otimes \,!\Lambda) & \xrightarrow{\mathrm{sbs} \otimes \,!(\mathrm{sbs})} & \Lambda \otimes \,!\Lambda \\
\| & & & & \downarrow \\
\mathcal{W}(\Lambda \otimes \,!\Lambda) \otimes \,!\Lambda & & & & \\
{\scriptstyle \mathcal{W}(\mathrm{app}) \otimes \mathrm{id}} \downarrow & & & & {\scriptstyle \mathrm{app}} \\
\mathcal{W}(\Lambda) \otimes \,!\Lambda & & \xrightarrow{\qquad\qquad \mathrm{sbs} \qquad\qquad} & & \Lambda
\end{array}
\qquad (9)
$$

In more conventional notation, reading $\mathrm{sbs}_n(M \otimes U) = M[U/v_{n+1}]$ where $M \in \Lambda$ and $U \subseteq \Lambda$ is finite, we can write these three diagrams (7) – (9) as:

$$
\mathrm{var}_{n+1}(V)[U/v_{n+1}] = \begin{cases} \mathrm{var}_n(V) & \text{if } v_{n+1} \notin V \\ \mathrm{var}_n(V - v_{n+1}) \vee \bigvee U & \text{if } v_{n+1} \in V \end{cases}
$$
$$
(\lambda v_{n+2}.\, M)[U/v_{n+1}] = \lambda v_{n+1}.\, M[v_{n+1}/v_{n+2}, v_{n+2}/v_{n+1}][U/v_{n+2}]
$$
$$
= \lambda v_{n+1}.\, M[v_{n+1}/v_{n+2}, U/v_{n+1}]
$$
$$
(M \cdot \{N_1, \ldots, N_k\})[U/v_{n+1}] = M[U/v_{n+1}] \cdot \{N_1[U/v_{n+1}], \ldots, N_k[U/v_{n+1}]\}.
$$

By construction the substitution map sbs is bilinear, so that:

$$\bot[U/v_{n+1}] = \bot \qquad (M \vee M')[U/v_{n+1}] = (M[U/v_{n+1}]) \vee (M'[U/v_{n+1}])$$
$$M[\emptyset/v_{n+1}] = \bot \qquad M[(U \cup U')/v_{n+1}] = (M[U/v_{n+1}]) \vee (M[U'/v_{n+1}]).$$

We see that the mere initiality of Λ in the presheaf category $\mathcal{EM}(\mathcal{P}_{\mathrm{fin}})^{\mathbb{N}}$ gives a lot of information about the term calculus involved. As usual, (initial) algebras of functors do not involve any equations. If needed, they will have to be imposed explicitly. For instance, in the current setting it makes sense to require the analogue of the familiar (β)-rule $(\lambda x. M)N = M[N/x]$ for (ordinary) lambda terms. Here it can be expressed diagrammatically as:

$$\mathcal{W}(\Lambda) \otimes !\Lambda \xrightarrow{\text{lam} \otimes \text{id}} \Lambda \otimes !\Lambda$$

with sbs and app mapping into Λ.

Another example is the (η)-rule: $\lambda x. Mx = M$, if x is not a free variable in M. In this setting it can be expressed as

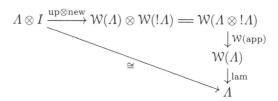

Here, $\text{new} \colon I \to \mathcal{W}(!\Lambda)$ denotes the generation of a free variable. It is defined as:

$$\text{new} = \left(I = !1 \xrightarrow{\delta} !!1 = !I \xrightarrow{!\kappa_2} !(\mathcal{F} + I) \cong !\mathcal{W}(\mathcal{F}) \xrightarrow{!\mathcal{W}(\text{var})} !\mathcal{W}(\Lambda)! = \mathcal{W}(!\Lambda) \right).$$

5.2   Weighted Lambda Calculus

We now consider a weighted version of the lambda calculus, by changing the monad in the above example. We fix a semiring S and consider the associated multiset monad $\mathcal{M} = \mathcal{M}_S$. Its Eilenberg-Moore algebras are semimodules over S. We write **SMod** $= \mathcal{EM}(\mathcal{M})$, with morphism given by linear maps. Let $\Lambda \in$ **SMod**$^{\mathbb{N}}$ be the initial algebra of the endofunctor on $\mathcal{EM}(\mathcal{M})^{\mathbb{N}} =$ **SMod**$^{\mathbb{N}}$ given by:

$$P \longmapsto \mathcal{F} + \mathcal{W}(P) + (P \otimes !P),$$

where $\mathcal{F}(n) = \mathcal{M}(n)$ and $!P(n) = \mathcal{M}(P(n))$. Notice that this functor is formally the same as for the non-deterministic lambda calculus (in the previous subsection).

As before, we describe this initial algebra as a co-triple of (linear) maps:

$$\mathcal{F} + \mathcal{W}(\Lambda) + (\Lambda \otimes !\Lambda) \xrightarrow[\cong]{[\text{var},\text{lam},\text{app}]} \Lambda$$

Elements of the set of terms $\Lambda(n) \in \mathbf{SMod}$ with variables from $\{v_1, \ldots, v_n\}$ are inductively given by:

- $\mathrm{var}_n(\varphi)$, where $\varphi \in \mathcal{F}(n) = \mathcal{M}(n) = \mathcal{M}(\{v_1, v_2, \ldots, v_n\})$; we will typically write φ (and, in general, elements of $\mathcal{M}(X)$) as a (finite) formal sum $\sum_i s_i v_i$.
- $\mathrm{lam}_n(N) = \lambda v_{n+1}. N$, where $N \in \Lambda(n+1)$;
- $\mathrm{app}(M, \sum_i s_i N_i) = M \cdot (\sum_i s_i N_i)$, where $M, N_i \in \Lambda(n)$;
- $0 \in \Lambda(n)$, and $s \bullet N \in \Lambda(n)$, for $N \in \Lambda(n)$ and $s \in S$, and $N + M \in \Lambda(n)$ for $N, M \in \Lambda(n)$. We write a fat dot $\bullet$ for scalar multiplication in order to distinguish it from application.

Slightly generalising the previous example, the non-standard features of this term calculus are the occurrences of *linear combinations* of variables φ in $\mathrm{var}_n(\varphi)$ and of *linear combinations* of terms as second argument in application $M \cdot (\sum_i s_i N_i)$. But using the linearity properties of the maps, they can be described also in terms of single variables/terms:

$$\mathrm{var}_n(\textstyle\sum_i s_i v_i) = \sum_i s_i \bullet \mathrm{var}_n(1 v_i)$$
$$M \cdot (\textstyle\sum_i s_i N_i) = \sum_i s_i \bullet (M \cdot 1 N_i).$$

(Recall that $1x = \eta(x) \in \mathcal{M}(X)$ is the singleton multiset.)

The substitution operation can now be defined by similar diagrams as in (7) – (9) above. For this example, we give it immediately in a more conventional notation:

$$\mathrm{var}_{n+1}(\textstyle\sum_{i \leq n+1} s_i v_i)[\psi/v_{n+1}] = \begin{cases} \mathrm{var}_n(\sum_{i \leq n} s_i v_i) & \text{if } s_{n+1} = 0 \\ \mathrm{var}_n(\sum_{i \leq n} s_i v_i) + s_{n+1} \bullet \psi & \text{if } s_{n+1} \neq 0 \end{cases}$$
$$(\lambda v_{n+2}. M)[\psi/v_{n+1}] = \lambda v_{n+1}. M[v_{n+1}/v_{n+2}, v_{n+2}/v_{n+1}][\psi/v_{n+2}]$$
$$= \lambda v_{n+1}. M[v_{n+1}/v_{n+2}, \psi/v_{n+1}]$$
$$(M \cdot (\textstyle\sum_i s_i N_i))[\psi/v_{n+1}] = M[\psi/v_{n+1}] \cdot (\sum_i s_k(N_i[\psi/v_{n+1}])).$$

5.3 Non-deterministic Automata

In this section, we present expressions with a fixed point operator denoting behaviours of non-deterministic automata. This formalises, using initial algebras, the work in [29]. We work again, first in the category $\mathcal{EM}(\mathcal{P}_{\mathrm{fin}})^{\mathbb{N}} = \mathbf{JSL}^{\mathbb{N}}$.

The presheaf of expressions $\mathsf{E} \in \mathbf{JSL}^{\mathbb{N}}$ is the initial algebra of the functor on $\mathbf{JSL}^{\mathbb{N}}$ given by:

$$P \longmapsto \mathcal{F} + \mathcal{W}(P) + 2 + A \cdot !P,$$

where $2 = \{\bot, \top\}$ is the constant presheaf, $\mathcal{F}(n) = \mathcal{P}_{\mathrm{fin}}(\{v_1, \ldots, v_n\})$, and $!P = \mathcal{P}_{\mathrm{fin}}(P)$.

Coalgebraically, non-deterministic automata are modelled using the functor $F(X) = 2 \times \mathcal{P}_{\mathrm{fin}}(X)^A$ (in $\mathbf{Sets}$). Both functors are tightly connected but further studying this connection is out of the scope of this paper.

We can describe this initial algebra as the following map in $\mathbf{JSL}^{\mathbb{N}}$.

$$\mathcal{F} + \mathcal{W}(\mathsf{E}) + 2 + A \cdot !\mathsf{E} \xrightarrow[\cong]{[\mathrm{var,fix,ops,pre}]} \mathsf{E}$$

Elements of the set of expressions $\mathsf{E}(n) \in \mathbf{JSL}$ with variables from $\{v_1, \ldots, v_n\}$ are inductively given by:

- $\mathrm{var}_n(V)$ where $V \subseteq \{v_1, \ldots, v_n\}$;
- $\mathrm{fix}_n(e) = \mu v_{n+1}.e$, where $e \in \mathsf{E}(n+1)$;
- $\mathbf{0} = \mathrm{ops}(\bot)$;
- $\mathbf{1} = \mathrm{ops}(\top)$;
- $a(\{e_1, \ldots e_k\}) = \mathrm{pre}(\kappa_a(\{e_1, \ldots e_k\}))$, for $a \in A$ and $e_i \in \mathsf{E}(n)$;
- $\bot$ and $e \vee e'$ for any $e, e' \in \mathsf{E}(n)$.

The intuition behind these expressions is best explained operationally. Even though we have not given at this stage any operational semantics to the above expressions, and hence they are just syntax, we depict below several non-deterministic automata and corresponding expressions.

The binder μ has a very similar role to the Kleene star $(-)^\star$. For instance, it is the intended semantics that $e_2 = \mu x.a(\{x\}) \vee 1 = a^\star$. Hence, semantically, we want μ to be a (least) fixed point. The fixed point equation $\mu x.e = e[\mu x.e/x]$ can be represented in this setting by the diagram:

Here the map sbs is defined in the same way as for the non-deterministic lambda calculus. In order to obtain a (sound and) complete calculus of these expressions with respect to bisimilarity, one would have to impose more equations, see [29].

In formal language theory, the prime equivalence used is language (or trace) equivalence, which is coarser than bisimilarity. Rabinovich [27] showed that in the case of a simple fixed point calculus like the one we derive here, it is enough to add one axiom to the bisimilarity calculus, namely the axiom expressing distributivity of prefixing over $\vee$, given as $a(\{e \vee e'\}) = a(\{e\}) \vee a(\{e'\})$.

Here, we can express this diagrammatically as:

$$
\begin{array}{c}
\xymatrix{
& & !!\mathsf{E} & & \\
!\mathsf{E} & A \cdot \mathsf{E} & & A \cdot !\mathsf{E} & !\mathsf{E}
}
\end{array}
\tag{10}
$$

$$
!\mathsf{E} \xleftarrow{\ \kappa_a\ } A \cdot \mathsf{E} \xrightarrow{\ \mathrm{pre}\ } \mathsf{E} \xleftarrow{\ \mathrm{pre}\ } A \cdot !\mathsf{E} \xleftarrow{\ \kappa_a\ } !\mathsf{E}
$$

with top maps ε and $!\varepsilon$ from $!!\mathsf{E}$.

This diagram is actually a more general formulation of the Rabinovich distributivity rule, since the starting point $!!\mathsf{E} = \mathcal{P}_{\mathrm{fin}}^2(\mathsf{E})$ at the top involves sets of sets of expressions. To recover the binary version above one should start from $\{\{e_1, e_2\}\} \in {!!\mathsf{E}}$. The commutativity of the above diagram says that $a(\{e_1 \vee e_2\}) = a(\{e_1, e_2\})$, and the right side of this equation rewrites to $a(\{e_1, e_2\}) = a(\{e_1\}) \vee a(\{e_2\})$, since pre is a **JSL** map. The relevance of capturing this law diagrammatically is that it provides a guideline to which law gives trace semantics in other settings. For instance, we show in the next subsection that for weighted automata, the same diagram can be used to axiomatise trace semantics.

5.4   Weighted Automata

We can obtain a syntax for weighted automata in a similar way by switching to the multiset monad $\mathcal{M}$, for a semiring S, and considering the initial algebra E of the functor on $\mathbf{SMod}^{\mathbb{N}}$:

$$
P \longmapsto \mathcal{F} + \mathcal{W}(P) + S + A \cdot {!P},
$$

where S denotes the constant presheaf $\Delta(S)$, and $\mathcal{F}(n) = \mathcal{M}(\{v_1, \ldots, v_n\})$, and $!P = \mathcal{M}(P)$.

Again, coalgebraically, weighted automata are modelled using the functor $F(X) = S \times \mathcal{M}(X)^A$ (in **Sets**). We can describe the above initial algebra as the following isomorphism.

$$
\mathcal{F} + \mathcal{W}(\mathsf{E}) + S + A \cdot {!\mathsf{E}} \xrightarrow[\cong]{\ [\mathrm{var},\mathrm{fix},\mathrm{val},\mathrm{pre}]\ } \mathsf{E}
$$

Elements of the set of expressions $\mathsf{E}(n) \in \mathbf{SMod}$ with variables from $\{v_1, \ldots, v_n\}$ are inductively given by:

- $\mathrm{var}_n(V)$ where $V \subseteq \{v_1, \ldots, v_n\}$;
- $\mathrm{fix}_n(e) = \mu v_{n+1}.e$, where $e \in \mathsf{E}(n+1)$;

- $\underline{s} = \mathrm{val}(s)$, for any $s \in S$;
- $a(\sum_i s_i e_i) = \mathrm{pre}(\kappa_a(\sum_i s_i e_i))$;
- 0, $s \bullet e$, and $e + e'$ for any $e, e' \in \mathsf{E}(n)$ and $s \in S$.

Intuitively, the expression $\underline{s}$ denotes that the output of a state is s and $a(\sum_{k \le m} s_k e_k)$ denotes a state with m transitions labelled by a, s_k to the states denoted by the expressions $e_1, \ldots, e_m$. We depict below two examples of expressions (over the semiring of integers) and the respective (intended) weighted automata they represent.

$e_1 = b(2 \bullet \underline{1})$	$e_2 = \mu x.a(3x + 2e_1) + \underline{1}$

Note that e_1 above has output 0 because the expression only specifies the b transition and no concrete output. Intuitively, the expression e_1 specifies a state which has a b transition with weight 2 to a state specified by the expression $\underline{1}$. The latter specifies a state with no transitions and with output 1. The use of $+$ in e_2 allows the combined specification of transitions and outputs and hence e_2 has output 1.

Examples of equations one might want to impose in this calculus include $a(r0) = 0$, which can be expressed using the diagram:

$$
\begin{array}{ccc}
I = \;!1 = S & \xrightarrow{\quad 0 \quad} & \\
{\scriptstyle (-)\bullet\eta(0)}\downarrow & & \searrow \\
!E & \xrightarrow[\kappa_a]{} A \cdot !E & \xrightarrow[\mathrm{pre}]{} E
\end{array}
$$

Recent papers [3,7] on generalising the work of Rabinovich, have proposed, in a rather *ad hoc* fashion, axiomatisations of trace semantics for weighted automata. The very same diagram (10) capturing Rabinovich's axiom for non-deterministic automata can now be interpreted in this weighted setting and it will result in the following axiom — which we present immediately a binary version, for readability.

$$a(s(e_1 + e_2)) = a(se_1) + a(se_2).$$

(We write se for the singleton multiset $s \bullet 1e$.)

Final Remarks

We have exploited initiality in categories of algebra-valued presheaves $\mathcal{EM}(T)^{\mathbb{N}}$, for a monad T on **Sets**, to modularly obtain a syntactic calculus of expressions as initial algebra of a given functor. The theory involved provides a systematic description of several non-trivial examples: non-deterministic and weighted lambda calculus and expressions for (weighted) automata.

The use of presheaves to describe calculi with binding operators has first been studied by Fiore, Plotkin and Turi. They used set-valued presheaves and did not explore the formalisation of relevant equations for concrete calculi. They have also not explored their theory to derive fixed point calculi. Non-deterministic versions of the lambda calculus have appeared in, for instance, [6,26]. The version we derive in this paper is however slightly different because we have linearity also in the second argument of application.

As future work, we wish to explore the formalisation of equations that we presented for the concrete examples. This opens the door to developing a systematic account of sound and complete axiomatisations for different equivalences (bisimilarity, trace, failure, *etc.*). Preliminary work for (generalised) regular expressions and bisimilarity has appeared in [15,31,30] and for trace semantics in [3].

Acknowledgments. Thanks to Sam Staton for suggesting several improvements.

References

1. Atkey, R., Johann, P., Ghani, N.: When is a type refinement an inductive type? In: Hofmann, M. (ed.) FOSSACS 2011. LNCS, vol. 6604, pp. 72–87. Springer, Heidelberg (2011)
2. Barr, M., Wells, C.: Toposes, Triples and Theories. Springer, Berlin (1985), Revized and corrected version available from
   www.cwru.edu/artsci/math/wells/pub/ttt.html
3. Bonsangue, M., Milius, S., Silva, A.: Sound and complete axiomatizations of coalgebraic language equivalence. ACM Trans. on Computational Logic 14(1) (2013)
4. Borceux, F.: Handbook of Categorical Algebra. Encyclopedia of Mathematics, vol. 50, 51 and 52. Cambridge Univ. Press (1994)
5. Coecke, B., Pavlović, D., Vicary, J.: A new description of orthogonal bases. Math. Struct. in Comp. Sci., 1–13 (2012)
6. de' Liguoro, U., Piperno, A.: Non-deterministic extensions of untyped λ-calculus. Inf. & Comp. 122, 149–177 (1995)
7. Ésik, Z., Kuich, W.: Free iterative and iteration k-semialgebras. CoRR, abs/1008.1507 (2010)
8. Fiore, M., Plotkin, G., Turi, D.: Abstract syntax and variable binding. In: LICS, pp. 193–202. IEEE Computer Society (1999)
9. Fiore, M., Staton, S.: Comparing operational models of name-passing process calculi. Inf. & Comp. 2004(4), 524–560 (2006)
10. Fiore, M., Turi, D.: Semantics of name and value passing. In: Logic in Computer Science, pp. 93–104. IEEE Computer Science Press (2001)
11. Goguen, J., Thatcher, J., Wagner, E., Wright, J.: Initial algebra semantics and continuous algebras. Journ. ACM 24(1), 68–95 (1977)
12. Hasuo, I., Jacobs, B., Sokolova, A.: Generic trace theory via coinduction. Logical Methods in Computer Science 3(4:11) (2007)
13. Jacobs, B.: Semantics of weakening and contraction. Ann. Pure & Appl. Logic 69(1), 73–106 (1994)
14. Jacobs, B.: Categorical Logic and Type Theory. North Holland, Amsterdam (1999)

15. Jacobs, B.: A bialgebraic review of deterministic automata, regular expressions and languages. In: Futatsugi, K., Jouannaud, J.-P., Meseguer, J. (eds.) Algebra, Meaning, and Computation. LNCS, vol. 4060, pp. 375–404. Springer, Heidelberg (2006)
16. Jacobs, B.: Bases as coalgebras. In: Corradini, A., Klin, B., Cîrstea, C. (eds.) CALCO 2011. LNCS, vol. 6859, pp. 237–252. Springer, Heidelberg (2011)
17. Jacobs, B.: Introduction to Coalgebra. Towards Mathematics of States and Observations (2012) Book, in preparation; version 2 available from http://www.cs.ru.nl/B.Jacobs/CLG/JacobsCoalgebraIntro.pdf
18. Joyal, A., Moerdijk, I.: Algebraic Set Theory. LMS, vol. 220. Cambridge Univ. Press (1995)
19. Klin, B.: Bialgebras for structural operational semantics: An introduction. Theor. Comp. Sci. 412(38), 5043–5069 (2011)
20. Kock, A.: Bilinearity and cartesian closed monads. Math. Scand. 29, 161–174 (1971)
21. Kock, A.: Closed categories generated by commutative monads. Journ. Austr. Math. Soc XII, 405–424 (1971)
22. Lambek, J.: A fixed point theorem for complete categories. Math. Zeitschr. 103, 151–161 (1968)
23. Manes, E.G.: Algebraic Theories. Springer, Berlin (1974)
24. Mac Lane, S.: Categories for the Working Mathematician. Springer, Berlin (1971)
25. Møgelberg, R.E., Staton, S.: Linearly-used state in models of call-by-value. In: Corradini, A., Klin, B., Cîrstea, C. (eds.) CALCO 2011. LNCS, vol. 6859, pp. 298–313. Springer, Heidelberg (2011)
26. Pagani, M., della Rocca, S.R.: Solvability in resource lambda-calculus. In: Ong, L. (ed.) FOSSACS 2010. LNCS, vol. 6014, pp. 358–373. Springer, Heidelberg (2010)
27. Rabinovich, A.M.: A complete axiomatisation for trace congruence of finite state behaviors. In: Main, M.G., Melton, A.C., Mislove, M.W., Schmidt, D., Brookes, S.D. (eds.) MFPS 1993. LNCS, vol. 802, pp. 530–543. Springer, Heidelberg (1994)
28. Rutten, J.: Universal coalgebra: a theory of systems. Theor. Comput. Sci. 249(1), 3–80 (2000)
29. Silva, A.: Kleene coalgebra. PhD thesis, Radboud University Nijmegen (2010)
30. Silva, A., Bonchi, F., Bonsangue, M., Rutten, J.: Generalizing the powerset construction, coalgebraically. In: Proc. FSTTCS 2010. LIPIcs, vol. 8, pp. 272–283 (2010)
31. Silva, A., Bonsangue, M., Rutten, J.: Non-deterministic kleene coalgebras. Logical Methods in Computer Science 6(3) (2010)
32. Uustalu, T., Vene, V., Pardo, A.: Recursion schemes from comonads. Nordic Journ. Comput. 8(3), 366–390 (2001)

Abstract Tensor Systems as Monoidal Categories

Aleks Kissinger

Department of Computer Science
University of Oxford

Abstract. The primary contribution of this paper is to give a formal, categorical treatment to Penrose's abstract tensor notation, in the context of traced symmetric monoidal categories. To do so, we introduce a typed, sum-free version of an abstract tensor system and demonstrate the construction of its associated category. We then show that the associated category of the free abstract tensor system is in fact the free traced symmetric monoidal category on a monoidal signature. A notable consequence of this result is a simple proof for the soundness and completeness of the diagrammatic language for traced symmetric monoidal categories.

Dedicated to Joachim Lambek on the occasion of his 90th birthday.

1 Introduction

This paper formalises the connection between monoidal categories and the abstract index notation developed by Penrose in the 1970s, which has been used by physicists directly, and category theorists implicitly, via the diagrammatic languages for traced symmetric monoidal and compact closed categories. This connection is given as a representation theorem for the free traced symmetric monoidal category as a syntactically-defined strict monoidal category whose morphisms are equivalence classes of certain kinds of terms called *Einstein expressions*. Representation theorems of this kind form a rich history of coherence results for monoidal categories originating in the 1960s [17,6]. Lambek's contribution [15,16] plays an essential role in this history, providing some of the earliest examples of syntactically-constructed free categories and most of the key ingredients in Kelly and Mac Lane's proof of the coherence theorem for closed monoidal categories [11]. Recently, Lambek has again become interested in the role of compact closed categories (a close relative of traced symmetric monoidal categories) in linguistics and physics, both contributing [14] and inspiring [2,4] ground-breaking new work there. The present work also aims to build a bridge between monoidal categories and theoretical physics, by formalising the use of a familiar language in physics within a categorical context. For these reasons, the author would like to dedicate this exposition to Lambek, on the occasion of his 90th birthday.

Tensors are a fundamental mathematical tool used in many areas of physics and mathematics with notable applications in differential geometry, relativity

C. Casadio et al. (Eds.): Lambek Festschrift, LNCS 8222, pp. 235–252, 2014.

theory, and high-energy physics. A (concrete) *tensor* is an indexed set of numbers in some field (any field will do, but we'll use $\mathbb{C}$).

$$\{\psi_{i_1,\ldots,i_m}^{j_1,\ldots,j_n} \in \mathbb{C}\}_{i_k,j_k \in \{1,\ldots,D\}} \tag{1}$$

One should think of the lower indices as *inputs* and the upper indices as *outputs*. Notable special cases are column vectors v^j, row vectors ξ_i, and matrices M_i^j. Tensors can be combined via the *tensor product* and *contraction*. The product of two tensors is defined as a new tensor whose elements are defined point-wise as products.

$$(\psi\phi)_{i,i'}^{j,j'} := \psi_i^j \phi_{i'}^{j'}$$

Contraction is a procedure by which a lower index is "plugged into" an upper index by summing them together.

$$\theta_i^j := \sum_{k=1}^{D} \psi_i^k \phi_k^j$$

Special cases of contraction are matrix composition, application of a matrix to a vector, and the trace of a matrix. It is customary to employ the *Einstein summation convention*, whereby repeated indices are assumed to be summed over.

$$\psi_{i,j}^k \phi_{k,l}^m \xi_m^i := \sum_{i,k,m} \psi_{i,j}^k \phi_{k,l}^m \xi_m^i$$

In other words, indices occurring only once are considered *free* in the tensor expression (and can be used in further contractions), whereas repeated indices are implicitly *bound* by the sum.

Abstract tensor notation was defined by Penrose in 1971 [21] to give an elegant way to describe various types of multi-linear maps without the encumbrance of fixing bases. It allows one to reason about much more general processes with many inputs and outputs *as if* they were just tensors. In that paper, he actually introduced two notations. He introduced a term-based notation, where the objects of interest are *abstract Einstein expressions*, and an equivalent (more interesting) notation that is diagrammatic. There, tensors are represented as *string diagrams*. This diagrammatic notation, as an elegant way of expressing tensor expressions, has appeared in popular science [20], theoretical physics [19], representation theory [5], and (in its more abstract form) foundations of physics [7].

Twenty years after Penrose's original paper, Joyal and Street formalised string diagrams as topological objects and showed that certain kinds of these diagrams can be used to form free monoidal categories [10]. From this work came a veritable zoo of diagrammatic languages [22] for describing various flavours of monoidal categories. These languages, as a calculational tool for monoidal categories, have played a crucial role in the development of categorical quantum mechanics [1,3] and the theory of quantum groups [24], as well as recently finding applications in computational linguistics [2,4].

While categorical string diagrams were developed very much in the same spirit as Penrose's notation, little work has been done formally relating abstract tensor systems to monoidal categories. This is the first contribution of this paper. In section 5, we show that it is possible to construct a traced symmetric monoidal category from any abstract tensor system in a natural way. Furthermore, we show that reasoning with abstract tensors is sound and complete with respect to traced symmetric monoidal categories by showing that the associated category of the free abstract tensor system is, in fact, the free traced SMC.

It is generally well known that string diagrams are sound an complete for traced symmetric monoidal categories. This fact was alluded to in Joyal and Street's original paper, but the authors stopped short of providing a proof for the case where string diagrams had "feedback" (i.e. traced structure). Subsequently, this fact is often stated without proof [22], sketched[1], or restricted to the case where the morphisms in the diagram have exactly one input and one output [12,23]. Thus, the second contribution of this paper is a proof of soundness and completeness of the diagrammatic language as a corollary to the main theorem about freeness of the category generated by the free abstract tensor system.

2   Abstract Tensor Systems

In this section, we shall define abstract tensor systems in a manner similar to Penrose's original definition in 1971 [21]. It will differ from the Penrose construction in two ways. First, we will not consider addition of tensors as a fundamental operation, but rather as extra structure that one could put on top (c.f. enriched category theory). Second, we will allow tensor inputs and outputs to have more than one type. In the previous section, we assumed for simplicity that the dimension of all indices was some fixed number D. We could also allow this to vary, as long as care is taken to only contract together indices of the same dimension. This yields a very simple example of a typed tensor system. Many other examples occur when we use tensors to study categories and operational physical theories.

For a set of types $\mathcal{U} = \{A, B, C, \ldots\}$, fix a set of labels:

$$\mathcal{L} = \{a_1, a_2, \ldots, b_1, b_2, \ldots\}$$

and a typing function $\tau : \mathcal{L} \to \mathcal{U}$. We will always assume there are at least countably many labels corresponding to each type in $\mathcal{U}$. For finite subsets $\mathbf{x}, \mathbf{y} \in \mathcal{P}_f(\mathcal{L})$ and $\{x_1, \ldots, x_N\} \subseteq \mathbf{x}$ let:

$$[x_1 \mapsto y_1, x_2 \mapsto y_2, \ldots, x_N \mapsto y_N] : \mathbf{x} \to \mathbf{y} \qquad (2)$$

be the function which sends each x_i to y_i and leaves all of the other elements of $\mathbf{x}$ fixed.

---

[1] The sketched proof of soundness/completeness for a diagrammatic language which the authors call *sharing graphs* in [9] is actually a special case of a theorem in Hasegawa's thesis [8], characterising the free cartesian-center traced symmetric monoidal category, which is proved in detail therein. Thanks to Masahito Hasegawa and Gordon Plotkin for pointing this out.

Definition 1. *A (small) abstract tensor system consists of:*

- *A set $\mathcal{T}(\mathbf{x}, \mathbf{y})$ for all $\mathbf{x}, \mathbf{y} \in \mathcal{P}_f(\mathcal{L})$ such that $\mathbf{x} \cap \mathbf{y} = \emptyset$,*
- *a tensor product operation:*

$$(- \cdot -) : \mathcal{T}(\mathbf{x}, \mathbf{y}) \times \mathcal{T}(\mathbf{x}', \mathbf{y}') \to \mathcal{T}(\mathbf{x} \cup \mathbf{x}', \mathbf{y} \cup \mathbf{y}')$$

defined whenever $(\mathbf{x} \cup \mathbf{y}) \cap (\mathbf{x}' \cup \mathbf{y}') = \emptyset$
- *a contraction function for $a \in \mathbf{x}, b \in \mathbf{y}$ such that $\tau(a) = \tau(b)$:*

$$\mathcal{K}_a^b : \mathcal{T}(\mathbf{x}, \mathbf{y}) \to \mathcal{T}(\mathbf{x} - \{a\}, \mathbf{y} - \{b\})$$

- *a bijection between sets of tensors, called a relabelling:*

$$r_f : \mathcal{T}(\mathbf{x}, \mathbf{y}) \overset{\sim}{\to} \mathcal{T}(f(\mathbf{x}), f(\mathbf{y}))$$

for every bijection $f : (\mathbf{x} \cup \mathbf{y}) \overset{\sim}{\to} \mathbf{z}$ such that $\tau(f(x)) = \tau(x)$ for all $x \in \mathbf{x} \cup \mathbf{y}$,
- *a chosen tensor called a δ-element $\delta_a^b \in \mathcal{T}(\{a\}, \{b\})$ for all a, b such that $\tau(a) = \tau(b)$, along with an "empty" δ-element $1 \in \mathcal{T}(\emptyset, \emptyset)$*

Before giving the axioms, we introduce some notation. Let $\psi\phi := \psi \cdot \phi$ and $\psi[f] := r_f(\psi)$. For $\psi \in \mathcal{T}(\mathbf{x}, \mathbf{y})$ let $L(\psi) = \mathbf{x} \cup \mathbf{y}$. If a label a is in $L(\psi)$, we say a is *free* in ψ. If a label occurs in a contraction, we say it is *bound*. Using this notation, the axioms of an abstract tensor system are as follows:

T1. $\mathcal{K}_a^b(\mathcal{K}_{a'}^{b'}(\psi)) = \mathcal{K}_{a'}^{b'}(\mathcal{K}_a^b(\psi))$
T2. $(\psi\phi)\xi = \psi(\phi\xi)$, $\psi 1 = \psi = 1\psi$, and $\psi\phi = \phi\psi$
T3. $\mathcal{K}_a^b(\psi\phi) = (\mathcal{K}_a^b(\psi))\phi$ for $a, b \notin L(\phi)$
T4. $\mathcal{K}_a^b(\delta_a^c\psi) = \psi[b \mapsto c]$ and $\mathcal{K}_b^c(\delta_a^c\psi) = \psi[b \mapsto a]$
L1. $\psi[f][g] = \psi[g \circ f]$ and $\psi[\mathrm{id}] = \psi$
L2. $(\psi[f])\phi = (\psi\phi)[f]$ where $cod(f) \cap L(\phi) = \emptyset$
L3. $\mathcal{K}_a^b(\psi)[f'] = \mathcal{K}_{f(a)}^{f(b)}(\psi[f])$ where f' is the restriction of f to $L(\psi) - \{a, b\}$
L4. $\delta_a^b[a \mapsto a', b \mapsto b'] = \delta_{a'}^{b'}$

Note that L3 implies in particular that the choice of bound labels in irrelevant to the value of a contracted tensor.

Lemma 1. *Let ψ be a tensor containing a lower label a and upper label b, and let a', b' be distinct labels not occurring in $L(\psi)$ such that $\tau(a) = \tau(a')$ and $\tau(b) = \tau(b')$. Then*

$$\mathcal{K}_a^b(\psi) = \mathcal{K}_{a'}^{b'}(\psi[a \mapsto a', b \mapsto b'])$$

Proof. Let $f = [a \mapsto a', b \mapsto b']$ and note that the restriction of f to $L(\psi) - \{a, b\}$ is the identity map. Then:

$$\mathcal{K}_{a'}^{b'}(\psi[a \mapsto a', b \mapsto b']) = \mathcal{K}_{f(a)}^{f(b)}(\psi[f]) = \mathcal{K}_a^b(\psi)[\mathrm{id}] = \mathcal{K}_a^b(\psi)$$

2.1   Einstein Notation and Free Abstract Tensor Systems

$\mathcal{T}(\mathbf{x}, \mathbf{y})$ is just an abstract set. Its elements should be thought of as "black boxes" whose inputs are labelled by the set $\mathbf{x}$ and whose outputs are labelled by the set $\mathbf{y}$. Despite this sparse setting, working with abstract tensors is no more difficult than working with concrete tensors, with the help of some suggestive notation.

First, let a *tensor symbol* $\Psi = (\psi, \boldsymbol{x}, \boldsymbol{y})$ be a triple consisting of a tensor $\psi \in \mathcal{T}(\mathbf{x}, \mathbf{y})$ and lists $\boldsymbol{x}, \boldsymbol{y}$ with no repetition such that the elements of $\boldsymbol{x}$ are precisely $\mathbf{x}$ and the elements of $\boldsymbol{y}$ are precisely $\mathbf{y}$. Equivalently, a tensor symbol is a tensor along with a total ordering on input and output labels.

Notation 1. *Let $\boldsymbol{x} = [x_1, \ldots, x_m]$ and $\boldsymbol{y} = [y_1, \ldots, y_n]$ be lists of labels. Then we write the tensor symbol $\Psi = (\psi, \boldsymbol{x}, \boldsymbol{y})$ as:*

$$\psi_{\boldsymbol{x}}^{\boldsymbol{y}} \quad \text{or} \quad \psi_{x_1, \ldots, x_m}^{y_1, \ldots, y_n}$$

If $m = n$ and $\tau(x_i) = \tau(y_i)$, then let:

$$\delta_{\boldsymbol{x}}^{\boldsymbol{y}} := \delta_{x_1}^{y_1} \ldots \delta_{x_n}^{y_n}$$

In particular, the above expression evaluates to $1 \in \mathcal{T}(\emptyset, \emptyset)$ when $\boldsymbol{x} = \boldsymbol{y} = []$.

An *alphabet* $\mathcal{A}$ for an abstract tensor system is a set of tensor symbols such that for all $\mathbf{x}, \mathbf{y}$ each element $\psi \in \mathcal{T}(\mathbf{x}, \mathbf{y})$ occurs at most once.

The fact that labels in a tensor symbol are ordered may seem redundant, given that the labels themselves identify inputs and outputs in ψ. However, it gives us a convenient (and familiar) way to express relabellings. Given $\psi_{x_1, \ldots, x_m}^{y_1, \ldots, y_n} \in \mathcal{A}$, we can express a relabelled tensor as:

$$[\![\psi_{a_1, \ldots, a_m}^{b_1, \ldots, b_n}]\!] = \psi[x_1 \mapsto a_1, \ldots, x_m \mapsto a_m, y_1 \mapsto b_1, \ldots, y_n \mapsto b_n] \qquad (3)$$

It will often be convenient to refer to arbitrary tensors $\psi_{x_1, \ldots, x_m}^{y_1, \ldots, y_n} \in \mathcal{T}(\mathbf{x}, \mathbf{y})$ using tensor symbols. In this case, we treat subsequent references to ψ (possibly with different labels) as tensors that have been relabelled according to (3).

Definition 2. *An Einstein expression over an alphabet $\mathcal{A}$ is a list of δ-elements and (possibly relabelled) tensor symbols, where each label is either distinct or occurs as a repeated upper and lower label.*

For an Einstein expression E, let $E_{[a \mapsto a']}$ and $E^{[a \mapsto a']}$ be the same expression, but with a lower or upper label replaced. Einstein expressions are interpreted as abstract tensors as follows. First, any repeated labels are replaced by new, fresh labels of the same type, along with contractions.

$$[\![E]\!] = \mathcal{K}_a^{\bar{a}}([\![E^{[a \mapsto \bar{a}]}]\!]) \qquad \text{where } a \text{ is repeated and } \bar{a} \text{ is not in } E$$

Once all labels are distinct, juxtaposition is treated as tensor product:

$$[\![EE']\!] = [\![E]\!][\![E']\!] \qquad \text{where } EE' \text{ has no repeated labels}$$

Single tensor symbols are evaluated as in equation (3), and empty expressions evaluate to $1 \in \mathcal{T}(\emptyset, \emptyset)$. We will often suppress the semantic brackets $[\![-]\!]$ when it is clear we are talking about equality of tensors rather than syntactic equality of Einstein expressions.

Theorem 2. An Einstein expression unambiguously represents an element of some set $\mathcal{T}(\mathbf{x}, \mathbf{y})$. Furthermore, any expression involving constants taken from the alphabet $\mathcal{A}$, labellings, tensor products, and contractions can be expressed this way.

Proof. First, we show that the expression E represents an abstract tensor without ambiguity. In the above prescription, the choices we are free to make are (1) the order in which contractions are performed, (2) the choice of fresh labels $\bar{a}$, and (3) the order in which tensor products are evaluated. However, (1) is irrelevant by axiom T1 of an abstract tensor system, (2) by Lemma 1, and (3) by axiom T2.

For the other direction, suppose e is an expression involving constants, relabellings, tensor product, and contraction. Then, we can use the axioms of an abstract tensor system to pull contractions to the outside and push labellings down to the level of constants. Then, by the axioms of an abstract tensor system, there is an equivalence of expressions:

$$e \equiv \mathcal{K}_{a_1}^{b_1}(\mathcal{K}_{a_1}^{b_1}(\ldots(\mathcal{K}_{a_n}^{b_n}(\psi_1[f_1]\psi_2[f_2]\ldots\psi_m[f_m])))\ldots)$$

Let Ψ_i be the tensor symbol corresponding to the relabelled constant $\psi_i[f_i]$. Then, there is an equality of tensors: $e = [\![(\Psi_1\Psi_2\ldots\Psi_m)^{[b_1 \mapsto a_1, \ldots, b_n \mapsto a_n]}]\!]$

By Lemma 1, the particular choice of bound labels in an Einstein expression is irrelevant. That is, $[\![E]\!] = [\![E_{[x \mapsto \bar{x}]}^{[x \mapsto \bar{x}]}]\!]$, for x a bound label and $\bar{x}$ a new fresh label such that $\tau(x) = \tau(\bar{x})$. Also note that, by axiom T4 of an abstract tensor system, it is sometimes possible to eliminate δ elements from Einstein expressions.

$$[\![E\delta_a^b]\!] = [\![E^{[a \mapsto b]}]\!] \qquad \text{if } E \text{ contains } a \text{ as an upper label}$$

and similarly:

$$[\![E\delta_a^b]\!] = [\![E_{[b \mapsto a]}]\!] \qquad \text{if } E \text{ contains } b \text{ as a lower label}$$

The only cases where such a reduction is impossible are (1) when neither label on δ is repeated, or (2) when the repeated label is on δ itself: δ_a^a. For reasons that will become clear in the graphical notation, case (1) is called a *bare wire* and case (2) is called a *circle*. If no δ-elimination is possible, an expression E is called a *reduced Einstein expression*. Up to permutation of tensor symbols and renaming of bound labels, this reduced form is unique.

We are now ready to define the *free abstract tensor system* over an alphabet $\mathcal{A}$. First, define an equivalence relation on Einstein expressions. Let $E \approx E'$ if E can be obtained from E' by permuting tensor symbols, adding or removing δ-elements as above, or renaming repeated labels. Let $|E|$ be the $\approx$-equivalence class of E. Then, the free abstract tensor system $\mathbf{Free}(\mathcal{A})$ is defined as follows.

- $\mathcal{T}(\mathbf{x}, \mathbf{y})$ is the set of $|E|$ where $\mathbf{x}$ and $\mathbf{y}$ occur as non-repeated lower and upper labels in E, respectively
- $|E| \cdot |E'| = |EE'|$, with E and E' chosen with no common labels
- $\mathcal{K}^b_a(|E|) = |E^{[b \mapsto a]}|$
- $|E|[f] = |E^{[f']}_{[f']}|$ where f' sends bound labels to new fresh labels (i.e. not in $\mathrm{cod}(f)$) of the correct type and acts as f otherwise

Since we assume infinitely many labels, it is always possible to find fresh labels. Furthermore, the three tensor operations do not depend on the choice of (suitable) representative. Note, it is also possible to define the free ATS in terms of reduced Einstein expressions, in closer analogy with free groups. However, it will be convenient in the proof of Theorem 7 to let $|E|$ contain non-reduced expressions as well.

3   Diagrammatic Notation

There is another, often more convenient, alternative to Einstein notation for writing tensor expressions: *string diagram* notation. Tensor symbols are represented as boxes with one input for each lower index and one output for each upper index.

These inputs and outputs are marked with both a label and the type of the label, but this data is often suppressed if it is irrelevant or clear from context. A repeated label is indicated by connecting an input of one box to an output of another. These connections are called *wires*.

$$\psi^c_{a,b}\phi^{b,e}_d \;\Rightarrow\; \qquad\qquad\qquad\qquad (4)$$

Repeated labels are not written, as they are irrelevant by Lemma 1. δ-elements are written as *bare wires*, i.e. wires that are not connected to any boxes. In particular, contracting the input and the output of a δ element together yields a circle. Also, connecting a box to a bare wire has no effect, which is consistent with axiom T4.

The most important feature of the diagrammatic notation is that only the connectivity of the diagram matters. Therefore, the value is invariant under topological deformations. For example:

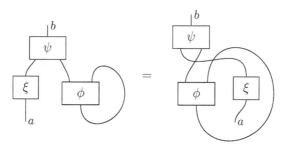

Theorem 3. For an alphabet $\mathcal{A}$, any tensor in **Free**$(\mathcal{A})$ can be unambiguously represented in the diagrammatic notation.

Proof. For a diagram D, form E as follows. First, chose a label that does not already occur in D for every wire connecting two boxes and for every circle. Then, let $E = \Psi_1 \ldots \Psi_n$, where each Ψ_i is a box from D, with labels taken from the input and output wires. Then, D represents the $\approx$-equivalence class of E defined in the previous section. By definition of $|E|$, the choice of labels for previously unlabelled wires in D and the order of the Ψ_i are irrelevant. Thus, D defines precisely one equivalence class $|E|$ in this manner.

4 Traced Symmetric Monoidal Categories

A monoidal category $(\mathcal{C}, \otimes, I, \alpha, \lambda, \rho)$ is a category that has a horizontal composition operation $\otimes : \mathcal{C} \times \mathcal{C} \to \mathcal{C}$ called the *monoidal product* that is associative and unital (up to isomorphism) and interacts well with the categorical (aka vertical) composition. A *strict* monoidal category is a monoidal category such that the natural isomorphisms α, λ, ρ are all identity maps. A *symmetric* monoidal category has an additional swap map $\sigma_{A,B} : A \otimes B \to B \otimes A$ such that $\sigma_{A,B} = \sigma_{B,A}$, and it interacts well with the rest of the monoidal structure. For full details, see e.g. [18].

Definition 3. *A traced symmetric monoidal category* $\mathcal{C}$ *is a symmetric monoidal category with a function*

$$Tr^X : \hom_{\mathcal{C}}(A \otimes X, B \otimes X) \to \hom_{\mathcal{C}}(A, B)$$

defined for all objects A, B, X, *satisfying the following five axioms:*[2]

1. $Tr^X((g \otimes X) \circ f \circ (h \otimes X)) = g \circ Tr^X(f) \circ h$
2. $Tr^Y(f \circ (A \otimes g)) = Tr^X((B \otimes g) \circ f)$

---

[2] Note that some structure isomorphisms have been suppressed for clarity. The coherence theorem for monoidal categories lets us do this without ambiguity.

3. $Tr^I(f) = f$ and $Tr^{X \otimes Y}(f) = Tr^X(Tr^Y(f))$
4. $Tr^X(g \otimes f) = g \otimes Tr^X(f)$
5. $Tr^X(\sigma_{X,X}) = 1_X$

Just as monoidal categories have strict and non-strict versions, so too do monoidal functors. *Strict* (traced, symmetric) monoidal functors preserve all of the categorical structure up to equality, whereas *strong* functors preserve all of the structure up to coherent natural isomorphism. The term "strong" is used by way of contrast with *lax* monoidal functors, which preserve the structure only up to (possibly non-invertible) natural transformations. Again, see [18] for full definitions.

Let **TSMC** be the category of traced symmetric monoidal categories and strong monoidal functors that preserve symmetry maps and the trace operation, and let **TSMC**$_s$ be the strict version.

4.1   The Free Traced Symmetric Monoidal Category

Two morphisms are equal in a free (symmetric, traced, compact closed, etc.) monoidal category *if and only if* their equality can be established only using the axioms of that category. Thus free monoidal categories are a powerful tool for proving theorems which hold in *all* categories of a particular kind. Free monoidal categories are defined over a collection of generators called a *monoidal signature*.

Notation 4. *For a set X, let X^* be the free monoid over X, i.e. the set of lists with elements taken from X where multiplication is concatenation and the unit is the empty list. For a function $f : X \to Y$, let $f^* : X^* \to Y^*$ be the lifting of f to lists: $f^*([x_1, \ldots, x_n]) = [f(x_1), \ldots, f(x_n)]$.*

Definition 4. *A (small, strict) monoidal signature $T = (O, M, \mathrm{dom}, \mathrm{cod})$ consists of a set of objects O, a set of morphisms M, and a pair of functions $\mathrm{dom} : M \to O^*$ and $\mathrm{cod} : M \to O^*$.*

The maps dom and cod should be interpreted as giving input and output types to a morphism $m \in M$. For instance, if $\mathrm{dom}(m) = [A, B, C]$ and $\mathrm{cod}(m) = [D]$, then m represents a morphism $m : A \otimes B \otimes C \to D$. The empty list is interpreted as the tensor unit I.

There is also a notion of a non-strict monoidal signature. In that case, O^* is replaced with the free $(\otimes, I)$-algebra over O. However, by the coherence theorem of monoidal categories, there is little difference between strict monoidal signatures and non-strict monoidal signatures with some fixed choice of bracketing.

Definition 5. *For monoidal signatures S, T, a monoidal signature homomorphism f consists of functions $f_O : O_S \to O_T$ and $f_M : M_S \to M_T$ such that $\mathrm{dom}_T \circ f_M = f_O^* \circ \mathrm{dom}_S$ and $\mathrm{cod}_T \circ f_M = f_O^* \circ \mathrm{cod}_S$. **MonSig** is the category of monoidal signatures and monoidal signature homomorphisms.*

A monoidal signature is essentially a strict monoidal category without composition or identity maps. A monoidal signature homomorphism is thus a strict monoidal functor, minus the condition that it respect composition and identity maps.

There is an evident forgetful functor from **TSMC**$_s$ into **MonSig**, by throwing away composition. If this forgetful functor has a left adjoint F, the image of a signature T under F is called the *free strict monoidal category* over T.

However, when considering the free non-strict category, the issue becomes a bit delicate. In particular, it is no longer reasonable to expect the lifted morphism $\widetilde{v}$ to be unique *on the nose*, but rather unique up to coherent natural isomorphism. Thus, the adjunction **MonSig** $\dashv$ **TSMC**$_s$ should be replaced with a pseudo-adjunction of some sort. To side-step such higher categorical issues, Joyal and Street simply state the appropriate correspondence between valuations of a signature and strong symmetric monoidal functors from the free category [10]. Here, we state the traced version of their definition. Let $[T, C]$ be the category of valuations of T in C and **TSMC**(C, D) be the category of strong traced symmetric monoidal functors from C to D and monoidal natural isomorphisms.

Definition 6. *For a monoidal signature $\boldsymbol{Sig}(A)$, a traced symmetric monoidal category $\mathbb{F}(T)$ is called the free traced SMC when, for any traced SMC C, there exists a valuation $\eta \in ob([\boldsymbol{Sig}(A), C])$ such that:*

$$(- \circ \eta) : \mathbf{TSMC}(\mathbb{F}(T), C) \to [T, C]$$

yields an equivalence of categories.

This equivalence of categories plays an analogous role to the isomorphism of hom-sets characterising an adjunction. For brevity, we omit the definitions of $[T, C]$ and $(- \circ \eta)$. The first represents the category of valuations of a monoidal signature T into a (possibly non-strict) monoidal category C and valuation morphisms (i.e. the valuation analogue of natural transformations). The latter represents the natural way to "compose" a valuation η with a strong monoidal functor to yield a new valuation. Details can be found in [10].

5   The Traced SMC of an Abstract Tensor System

In this section, we construct the associated traced symmetric monoidal category of an abstract tensor system. We shall see that an abstract tensor system and a traced SMC are essentially two pictures of the same thing. However, these two pictures vary in how they refer to the inputs and outputs of maps. On the one hand, traced SMCs take the input and output ordering to be fixed, and rely on structural isomorphisms for shuffling inputs and outputs around. On the other, abstract tensor systems refer to inputs and outputs by *labels*, but care must be taken to make sure these labels are given in a consistent manner.

Let $\mathbb{N}$ be the natural numbers and $\mathbb{B} = \{0, 1\}$. From hence forth, we will assume that the set $\mathcal{L}$ of labels has a special subset $\mathcal{L}_c \cong \mathcal{U} \times \mathbb{N} \times \mathbb{B}$ called the *canonical* labels. We write the elements $(X, i, 0)$ and $(X, i, 1)$ as $x_i^{(0)}$ and $x_i^{(1)}$, respectively. Then, let:

$$\tau(x_i^{(0)}) = \tau(x_i^{(1)}) = X$$

As we shall see in definition 7, using canonical labels allows us to impose input and output ordering on a tensor in order to treat it as a morphism in a monoidal category. It also yields a natural choice of free labels in the monoidal product of two tensors, which is an important consideration when the labels of the two tensors being combined are not disjoint.

Notation 5. *For $X = [X_1, X_2, \ldots, X_N]$ a list of types, define the following list of labels for $1 \leq m < n \leq N$ and $i = 0, 1$:*

$$x_{m..n}^{(i)} := [x_m^{(i)}, x_{m+1}^{(i)}, \ldots, x_{n-1}^{(i)}, x_n^{(i)}]$$

The set containing the above elements is denoted $\mathbf{x}_{m..n}^{(i)}$. In the case where $m = 1$ and $n = length(X)$, we often omit the subscripts, writing simply $x^{(i)}$ and $\mathbf{x}^{(i)}$.

Definition 7. *Let $\mathcal{S} = (\mathcal{U}, \mathcal{L}, \mathcal{T}(-, -))$ an abstract tensor system with a choice of canonical labels $\mathcal{L}_c \subseteq \mathcal{L}$. Then $\mathbb{C}[\mathcal{S}]$ is the traced symmetric monoidal category defined as follows:*

$$ob(\mathbb{C}[\mathcal{S}]) = \mathcal{U}^*$$

$$\hom_{\mathbb{C}[\mathcal{S}]}(X, Y) = \mathcal{T}(\mathbf{x}^{(0)}, \mathbf{y}^{(1)})$$

$$X \otimes Y = XY \quad (I = [\,])$$

For $\psi : X \to Y$, $\phi : Y \to Z$, $\tilde{\psi} : U \to V$, and $\xi : U \otimes X \to V \otimes X$, the rest of the structure is defined as:

$$\phi_{\mathbf{y}^{(0)}}^{\mathbf{z}^{(1)}} \circ \psi_{\mathbf{x}^{(0)}}^{\mathbf{y}^{(1)}} = \psi_{\mathbf{x}^{(0)}}^{\mathbf{y}'} \phi_{\mathbf{y}'}^{\mathbf{z}^{(1)}}$$

$$\psi_{\mathbf{x}^{(0)}}^{\mathbf{y}^{(1)}} \otimes \tilde{\psi}_{\mathbf{u}^{(0)}}^{\mathbf{v}^{(1)}} = \psi_{\mathbf{x}_{1..m}^{(0)}}^{\mathbf{y}_{1..n}^{(1)}} \tilde{\psi}_{\mathbf{u}_{m+1..m+m'}^{(0)}}^{\mathbf{v}_{n+1..n+n'}^{(1)}}$$

$$id_X = \delta_{\mathbf{x}^{(0)}}^{\mathbf{x}^{(1)}}$$

$$\sigma_{X,Y} = \delta_{\mathbf{x}_{1..m}^{(0)}}^{\mathbf{x}_{n+1..n+m}^{(1)}} \delta_{\mathbf{y}_{m+1..m+n}^{(0)}}^{\mathbf{y}_{1..n}^{(1)}}$$

$$Tr^X(\xi_{\mathbf{u}_{1..m}^{(0)} \mathbf{x}_{m+1..m+k}^{(0)}}^{\mathbf{v}_{1..n}^{(1)} \mathbf{x}_{n+1..n+k}^{(1)}}) = \xi_{\mathbf{u}_{1..m}^{(0)} \mathbf{x}'}^{\mathbf{v}_{1..n}^{(1)} \mathbf{x}'}$$

where $\mathbf{x}'$ and $\mathbf{y}'$ are chosen as fresh (possibly non-canonical) labels.

Theorem 6. $\mathbb{C}[\mathcal{S}]$ *is a strict, traced symmetric monoidal category.*

Proof. Associativity follows from ATS axioms (used implicitly in the Einstein notation):

$$(\xi \circ \phi) \circ \psi = \psi_{\mathbf{x}^{(0)}}^{\mathbf{y}'} \phi_{\mathbf{y}'}^{\mathbf{z}'} \xi_{\mathbf{z}'}^{\mathbf{w}^{(1)}} = \xi \circ (\phi \circ \psi)$$

and similarly for identity maps. Associativity and unit laws of the monoidal product follow straightforwardly from associativity of $(- \cdot -)$. The interchange law can be shown as:

$$(\tilde{\psi}^{z(1)}_{y(0)} \otimes \tilde{\phi}^{w(1)}_{v(0)}) \circ (\psi^{y(1)}_{x(0)} \otimes \phi^{v(1)}_{u(0)}) = \psi^{w'}_{x(0)} \phi^{x'}_{u(0)} \tilde{\psi}^{y(1)}_{w'}{}_{1..n} \tilde{\phi}^{z(1)}_{x'}{}_{n+1..n+n'}$$

$$= \psi^{w'}_{x(0)} \tilde{\psi}^{y(1)}_{w'}{}_{1..n} \phi^{x'}_{u(0)} \tilde{\phi}^{z(1)}_{x'}{}_{n+1..n+n'}$$

$$= (\tilde{\psi} \circ \psi) \otimes (\tilde{\phi} \circ \phi)$$

Verification of the symmetry and trace axioms is a routine application of the ATS axioms.

6   The Free ATS and the Free Traced SMC

In this section, we will show that the free abstract tensor system over an alphabet induces a free traced symmetric monoidal category. We assume for the remainder of this section that tensor symbols in an alphabet $\mathcal{A}$ are canonically labelled. That is, they are of the form $\psi^{y(1)}_{x(0)} \in \mathcal{A}$. As the labels have no *semantic* content, we can always replace an arbitrary alphabet with a canonically labelled one.

Also note that canonically labelled alphabets and monoidal signatures are essentially the same thing. Let $\mathrm{Sig}(\mathcal{A})$ be the monoidal signature with morphisms $\psi^{y(1)}_{x(0)} \in \mathcal{A}$ and the dom and cod maps defined by:

$$\mathrm{dom}(\psi^{y(1)}_{x(0)}) = X \qquad \mathrm{cod}(\psi^{y(1)}_{x(0)}) = Y$$

For any signature $S = (\mathcal{O}, \mathcal{M}, \mathrm{dom}, \mathrm{cod})$, it is always possible to define an alphabet $\mathcal{A}$ such that $S = \mathrm{Sig}(\mathcal{A})$. Thus, we will often use the designation $\mathrm{Sig}(\mathcal{A})$ to refer to an *arbitrary* monoidal signature.

For $\mathrm{Free}(\mathcal{A})$ the free ATS over $\mathcal{A}$, we will show that $\mathbb{C}[\mathrm{Free}(\mathcal{A})]$ is the free traced SMC over $\mathrm{Sig}(\mathcal{A})$. We will do this by first considering the strict case, where we construct the unique strict traced symmetric monoidal functor $\tilde{v}$ that completes the following diagram, for signature homomorphisms η, v:

$$\mathrm{Sig}(\mathcal{A}) \xrightarrow{\eta} \mathbb{C}[\mathrm{Free}(\mathcal{A})]$$

$$\searrow_{v} \qquad \downarrow^{\tilde{v}}$$

$$\mathcal{C}$$

(5)

Before we get to the bulk of the proof, we introduce some notation. The first thing we introduce is the notion of *labelling* a morphism.

Definition 8. *For a set $\mathcal{L}$ of labels and a function $\mu : \mathcal{L} \to ob(\mathcal{C})$, an object is called μ-labelled if it is equipped with a list i such that:*

$$X = \mu(i_1) \otimes \mu(i_2) \otimes \ldots \otimes \mu(i_n)$$

A morphism is called μ-labelled if its domain and codomain have μ-labellings for disjoint lists $\boldsymbol{i}$, $\boldsymbol{j}$.

To simplify notation, we write μ labelled objects as follows:

$$X = X_{i_1} \otimes X_{i_2} \otimes \ldots \otimes X_{i_n}$$

where $X_{i_k} = \mu(i_k)$. For a μ-labelled object $(X, \boldsymbol{i})$ and a label $i \in \boldsymbol{i}$, $\sigma_{X:i}$ is the (unique) symmetry map that permutes the object X_i to the end of the list and leaves the other objects fixed.

In any traced SMC, we can define a contraction operator $C_i^j(-)$ which "traces together" the i-th input with the j-th output on a labelled morphism.

Definition 9. *Let $f : X_{i_1} \otimes \ldots \otimes X_{i_M} \to Y_{j_1} \otimes \ldots \otimes Y_{j_N}$ be a labelled morphism in a traced symmetric monoidal category such that for labels $i \in \{i_1, \ldots, i_M\}$ and $j \in \{j_1, \ldots, j_N\}$, $X_i = Y_j$. Then we define the trace contraction $C_i^j(f)$ as follows:*

$$C_i^j(f) := Tr^{X_i = Y_j}(\sigma_{Y:j} \circ f \circ \sigma_{X:i}^{-1})$$

Note that a contraction of a labelled morphism yields another labelled morphism, by deleting the contracted objects from the label lists. Thus we can contract many times, and the resulting morphism does not depend on the order in which we perform contractions.

Lemma 2. Contractions are commutative. For a labelled morphism f distinct indices i, i' and j, j':

$$C_i^j(C_{i'}^{j'}(f)) = C_{i'}^{j'}(C_i^j(f))$$

Definition 10. *For a strict traced symmetric monoidal category $\mathcal{C}$, define a set M of atomic morphisms, such that any morphism in $\mathcal{C}$ can be obtained from those morphisms and the traced symmetric structure. An labelled morphism is called disconnected if it is of the form $f = f_1 \otimes \ldots \otimes f_K$, where each f_k is a labelled morphism in M:*

$$f_k : X_{i_{k,1}} \otimes \ldots X_{i_{k,M_k}} \to Y_{j_{k,1}} \otimes \ldots \otimes Y_{j_{k,N_k}}$$

Definition 11. *Let $f = f_1 \otimes \ldots \otimes f_M$ be a disconnected labelled morphism. For distinct indices $\{i_1, \ldots, i_P\} \subseteq \{i_{1,1}, \ldots i_{K,M_K}\}$ and $\{j_1, \ldots, j_P\} \subseteq \{j_{1,1}, \ldots j_{K,N_K}\}$, a map f' is said to be in contraction normal form (CNF) if:*

$$f' = C_{i_1}^{j_1}(C_{i_2}^{j_2}(\ldots(C_{i_P}^{j_P}(f))\ldots))$$

Definition 12. *Let f and f' be given as in Definition 11. A component f_k of f is said to be totally contracted if the indices of all of its inputs occur in $\{i_1, \ldots, i_P\}$ and the indices of all of its outputs occur in $\{j_1, \ldots, j_P\}$.*

Lemma 3. For f and f' from Definition 11, totally contracted components of f can be re-ordered arbitrarily by relabelling.

Lemma 4. Let f, f' be Defined as in 11. If $f_k = 1_{X_{i_{k,1}}} = 1_{Y_{j_{k,1}}}$ is a totally contracted identity map that is not a circle (i.e. it is not contracted to itself), then it can be removed by relabelling.

For full proofs of lemmas 2, 3, and 4, see [13]. The final ingredient we need for the main theorem is the correspondence between the operations C_i^j and $\mathcal{K}_i^j$. First, note that labelled morphisms in $\mathbb{C}[\mathbf{Free}(\mathcal{A})]$ are in 1-to-1 correspondence with tensors in $\mathbf{Free}(\mathcal{A})$. That is, a morphism $\psi^{\boldsymbol{y}\,(1)}_{\boldsymbol{x}\,(0)} : \boldsymbol{X} \to \boldsymbol{Y}$ labelled by $(\boldsymbol{i}, \boldsymbol{j})$ defines the tensor $\psi_{\boldsymbol{i}}^{\boldsymbol{j}}$. By abuse of notation, we will write $\psi_{\boldsymbol{i}}^{\boldsymbol{j}}$ for both the tensor and the corresponding labelled morphism.

Lemma 5. For some fixed objects $\boldsymbol{X}, \boldsymbol{Y} \in \mathbb{C}[\mathbf{Free}(\mathcal{A})]$, a labelled morphism $\psi_{\boldsymbol{i}}^{\boldsymbol{j}} : \boldsymbol{X} \to \boldsymbol{Y}$ in $\mathbb{C}[\mathbf{Free}(\mathcal{A})]$, and labels $i \in \boldsymbol{i}, j \in \boldsymbol{j}$:

$$C_i^j(\psi_{\boldsymbol{i}}^{\boldsymbol{j}}) = \mathcal{K}_i^j(\psi_{\boldsymbol{i}}^{\boldsymbol{j}})$$

With the help of these lemmas, we are now ready to prove the main theorem.

Theorem 7. $\mathbb{C}[\mathbf{Free}(\mathcal{A})]$ is the free strict traced symmetric monoidal category over $\mathbf{Sig}(\mathcal{A})$.

Proof. Let η be the monoidal signature homomorphism that is identity on objects and sends morphisms $\psi^{\boldsymbol{y}\,(1)}_{\boldsymbol{x}\,(0)} \in \mathcal{A}$ to themselves, considered as Einstein expressions with one tensor symbol.

$$\psi^{\boldsymbol{y}\,(1)}_{\boldsymbol{x}\,(0)} \in \mathcal{T}(\mathbf{x}^{(0)}, \mathbf{y}^{(1)}) = \hom_{\mathbb{C}[\mathbf{Free}(\mathcal{A})]}(\boldsymbol{X}, \boldsymbol{Y})$$

Now, supposing we are given another monoidal signature homomorphism v from $\mathrm{Sig}(\mathcal{A})$ into a strict traced symmetric monoidal category $\mathcal{C}$. Our goal is to build a traced symmetric monoidal functor $\widetilde{v} : \mathbb{C}[\mathbf{Free}(\mathcal{A})] \to \mathcal{C}$ such that $\widetilde{v}\eta = v$. On objects:

$$\widetilde{v}([X_1, X_2, \ldots, X_n]) = v(X_1) \otimes v(X_2) \otimes \ldots \otimes v(X_n)$$

Let $|E| : \boldsymbol{X} \to \boldsymbol{Y}$ be morphism in $\mathbb{C}[\mathbf{Free}(\mathcal{A})]$. In other words, it is an equivalence class of Einstein expressions, up to permutation of tensor symbols, renaming of bound labels, and δ-contraction. Choose some representative E of $|E|$ such that E is of the form:

$$E = \delta^{\boldsymbol{x}'}_{\boldsymbol{x}\,(0)} \delta^{\boldsymbol{y}\,(1)}_{\boldsymbol{y}'} \Psi_1 \Psi_2 \ldots \Psi_M \tag{6}$$

where Ψ_i are tensor symbols (or δ-elements) and $E' = \Psi_1\Psi_2\ldots\Psi_M$ is an Einstein expression with upper labels $\boldsymbol{x}'$ and lower labels $\boldsymbol{y}'$. In other words, E' contains no free labels, and $\boldsymbol{x}'$ and $\boldsymbol{y}'$ are disjoint. Form an expression F from E by choosing a fresh j_k for each repeated label i_k. Reading the repeated indices in E from left to rewrite, the choice of E fixes a unique expression:

$$E = \mathcal{K}_{i_1}^{j_1}(\mathcal{K}_{i_2}^{j_2}(\ldots(\mathcal{K}_{i_N}^{j_N}(F))\ldots)$$

which, by Lemma 5 can be expressed:

$$E = C_{i_1}^{j_1}(C_{i_2}^{j_2}(\ldots(C_{i_N}^{j_N}(F)\ldots)$$

up to bound labels j_k. Since F contains no repeated labels,

$$E = C_{i_1}^{j_1}(C_{i_2}^{j_2}(\ldots(C_{i_N}^{j_N}(1_{\boldsymbol{X}} \otimes 1_{\boldsymbol{Y}} \otimes \Psi_1 \otimes \ldots \otimes \Psi_n)\ldots)$$

Then, $\tilde{v}$ must respect v and preserve the traced symmetric monoidal structure. In particular, it must preserve C_i^j. So, the only possible value for $\tilde{v}(\widehat{E})$ is:

$$\tilde{v}(\widehat{E}) = C_{i_1}^{j_1}(C_{i_2}^{j_2}(\ldots(C_{i_N}^{j_N}(1_{v(\boldsymbol{X})} \otimes 1_{v(\boldsymbol{Y})} \otimes v(\Psi_1) \otimes \ldots \otimes v(\Psi_n))\ldots)$$

where the labelling on the argument is inherited from the labelling on F.

Then, since all of the non-canonical labels in E are contracted, $\tilde{v}(\widehat{E})$ is indeed a morphism from $\tilde{v}(\boldsymbol{X})$ to $\tilde{v}(\boldsymbol{Y})$. For this to be well-defined, we need to show it does not depend on the choice of E. First, note that the choice of bound labels is irrelevant because the operation C_i^j is defined in terms of (pairs of) *positions* of labels, and does not depend on the choice of labels themselves. Next, consider the form (6) we have fixed for E. The order of tensor symbols is fixed except for in $\Psi_1\Psi_2\ldots\Psi_M$. But then, all of the symbols in $\Psi_1\Psi_2\ldots\Psi_M$ must be totally contracted, so by Lemma 3, the order of the corresponding $v(\Psi_i)$ are irrelevant. Furthermore, δ expansion or removal will not affect the value of $\tilde{v}(\widehat{E})$ by corollary 4. Thus $\tilde{v}$ is well-defined.

Next, we show that $\tilde{v}$ is a traced symmetric monoidal functor. It follows immediately from the definition that $\tilde{v}$ preserves the C_i^j operation:

$$C_i^j(\tilde{v}(f)) = \tilde{v}(C_i^j(f))$$

where $\tilde{v}(f)$ inherits its labelling from f. The fact that $\tilde{v}$ preserves the monoidal product follows from the definition of C_i^j and the trace axioms. Then, since all of the rest of the traced symmetric monoidal structure can be defined in terms of C_i^j and $\otimes$, $\tilde{v}$ must preserve it.

This suffices to establish that $\mathbb{C}[\mathbf{Free}(\mathcal{A})]$ is the free *strict* traced symmetric monoidal category. The extension to the non-strict case is now routine.

Corollary 1. $\mathbb{C}[\mathbf{Free}(\mathcal{A})]$ is the free traced symmetric monoidal category over $\mathbf{Sig}(\mathcal{A})$.

Proof. Theorem 7 establishes the correspondence between valuations and traced symmetric monoidal functors when $\mathcal{C}$ is strict. The remainder of the proof is similar to that of theorem 1.2 in [10].

6.1   The Diagrammatic Free Category

In [10], Joyal and Street defined the free symmetric monoidal category in terms of *anchored diagrams* with valuations.

Definition 13. *A generalised topological graph is a topological space G and a distinguished finite discrete subspace G_0 such that $G - G_0$ is isomorphic to a disjoint union of finitely many copies of the open interval $(0,1)$ or circles S^1.*

The points in G_0 are called nodes and the components of $G - G_0$ are called wires. An anchored diagram is then a generalised topological graph with three extra pieces of data:

1. A choice of orientation for each wire.
2. For each $n \in G_0$, a total ordering for the input and output wires of n, where inputs and outputs are distinguished using the orientation of wires.
3. A total ordering on the inputs and outputs of G as a whole, i.e. the ends of wires that are not connected to nodes.

A valuation v of G is then an assignment of objects to wires and morphisms to the points in G_0. The data (G, v) can be pictured as follows:

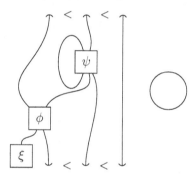

Let $\mathbf{Anchor}(\mathbf{Sig}(\mathcal{A}))$ be the category whose objects are lists of objects from $\mathbf{Sig}(\mathcal{A})$ and whose morphisms are isomorphism classes of anchored diagrams with valuations. Composition is defined by plugging the outputs of one diagram into the inputs of another and monoidal product as diagram juxtaposition. Let $\mathbf{Anchor}_p(\mathbf{Sig}(\mathcal{A}))$ be the restriction of morphisms to *progressive* diagrams, i.e. diagrams with no feedback. For details on this construction, see [10].

Theorem 8 ([10]). $\mathbf{Anchor}_p(\mathbf{Sig}(\mathcal{A}))$ *is the free symmetric monoidal category on* $\mathbf{Sig}(\mathcal{A})$.

If we drop the progressive constraint, $\mathbf{Anchor}(\mathbf{Sig}(\mathcal{A}))$ admits a trace operation in the obvious way. So, we can now state the above result for traced symmetric monoidal categories as a corollary to Theorem 7. This is due to the close relationship between Einstein expressions in diagrams demonstrated in section 3.

Corollary 2. Anchor(Sig($\mathcal{A}$)) is the free traced symmetric monoidal category on **Sig($\mathcal{A}$)**.

Proof. We can prove this by demonstrating a (traced symmetric monoidal) isomorphism of categories between **Anchor(Sig($\mathcal{A}$))** and $\mathbb{C}[\mathbf{Free}(\mathcal{A})]$. These categories have the same objects, so it suffices to show an isomorphism of hom-sets. By a construction similar to that described in section 3, a tensor in $\mathbb{C}[\mathbf{Free}(\mathcal{A})]$ defines a unique anchored diagram, where the total ordering on inputs and outputs is defined using the partial ordering on canonical labels: $x_i \leq y_j \Leftrightarrow i \leq j$. Furthermore, any anchored diagram can be expressed this way.

References

1. Abramsky, S., Coecke, B.: A categorical semantics of quantum protocols. In: Proceedings from LiCS (2004) arXiv:quant-ph/0402130v5
2. Coecke, B., Grefenstette, E., Sadrzadeh, M.: Lambek vs. Lambek: Functorial Vector Space Semantics and String Diagrams for Lambek Calculus (2013) arXiv: 1302.0393 [math.LO]
3. Coecke, B., Duncan, R.: Interacting Quantum Observables: Categorical Algebra and Diagrammatics (2009) arXiv:0906.4725v1 [quant-ph]
4. Coecke, B., Sadrzadeh, M., Clark, S.: Mathematical Foundations for a Compositional Distributional Model of Meaning (2010) arXiv:1003.4394v1 [cs.CL]
5. Cvitanovic, P.: Group Theory: Birdtracks, Lie's, and Exceptional Groups. Princeton University Press (2008)
6. Epstein, D.: Functors between tensored categories. Inventiones Mathematicae 1(3), 221–228 (1966)
7. Hardy, L.: A Formalism-local framework for general probabilistic theories including quantum theory (2010)
8. Hasegawa, M.: Models of Sharing Graphs (A Categorical Semantics of Let and Letrec). PhD thesis, University of Edinburgh (1997)
9. Hasegawa, M., Hofmann, M.O., Plotkin, G.: Finite Dimensional Vector Spaces Are Complete for Traced Symmetric Monoidal Categories. In: Avron, A., Dershowitz, N., Rabinovich, A. (eds.) Pillars of Computer Science. LNCS, vol. 4800, pp. 367–385. Springer, Heidelberg (2008),
   http://dx.doi.org/10.1007/978-3-540-78127-1_20
10. Joyal, A., Street, R.: The geometry of tensor calculus I. Advances in Mathematics 88, 55–113 (1991)
11. Kelly, G., Maclane, S.: Coherence in closed categories. Journal of Pure and Applied Algebra 1(1), 97–140 (1971)
12. Kelly, M., Laplaza, M.L.: Coherence for Compact Closed Categories. Journal of Pure and Applied Algebra 19, 193–213 (1980)
13. Kissinger, A.: Pictures of Processes: Automated Graph Rewriting for Monoidal Categories and Applications to Quantum Computing. PhD thesis, University of Oxford (2011)
14. Lambek, J.: Compact monoidal categories from linguistics to physics. In: New Structures for Physics. Lecture Notes in Physics, ch. 8. Springer (2011)
15. Lambek, J.: Deductive systems and categories I: Syntactic calculus and residuated categories. Math. Systems Theory 2, 287–318 (1968)

16. Lambek, J.: Deductive systems and categories ii: Standard constructions and closed categories. Lecture Notes in Mathematics, vol. 86. Springer (1969)
17. Mac Lane, S.: Natural associativity and commutativity. Rice University Studies 49, 28–46 (1963)
18. Mac Lane, S.: Categories for the working mathematician. Springer (1998)
19. Penrose, R.: Spinors and Space-Time: Two-Spinor Calculus and Relativistic Fields, vol. 1. Cambridge Monographs on Mathematical Physics (1997)
20. Penrose, R.: The Road to Reality: A Complete Guide to the Laws of the Universe. Vintage Books (2004)
21. Penrose, R.: Applications of negative dimensional tensors. In: Combinatorial Mathematics and its Applications, pp. 221–244. Academic Press (1971)
22. Selinger, P.: A survey of graphical languages for monoidal categories. New Structures for Physics, 275–337 (2009)
23. Shum, M.: Tortile tensor categories. Journal of Pure and Applied Algebra 93(1), 57–110 (1994)
24. Turaev, V.G.: Quantum invariants of knots and 3-manifolds. de Gruyter Studies in Mathematics, vol. 18. Walter de Gruyter (1994)

On Canonical Embeddings
of Residuated Groupoids

Mirosława Kołowska-Gawiejnowicz

Adam Mickiewicz University in Poznań
Faculty of Mathematics and Computer Science
e-mail: mkolowsk@amu.edu.pl

Abstract. We prove that every symmetric residuated groupoid is embeddable in a boolean double residuated groupoid. Analogous results are obtained for other classes of algebras, e.g. (commutative) symmetric residuated semigroups, symmetric residuated unital groupoids, cyclic bilinear algebras. We also show that powerset algebras constructed in the paper preserve some Grishin axioms.

Keywords: representation theorems, residuated groupoids, symmetric residuated groupoids, boolean double residuated groupoid.

1 Introduction

Residuated groupoids with operations $\otimes, \backslash, /$ are models of Nonassociative Lambek Calculus (NL) [11] and other weak substructural logics [7]. Symmetric residuated groupoids with operations $\otimes, \backslash, /$ and dual operations $\oplus, \oslash, \obslash$ are models of certain symmetric substructural logics, as e.g. Grishin's extensions of the Lambek calculus [8]. In particular, Moortgat [16] studies a nonassociative symmetric substructural logic, called Lambek-Grishin calculus (LG), as a type logic for Type-Logical Grammar. Let us recall the calculus LG. Types are formed out of atomic types $p, q, r, \ldots$ by means of the following formation rule: if A, B are types, then also $A \otimes B, A \backslash B, A / B, A \oplus B, A \oslash B, A \obslash B$ are types. The minimal LG is given by the preorder axioms:
$$A \to A; \text{ if } A \to B \text{ and } B \to C \text{ then } A \to C,$$
together with the residuation and dual residuation laws:

$$A \to C/B \quad \text{iff} \quad A \otimes B \to C \quad \text{iff} \quad B \to A \backslash C,$$
$$C \oslash B \to A \quad \text{iff} \quad C \to A \oplus B \quad \text{iff} \quad A \obslash C \to B.$$

Algebraic models of this calculus are symmetric residuated groupoids.

Interesting extensions of this calculus can be obtained by adding Grishin axioms (see Section 5). Other well-known logics of that kind are Multiplicative Linear Logics, corresponding to commutative involutive symmetric residuated semigroups, and their noncommutative and nonassociative variants, e.g. InFL, InGL (see e.g. [1,7]).

C. Casadio et al. (Eds.): Lambek Festschrift, LNCS 8222, pp. 253–267, 2014.
© Springer-Verlag Berlin Heidelberg 2014

There are many natural constructions of multiple residuated groupoids, i.e. residuated groupoids with several residuation triples (see e.g. [5,9]). Dual residuated groupoids (satisfying the residuation law with respect to dual ordering $\geq$) can be constructed by using an involutive negation, e.g. set complementation $\sim$, and defining the dual residuation triple:

$$X \oplus Y = (X^\sim \otimes Y^\sim)^\sim, X \oslash Y = (X^\sim \backslash Y^\sim)^\sim, X \oslash Y = (X^\sim / Y^\sim)^\sim.$$

Białynicki-Birula and Rasiowa [2] show that every quasi-boolean algebra (i.e. a distributive lattice with an involutive negation, satisfying Double Negation and Transposition, or equivalently: Double Negation and one of De Morgan laws) is embeddable into a quasi-field of sets (i.e. a family of sets, closed under $\cup$, $\cap$ and a quasi-complement $\sim_g X = g[X]^\sim$, where g is an involutive mapping).

In this paper we prove similar results for symmetric residuated groupoids and related algebras. Our embedding preserves the residuated groupoid operations and negation(s). The target algebra is a field or a quasi-field of sets with additional operations of a symmetric residuated groupoid.

We prove that every symmetric residuated groupoid is a subalgebra of an algebra of the above form. As in [5], by a boolean residuated algebra one means a residuated algebra with additional boolean operations $\sim$, $\cup$, $\cap$, and similarly for a quasi-boolean residuated algebra. More precisely, we show that every symmetric residuated groupoid can be embedded in a boolean double residuated groupoid, which is a field of sets with additional operations $\otimes_1$, $\backslash_1$, $/_1$, $\otimes_2$, $\backslash_2$, $/_2$ and $\sim$ (dual operations are defined from $\otimes_2$, $\backslash_2$, $/_2$ as above). Analogous results are obtained for (commutative) symmetric residuated semigroups and other algebras of this kind. Furthermore, the target algebra always consists of subsets of some set, and the involutive negation is set complementation. The results elaborate final remarks of Buszkowski [5] who considers general residuated algebras, but does not provide any details of the representation. Let us notice that in [5] symmetric residuated algebras are called double residuated algebras.

We also show that the target algebra is a (commutative) semigroup, if the source algebra is so. Units 1 and 0 (for $\otimes$ and $\oplus$, respectively) are preserved, if the target algebra is restricted to a family of upsets. The latter is a quasi-field of sets, if the source algebra admits an involutive negation '$-$', and the embedding sends '$-$' to a quasi-complement. The target algebra is a cyclic bilinear algebra, if the source algebra is so.

Some ideas of our proofs are similar to those of Kurtonina and Moortgat [10] in their proof of the completeness of the minimal LG with respect to Kripke semantics. Our algebraic approach, however, reveals more uniformity of the whole construction, i.e. its two-level form where the first level is related to the ground level in the same way as the second one to the first one.

The paper is organized as follows. In Section 2 we discuss some basic notions. Section 3 contains some powerset constructions of residuated groupoids, dual residuated groupoids and symmetric residuated groupoids. The main result, a representation theorem for symmetric residuated groupoids, is proved in Section 4. In Section 5 we provide similar representation theorems for symmetric residuated semigroups, symmetric residuated unital groupoids and cyclic bilinear algebras. At the end of this section we consider some Grishin axioms.

2 Preliminaries

We begin this section with the definitions of some basic notions.

Let us recall that a structure $(M, \leq, \otimes)$ is *a partially ordered groupoid (p.o. groupoid)*, if $\leq$ is a partial order and $\otimes$ is a binary operation monotone in both arguments i.e. $a \leq b$ implies $a \otimes c \leq b \otimes c$ and $c \otimes a \leq c \otimes b$, for $a, b, c \in M$.

A *residuated groupoid* is a structure $(M, \leq, \otimes, \backslash, /)$ such that $(M, \leq)$ is a poset, $(M, \otimes)$ is a groupoid, and $\otimes, \backslash, /$ satisfy the residuation law:

$$a \leq c/b \quad \text{iff} \quad a \otimes b \leq c \quad \text{iff} \quad b \leq a\backslash c,$$

for all $a, b, c \in M$. It is easy to show that if $(M, \leq, \otimes, \backslash, /)$ is a residuated groupoid, then $(M, \leq, \otimes)$ is a p.o. groupoid.

A *dual residuated groupoid* is defined as a structure $(M, \leq, \oplus, \oslash, \obslash)$ such that $(M, \leq)$ is a poset, $(M, \oplus)$ is a groupoid, and $\oplus, \oslash, \obslash$ satisfy the dual residuation law:

$$c \obslash b \leq a \quad \text{iff} \quad c \leq a \oplus b \quad \text{iff} \quad a \oslash c \leq b$$

for all $a, b, c \in M$. Again $(M, \leq, \oplus)$ is a p.o. groupoid.

A structure $\boldsymbol{M} = (M, \leq, \otimes, \backslash, /, \oplus, \oslash, \obslash)$ is called *a symmetric residuated groupoid* iff the $(\leq, \otimes, \backslash, /)$-reduct of $\boldsymbol{M}$ and the $(\leq, \oplus, \oslash, \obslash)$-reduct of $\boldsymbol{M}$ are a residuated groupoid and a dual residuated groupoid, respectively.

An *involutive residuated groupoid* is a structure which arises from a residuated groupoid by adding a unary operation $-$ (we call it *an involutive negation*) which satisfies the following two conditions:

$$- - a = a \qquad \text{(Double Negation)}$$
$$a \leq b \Rightarrow -b \leq -a \quad \text{(Transposition)}$$

for all elements a, b. In a similar way we define involutive dual residuated groupoids, involutive symmetric residuated groupoids etc. Given lattice operations $\vee, \wedge$, the second condition is equivalent to $-(a \vee b) = (-a) \wedge (-b)$. Hence our involutive negation corresponds to a quasi-complement in the sense of [2] and a De Morgan negation (assuming Double Negation) in the sense of [6]. It is also called a cyclic negation in the literature on substructural logics (cf. [7]).

A *multiple p.o. groupoid* is an ordered algebra $\boldsymbol{M} = (M, \leq, \{\otimes\}_{i \in I})$ such that, for any $i \in I$, $(M, \leq, \otimes_i)$ is a p.o. groupoid. By a *multiple residuated groupoid* we mean a structure $\boldsymbol{M} = (M, \leq, \{\odot_i, \backslash_i, /_i\}_{i=1,\dots,n})$ such that the $(\leq, \odot_i, \backslash_i, /_i)$-reducts of $\boldsymbol{M}$ for $i = 1, 2, \dots n$ are residuated groupoids.

In this paper we only consider *double residuated groupoids*, i.e. multiple residuated groupoids for $i = 1, 2$.

Let M be an involutive residuated groupoid. We define the structure $M^\ominus = (M, \leq, \oplus, \oslash, \otimes, -)$ such that

$$a \oplus b = -((-a) \otimes (-b)),$$
$$a \oslash b = -((-a)/(-b)),$$
$$a \otimes b = -((-a)\backslash(-b)).$$

Now, let M denote a dual involutive residuated groupoid. We define the structure $M^- = (M, \leq, \otimes, /, \backslash, -)$ such that

$$a \otimes b = -((-a) \oplus (-b)),$$
$$a/b = -((-a) \oslash (-b)),$$
$$a\backslash b = -((-a) \otimes (-b)).$$

Lemma 1. *If M is an involutive residuated groupoid, then $M^\ominus$ is an involutive dual residuated groupoid. If M is an involutive dual residuated groupoid, then M^- is an involutive residuated groupoid.*

Proof. Assume that M is an involutive residuated groupoid. We show

$$c \oslash b \leq a \text{ iff } c \leq a \oplus b \text{ iff } a \otimes c \leq b.$$

We prove the first equivalence:
$c \leq a \oplus b$ iff $c \leq -((-a) \otimes (-b))$ iff $(-a) \otimes (-b) \leq -c$ iff
$-a \leq (-c)/(-b)$ iff $-((-c)/(-b)) \leq a$ iff $c \oslash b \leq a$
The second equivalence can be proved in an analogous way.

Assuming that M is an involutive dual residuated groupoid, the equivalences $a \leq c/b$ iff $a \otimes b \leq c$ iff $b \leq a\backslash c$ can be proved in an analogous way to the one above. $\qquad\square$

Observe that $M^{\ominus-} = M$ and $M^{-\ominus} = M$.

It is easy to show that for symmetric residuated groupoids the following conditions hold:

$$a \otimes (a\backslash b) \leq b, \quad (b/a) \otimes a \leq b,$$
$$b \leq a \oplus (a \otimes b), \quad b \leq (b \oslash a) \oplus a,$$
$$a \leq b \Rightarrow c\backslash a \leq c\backslash b, \quad a/c \leq b/c, \quad b\backslash c \leq a\backslash c, \quad c/b \leq c/a;$$
$$a \leq b \Rightarrow c \otimes a \leq c \otimes b, \quad a \oslash c \leq b \oslash c, \quad b \otimes c \leq a \otimes c, \quad c \oslash b \leq c \oslash a.$$

3   A Powerset Construction

Concrete residuated groupoids can be constructed in various ways. A basic construction is the powerset residuated groupoid.

Given a groupoid $M = (M, \otimes)$, we consider the powerset $\mathcal{P}(M)$ with operations defined as follows:

$$X \otimes Y = \{a \otimes b : a \in X, \ b \in Y\},$$
$$X\backslash Z = \{c \in M : \forall a \in X \ a \otimes c \in Z\},$$
$$Z/Y = \{c \in M : \forall b \in Y \ c \otimes b \in Z\}.$$

Then, $(\mathcal{P}(M), \subset, \otimes, \backslash, /)$ is a residuated groupoid; we denote this algebra by $\mathcal{P}(M)$. Every residuated groupoid can be embedded in a structure of this form as shown in [9]. In this paper, we apply a more general construction.

Starting from a p.o. groupoid $M = (M, \leq, \otimes)$ one can define another powerset algebra which will be denoted by $\mathcal{P}^{\leq}(M)$. For $X, Y, Z \subset M$, we define operations:

$$X \widehat{\otimes} Y = \{c \in M : \exists a \in X \; \exists b \in Y \; a \otimes b \leq c\},$$
$$X \widehat{\backslash} Z = \{b \in M : \forall a \in X \; \forall c \in M \; (a \otimes b \leq c \Rightarrow c \in Z)\},$$
$$Z \widehat{/} Y = \{a \in M : \forall b \in Y \; \forall c \in M \; (a \otimes b \leq c \Rightarrow c \in Z)\}.$$

The following lemma holds.

Lemma 2. $\mathcal{P}^{\leq}(M) = (\mathcal{P}(M), \subset, \widehat{\otimes}, \widehat{\backslash}, \widehat{/})$ is a residuated groupoid.

Proof. We prove that the residuation law holds, i.e.

$$Y \subset X \widehat{\backslash} Z \quad \text{iff} \quad X \widehat{\otimes} Y \subset Z \quad \text{iff} \quad X \subset Z \widehat{/} Y$$

for every $X, Y, Z \in \mathcal{P}(M)$.

Assume $Y \subset X \widehat{\backslash} Z$. Let $c \in X \widehat{\otimes} Y$. By the definition of operation $\widehat{\otimes}$, there exist $a \in X$ and $b \in Y$ such that $a \otimes b \leq c$. Since $b \in Y$, then $b \in X \widehat{\backslash} Z$. Hence, by the definition of operation $\widehat{\backslash}$, $c \in Z$.

Assume $X \widehat{\otimes} Y \subset Z$. Let $b \in Y$. Let $a \in X$, $c \in M$ and $a \otimes b \leq c$. By the definition of operation $\widehat{\otimes}$, we have $c \in X \widehat{\otimes} Y$, so $c \in Z$. Finally, by the definition of operation $\widehat{\backslash}$, $b \in X \widehat{\backslash} Z$.

The proof of the second equivalence is analogous. $\qquad\square$

The same construction can be performed with the reverse ordering $\geq$. Starting from a p.o. groupoid $M = (M, \leq, \oplus)$, we define a dual powerset algebra $\mathcal{P}^{\geq}(M)$. For $X, Y, Z \subset M$, we define operations:

$$X \widehat{\oplus} Y = \{c \in M : \exists a \in X \; \exists b \in Y \; c \leq a \oplus b\},$$
$$X \widehat{\oslash} Z = \{b \in M : \forall a \in X \; \forall c \in M \; (c \leq a \oplus b \Rightarrow c \in Z)\},$$
$$Z \widehat{\oslash} Y = \{a \in M : \forall b \in Y \; \forall c \in M \; (c \leq a \oplus b \Rightarrow c \in Z)\}.$$

Lemma 3. $\mathcal{P}^{\geq}(M) = (\mathcal{P}(M), \subset, \widehat{\oplus}, \widehat{\oslash}, \widehat{\oslash})$ is a residuated groupoid.

Proof. Observe that $M' = (M, \geq, \oplus)$ is a p.o. groupoid. $\mathcal{P}^{\geq}(M)$ is exactly the algebra considered in Lemma 2 for M'. $\qquad\square$

In all cases we obtained some powerset residuated groupoids. Dual residuated groupoids can be constructed from them in the way described in Lemma 1. Of course, $\mathcal{P}^{\leq}(M)$ and $\mathcal{P}^{\geq}(M)$ can be expanded by the set complementation:

$$X^{\sim} = \{a \in M : a \notin X\}.$$

Clearly, $\sim$ is an involutive negation on $\mathcal{P}(M)$. We can define dual operations on $\mathcal{P}^{\geq}(M)$ as follows:

$$X \widehat{\oplus} Y = (X^{\sim} \bar{\oplus} Y^{\sim})^{\sim},$$
$$X \widehat{\oslash} Z = (X^{\sim} \bar{\oslash} Z^{\sim})^{\sim},$$
$$Z \widehat{\oslash} Y = (Z^{\sim} \bar{\oslash} Y^{\sim})^{\sim}.$$

Lemma 4. $\mathcal{P}_d^{\geq}(M) = (\mathcal{P}(M), \subset, \widehat{\oplus}, \widehat{\oslash}, \widehat{\oslash})$ *is a dual residuated groupoid.*

Proof. It is an easy consequence of Lemma 1 and Lemma 3.   □

The next lemma shows an alternative way of defining operations $\widehat{\oplus}, \widehat{\oslash}, \widehat{\oslash}$.

Lemma 5. *The operations* $\widehat{\oplus}, \widehat{\oslash}, \widehat{\oslash}$ *can also be defined as follows:*

$$X \widehat{\oplus} Y = \{c \in M : \forall a, b \in M \ (c \leq a \oplus b \Rightarrow (a \in X \vee b \in Y))\},$$
$$X \widehat{\oslash} Z = \{b \in M : \exists a \notin X \ \exists c \in Z \ c \leq a \oplus b\},$$
$$Z \widehat{\oslash} Y = \{a \in M : \exists b \notin Y \ \exists c \in Z \ c \leq a \oplus b\}.$$

Let $M = (M, \leq, \otimes, \backslash, /, \oplus, \oslash, \oslash)$ be a symmetric residuated groupoid. By $\mathcal{P}_{\leq}(M)$ we denote the algebra $(\mathcal{P}(M), \subset, \widehat{\otimes}, \widehat{\backslash}, \widehat{/}, \widehat{\oplus}, \widehat{\oslash}, \widehat{\oslash})$, where $\widehat{\otimes}, \widehat{\backslash}, \widehat{/}$ and $\widehat{\oplus}, \widehat{\oslash}, \widehat{\oslash}$ are defined as for $\mathcal{P}^{\leq}(M)$ and for $\mathcal{P}_d^{\geq}(M)$, respectively.

Lemma 6. *For any symmetric residuated groupoid M, $\mathcal{P}_{\leq}(M)$ is a symmetric residuated groupoid.*

Proof. It is an immediate consequence of Lemma 2 and Lemma 4.   □

Let $(M, \leq)$ be a poset. *An upset* is a set $X \subset M$ such that, if $x \in X$ and $x \leq y$, then $y \in X$, for all $x, y \in M$. *A downset* is a set $X \subset M$ such that, if $x \in X$ and $y \leq x$, then $y \in X$, for all $x, y \in M$.

By *a principal upset (downset) generated by* $a \in M$ we mean the set of all $b \in M$ such that $a \leq b$ ($b \leq a$). We denote it $\lceil a \rceil$ ($\lfloor a \rfloor$).

Observe that for any $X, Y \subset M$, $X \widehat{\otimes} Y$, $X \widehat{\backslash} Y$, $Y \widehat{/} X$ are upsets on $(M, \leq)$. Similarly, $X \bar{\oplus} Y$, $X \bar{\oslash} Y$, $Y \bar{\oslash} X$ are downsets. Consequently, $X \widehat{\oplus} Y$, $X \widehat{\oslash} Y$, $Y \widehat{\oslash} X$ are upsets.

Let us denote by U_M the set $\{X \subset M : X \text{ is an upset}\}$. Let us denote by $\boldsymbol{U_M}$ the partially ordered algebra $(U_M, \subset, \widehat{\otimes}, \widehat{\backslash}, \widehat{/}, \widehat{\oplus}, \widehat{\oslash}, \widehat{\oslash})$. Observe that $\boldsymbol{U_M}$ is a subalgebra of $\mathcal{P}_{\leq}(M)$. Clearly, $\boldsymbol{U_M}$ is a symmetric residuated groupoid. Let us denote $D_M = \{X \subset M : X \text{ is a downset}\}$ and $\boldsymbol{D_M} = (D_M, \subset, \bar{\oplus}, \bar{\oslash}, \bar{\oslash})$. Observe that $\boldsymbol{D_M}$ is a subalgebra of $\mathcal{P}^{\geq}(M)$, where $M = (M, \leq, \oplus)$ is a p.o. groupoid.

Unfortunately, we know no embedding of the symmetric residuated groupoid M into $\mathcal{P}_{\leq}(M)$. The values of such an embedding should be upsets. Neither $h(a) = \lceil a \rceil$, nor $h(a) = \lfloor a \rfloor^{\sim}$ satisfies the homomorphism conditions for all operations $\otimes, \backslash, /, \oplus, \oslash, \oslash$. For instance, the first does not satisfy $h(a \backslash b) = h(a) \widehat{\backslash} h(b)$.

We construct the higher-level algebra $\mathcal{P}_\le(U_M)$. In this algebra the operations are denoted by $\otimes, \backslash, /, \oplus, \oslash, \oslash$. They can explicitly be defined as follows:

$$\mathcal{X} \otimes \mathcal{Y} = \{Z \in U_M : \exists X \in \mathcal{X}\ \exists Y \in \mathcal{Y}\ X \widehat{\otimes} Y \subset Z\},$$
$$\mathcal{X} \backslash \mathcal{Z} = \{Y \in U_M : \forall X \in \mathcal{X}\ \forall Z \in U_M\ (X \widehat{\otimes} Y \subset Z \Rightarrow Z \in \mathcal{Z})\},$$
$$\mathcal{Z} / \mathcal{Y} = \{X \in U_M : \forall Y \in \mathcal{Y}\ \forall Z \in U_M\ (X \widehat{\otimes} Y \subset Z \Rightarrow Z \in \mathcal{Z})\},$$
$$\mathcal{X} \oplus \mathcal{Y} = \{Z \in U_M : \forall X \in U_M\ \forall Y \in U_M\ (Z \subset X \widehat{\oplus} Y \Rightarrow (X \in \mathcal{X} \vee Y \in \mathcal{Y}))\},$$
$$\mathcal{X} \oslash \mathcal{Z} = \{Y \in U_M : \exists X \notin \mathcal{X}\ \exists Z \in \mathcal{Z}\ Z \subset X \widehat{\oplus} Y\},$$
$$\mathcal{Z} \oslash \mathcal{Y} = \{X \in U_M : \exists Y \notin \mathcal{Y}\ \exists Z \in \mathcal{Z}\ Z \subset X \widehat{\oplus} Y\},$$

for all $\mathcal{X}, \mathcal{Y}, \mathcal{Z} \subset U_M$.

The following lemma holds.

Lemma 7. $\mathcal{P}_\le(U_M) = (\mathcal{P}(U_M), \subset, \otimes, \backslash, /, \oplus, \oslash, \oslash)$ *is a symmetric residuated groupoid.*

Proof. It is an immediate consequence of Lemma 6.   □

Clearly, $\mathcal{P}_\le(U_M)$ with complementation $\sim$ is an involutive symmetric residuated groupoid. Further, $\mathcal{P}_\le(U_M)$ is a boolean symmetric residuated groupoid, since $\mathcal{P}(U_M)$ is a boolean algebra (a field of all subsets of a set).

4   Main Theorem

In this section, we prove the main result of the paper.

Theorem 1. *Every symmetric residuated groupoid M is embeddable into the boolean symmetric residuated groupoid $\mathcal{P}_\le(U_M)$.*

Proof. We define a function $h : M \to \mathcal{P}(U_M)$ by setting: $h(a) = \{X \in U_M : a \in X\}$. First, we show that h preserves the order, i.e.

$$a \le b \quad \text{iff} \quad h(a) \subset h(b), \quad \text{for all } a, b \in M.$$

($\Rightarrow$) Suppose $a \le b$. Let $X \in h(a)$. By the definition of h, $a \in X$. X is an upset, hence $a \in X$ and $a \le b$ imply $b \in X$. Thus $X \in h(b)$.

($\Leftarrow$) Suppose $h(a) \subset h(b)$. We have $a \in \lceil a \rceil \in h(a)$. Hence, $\lceil a \rceil \in h(b)$. By the definition of h, $b \in \lceil a \rceil$, it means that $a \le b$.

We show that h preserves all operations.

First, we show that $h(a \otimes b) = h(a) \otimes h(b)$.

($\subseteq$) Let $Z \in h(a \otimes b)$. We have then $a \otimes b \in Z$. Since $Z \in U_M$, then by the definition of operation $\widehat{\otimes}$, $\lceil a \rceil \widehat{\otimes} \lceil b \rceil \subset Z$. We have $\lceil a \rceil \in h(a)$, $\lceil b \rceil \in h(b)$. Then, by the definition of operation $\otimes$, we obtain $Z \in h(a) \otimes h(b)$.

($\supseteq$) Let $Z \in h(a) \otimes h(b)$. By the definition of operation $\otimes$, there exist $X \in h(a)$ and $Y \in h(b)$ such that $X \widehat{\otimes} Y \subset Z$. By the definition of h, $a \in X$ and $b \in Y$. Hence by the definition of operation $\widehat{\otimes}$, $a \otimes b \in X \widehat{\otimes} Y$. Thus, $a \otimes b \in Z$, and finally $Z \in h(a \otimes b)$.

Now, we show that $h(a\backslash b) = h(a)\backslash h(b)$.

($\subseteq$) Let $Y \in h(a\backslash b)$. We have then $a\backslash b \in Y$. Take $X \in h(a)$, $Z \in U_M$ such that $X\widehat{\otimes}Y \subset Z$. Since $a \in X$, $a \otimes (a\backslash b) \leq b$, so $b \in X\widehat{\otimes}Y$. Hence $b \in Z$. Thus $Z \in h(b)$ and $Y \in h(a)\backslash h(b)$.

($\supseteq$) Let $Y \in h(a)\backslash h(b)$. We have $\lceil a \rceil \in h(a)$. By the definition of operation $\backslash$, for all $Z \in U_M$ the following implication holds: if $\lceil a \rceil \widehat{\otimes}Y \subset Z$, then $Z \in h(b)$. We have then $\lceil a \rceil \widehat{\otimes}Y \in h(b)$, and hence $b \in \lceil a \rceil \widehat{\otimes}Y$. By the definition of operation $\widehat{\otimes}$, there exist $a' \in \lceil a \rceil$ and $y \in Y$ such that $a' \otimes y \leq b$. Hence $y \leq a'\backslash b \leq a\backslash b$, so $a\backslash b \in Y$. Thus $Y \in h(a\backslash b)$.

One proves $h(a/b) = h(a)/h(b)$ in an analogous way.

Now, we show that $h(a \oplus b) = h(a) \oplus h(b)$.

($\subseteq$) Let $Z \in h(a \oplus b)$. We have then $a \oplus b \in Z$. Let $X \in U_M$, $Y \in U_M$ be such that $Z \subset X\widehat{\oplus}Y$. Then $a \oplus b \in X\widehat{\oplus}Y$. By the definition of operation $\widehat{\oplus}$, we have $a \in X$ or $b \in Y$, so $X \in h(a)$ or $Y \in h(b)$. By the definition of operation $\oplus$, we obtain $Z \in h(a) \oplus h(b)$.

($\supseteq$) Let $Z \in h(a) \oplus h(b)$. By the definition of operation $\oplus$, for all $X \in U_M$, $Y \in U_M$, if $Z \subset X\widehat{\oplus}Y$ and $X \notin h(a)$, then $Y \in h(b)$. Let X be $\lfloor a \rfloor^{\sim}$ and let Y be $\lfloor a \rfloor^{\sim}\widehat{\otimes}Z$. We have then $Z \subset \lfloor a \rfloor^{\sim}\widehat{\oplus}(\lfloor a \rfloor^{\sim}\widehat{\otimes}Z)$. Since $\lfloor a \rfloor^{\sim} \notin h(a)$, therefore $\lfloor a \rfloor^{\sim}\widehat{\otimes}Z \in h(b)$, so $b \in \lfloor a \rfloor^{\sim}\widehat{\otimes}Z$. By the definition of operation $\widehat{\otimes}$, there exist $a' \notin \lfloor a \rfloor^{\sim}$ and $c \in Z$ such that $c \leq a' \oplus b$. Since $a' \leq a$, so $c \leq a \oplus b$. Hence $a \oplus b \in Z$ and $Z \in h(a \oplus b)$.

Finally, we show that $h(a \otimes b) = h(a) \otimes h(b)$.

($\subseteq$) Let $Y \in h(a \otimes b)$. We have then $a \otimes b \in Y$. We know that $\lfloor a \rfloor^{\sim} \notin h(a)$ and $\lceil b \rceil \in h(b)$. We show $\lceil b \rceil \subset \lfloor a \rfloor^{\sim}\widehat{\oplus}Y$. Let $d \in \lceil b \rceil$, so $b \leq d$. Let $d \leq x \oplus y$ and $x \notin \lfloor a \rfloor^{\sim}$. So $x \leq a$, and then $d \leq a \oplus y$. We obtain $a \otimes d \leq y$, so $a \otimes b \leq y$. Hence $y \in Y$. Consequently, $d \in \lfloor a \rfloor^{\sim}\widehat{\oplus}Y$. Therefore, by the definition of operation $\otimes$, $Y \in h(a) \otimes h(b)$.

($\supseteq$) Let $Y \in h(a) \otimes h(b)$. By the definition of operation $\otimes$, there exist $X \notin h(a)$ and $Z \in h(b)$ such that $Z \subset X\widehat{\oplus}Y$. We have then $a \notin X$ and $b \in Z$. Since $X \in U_M$, $X \subset \lfloor a \rfloor^{\sim}$, so $Z \subset \lfloor a \rfloor^{\sim}\widehat{\oplus}Y$, and hence $b \in \lfloor a \rfloor^{\sim}\widehat{\oplus}Y$. Since $b \leq a \oplus (a \otimes b)$ and $a \notin \lfloor a \rfloor^{\sim}$, then $a \otimes b \in Y$. Thus $Y \in h(a \otimes b)$.

One proves $h(a \oslash b) = h(a) \oslash h(b)$ in an analogous way. $\square$

It is easy to deduce from Theorem 1 that the Lambek-Grishin calculus is a conservative fragment of the Boolean Generalized Lambek Calculus from [5].

Representation theorems are studied by many authors. Bimbó and Dunn in [3] prove representation theorems for some types of generalized Galois logics (gaggles) such as boolean, distributive and partial (multi-)gaggles. To preserve operations, the set of upsets U_M in our case is replaced by the set of ultrafilters on M for boolean gaggles and by the set of prime filters on M for distributive lattices in [3].

5 Variants

In this section, based on the main result of the paper, we discuss certain variants of the representation theorem.

Let M be a symmetric residuated groupoid.

Fact 1. *If operation $\otimes$ (resp. $\oplus$) is associative in M, then operation $\widehat{\otimes}$ (resp. $\widehat{\oplus}$) is associative in $\mathcal{P}_{\leq}(M)$.*

Proof. We show $(X\widehat{\otimes}Y)\widehat{\otimes}Z \subset X\widehat{\otimes}(Y\widehat{\otimes}Z)$. Let $x \in (X\widehat{\otimes}Y)\widehat{\otimes}Z$. Then there exist $a \in X\widehat{\otimes}Y$ and $b \in Z$ such that $a \otimes b \leq x$, and next, there exist $c \in X$, $d \in Y$ such that $c \otimes d \leq a$. Hence $(c \otimes d) \otimes b \leq x$. By the associativity of $\otimes$ in M, $c \otimes (d \otimes b) \leq x$. Consequently, $x \in X\widehat{\otimes}(Y\widehat{\otimes}Z)$. The reverse inclusion can be proved in an analogous way.

In order to prove the associativity of operation $\widehat{\oplus}$, let us observe that operation $\bar{\oplus}$ is associative in the residuated groupoid $\mathcal{P}^{\geq}(M)$, where $M = (M, \leq, \oplus)$. The latter fact can be proved in a similar way as above. Thus, $(X\widehat{\oplus}Y)\widehat{\oplus}Z =$
$= ((X\widehat{\oplus}Y)^{\sim}\bar{\oplus}Z^{\sim})^{\sim} = ((X^{\sim}\bar{\oplus}Y^{\sim})\bar{\oplus}Z^{\sim})^{\sim} = (X^{\sim}\bar{\oplus}(Y^{\sim}\bar{\oplus}Z^{\sim}))^{\sim} = X\widehat{\oplus}(Y\widehat{\oplus}Z)$. $\square$

Observe that the associativity of operation $\widehat{\otimes}$ (resp. $\widehat{\oplus}$) implies the associativity of operation $\otimes$ (resp. $\oplus$) in $\mathcal{P}_{\leq}(U_M)$.

Fact 2. *If operation $\otimes$ (resp. $\oplus$) is commutative in M, then operation $\widehat{\otimes}$ (resp. $\widehat{\oplus}$) is commutative in $\mathcal{P}_{\leq}(M)$.*

Proof. Assume that operation $\otimes$ is commutative in M. Then $X\widehat{\otimes}Y = \{c \in M : \exists a \in X\, \exists b \in Y\, a \otimes b \leq c\} = \{c \in M : \exists b \in Y\, \exists a \in X\, b \otimes a \leq c\} = Y\widehat{\otimes}X$.

Assuming the commutativity of operation $\oplus$ in M, we can show in a similar way that $X\bar{\oplus}Y = Y\bar{\oplus}X$. Thus, $X\widehat{\oplus}Y = (X^{\sim}\bar{\oplus}\,Y^{\sim})^{\sim} = (Y^{\sim}\bar{\oplus}\,X^{\sim})^{\sim} = Y\widehat{\oplus}X$. $\square$

Observe that the commutativity of operation $\widehat{\otimes}$ (resp. $\widehat{\oplus}$) implies the commutativity of operation $\otimes$ (resp. $\oplus$) in $\mathcal{P}_{\leq}(U_M)$.

The above facts and observations allow us to state the following representation theorem for semigroups and commutative semigroups.

Theorem 2. *Every (commutative) symmetric residuated semigroup can be embedded into the (commutative) boolean symmetric residuated semigroup.*

A *unital groupoid* is an algebra $(M, \otimes, 1)$ such that $(M, \otimes)$ is a groupoid and 1 is a unit element for $\otimes$. A *symmetric residuated unital groupoid* is a structure $M = (M, \leq, \otimes, \backslash, /, 1, \oplus, \oslash, \oslash, 0)$ such that the $(\leq, \otimes, \backslash, /, \oplus, \oslash, \oslash)$-reduct of M is a symmetric residuated groupoid, 1 is a unit element for $\otimes$ and 0 is a unit element for $\oplus$. A *monoid* is a unital semigroup and a *symmetric residuated monoid* is a symmetric residuated unital semigroup.

Let M be a symmetric residuated unital groupoid. In U_M there exists a unit element $\mathbb{1}$ satisfying $X\widehat{\otimes}\,\mathbb{1} = X = \mathbb{1}\,\widehat{\otimes}X$, namely $\mathbb{1} = \lceil 1 \rceil$. If X is an upset, then

$X \widehat{\otimes} \lceil 1 \rceil = \{c \in M : \exists a \in X \ \exists b \in \lceil 1 \rceil \ a \otimes b \leq c\} = \{c \in M : \exists a \in X \ a \leq c\} = X = \lceil 1 \rceil \widehat{\otimes} X$. In D_M there exists a zero element $\underline{0}$ satisfying $X \widehat{\oplus} \underline{0} = X = \underline{0} \widehat{\oplus} X$, namely $\underline{0} = \lfloor 0 \rfloor$. If X is a downset, then $X \widehat{\oplus} \lfloor 0 \rfloor = \{c \in M : \exists a \in X \ \exists b \in \lfloor 0 \rfloor \ c \leq a \oplus b\} = \{c \in M : \exists a \in X \ c \leq a\} = X = \lfloor 0 \rfloor \widehat{\oplus} X$.

The zero element $\mathbb{O} \in \mathcal{P}_{\leq}(M)$ satisfying $X \widehat{\oplus} \mathbb{O} = X = \mathbb{O} \widehat{\oplus} X$ is $\lfloor 0 \rfloor^{\sim}$. We have $X \widehat{\oplus} \lfloor 0 \rfloor^{\sim} = (X^{\sim} \widehat{\oplus} \lfloor 0 \rfloor)^{\sim} = (X^{\sim})^{\sim} = X = \lfloor 0 \rfloor^{\sim} \widehat{\oplus} X$.

Now, we pass to $\mathcal{P}_{\leq}(U_M)$. Notice that, for any $\mathcal{X}, \mathcal{Y} \subset U_M$, the sets $\mathcal{X} \otimes \mathcal{Y}$, $\mathcal{X} \backslash \mathcal{Y}$, $\mathcal{Y} / \mathcal{X}$, $\mathcal{X} \oplus \mathcal{Y}$, $\mathcal{X} \otimes \mathcal{Y}$, $\mathcal{Y} \oslash \mathcal{X}$ are upsets with respect to $\subset$ on $\mathcal{P}(U_M)$. Consequently, the set $\mathcal{U}_{\mathcal{P}(U_M)}$, of all upsets on $\mathcal{P}(U_M)$, is a subalgebra of $\mathcal{P}_{\leq}(U_M)$. The unit element and the zero element can be defined as follows:

$\mathbf{1} = \{X \in U_M : 1 \in X\} = h(1)$,
$\mathbf{0} = \{X \in U_M : 0 \in X\} = h(0)$.

We have $\mathcal{X} \otimes \mathbf{1} = \{Z \in U_M : \exists X \in \mathcal{X} \ \exists Y \in \mathbf{1} \ X \widehat{\otimes} Y \subset Z\} = \{Z \in U_M : \exists X \in \mathcal{X} \ X \widehat{\otimes} \lceil 1 \rceil \subset Z\} = \{Z \in U_M : \exists X \in \mathcal{X} \ X \subset Z\} = \mathcal{X} = \mathcal{X} \otimes \mathbf{1}$.

We have for all $Y \in U_M, 0 \notin Y$ if, and only if, $Y \subset \lfloor 0 \rfloor^{\sim}$. In other words, $\lfloor 0 \rfloor^{\sim}$ is the greatest upset Y such that $0 \notin Y$. We prove $\mathcal{X} = \mathcal{X} \oplus \mathbf{0}$ for any $\mathcal{X} \in \mathcal{U}_{\mathcal{P}(U_M)}$. We show $\mathcal{X} \subset \mathcal{X} \oplus \mathbf{0}$. Assume $Z \in \mathcal{X}$. Let $Z \subset X \widehat{\oplus} Y$. Hence $X \widehat{\oplus} Y \in \mathcal{X}$. Assume $Y \notin \mathbf{0}$, hence $0 \notin Y$. Since $Y \subset \lfloor 0 \rfloor^{\sim}$, then $X \widehat{\oplus} Y \subset X \widehat{\oplus} \lfloor 0 \rfloor^{\sim} = X$. Consequently, $X \in \mathcal{X}$, which yields $Z \in \mathcal{X} \oplus \mathbf{0}$. Now, we show $\mathcal{X} \oplus \mathbf{0} \subset \mathcal{X}$. Assume $Z \in \mathcal{X} \oplus \mathbf{0}$. We have $Z \subset Z \widehat{\oplus} \lfloor 0 \rfloor^{\sim}$ and $0 \notin \lfloor 0 \rfloor^{\sim}$, so $\lfloor 0 \rfloor^{\sim} \notin \mathbf{0}$. It yields $Z \in \mathcal{X}$. $\mathcal{X} = \mathbf{0} \oplus \mathcal{X}$, for $\mathcal{X} \in \mathcal{U}_{\mathcal{P}(U_M)}$, can be proved in a similar way. We have then, $\mathcal{X} \oplus \mathbf{0} = \mathcal{X} = \mathbf{0} \oplus \mathcal{X}$.

$\mathcal{U}_{\mathcal{P}(U_M)}$ is a subalgebra of $\mathcal{P}_{\leq}(U_M)$. We have shown above that h embeds $M = (M, \leq, \otimes, \backslash, /, 1, \oplus, \otimes, \oslash, 0)$ into the algebra $\mathcal{U}_{\mathcal{P}(U_M)}$ and $h(1) = \mathbf{1}, h(0) = \mathbf{0}$. Notice that $\mathcal{U}_{\mathcal{P}(U_M)}$ is not closed under $\sim$, in general (similarly, U_M is not closed under $\sim$).

If M is an involutive symmetric residuated groupoid, then U_M (resp. $\mathcal{U}_{\mathcal{P}(U_M)}$) is closed under an involutive negation (a quasi-complement in the sense of [2]). We define $g : \mathcal{P}(M) \mapsto \mathcal{P}(M)$ as follows:

$$g(X) = (-X)^{\sim},$$

where $-X = \{-a : a \in X\}$. Clearly, $(-X)^{\sim} = -(X^{\sim})$, hence $g(g(X)) = X$, and $X \subset Y$ entails $g(Y) \subset g(X)$. Consequently, g is an involutive negation on $\mathcal{P}(M)$. Further, U_M is closed under g, so $(U_M, \subset, \widehat{\otimes}, \widehat{\backslash}, \widehat{/}, \widehat{\oplus}, \widehat{\otimes}, \widehat{\oslash}, g)$ is an involutive symmetric residuated groupoid.

We define an involutive negation $\sim_g$ on $\mathcal{P}(U_M)$ as follows:

$$\sim_g (\mathcal{X}) = g [\mathcal{X}]^{\sim} .$$

Clearly, $\sim_g$ arises from g in the same way as g arises from $-$. Consequently, $\sim_g$ is an involutive negation on $\mathcal{P}(U_M)$, and $\mathcal{U}_{\mathcal{P}(U_M)}$ is closed under $\sim_g$. We show that $h(-a) = \sim_g h(a)$ for all $a \in X$, for the mapping h defined above.

We have to show that $X \in \sim_g h(a)$ iff $X \in h(-a)$. The following equivalences hold: $X \in \sim_g h(a)$ iff $X \in g [h(a)]^{\sim}$ iff $X \notin g [h(a)]^{\sim}$ iff $g(X) \notin h(a)$ iff $(-X)^{\sim} \notin h(a)$ iff $a \notin (-X)^{\sim}$ iff $a \in -X$ iff $-a \in X$ iff $X \in h(-a)$.

Buszkowski [4] proved that each residuated semigroup with De Morgan negation is isomorphically embeddable into some residuated semigroup of cones with quasi-boolean complement. The following theorem yields a related result.

Theorem 3. *Every involutive symmetric residuated (unital) groupoid is embeddable into a quasi-boolean symmetric residuated (unital) groupoid, and similarly for involutive residuated (commutative) semigroups and monoids.*

A *bilinear algebra* can be defined as a symmetric residuated monoid with two negations $\sim$, $-$, satisfying:

$$\sim -a = a = -\sim a,$$
$$\sim (a \otimes b) = (\sim b) \oplus (\sim a),$$
$$-(a \otimes b) = (-b) \oplus (-a),$$
$$\sim a = a \backslash 0,$$
$$-a = 0/a,$$

for all elements a, b. An equivalent notion was defined in Lambek [14,15] as an algebra corresponding to Bilinear Logic. Bilinear Logic is equivalent to the multiplicative fragment of Noncommutative MALL of Abrusci [1]. Some lattice models of this logic are discussed by Lambek in [13]. Cyclic Noncommutative MALL of Yetter [17] gives rise to cyclic bilinear algebras.

A *cyclic bilinear algebra* is a bilinear algebra $\boldsymbol{M}$ such that $\sim a = -a$; equivalently $\boldsymbol{M}$ is an involutive symmetric residuated monoid, satisfying:

$$-(a \otimes b) = (-b) \oplus (-a),$$
$$-a = a \backslash 0 = 0/a,$$

for all $a, b \in M$.

Let $\boldsymbol{M} = (M, \leq, \otimes, \backslash, /, 1, \oplus, \oslash, \oslash, 0, -)$ be a cyclic bilinear algebra. We show that the involutive function g defined above satisfies:

$$g(X \widehat{\otimes} Y) = g(Y) \widehat{\oplus} g(X),$$
$$g(X) = X \widehat{\backslash} \lfloor 0 \rfloor^{\sim} = \lfloor 0 \rfloor^{\sim} \widehat{/} X,$$

for all $X, Y \in U_{\boldsymbol{M}}$.

We show the first equation. Assume $-(a \otimes b) = (-b) \oplus (-a)$. We have $g(X \widehat{\otimes} Y) = g(Y) \widehat{\oplus} g(X)$ iff $-(X \widehat{\otimes} Y)^{\sim} = (-Y)^{\sim} \widehat{\oplus} (-X)^{\sim}$ iff $-(X \widehat{\otimes} Y) = (-Y) \widehat{\oplus} (-X)$. The following equivalences hold: $c \in -(X \widehat{\otimes} Y)$ iff $-c \in X \widehat{\otimes} Y$ iff there exist $a \in X$ and $b \in Y$ such that $a \otimes b \leq -c$ iff there exist $a \in X$ and $b \in Y$ such that $c \leq -(a \otimes b) = (-b) \oplus (-a)$ iff there exist $a' \in -X$ and $b' \in -Y$ such that $c \leq b' \oplus a'$ iff $c \in (-Y) \widehat{\oplus} (-X)$. So, we have shown $g(X \widehat{\otimes} Y) = g(Y) \widehat{\oplus} g(X)$.

Now, we prove $g(X) = X \widehat{\backslash} \lfloor 0 \rfloor^{\sim}$ i.e. $(-X)^{\sim} = X \widehat{\backslash} \lfloor 0 \rfloor^{\sim}$, or equivalently $-X = (X \widehat{\backslash} \lfloor 0 \rfloor^{\sim})^{\sim}$. We have $b \in (X \widehat{\backslash} \lfloor 0 \rfloor^{\sim})^{\sim}$ iff $b \notin X \widehat{\backslash} \lfloor 0 \rfloor^{\sim}$ iff there exist $a \in X$ and $c \in M$ such that $a \otimes b \leq c$ and $c \notin \lfloor 0 \rfloor^{\sim}$ (i.e. $c \leq 0$) iff there exists $a \in X$ such that $a \otimes b \leq 0$ iff there exists $a \in X$ such that $a \leq 0/b = -b$ iff $-b \in X$ iff $b \in -X$.

One proves $g(X) = \lfloor 0 \rfloor^{\sim} \widehat{/} X$ in an analogous way.

Since $\sim_g$ arises from g in the same way as g arises from $-$, one can analogously show that the involutive negation $\sim_g$ satisfies:

$$\sim_g (\mathcal{X} \otimes \mathcal{Y}) = \sim_g (\mathcal{Y}) \oplus \sim_g (\mathcal{X}),$$
$$\sim_g (\mathcal{X}) = \mathcal{X} \backslash \mathbf{0} = \mathbf{0}/\mathcal{X},$$

for all $\mathcal{X}, \mathcal{Y} \in \mathcal{U}_{\mathcal{P}(U_M)}$. We obtain the following theorem:

Theorem 4. *Every cyclic bilinear algebra M is embeddable into the quasi-boolean cyclic bilinear algebra $\mathcal{U}_{\mathcal{P}(U_M)}$, which is a quasi-field of sets.*

An analogous result can be proved for bilinear algebras, but then the target algebra $\mathcal{U}_{\mathcal{P}(U_M)}$ is a weak quasi-field of sets with two weak quasi-complements $\sim_g \mathcal{X} = g[\mathcal{X}]^\sim$, $-_f \mathcal{X} = f[\mathcal{X}]^\sim$, where $g(X) = (\sim X)^\sim$, $f(X) = (-X)^\sim$. We also have $h(\sim a) = \sim_g h(a)$, $h(-a) = -_f h(a)$. If the associativity of $\otimes$ and $\oplus$ is not assumed, then similar results can be obtained for cyclic InGL-algebras (without lattice operations) in the sense of [7].

Grishin [8] considered a formal system whose algebraic models are symmetric residuated monoids with 0 additionally satisfying the laws of mixed associativity:

$$1.\ a \otimes (b \oplus c) \leq (a \otimes b) \oplus c$$
$$2.\ (a \oplus b) \otimes c \leq a \oplus (b \otimes c)$$

We propose to call such structures *associative Lambek-Grishin algebras* (associative LG-algebras). Omitting the associativity of $\otimes, \oplus$, one obtains a more general class of LG-algebras.

Moortgat [16] and other authors consider systems admitting so-called Grishin axioms. Some axioms of that kind are listed below.

	Associativity		Commutativity	
Group I	$1_a.\ a \otimes (b \oplus c) \leq (a \otimes b) \oplus c$		$1_c.\ a \otimes (b \oplus c) \leq b \oplus (a \otimes c)$	
	$2_a.\ (a \oplus b) \otimes c \leq a \oplus (b \otimes c)$		$2_c.\ (a \oplus b) \otimes c \leq (a \otimes c) \oplus b$	
Group II	$1_a.\ (a \otimes b) \otimes c \leq a \otimes (b \otimes c)$		$1_c.\ a \otimes (b \otimes c) \leq b \otimes (a \otimes c)$	
	$2_a.\ a \otimes (b \otimes c) \leq (a \otimes b) \otimes c$		$2_c.\ (a \otimes b) \otimes c \leq (a \otimes c) \otimes b$	
Group III	$1_a.\ (a \oplus b) \oplus c \leq a \oplus (b \oplus c)$		$1_c.\ a \oplus (b \oplus c) \leq b \oplus (a \oplus c)$	
	$2_a.\ a \oplus (b \oplus c) \leq (a \oplus b) \oplus c$		$2_c.\ (a \oplus b) \oplus c \leq (a \oplus c) \oplus b$	
Group IV	$1_a.\ (a \backslash b) \oslash c \leq a \backslash (b \oslash c)$		$1_c.\ a \oslash (b \backslash c) \leq b \backslash (a \oslash c)$	
	$2_a.\ a \oslash (b/c) \leq (a \oslash b)/c$		$2_c.\ (a/b) \oslash c \leq (a \oslash c)/b$	

Some axiomatization of a bilinear algebra obtained by adding selected Grishin axioms was described by Lambek in [12].

We show that the powerset algebras, defined above, preserve axioms from Groups I-IV. We denote by I.1_a the first axiom from Group I of Associativity, and similarly for the other axioms.

Let A be a symmetric residuated groupoid.

Proposition 1. *If the axiom I.1$_a$ (resp. IV.1$_a$, I.2$_a$, IV.2$_a$, I.1$_c$, IV.1$_c$, I.2$_c$, IV.2$_c$) is valid in the basic algebra $\boldsymbol{A}$, then the corresponding axiom IV.1$_a$ (resp. I.1$_a$, IV.2$_a$, I.2$_a$, IV.1$_c$, I.1$_c$, IV.2$_c$, I.2$_c$) is valid in the algebra $\mathcal{P}_\leq(\boldsymbol{A})$.*

Proof. Assume that the mixed associativity law I.1$_a$ is valid in algebra $\boldsymbol{A}$. We show that the appropriate law IV.1$_a$ is valid in $\mathcal{P}_\leq(\boldsymbol{A})$: $(A\widehat{\backslash}B)\widehat{\oslash}C \subseteq A\widehat{\backslash}(B\widehat{\oslash}C)$.

Let $x \in (A\widehat{\backslash}B)\widehat{\oslash}C$. By the definition of operation $\widehat{\oslash}$ there exist $c \notin C$ and $b \in A\widehat{\backslash}B$ such that $b \leq x \oplus c$. We fix $a \in A$. By monotonicity of $\otimes$ we obtain $a \otimes b \leq a \otimes (x \oplus c)$. By assumption $a \otimes b \leq (a \otimes x) \oplus c$. Since $a \in A$ and $b \in A\widehat{\backslash}B$, then $a \otimes b \in B$ by the definition of operation $\widehat{\backslash}$. Since $c \notin C$ and $a \otimes b \in B$, so $a \otimes x \in B\widehat{\oslash}C$. Consequently, $x \in A\widehat{\backslash}(B\widehat{\oslash}C)$.

Assume now that the mixed associativity law IV.1$_a$ is valid in $\boldsymbol{A}$. We show that the appropriate law I.1$_a$ is valid in $\mathcal{P}_\leq(\boldsymbol{A})$: $A\widehat{\otimes}(B\widehat{\oplus}C) \subseteq (A\widehat{\otimes}B)\widehat{\oplus}C$.

Let $x \in A\widehat{\otimes}(B\widehat{\oplus}C)$. By the definition of operation $\widehat{\otimes}$ there exist $a \in A$ and $b \in B\widehat{\oplus}C$ such that $a \otimes b \leq x$. We claim that $x \in (A\widehat{\otimes}B)\widehat{\oplus}C$ i.e. for all u, v: if $x \leq u \oplus v$ then $u \in A\widehat{\otimes}B$ or $v \in C$. Assume that $x \leq u \oplus v$. Suppose that $v \notin C$. We show that $u \in A\widehat{\otimes}B$. By the residuation law we have $x \oslash v \leq u$. Take $a \in A$. By monotonicity of $\backslash$ we obtain $a\backslash(x \oslash v) \leq a\backslash u$ and by assumption $(a\backslash x) \oslash v \leq a\backslash u$. By the residuation we have $a\backslash x \leq (a\backslash u) \oplus v$. Since $b \leq a\backslash x$ then $b \leq (a\backslash u) \oplus v$. Since $b \in B\widehat{\oplus}C$ and $v \notin C$, so $a\backslash u \in (B\widehat{\oplus}C)\widehat{\oslash}C \subseteq B$. We have $a \in A$ and $a\backslash u \in B$. Consequently, $u \in A\widehat{\otimes}B$.

Assume that the mixed (weak)-commutativity law I.1$_c$ is valid in $\boldsymbol{A}$. We show that the appropriate law IV.1$_c$ is valid in $\mathcal{P}_\leq(\boldsymbol{A})$: $A\widehat{\oslash}(B\widehat{\backslash}C) \subseteq B\widehat{\backslash}(A\widehat{\oslash}C)$.

Let $x \in A\widehat{\oslash}(B\widehat{\backslash}C)$. There exist $a \notin A$ and $c \in B\widehat{\backslash}C$ such that $c \leq a \oplus x$. We fix $b \in B$. By monotonicity of $\otimes$ we obtain $b \otimes c \leq b \otimes (a \oplus x)$. By assumption $b \otimes c \leq a \oplus (b \otimes x)$. Since $b \in B$ and $c \in B\widehat{\backslash}C$, then $b \otimes c \in C$. We have $a \notin A$ and $b \otimes c \in C$, so $b \otimes x \in A\widehat{\oslash}C$. Consequently, $x \in B\widehat{\backslash}(A\widehat{\oslash}C)$.

Assume now that the mixed (weak)-commutativity law IV.1$_c$ is valid in $\boldsymbol{A}$. We show that the law I.1$_c$ is valid in $\mathcal{P}_\leq(\boldsymbol{A})$: $A\widehat{\otimes}(B\widehat{\oplus}C) \subseteq B\widehat{\oplus}(A\widehat{\otimes}C)$.

Let $x \in A\widehat{\otimes}(B\widehat{\oplus}C)$. There exist $a \in A$ and $b \in B\widehat{\oplus}C$ such that $a \otimes b \leq x$. We claim that $x \in B\widehat{\oplus}(A\widehat{\otimes}C)$ i.e. for all u, v: if $x \leq u \oplus v$ then $u \in B$ or $v \in A\widehat{\otimes}C$. Assume that $x \leq u \oplus v$. Suppose that $u \notin B$. We show that $v \in A\widehat{\otimes}C$. By the residuation law we have $u \oslash x \leq v$. Take $a \in A$. By monotonicity of $\backslash$ we obtain $a\backslash(u \oslash x) \leq a\backslash v$ and by assumption $u \oslash (a\backslash x) \leq a\backslash v$. By the residuation we have $a\backslash x \leq u \oplus (a\backslash v)$. Since $b \leq a\backslash x$ then $b \leq u \oplus (a\backslash v)$. Since $b \in B\widehat{\oplus}C$ and $u \notin B$, so $a\backslash v \in B\widehat{\oslash}(B\widehat{\oplus}C) \subseteq C$. We have $a \in A$ and $a\backslash v \in C$. Consequently, $v \in A\widehat{\otimes}C$.

The cases for axioms I.2$_a$, IV.2$_a$, I.2$_c$ and IV.2$_c$ are proved in a similar way. $\qquad\square$

Proposition 2. *If the axiom II.1$_a$ (resp. II.2$_a$, III.1$_a$, III.2$_a$, II.1$_c$, II.2$_c$, III.1$_c$, III.2$_c$) is valid in the basic algebra $\boldsymbol{A}$, then the corresponding axiom II.2$_a$ (resp. II.1$_a$, III.2$_a$, III.1$_a$, II.2$_c$, II.1$_c$, III.2$_c$, III.1$_c$) is valid in the algebra $\mathcal{P}_\leq(\boldsymbol{A})$.*

We omit an easy proof of this proposition.

Corollary 1. *If A satisfies an axiom from the above list, then $\mathcal{P}_\leq(U_A)$ satisfies the same axiom.*

This corollary yields the following proposition.

Proposition 3. *If A is an (resp. associative) LG-algebra, then $\mathcal{U}_{\mathcal{P}(U_A)}$ is an (resp. associative) LG-algebra.*

Acknowledgement. I thank Wojciech Buszkowski and two anonymous referees for helpful comments.

References

1. Abrusci, V.M.: Phase semantics and sequent calculus for pure noncommutative classical linear propositional logic. The Journal of Symbolic Logic 56(4), 1403–1451 (1991)
2. Białynicki-Birula, A., Rasiowa, H.: On the Representation of Quasi-Boolean Algebras. Bulletin de l'Académie Polonaise des Sciences 5, 259–261 (1957)
3. Bimbó, K., Dunn, J.M.: Generalized Galois Logics: Relational Semantics of Nonclassical Logical Calculi. CSLI Lecture Notes 188, Stanford (2008)
4. Buszkowski, W.: Categorial Grammars with Negative Information. In: Wansing, H. (ed.) Negation. A Notion in Focus, pp. 107–126. W. de Gruyter, Berlin (1996)
5. Buszkowski, W.: Interpolation and FEP for Logics of Residuated Algebras. Logic Journal of the IGPL 19(3), 437–454 (2011)
6. Dunn, J.M.: Generalized Ortho Negation. In: Wansing, H. (ed.) Negation. A Notion in Focus, pp. 3–26. W. de Gruyter, Berlin (1996)
7. Galatos, N., Jipsen, P., Kowalski, T., Ono, H.: Residuated Lattices: An Algebraic Glimpse at Substructural Logics, vol. 151. Elsevier, Amsterdam (2007)
8. Grishin, V.N.: On a generalization of the Ajdukiewicz-Lambek system. In: Studies in Non-Commutative Logics and Formal Systems, Russian, Nauka, Moscow, pp. 315–343 (1983)
9. Kołowska-Gawiejnowicz, M.: Powerset Residuated Algebras and Generalized Lambek Calculus. Mathematical Logic Quarterly 43, 60–72 (1997)
10. Kurtonina, N., Moortgat, M.: Relational semantics for the Lambek-Grishin calculus. In: Ebert, C., Jäger, G., Michaelis, J. (eds.) MOL 10. LNCS, vol. 6149, pp. 210–222. Springer, Heidelberg (2010)
11. Lambek, J.: On the calculus of syntactic types. In: Jacobson, R. (ed.) Structure of Language and Its Mathematical Aspects, pp. 166–178. AMS, Providence (1961)
12. Lambek, J.: From Categorial Grammar to Bilinear Logic. In: Došen, K., Schröder-Heister, P. (eds.) Substructural Logics, pp. 207–237. Oxford University Press (1993)
13. Lambek, J.: Some lattice models of bilinear logic. Algebra Universalis 34, 541–550 (1995)
14. Lambek, J.: Bilinear logic in algebra and linguistics. In: Girard, J.-Y., Lafont, Y., Regnier, L. (eds.) Advances in Linear Logic, pp. 43–59. Cambridge University Press, Cambridge (1995)
15. Lambek, J.: Bilinear logic and Grishin algebras. In: Orłowska, E. (ed.) Logic at Work, pp. 604–612. Physica-Verlag, Heidelberg (1999)

16. Moortgat, M.: Symmetries in natural language syntax and semantics: Lambek-Grishin calculus. In: Leivant, D., de Queiroz, R. (eds.) WoLLIC 2007. LNCS, vol. 4576, pp. 264–284. Springer, Heidelberg (2007)
17. Yetter, D.N.: Quantales and (non-commutative) linear logic. The Journal of Symbolic Logic 55(1), 41–64 (1990)

L-Completeness of the Lambek Calculus with the Reversal Operation Allowing Empty Antecedents

Stepan Kuznetsov

Moscow State University
`sk@lpcs.math.msu.su`

Abstract. In this paper we prove that the Lambek calculus allowing empty antecedents and enriched with a unary connective corresponding to language reversal is complete with respect to the class of models on subsets of free monoids (L-models).

1   The Lambek Calculus with the Reversal Operation

We consider the calculus L, introduced in [4]. The set $\mathrm{Pr} = \{p_1, p_2, p_3, \dots\}$ is called the set of *primitive types*. *Types* of L are built from primitive types using three binary connectives: $\backslash$ *(left division)*, $/$ *(right division)*, and $\cdot$ *(multiplication)*; we shall denote the set of all types by Tp. Capital letters $(A, B, \dots)$ range over types. Capital Greek letters (except Σ) range over finite (possibly empty) sequences of types; Λ stands for the empty sequence. Expressions of the form $\Gamma \to C$, where $\Gamma \neq \Lambda$, are called *sequents* of L.

Axioms: $A \to A$.

Rules:

$$\frac{A\Pi \to B}{\Pi \to A \backslash B} \; (\to \backslash), \; \Pi \neq \Lambda \qquad \frac{\Pi \to A \quad \Gamma B \Delta \to C}{\Gamma \Pi (A \backslash B) \Delta \to C} \; (\backslash \to)$$

$$\frac{\Pi A \to B}{\Pi \to B / A} \; (\to /), \; \Pi \neq \Lambda \qquad \frac{\Pi \to A \quad \Gamma B \Delta \to C}{\Gamma (B / A) \Pi \Delta \to C} \; (/ \to)$$

$$\frac{\Pi \to A \quad \Delta \to B}{\Pi \Delta \to A \cdot B} \; (\to \cdot) \qquad \frac{\Gamma A B \Delta \to C}{\Gamma (A \cdot B) \Delta \to C} \; (\cdot \to)$$

$$\frac{\Pi \to A \quad \Gamma A \Delta \to C}{\Gamma \Pi \Delta \to C} \; (\text{cut})$$

The (cut) rule is eliminable [4].

We also consider an extra unary connective ^{R} (written in the postfix form, A^{R}). The extended set of types is denoted by Tp^{R}. For a sequence of types $\Gamma = A_1 A_2 \dots A_n$ let $\Gamma^{\mathrm{R}} \rightleftharpoons A_n^{\mathrm{R}} \dots A_2^{\mathrm{R}} A_1^{\mathrm{R}}$ ("$\rightleftharpoons$" here and further means "equal by definition").

The calculus L^{R} is obtained from L by adding three rules for ^{R}:

$$\frac{\Gamma \to C}{\Gamma^{\mathrm{R}} \to C^{\mathrm{R}}} \; (\mathrm{R} \to {}^{\mathrm{R}}) \qquad \frac{\Gamma A^{\mathrm{RR}} \Delta \to C}{\Gamma A \Delta \to C} \; (\mathrm{RR} \to)_{\mathrm{E}} \qquad \frac{\Gamma \to C^{\mathrm{RR}}}{\Gamma \to C} \; (\to {}^{\mathrm{RR}})_{\mathrm{E}}$$

C. Casadio et al. (Eds.): Lambek Festschrift, LNCS 8222, pp. 268–278, 2014.
© Springer-Verlag Berlin Heidelberg 2014

Dropping the $\Pi \neq \Lambda$ restriction on the $(\rightarrow \backslash)$ and $(\rightarrow /)$ rules of L leads to *the Lambek calculus allowing empty antecedents* called L^*. The calculus L^{*R} is obtained from L^* by changing the type set from Tp to Tp^R and adding the $(^R \rightarrow {}^R)$, $(^{RR} \rightarrow)_E$, and $(\rightarrow {}^{RR})_E$ rules.

Unfortunately, no cut elimination theorem is known for L^R and L^{*R}. Nevertheless, L^R is a conservative extension of L, and L^{*R} is a conservative extension of L^*:

Lemma 1. *A sequent formed of types from* Tp *is provable in* L^R *(L^{*R}) if and only if it is provable in* L *(resp.,* L^**).*

This lemma will be proved later via a semantic argument.

2   Normal Form for Types

The ^{R} connective in the Lambek calculus and linear logic was first considered in [5] (there it is denoted by $^\smile$). In [5], this connective is axiomatised using Hilbert-style axioms:

$$A^{RR} \leftrightarrow A \qquad \text{and} \qquad (A \cdot B)^R \leftrightarrow B^R \cdot A^R.$$

Here $F \leftrightarrow G$ ("F is *equivalent* to G") is a shortcut for two sequents: $F \rightarrow G$ and $G \rightarrow F$. The relation $\leftrightarrow$ is reflexive, symmetric, and transitive (due to the rule (cut)). Using (cut) one can prove that if $L^R \vdash F_1 \rightarrow G_1$, $F_1 \leftrightarrow F_2$, and $G_1 \leftrightarrow G_2$, then $L^R \vdash F_2 \rightarrow G_2$. Also, $\leftrightarrow$ is a *congruence relation*, in the following sense: if $A_1 \leftrightarrow A_2$ and $B_1 \leftrightarrow B_2$, then $A_1 \cdot B_1 \leftrightarrow A_2 \cdot B_2$, $A_1 \backslash B_1 \leftrightarrow A_2 \backslash B_2$, $B_1 / A_1 \leftrightarrow B_2 / A_2$, $A_1^R \leftrightarrow A_2^R$.

These axioms are provable in L^R and, vice versa, adding them to L yields a calculus equivalent to L^R. The same is true for L^{*R} and L^* respectively.

Furthermore, the following two equivalences hold in L^R and L^{*R}:

$$(A \backslash B)^R \leftrightarrow B^R / A^R \qquad \text{and} \qquad (B / A)^R \leftrightarrow A^R \backslash B^R.$$

Using the four equivalences above one can prove by induction that any type $A \in Tp^R$ is equivalent to its *normal form* $tr(A)$, defined as follows:

1. $tr(p_i) \coloneqq p_i$;
2. $tr(p_i^R) \coloneqq p_i^R$;
3. $tr(A \cdot B) \coloneqq tr(A) \cdot tr(B)$;
4. $tr(A \backslash B) \coloneqq tr(A) \backslash tr(B)$;
5. $tr(B / A) \coloneqq tr(B) / tr(A)$;
6. $tr((A \cdot B)^R) \coloneqq tr(B^R) \cdot tr(A^R)$;
7. $tr((A \backslash B)^R) \coloneqq tr(B^R) / tr(A^R)$;
8. $tr((B / A)^R) \coloneqq tr(A^R) \backslash tr(B^R)$;
9. $tr(A^{RR}) \coloneqq tr(A)$.

In the normal form, the ^{R} connective can appear only on occurrences of primitive types. Obviously, $tr(tr(A)) = tr(A)$ for every type A.

We also consider variants of L and L* with $\mathrm{Tp} \cup \{p^{\mathrm{R}} \mid p \in \mathrm{Tp}\}$ instead of Tp as the set of primitive types. These calculi will be called L′ and L*′ respectively. Obviously, if a sequent is provable in L′, then all its types are in normal form and this sequent is provable in L^{R} (and the same for L*′ and L*R). Later we shall prove the converse statement:

Lemma 2. *A sequent $F_1 \ldots F_n \to G$ is provable in L^{R} (resp., L*R) if and only if the sequent $tr(F_1) \ldots tr(F_n) \to tr(G)$ is provable in L′ (resp., L*′).*

3   L-Models

Now let Σ be an alphabet (an arbitrary nonempty set, finite or countable). By Σ^+ we denote the set of all nonempty words over Σ; the set of all words over Σ, including the empty word, is denoted by Σ^*. The set Σ^* with the operation of word concatenation is the *free monoid* generated by Σ; the empty word ϵ is the unit of this monoid. Subsets of Σ^* are called *languages* over Σ. The set Σ^+ with the same operation is the *free semigroup* generated by Σ. Its subsets are *languages without the empty word*.

The set $\mathcal{P}(\Sigma^*)$ of all languages is also a monoid: if $M, N \subseteq \Sigma^*$, then let $M \cdot N$ be $\{uv \mid u \in M, v \in N\}$; the singleton $\{\epsilon\}$ is the unit. Likewise, the set $\mathcal{P}(\Sigma^+)$ is a semigroup with the same multiplication operation.

On these two structures one can also define two *division* operations: $M \setminus N \leftrightharpoons \{u \in \Sigma^* \mid (\forall v \in M)\, vu \in N\}$, $N / M \leftrightharpoons \{u \in \Sigma^* \mid (\forall v \in M)\, uv \in N\}$ for $\mathcal{P}(\Sigma^*)$, and $M \setminus N \leftrightharpoons \{u \in \Sigma^* \mid (\forall v \in M)\, vu \in N\}$, $N / M \leftrightharpoons \{u \in \Sigma^+ \mid (\forall v \in M)\, uv \in N\}$ for $\mathcal{P}(\Sigma^+)$. Note that, unlike multiplication, the $\mathcal{P}(\Sigma^*)$ version of division operations does not coincide with the $\mathcal{P}(\Sigma^+)$ one even for languages without the empty word. For example, if $M = N = \{a\}$ ($a \in \Sigma$), then $M \setminus N$ is $\{\epsilon\}$ in $\mathcal{P}(\Sigma^*)$ and empty in $\mathcal{P}(\Sigma^+)$.

These three operations on languages naturally correspond to three connectives of the Lambek calculus, thus giving an interpretation for Lambek types and sequents. An *L-model* is a pair $\mathcal{M} = \langle \Sigma, w \rangle$, where Σ is an alphabet and w is a function that maps Lambek calculus types to languages over Σ, such that $w(A \cdot B) = w(A) \cdot w(B)$, $w(A \setminus B) = w(A) \setminus w(B)$, and $w(B / A) = w(B) / w(A)$ for all $A, B \in \mathrm{Tp}$. One can consider models either with or without the empty word, depending on what set of languages ($\mathcal{P}(\Sigma^*)$ or $\mathcal{P}(\Sigma^+)$), and, more importantly, what version of the division operations is used. Models with and without the empty word are similar but different (in particular, models with the empty word are not a generalisation of models without it). Obviously, w can be defined on primitive types in an arbitrary way, and then it is uniquely propagated to all types.

A sequent $F_1 \ldots F_n \to G$ is considered *true* in a model $\mathcal{M}$ ($\mathcal{M} \vDash F_1 \ldots F_n \to G$) if $w(F_1) \cdot \ldots \cdot w(F_n) \subseteq w(G)$. If the sequent has an empty antecedent ($n = 0$), i. e., is of the form $\to G$, then it is considered true if $\epsilon \in w(G)$. This implies that such sequents are never true in L-models without the empty word. L-models give sound and complete semantics for L and L*, due to the following theorem:

Theorem 1. *A sequent is provable in* L *if and only if it is true in all L-models without the empty word. A sequent is provable in* L* *if and only if it is true in all L-models with the empty word.*

This theorem is proved in [8] for L and in [9] for L*; its special case for the product-free fragment (where we keep only types without multiplication) is much easier and appears in [1].

Note that for L and L-models without the empty word it is sufficient to consider only sequents with one type in the antecedent, since $L \vdash F_1 F_2 \ldots F_n \to G$ if and only if $L \vdash F_1 \cdot F_2 \cdot \ldots \cdot F_n \to G$. For L* and L-models with the empty word it is sufficient to consider only sequents with empty antecedent, since $L^* \vdash F_1 \ldots F_{n-1} F_n \to G$ if and only if $L^* \vdash \to F_n \backslash (F_{n-1} \backslash \ldots \backslash (F_1 \backslash G) \ldots))$.

4   L-Models with the Reversal Operation

The new $^\mathrm{R}$ connective corresponds to the *language reversal* operation. For $u = a_1 a_2 \ldots a_n$ $(a_1, \ldots, a_n \in \Sigma, n \geq 1)$ let $u^\mathrm{R} \leftrightharpoons a_n \ldots a_2 a_1$; $\epsilon^\mathrm{R} \leftrightharpoons \epsilon$. For a language M let $M^\mathrm{R} \leftrightharpoons \{u^\mathrm{R} \mid u \in M\}$. The notion of L-model is easily modified to deal with the new connective by adding additional constraints on w: $w(A^\mathrm{R}) = w(A)^\mathrm{R}$ for every type A.

One can easily show that the calculi L^R and $\mathrm{L}^{*\mathrm{R}}$ are sound with respect to L-models with the reversal operation (without and with the empty word respectively). Now, using this soundness statement and Pentus' completeness theorem (Theorem 1), we can prove Lemma 1 (conservativity of L^R over L and $\mathrm{L}^{*\mathrm{R}}$ over L*): if a sequent is provable in L^R (resp., $\mathrm{L}^{*\mathrm{R}}$) and does not contain the $^\mathrm{R}$ connective, then it is true in all L-models without the empty word (resp., with the empty word). Moreover, in these L-models the language reversal operation is never used. Therefore, the sequent involved is provable in L (resp., L*) due to the completeness theorem.

The completeness theorem for L^R is proved in [3] (the product-free case is again easy and is handled in [6] using Buszkowski's argument [1]):

Theorem 2. *A sequent is provable in* L^R *if and only if it is true in all L-models with the reversal operation and without the empty word.*

In this paper we present a proof for the $\mathrm{L}^{*\mathrm{R}}$ version of this theorem:

Theorem 3. *A sequent is provable in* $\mathrm{L}^{*\mathrm{R}}$ *if and only if it is true in all L-models with the reversal operation and without the empty word.*

The proof basically duplicates the proof of Theorem 2 from [3]; changes are made to handle the empty word cases.

The main idea is as follows: if a sequent in normal form is not provable in $\mathrm{L}^{*\mathrm{R}}$, then it is not provable in $\mathrm{L}^{*\prime}$. Therefore, by Theorem 1, there exists a model in which this sequent is not true, but this model does not necessarily satisfy all of the conditions $w(A^\mathrm{R}) = w(A)^\mathrm{R}$. We want to modify our model by adding $w(A^\mathrm{R})^\mathrm{R}$ to $w(A)$. For L^R [3], we can first make the sets $w(A^\mathrm{R})^\mathrm{R}$ and $w(A)$

disjoint by replacing every letter $a \in \Sigma$ by a long word $a^{(1)} \dots a^{(N)}$ ($a^{(i)}$ are symbols from a new alphabet); then the new interpretation for A is going to be $w(A) \cup w(A^R)^R \cup T$ with an appropriate "trash heap" set T. For L^{*R}, we cannot do this directly, because ϵ will still remain the same word after the substitution of long words for letters. Fortunately, the model given by Theorem 1 enjoys a sort of weak universal property: if a type A is a subtype of our sequent, then $\epsilon \in w(A)$ if *and only if* $L^{*\prime} \vdash \to A$. Hence, if $\epsilon \in w(A)$, then $\epsilon \in w(A^R)$, and vice versa, so the empty word does not do any harm here.

Note that essentially here we need only the fact that our sequent is not derivable in $L^{*\prime}$, but not L^{*R}, and from this assumption we prove the existence of a model falsifying it. Hence, the sequent is not provable in L^{*R}. Therefore, we have proved Lemma 2.

5   L-Completeness of L^{*R} (Proof)

Let $L^{*R} \not\vdash \to G$ (as mentioned earlier, it is sufficient to consider sequents with empty antecedent). Also let G be in normal form (otherwise replace it by $tr(G)$).

Since $L^{*R} \not\vdash \to G$, $L^{*\prime} \not\vdash \to G$. The calculus $L^{*\prime}$ is essentially the same as L^*, therefore Theorem 1 gives us a structure $\mathcal{M} = \langle \Sigma, w \rangle$ such that $\epsilon \notin w(G)$. The structure $\mathcal{M}$ indeed falsifies $\to G$, but it is not a model in the sense of our new language: some of the conditions $w(p_i^R) = w(p_i)^R$ might be not satisfied.

Let Φ be the set of all subtypes of G (including G itself; the notion of subtype is understood in the sense of L^R).

The construction of $\mathcal{M}$ (see [9]) guarantees that the following two statements hold for every $A \in \Phi$:

  1. $w(A) \neq \varnothing$;
  2. $\epsilon \in w(A) \iff L^{*\prime} \vdash \to A$.

We introduce an inductively defined counter $f(A)$, $A \in \Phi$: $f(p_i) \leftharpoondown 1$, $f(p_i^R) \leftharpoondown 1$, $f(A \cdot B) \leftharpoondown f(A) + f(B) + 10$, $f(A \backslash B) \leftharpoondown f(B)$, $f(B / A) \leftharpoondown f(B)$. Let $K \leftharpoondown \max\{f(A) \mid A \in \Phi\}$, $N \leftharpoondown 2K + 25$ (N should be odd, greater than K, and big enough itself).

Let $\Sigma_1 \leftharpoondown \Sigma \times \{1, \dots, N\}$. We shall denote the pair $\langle a, j \rangle \in \Sigma_1$ by $a^{(j)}$. Elements of Σ and Σ_1 will be called *letters* and *symbols* respectively. A symbol can be *even* or *odd* depending on the parity of the superscript. Consider a homomorphism $h \colon \Sigma^* \to \Sigma_1^*$, defined as follows: $h(a) \leftharpoondown a^{(1)} a^{(2)} \dots a^{(N)}$ ($a \in \Sigma$), $h(a_1 \dots a_n) \leftharpoondown h(a_1) \dots h(a_n)$, $h(\epsilon) = \epsilon$. Let $P \leftharpoondown h(\Sigma^+)$. Note that h is a bijection between Σ^* and $P \cup \{\epsilon\}$ and between Σ^+ and P.

Lemma 3. *For all $M, N \subseteq \Sigma^*$ we have*

  1. $h(M \cdot N) = h(M) \cdot h(N)$;
  2. *if $M \neq \varnothing$, then $h(M \backslash N) = h(M) \backslash h(N)$ and $h(N / M) = h(N) / h(M)$.*

Proof

  1. By the definition of a homomorphism.

2. $\boxed{\subseteq}$ Let $u \in h(M \setminus N)$. Then $u = h(u')$ for some $u' \in M \setminus N$. For all $v' \in M$ we have $v'u' \in N$. Take an arbitrary $v \in h(M)$, $v = h(v')$ for some $v' \in M$. Since $u' \in M \setminus N$, $v'u' \in N$, whence $vu = h(v')h(u') = h(v'u') \in h(N)$. Therefore $u \in h(M) \setminus h(N)$.

$\boxed{\supseteq}$ Let $u \in h(M) \setminus h(N)$. First we claim that $u \in P \cup \{\epsilon\}$. Suppose the contrary: $u \notin P \cup \{\epsilon\}$. Take $v' \in M$ (M is nonempty by assumption). Since $v = h(v') \in P \cup \{\epsilon\}$, $vu \notin P \cup \{\epsilon\}$. On the other hand, $vu \in h(N) \subseteq P \cup \{\epsilon\}$. Contradiction. Now, since $u \in P \cup \{\epsilon\}$, $u = h(u')$ for some $u' \in \Sigma^+$. For an arbitrary $v' \in M$ and $v \leftrightharpoons h(v')$ we have $h(v'u') = vu \in h(N)$, whence $v'u' \in N$, whence $u' \in M \setminus N$. Therefore, $u = h(u') \in h(M \setminus N)$. The $/$ case is handled symmetrically.

We construct a new model $\mathcal{M}_1 = \langle \Sigma_1, w_1 \rangle$, where $w_1(z) \leftrightharpoons h(w(z))$ ($z \in \mathrm{Pr}'$). Due to Lemma 3, $w_1(A) = h(w_1(A))$ for all $A \in \Phi$, whence $w_1(F) = h(w(F)) \not\subseteq h(w(G)) = w_1(G)$ ($\mathcal{M}_1$ is also a countermodel in the language without $^\mathrm{R}$). Note that $w_1(A) \subseteq P \cup \{\epsilon\}$ for any type A; moreover, if $A \in \Phi$, then $\epsilon \in w_1(A)$ if and only if $\mathrm{L}^{*\prime} \vdash \to A$.

Now we introduce several auxiliary subsets of Σ_1^+ (by $\mathrm{Subw}(M)$ we denote the set of all nonempty subwords of words from M, i.e. $\mathrm{Subw}(M) \leftrightharpoons \{u \in \Sigma_1^+ \mid (\exists v_1, v_2 \in \Sigma_1^*)\, v_1 u v_2 \in M\}$):

$T_1 \leftrightharpoons \{u \in \Sigma_1^+ \mid u \notin \mathrm{Subw}(P \cup P^\mathrm{R})\}$;

$T_2 \leftrightharpoons \{u \in \mathrm{Subw}(P \cup P^\mathrm{R}) \mid \text{the first or the last symbol of } u \text{ is even}\}$;

$E \leftrightharpoons \{u \in \mathrm{Subw}(P \cup P^\mathrm{R}) - (P \cup P^\mathrm{R}) \mid \text{both the first symbol and the last symbol of } u \text{ are odd}\}$.

The sets P, P^R, T_1, T_2, and E form a partition of Σ_1^+ into nonintersecting parts. The set Σ_1^* is now split into six disjoint subsets: P, P^R, T_1, T_2, E, and $\{\epsilon\}$. For example, $a^{(1)}b^{(10)}a^{(2)} \in T_1$, $a^{(N)}b^{(1)} \ldots b^{(N-1)} \in T_2$, $a^{(7)}a^{(6)}a^{(5)} \in E$ ($a, b \in \Sigma$). Let $T \leftrightharpoons T_1 \cup T_2$, $T_i(k) \leftrightharpoons \{u \in T_i \mid |u| \geq k\}$ ($i = 1, 2$, $|u|$ is the length of u), $T(k) \leftrightharpoons T_1(k) \cup T_2(k) = \{u \in T \mid |u| \geq k\}$. Note that if the first or the last symbol of u is even, then $u \in T$, no matter whether it belongs to $\mathrm{Subw}(P \cup P^\mathrm{R})$. The index k (possibly with subscripts) here and further ranges from 1 to K. For all k we have $T(k) \supseteq T(K)$.

Lemma 4

1. $P \cdot P \subseteq P$, $P^\mathrm{R} \cdot P^\mathrm{R} \subseteq P^\mathrm{R}$;
2. $T^\mathrm{R} = T$, $T(k)^\mathrm{R} = T(k)$;
3. $P \cdot P^\mathrm{R} \subseteq T(K)$, $P^\mathrm{R} \cdot P \subseteq T(K)$;
4. $P \cdot T \subseteq T(K)$, $T \cdot P \subseteq T(K)$;
5. $P^\mathrm{R} \cdot T \subseteq T(K)$, $T \cdot P^\mathrm{R} \subseteq T(K)$;
6. $T \cdot T \subseteq T$.

Proof

1. Obvious.
2. Directly follows from our definitions.
3. Any element of $P \cdot P^\mathrm{R}$ or $P^\mathrm{R} \cdot P$ does not belong to $\mathrm{Subw}(P \cup P^\mathrm{R})$ and its length is at least $2N > K$. Therefore it belongs to $T_1(K) \subseteq T(K)$.

4. Let $u \in P$ and $v \in T$. If $v \in T_1$, then uv is also in T_1. Let $v \in T_2$. If the last symbol of v is even, then $uv \in T$. If the last symbol of v is odd, then $uv \notin \text{Subw}(P \cup P^R)$, whence $uv \in T_1 \subseteq T$. Since $|uv| > |u| \geq N > K$, $uv \in T(K)$.

The claim $T \cdot P \subseteq T$ is handled symmetrically.

5. $P^R \cdot T = P^R \cdot T^R = (T \cdot P)^R \subseteq T(K)^R = T(K)$. $T \cdot P^R = T^R \cdot P^R = (P \cdot T)^R \subseteq T(K)^R = T(K)$.

6. Let $u, v \in T$. If at least one of these two words belongs to T_1, then $uv \in T_1$. Let $u, v \in T_2$. If the first symbol of u or the last symbol of v is even, then $uv \in T$. In the other case u ends with an even symbol, and v starts with an even symbol. But then we have two consecutive even symbols in uv, therefore $uv \in T_1$.

Let us call words of the form $a^{(i)}a^{(i+1)}a^{(i+2)}$, $a^{(N-1)}a^{(N)}b^{(1)}$, and $a^{(N)}b^{(1)}b^{(2)}$ $(a, b \in \Sigma, 1 \leq i \leq N - 2)$ *valid triples of type I* and their reversals (namely, $a^{(i+2)}a^{(i+1)}a^{(i)}$, $b^{(1)}a^{(N)}a^{(N-1)}$, and $b^{(2)}b^{(1)}a^{(N)}$) *valid triples of type II*. Note that valid triples of type I (resp., of type II) are the only possible three-symbol subwords of words from P (resp., P^R).

Lemma 5. *A word u of length at least three is a subword of a word from $P \cup P^R$ if and only if any three-symbol subword of u is a valid triple of type I or II.*

Proof. The nontrivial part is "if". We proceed by induction on $|u|$. Induction base ($|u| = 3$) is trivial. Let u be a word of length $m + 1$ satisfying the condition and let $u = u'x$ ($x \in \Sigma_1$). By induction hypothesis ($|u'| = m$), $u' \in \text{Subw}(P \cup P^R)$. Let $u' \in \text{Subw}(P)$ (the other case is handled symmetrically); u' is a subword of some word $v \in P$. Consider the last three symbols of u. Since the first two of them also belong to u', this three-symbol word is a valid triple of type I, not type II. If it is of the form $a^{(i)}a^{(i+1)}a^{(i+2)}$ or $a^{(N)}b^{(1)}b^{(2)}$, then x coincides with the symbol next to the occurrence of u' in v, and therefore $u = u'x$ is also a subword of v. If it is of the form $a^{(N-1)}a^{(N)}b^{(1)}$, then, provided $v = v_1u'v_2$, v_1u' is also an element of P, and so is the word $v_1u'b^{(1)}b^{(2)} \ldots b^{(N)}$, which contains $u = u'b^{(1)}$ as a subword. Thus, in all cases $u \in \text{Subw}(P)$.

Now we construct one more model $\mathcal{M}_2 = \langle \Sigma_1, w_2 \rangle$, where $w_2(p_i) \rightleftharpoons w_1(p_i) \cup w_1(p_i^R)^R \cup T$, $w_2(p_i^R) \rightleftharpoons w_1(p_i)^R \cup w_1(p_i^R) \cup T$. This model is a model even in the sense of the enriched language. To finish the proof, we need to check that $\mathcal{M}_2 \not\vdash \to G$, e.g. $w_2(G) \not\ni \epsilon$.

Lemma 6. *For any $A \in \Phi$ the following holds:*

1. $w_2(A) \subseteq P \cup P^R \cup \{\epsilon\} \cup T$;
2. $w_2(A) \supseteq T(f(A))$;
3. $w_2(A) \cap (P \cup \{\epsilon\}) = w_1(A)$ *(in particular, $w_2(A) \cap (P \cup \{\epsilon\}) \neq \varnothing$)*;
4. $w_2(A) \cap (P^R \cup \{\epsilon\}) = w_1(tr(A^R))^R$ *(in particular, $w_2(A) \cap (P^R \cup \{\epsilon\}) \neq \varnothing$)*;
5. $\epsilon \in w_2(A) \iff \mathbf{L}^{*\prime} \vdash \to A$.

Proof. We prove statements 1–4 simultaneously by induction on type A.

The induction base is trivial. Further we shall refer to the i-th statement of the induction hypothesis ($i = 1, 2, 3, 4$) as "IH-i".

1. Consider three possible cases.

a) $A = B \cdot C$. Then $w_2(A) = w_2(B) \cdot w_2(C) \subseteq (P \cup P^{\mathrm{R}} \cup \{\epsilon\} \cup T) \cdot (P \cup P^{\mathrm{R}} \cup \{\epsilon\} \cup T) \subseteq P \cup P^{\mathrm{R}} \cup \{\epsilon\} \cup T$ (Lemma 4).

b) $A = B \setminus C$. Suppose the contrary: in $w_2(A)$ there exists an element $u \in E$. Then $vu \in w_2(C)$ for any $v \in w_2(B)$. We consider several subcases and show that each of those leads to a contradiction.

i) $u \in \mathrm{Subw}(P)$, and the superscript of the first symbol of u (as $\epsilon \notin E$, u contains at least one symbol) is not 1. Let the first symbol of u be $a^{(i)}$. Note that i is odd and $i > 2$. Take $v = a^{(3)} \ldots a^{(N)} a^{(1)} \ldots a^{(i-1)}$. The word v has length at least $N \geq K$ and ends with an even symbol, therefore $v \in T(K) \subseteq T(f(B)) \subseteq w_2(B)$ (IH-2). On the other hand, $vu \in \mathrm{Subw}(P)$ and the first symbol and the last symbol of vu are odd. Therefore, $vu \in E$ and $vu \in w_2(C)$, but $w_2(C) \cap E = \varnothing$ (IH-1). Contradiction.

ii) $u \in \mathrm{Subw}(P)$, and the first symbol of u is $a^{(1)}$ (then the superscript of the last symbol of u is not N, because otherwise $u \in P$). Take $v \in w_2(B) \cap (P \cup \{\epsilon\})$ (this set is nonempty due to IH-3). If $v = \epsilon$, then $vu = u \in E$. Otherwise the first and the last symbol of vu are odd, and $vu \in \mathrm{Subw}(P) - P$, and again we have $vu \in E$. Contradiction.

iii) $u \in \mathrm{Subw}(P^{\mathrm{R}})$, and the superscript of the first symbol of u is not N (the first symbol of u is $a^{(i)}$, i is odd). Take $v = a^{(N-2)} \ldots a^{(1)} a^{(N)} \ldots a^{(i+1)} \in T(K) \subseteq w_2(B)$. Again, $vu \in E$.

iv) $u \in \mathrm{Subw}(P^{\mathrm{R}})$, and the first symbol of u is $a^{(N)}$. Take $v \in w_2(B) \cap (P^{\mathrm{R}} \cup \{\epsilon\})$ (nonempty due to IH-4). $vu \in E$.

c) $A = C \,/\, B$. Proceed symmetrically.

2. Consider three possible cases.

a) $A = B \cdot C$. Let $k_1 \leftrightharpoons f(B)$, $k_2 \leftrightharpoons f(C)$, $k \leftrightharpoons k_1 + k_2 + 10 = f(A)$. Due to IH-2, $w_2(B) \supseteq T(k_1)$ and $w_2(C) \supseteq T(k_2)$. Take $u \in T(k)$. We have to prove that $u \in w_2(A)$. Consider several subcases.

i) $u \in T_1(k)$. By Lemma 5 ($|u| \geq k > 3$ and $u \notin \mathrm{Subw}(P \cup P^{\mathrm{R}})$) in u there is a three-symbol subword xyz that is not a valid triple of type I or II. Divide the word u into two parts, $u = u_1 u_2$, such that $|u_1| \geq k_1 + 5$, $|u_2| \geq k_2 + 5$. If needed, shift the border between parts by one symbol to the left or to the right, so that the subword xyz lies entirely in one part. Let this part be u_2 (the other case is handled symmetrically). Then $u_2 \in T_1(k_2)$. If u_1 is also in T_1, then the proof is finished. Consider the other case. Note that in any word from $\mathrm{Subw}(P \cup P^{\mathrm{R}})$ among any three consecutive symbols at least one is even. Shift the border to the left by at most 2 symbols to make the last symbol of u_1 even. Then $u_1 \in T(k_1)$, and u_2 remains in $T_1(k_2)$. Thus $u = u_1 u_2 \in T(k_1) \cdot T(k_2) \subseteq w_2(B) \cdot w_2(C) = w_2(A)$.

ii) $u \in T_2(k)$. Let u end with an even symbol (the other case is symmetric). Divide the word u into two parts, $u = u_1 u_2$, $|u_1| \geq k_1 + 5$, $u_2 \geq k_2 + 5$, and shift the border (if needed), so that the last symbol of u_1 is even. Then both u_1 and u_2 end with an even symbol, and therefore $u_1 \in T(k_1)$ and $u_2 \in T(k_2)$.

b) $A = B \setminus C$. Let $k = f(C) = f(A)$. By IH-2, $w_2(C) \supseteq T(k)$. Take $u \in T(k)$ and an arbitrary $v \in w_2(B) \subseteq P \cup P^{\mathrm{R}} \cup \{\epsilon\} \cup T$. By Lemma 4, statements 4–6, $vu \in (P \cup P^{\mathrm{R}} \cup \{\epsilon\} \cup T) \cdot T \subseteq T$, and since $|vu| \geq |u| \geq k$, $vu \in T(k) \subseteq w_2(C)$. Thus $u \in w_2(A)$.

c) $A = C \,/\, B$. Symmetrically.

3. Consider three possible cases.

a) $A = B \cdot C$.

$\boxed{\supseteq}$ $u \in w_1(A) = w_1(B) \cdot w_1(C) \subseteq w_2(B) \cdot w_2(C) = w_2(A)$ (IH-3); $u \in P \cup \{\epsilon\}$.

$\boxed{\subseteq}$ Suppose $u \in P$ and $u \in w_2(A) = w_2(B) \cdot w_2(C)$. Then $u = u_1 u_2$, where $u_1 \in w_2(B)$ and $u_2 \in w_2(C)$. First we claim that $u_1 \in P \cup \{\epsilon\}$. Suppose the contrary. By IH-1, $u_1 \in P^{\mathrm{R}} \cup T$, $u_2 \in P \cup P^{\mathrm{R}} \cup \{\epsilon\} \cup T$, and therefore $u = u_1 u_2 \in (P^{\mathrm{R}} \cup T) \cdot (P \cup P^{\mathrm{R}} \cup \{\epsilon\} \cup T) \subseteq P^{\mathrm{R}} \cup T$ (Lemma 4, statements 1, 3–6). Hence $u \notin P \cup \{\epsilon\}$. Contradiction. Thus, $u_1 \in P \cup \{\epsilon\}$. Similarly, $u_2 \in P \cup \{\epsilon\}$, and by IH-3 we obtain $u_1 \in w_1(B)$ and $u_2 \in w_1(C)$, whence $u = u_1 u_2 \in w_1(A)$.

b) $A = B \setminus C$.

$\boxed{\supseteq}$ Take $u \in w_1(B \setminus C) \subseteq P \cup \{\epsilon\}$. First we consider the case where $u = \epsilon$. Then we have $\mathrm{L}^{*\prime} \vdash \rightarrow B \setminus C$, whence $u = \epsilon \in w_2(B \setminus C)$. Now let $u \in P$. For any $v \in w_1(B)$ we have $vu \in w_1(C)$. We claim that $u \in w_2(B \setminus C)$. Take $v \in w_2(B) \subseteq P \cup P^{\mathrm{R}} \cup \{\epsilon\} \cup T$ (IH-1). If $v \in P \cup \{\epsilon\}$, then $v \in w_1(B)$ (IH-3), and $vu \in w_1(C) \subseteq w_2(C)$ (IH-3). If $v \in P^{\mathrm{R}} \cup T$, then $vu \in (P^{\mathrm{R}} \cup T) \cdot P \subseteq T(K) \subseteq w_2(C)$ (Lemma 4, statements 3 and 4, and IH-2). Therefore, $u \in w_2(B) \setminus w_2(C) = w_2(B \setminus C)$.

$\boxed{\subseteq}$ If $u \in w_2(B \setminus C)$ and $u \in P \cup \{\epsilon\}$, then for any $v \in w_1(B) \subseteq w_2(B)$ we have $vu \in w_2(C)$. Since $v, u \in P \cup \{\epsilon\}$, $vu \in P \cup \{\epsilon\}$. By IH-3, $vu \in w_1(C)$. Thus $u \in w_1(B \setminus C)$.

c) $A = C \,/\, B$. Symmetrically.

4. Consider three cases.

a) $A = B \cdot C$. Then $tr(A^{\mathrm{R}}) = tr(C^{\mathrm{R}}) \cdot tr(B^{\mathrm{R}})$.

$\boxed{\supseteq}$ $u \in w_1(tr(A^{\mathrm{R}}))^{\mathrm{R}} = w_1(tr(C^{\mathrm{R}}) \cdot tr(B^{\mathrm{R}}))^{\mathrm{R}} = \left(w_1(tr(C^{\mathrm{R}})) \cdot w_1(tr(B^{\mathrm{R}}))\right)^{\mathrm{R}} = w_1(tr(B^{\mathrm{R}}))^{\mathrm{R}} \cdot w_1(tr(C^{\mathrm{R}}))^{\mathrm{R}} \subseteq w_2(B) \cdot w_2(C) = w_2(A)$ (IH-4); $u \in P^{\mathrm{R}}$.

$\boxed{\subseteq}$ Let $u \in P^{\mathrm{R}}$ and $u \in w_2(A) = w_2(B) \cdot w_2(C)$. Then $u = u_1 u_2$, where $u_1 \in w_2(B)$, $u_2 \in w_2(C)$. We claim that $u_1, u_2 \in P^{\mathrm{R}} \cup \{\epsilon\}$. Suppose the contrary. By IH-1, $u_1 \in P \cup T$, $u_2 \in P \cup P^{\mathrm{R}} \cup \{\epsilon\} \cup T$, whence $u = u_1 u_2 \in (P \cup T) \cdot (P \cup P^{\mathrm{R}} \cup \{\epsilon\} \cup T) \subseteq P \cup T$. Contradiction. Thus, $u_1 \in P^{\mathrm{R}} \cup \{\epsilon\}$, and therefore $u_2 \in P^{\mathrm{R}} \cup \{\epsilon\}$, and, using IH-4, we obtain $u_1 \in w_1(tr(B^{\mathrm{R}}))^{\mathrm{R}}$, $u_2 \in w_1(tr(C^{\mathrm{R}}))^{\mathrm{R}}$. Hence $u = u_1 u_2 \in w_1(tr(B^{\mathrm{R}}))^{\mathrm{R}} \cdot w_1(tr(C^{\mathrm{R}}))^{\mathrm{R}} = \left(w_1(tr(C^{\mathrm{R}})) \cdot w_1(tr(B^{\mathrm{R}}))\right)^{\mathrm{R}} = w_1(tr(C^{\mathrm{R}}) \cdot tr(B^{\mathrm{R}}))^{\mathrm{R}} = w_1(tr(A^{\mathrm{R}}))^{\mathrm{R}}$.

b) $A = B \setminus C$. Then $tr(A^{\mathrm{R}}) = tr(C^{\mathrm{R}}) \,/\, tr(B^{\mathrm{R}})$.

$\boxed{\supseteq}$ Let $u \in w_1(tr(C^{\mathrm{R}}) \,/\, tr(B^{\mathrm{R}}))^{\mathrm{R}} = w_1(tr(B^{\mathrm{R}}))^{\mathrm{R}} \setminus w_1(tr(C^{\mathrm{R}}))^{\mathrm{R}}$, First we consider the case where $u = \epsilon$. Then $\mathrm{L}^{*\prime} \vdash \rightarrow tr(C^{\mathrm{R}}) \,/\, tr(B^{\mathrm{R}})$, whence $\epsilon \in w_2(tr(C^{\mathrm{R}}) \,/\, tr(B^{\mathrm{R}})) = w_2(tr(A^{\mathrm{R}}))$. Therefore, $u \in w_2(tr(A^{\mathrm{R}}))^{\mathrm{R}}$. Now let $u \in$

P^R. For every $v \in w_1(tr(B^R))^R$ we have $vu \in w_1(tr(C^R))^R$. We claim that $u \in w_2(B \setminus C)$. Take an arbitrary $v \in w_2(B) \subseteq P \cup P^R \cup \{\epsilon\} \cup T$ (IH-1). If $v \in P^R \cup \{\epsilon\}$, then $v \in w_1(tr(B^R))^R$ (IH-4), whence $vu \in w_1(tr(C^R))^R \subseteq w_2(C)$. If $v \in P \cup T$, then (since $u \in P^R$) we have $vu \in (P \cup T) \cdot P^R \subseteq T(K) \subseteq w_2(C)$ (Lemma 4 and IH-2).

$\boxed{\subseteq}$ If $u \in w_2(B \setminus C)$ and $u \in P^R \cup \{\epsilon\}$, then for any $v \in w_1(tr(B^R))^R \subseteq w_2(B)$ we have $vu \in w_2(C)$. Since $v, u \in P^R \cup \{\epsilon\}$, $vu \in P^R \cup \{\epsilon\}$, therefore $vu \in w_1(tr(C^R))^R$ (IH-4). Thus $u \in w_1(tr(B^R))^R \setminus w_1(tr(C^R))^R = w_1(A^R)^R$.

c) $A = C / B$. Symmetrically.

This completes the proof of statements 1–4 of Lemma 6. Statement 5 follows from statement 3 and immediately yields Theorem 3 ($L^{*\prime} \not\vdash \to G$, whence $\epsilon \notin w_2(G)$).

6   Grammars and Complexity

The Lambek calculus and its variants are used for describing formal languages via Lambek categorial grammars. An L^*-*grammar* is a triple $\mathcal{G} = \langle \Sigma, H, \rhd \rangle$, where Σ is a finite alphabet, $H \in \mathrm{Tp}$, and $\rhd$ is a finite correspondence between Tp and Σ ($\rhd \subset \mathrm{Tp} \times \Sigma$). The *language generated by* $\mathcal{G}$ is the set of all nonempty words $a_1 \ldots a_n$ over Σ for which there exist types $B_1, \ldots, B_n$ such that $L^* \vdash B_1 \ldots B_n \to H$ and $B_i \rhd a_i$ for all $i \le n$. We denote this language by $\mathfrak{L}(\mathcal{G})$. The notion of L-*grammar* is defined in a similar way. These class of grammars are *weakly equivalent* to the classes of context-free grammars with and without ϵ-rules in the following sense:

Theorem 4. *A formal language is context-free if and only if it is generated by some* L^**-grammar. A formal language without the empty word is context-free if and only if it is generated by some* L*-grammar.* [7] [2]

By modifying our definition in a natural way one can introduce the notion of L^{*R}-grammar and L^R-grammar. These grammars also generate precisely all context-free languages (resp., context-free languages without the empty word):

Theorem 5. *A formal language is context-free if and only if it is generated by some* L^{*R}*-grammar. A formal language without the empty word is context-free if and only if it is generated by some* L^R*-grammar.*

Proof. The "only if" part follows directly from Theorem 4 due to the conservativity of L^{*R} over L^* and L^R over L (Lemma 1).

The "if" part is proved by replacing all types in an L^{*R}-grammar (L^*-grammar) by their normal forms and applying Lemma 2.

Since A / B is equivalent in L^R and L^{*R} to $(B^R \setminus A^R)^R$, and the derivability problem in Lambek calculus with two division operators is NP-complete [10] (this holds both for L and L^*), the derivability problem is NP-complete even for the fragment of L^R (L^{*R}) with one division.

Acknowledgments. I am grateful to Prof. Mati Pentus for fruitful discussions and constant attention to my work. I am also grateful to Prof. Sergei Adian for inspiring techniques of working with words over an alphabet given in his lectures and papers.

This research was supported by the Russian Foundation for Basic Research (grants 11-01-00281-a and 12-01-00888-a), by the Presidential Council for Support of Leading Research Schools (grant NSh-65648.2010.1) and by the Scientific and Technological Cooperation Programme Switzerland–Russia (STCP-CH-RU, project "Computational Proof Theory").

References

1. Buszkowski, W.: Compatibility of categorial grammar with an associated category system. Zeitschr. für Math. Logik und Grundl. der Math. 28, 229–238 (1982)
2. Kuznetsov, S.: Lambek grammars with one division and one primitive type. Logic Journal of the IGPL 28(1), 207–221 (2012)
3. Kuznetsov, S.: L-completeness of the Lambek calculus with the reversal operation. In: Béchet, D., Dikovsky, A. (eds.) Logical Aspects of Computational Linguistics. LNCS, vol. 7351, pp. 151–160. Springer, Heidelberg (2012)
4. Lambek, J.: The mathematics of sentence structure. American Math. Monthly 65(3), 154–170 (1958)
5. Lambek, J.: From categorial grammar to bilinear logic. In: Došen, K., Schroeder-Heister, P. (eds.) Substructural Logics. Studies in Logic and Computation, vol. 2, pp. 128–139. Clarendon Press, Oxford (1993)
6. Minina, V.A.: Completeness of the Lambek syntactic calculus with the involution operation. Diploma paper, Dept. of Math. Logic and Theory of Algorithms, Moscow State University (2001) (in Russian)
7. Pentus, M.: Lambek grammars are context free. In: 8th Annual IEEE Symposium on Logic in Computer Science, pp. 429–433. IEEE Computer Society Press, Los Alamitos (1993)
8. Pentus, M.: Models for the Lambek calculus. Annals of Pure and Applied Logic 75(1-2), 179–213 (1995)
9. Pentus, M.: Free monoid completeness of the Lambek calculus allowing empty premises. In: Larrazabal, J.M., Lascar, D., Mints, G. (eds.) Logic Colloquium 1996. LNL, vol. 12, pp. 171–209. Springer, Berlin (1998)
10. Savateev, Y.: Product-free Lambek calculus is NP-complete. In: Artemov, S., Nerode, A. (eds.) LFCS 2009. LNCS, vol. 5407, pp. 380–394. Springer, Heidelberg (2008)

A Note on Multidimensional Dyck Languages

Michael Moortgat

Utrecht Institute of Linguistics

M.J.Moortgat@uu.nl

Abstract. Multidimensional Dyck languages generalize the language of balanced brackets to alphabets of size > 2. Words contain each alphabet symbol with the same multiplicity. In addition, reading a word from left to right, there are always at least as many a_i as $a_{(i+1)}$, where a_j is the jth alphabet symbol in the lexicographic ordering.

We compare the Dyck languages with MIX languages, where the multiplicity constraint is respected, but the order of the symbols is free. To understand the combinatorics of the Dyck languages, we study the bijection with standard Young tableaux of rectangular shape, and, for the three-dimensional case, with planar webs for combinatorial spider categories. We present a typelogical analysis of Dyck languages with an alphabet of size d in terms of a construction that aligns $(d-1)$ grammars for the two-symbol case.

This paper is dedicated to Jim Lambek on the occasion of his 90th birthday. More than 50 years ago, the deductive method of his 'Mathematics of Sentence Structure' raised the standards for doing computational linguistics to a new level. His work has been a continuing source of inspiration.

1 Introduction

Typelogical grammar, ever since the rediscovery of Lambek's Syntactic Calculus in the 1980s, has been struggling to come to grips with *discontinuity*: information flow between physically detached parts of an utterance. Emmon Bach, in a number of papers [1,2], discussed generalizations of categorial grammar that would be able to deal with 'scramble' languages: languages that are the free permutations of arbitrary context-free languages. As an abstract example of a language with total word order freedom, Bach proposed MIX: the words of this language are strings over a three-letter alphabet with an equal number of occurrences of each letter but no constraints on their order. MIX, in other words, is the scramble of context-free (actually, regular) $(abc)^+$.

A natural generalization of MIX allows variation in the size of the alphabet, keeping the multiplicity constraint on the letters. Let us call the size of the alphabet the *dimension* of the language. For positive d and n, let us write M_n^d for the (finite) language consisting of words of length dn over an alphabet with d symbols where each symbol occurs with multiplicity n.

$$M_n^d = \{w \in \{a_1, \ldots, a_d\}^+ \mid |w|_{a_1} = \ldots = |w|_{a_d} = n\}$$

C. Casadio et al. (Eds.): Lambek Festschrift, LNCS 8222, pp. 279–296, 2014.

Dropping the subscript, we write M^d for $\bigcup_{0<n} M_n^d$.

The restriction to positive multiplicity is characteristic of a categorial view on these languages: categorial grammars, being fully lexicalized, do not recognize the empty word. In the context of rewriting grammars, it is natural to also allow $n = 0$. In the degenerate case of $d = 1$, the language consists of the nth powers of the single alphabet symbol; there is nothing to scramble then.

I this paper, I consider a similar multidimensional generalization for Dyck languages. These are d-dimensional MIX languages with an extra prefix constraint: reading a word from left to right, there are always at least as many letters a_i as a_{i+1}, where a_j is the j-th symbol in the lexicographic order of the alphabet[1]

$$D_n^d = \{w \in M_n^d \mid \text{ for every prefix } u \text{ of } w, |u|_{a_i} \geq |u|_{a_{i+1}}, 1 \leq i < d\} \quad (1)$$

As before, we write D^d for $\bigcup_{0<n} D_n^d$. In the two-dimensional case D^2, we have the familiar language of well-balanced brackets consisting of words over an alphabet $\{a, b\}$ (with $a < b$) reading a as the opening and b as the closing bracket symbol. Whereas MIX stands for total word order freedom, the generalized Dyck languages represent discontinuity of the *interleaving* type: words of D_n^d merge n copies of the word $a_1 \cdots a_d$ in such a way that the order of the letters is respected. Interleaving (or shuffling) has been proposed for the analysis of semifree word order and bounded discontinuous constituency, for example in the 'sequence union' operation of [21]. Complexity of parsing of extended context-free formalisms with interleaving and concatenation operations has been studied in [18]. For applications of interleaving in various areas of computer science, see [3].

As for the position of M^d and D^d in the extended Chomsky hierarchy, much remains to be settled. Both M^2 and D^2 are context-free. From $d > 2$ on, one goes beyond the context-free languages. In the 3D case, for example, intersection with regular $a^+b^+c^+$ produces $a^nb^nc^n$, which is not context-free. Recent work by Salvati and Kanazawa [22,12] shows that M^3 is a 2-MCFG (Multiple Context Free Language [24]), but a non-wellnested one, which means M^3 is not a TAL (Tree Adjoining Language [10]). For higher dimensions, the challenge, for the MIX and the Dyck languages alike, would be to relate the size of their alphabet to the minimal degree (or fan-out) of the k-MCFG required for their recognition. Similarly, from the categorial point of view, one would like to see a hierarchy of stepwise generalizations of Lambek's Syntactic Calculus in correspondence with the rise in dimensionality.

This short note does not aim to answer these foundational questions in their full generality. We *do* hope to clarify the nature of the discontinuity exhibited by multidimensional Dyck languages by studying their correspondence with some well-understood combinatorial objects (sections §3 to §5). In §6 we present a typelogical anaysis of d-symbol Dyck languages in terms of an alignment of $(d-1)$ grammars for the two-symbol case. In §2, we set the stage with a comparison of the cardinalities of the Dyck and MIX languages.

---

[1] D^d should not be confused with the generalization that considers k different *pairs* of opening/closing parentheses $[_1,]_1, \ldots, [_k,]_k$.

2   Counting Words

Table 1 compares the size of M_n^d and D_n^d for small values of $d \geq 2, n \geq 0$, and gives the identifiers of these number sequences in $OEIS$, the Online Encyclopedia of Integer Sequences.

Table 1. Cardinality of M_n^d and D_n^d (grayscale) for small values of $d \geq 2, n \geq 0$

$d\backslash n$	0	1	2	3	4	5	6	OEIS
2	1	2	6	20	70	252	924	A000984
	1	1	2	5	14	42	132	A000108
3	1	6	90	1680	34650	756756	17153136	A006480
	1	1	5	42	462	6006	87516	A005789
4	1	24	2520	369600	63063000	11732745024	2308743493056	A008977
	1	1	14	462	24024	1662804	140229804	A005790

A useful graphical representation highlighting the differences between M_n^d and D_n^d pictures the words of these languages as *monotone paths* along the edges of a d-dimensional grid, with sides measuring n unit steps: these are paths starting at the origin $(0, \ldots, 0)$ and ending at $(n, \ldots, n)$, and consisting entirely of forward steps along one of the dimensions. All such paths are legitimate for the MIX words. For Dyck words, the prefix constraint translates into the requirement that the coordinates of the points visited are weakly decreasing $x_1 \geq x_2 \geq \cdots \geq x_d$. In the two-dimensional case, this means a path can touch but not cross the diagonal. Compare the two words over the alphabet $\{a, b\}$ with letter multiplicity 3 in (2). Paths start at the lower left corner and end at the upper right corner. The letter a is interpreted as a step to the right, the letter b as a step up. The path in (b) respects the prefix constraint; the path in (a) does not.

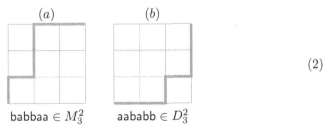

$$(2)$$

$$\text{babbaa} \in M_3^2 \qquad \text{aababb} \in D_3^2$$

A comparable contrast for the 3-dimensional case is given in (3). All points visited by the path in (3b) have coordinates $x \geq y \geq z$:

$$(0,0,0) \xrightarrow{a} (1,0,0) \xrightarrow{a} (2,0,0) \xrightarrow{b} (2,1,0) \xrightarrow{c} (2,1,1) \xrightarrow{b} (2,2,1) \xrightarrow{c} (2,2,2)$$

whereas the path in (3b) violates the Dyck prefix constraint at the first step.

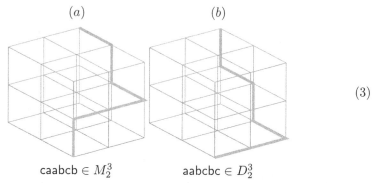

$$\text{caabcb} \in M_2^3 \qquad\qquad \text{aabcbc} \in D_2^3$$

The cardinality of the MIX languages is easily determined, see (4). There are $\binom{dn}{n}$ possible choices for the position of the n copies of the first alphabet symbol among dn candidate positions, then $\binom{(d-1)n}{n}$ possibilities for the n copies of the second alphabet symbol among the remaining $(d-1)n$ positions, and so on until one reaches the dth symbol for which no alternatives remain. In other words, (4) counts the number of distinct permutations of a multiset with d distinct elements each occurring with multiplicity n.

$$|M_n^d| \;=\; \prod_{k=1}^{d} \binom{kn}{n} \;=\; \frac{(dn)!}{(n!)^d} \tag{4}$$

The cardinality of the Dyck languages is given by the multidimensional Catalan numbers C_n^d which are given with the formula (5) in [7]. We will present an alternative formula with a direct combinatorial interpretation once we have discussed the connection between Dyck words and rectangular standard Young tableaux.

$$|D_n^d| = C_n^d = (dn)! \times \prod_{k=0}^{d-1} \frac{k!}{(n+k)!} \tag{5}$$

In the two-dimensional case, (5) gives the Catalan numbers $1, 1, 2, 5, 14, 42, 132, \ldots$ Note also that $|D_n^d| = |D_d^n|$; the triangular arrangement for D_n^d brings this out clearly. (Values for $n \leq 2, d \leq 2$. Multiplicity n increases along the $\searrow$ diagonal, dimensionality d along the $\nearrow$ diagonal.)

$$
\begin{array}{ccccccccc}
 & & & & 2 & & & & \\
 & & & 5 & & 5 & & & \\
 & & 14 & & 42 & & 14 & & \\
 & 42 & & 462 & & 462 & & 42 & \\
132 & & 6006 & & 24024 & & 6006 & & 132
\end{array}
$$

3 Dyck Words and Rectangular Standard Young Tableaux

Young tableaux are rich combinatorial objects, widely used in representation theory of the symmetric and general linear groups and in algebraic geometry.

They are less known in computational linguistics, so let us start with some definitions, following [5]. Let $\lambda = (\lambda_1, \ldots, \lambda_k)$ be a partition of an integer n, i.e. a multiset of positive integers the sum of which is n, and let us list the k parts in weakly decreasing order. With such a partition, we associate a *Young diagram*: an arrangement of n boxes into left-aligned rows of length $\lambda_1 \geq \lambda_2 \geq \cdots \geq \lambda_k$. A *standard Young tableau* of shape λ is obtained by placing the integers 1 through n in these boxes in such a way that the entries are strictly increasing from left to right in the rows and from top to bottom in the columns. Henceforth, when we say tableau, we mean standard Young tableau.

Given a partition of n with a diagram of shape λ, the number of possible tableaux with that shape is given by the Frame-Robinson-Thrall [4] *hook length formula*, which is $n!$ divided by the product of the *hook lengths* of the n boxes; the hook length of a box is the number of boxes to the right in its row and below it in its column, plus one for the box itself. Given a tableau T, let T' be the tableau obtained from T by a reflection through the main diagonal, taking rows to columns and columns to rows. The diagrams for T and T', clearly, produce the same result for the hook length formula.

Given a standard Young tableau one can read off its *Yamanouchi word*. The Yamanouchi word for a tableau T of shape $\lambda = (\lambda_1, \ldots, \lambda_k)$ is a word $w = w_1 \cdots w_n$ over a k-symbol alphabet $\{1, 2, \ldots, k\}$ such that w_i is the row that contains the integer i in T. Conversely, given a word $w = w_1 \cdots w_n$ over a k-symbol alphabet $\{1, 2, \ldots, k\}$ with the property that, reading w from left to right, there are never fewer letters i than letters $(i+1)$, we can recover a standard Young tableau with k rows.

We illustrate these definitions in (6) below. The partition $(3, 3, 2, 1)$ of the integer 9 has the Young diagram in (a). Distributing the integers $1, \ldots, 9$ over the nine boxes as in (b) yields a well-formed standard Young tableau with strictly increasing rows and columns. The Yamanouchi word for this tableau is given in (c). The fifth letter of the word is 4, because the integer 5 appears on the fourth row in the tableau. In (d), we give the hook lengths of the boxes of diagram (a). The hook length formula then says there are 168 standard Young tableaux of that shape: 9! divided by $(6 \cdot 4 \cdot 2 \cdot 5 \cdot 3 \cdot 1 \cdot 3 \cdot 1 \cdot 1)$.

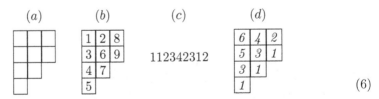

$$(6)$$

The definitions above apply to standard Young tableaux in general. To obtain the multidimensional Dyck words we are interested in, we restrict to *rectangular* tableaux of shape $d \times n$ (d rows by n columns). These tableaux correspond to *balanced* Yamanouchi words of length dn— words in which the constituting letters appear with equal multiplicity. The prefix constraint of the generalized Dyck languages is captured by the tableau constraint that the box entries are strictly increasing from top to bottom in the columns; the restriction to rows of equal length corresponds to the letter multiplicity constraint on Dyck (and MIX) words. Note that, in accordance with the practice in formal linguistics, we use the alphabet a < b < c < $\cdots$, rather than the alphabet 1 < 2 < 3 < $\cdots$ of (6c).

For rectangular tableaux, the hook length formula takes on the simple form of (7) below.

$$\frac{dn!}{\prod_{k=1}^{n} k^{\overline{d}}} \tag{7}$$

(writing $n^{\overline{m}}$ for rising factorial powers $n(n+1)\cdots(n+m-1)$) which then counts the words of D_n^d for positive n. We saw in Table 1 that D_n^d and D_d^n have the same cardinality; indeed, the hook length formula for the D_n^d and for the D_d^n diagrams produces the same result. On the level of individual words, we have a bijection between words of D_n^d and words of D_d^n, reading them as the Yamanouchi words of a $d \times n$ tableau T and of the transposed $n \times d$ tableau T' respectively.

We illustrate with D_2^3, words over a three-letter alphabet with letter multiplicity 2. On the left are the hook lengths of the boxes in a diagram of shape $(3,2)$. By the hook length formula (7), there are $6!/(1 \cdot 2^2 \cdot 3^2 \cdot 4) = 720/144 = 5$ standard Young tableaux of that shape. They are listed in (a) through (e) together with their Yamanouchi words $y(T)$. Finally, $y(T')$ gives the words corresponding to the transposed tableaux T'. These are the five words of D_3^2.

$$\begin{array}{ccccc}
(a) & (b) & (c) & (d) & (e)
\end{array}$$

$$\begin{array}{|c|c|}\hline 4 & 3 \\\hline 3 & 2 \\\hline 2 & 1 \\\hline\end{array} \qquad T: \begin{array}{|c|c|}\hline 1 & 4 \\\hline 2 & 5 \\\hline 3 & 6 \\\hline\end{array} \quad \begin{array}{|c|c|}\hline 1 & 3 \\\hline 2 & 4 \\\hline 5 & 6 \\\hline\end{array} \quad \begin{array}{|c|c|}\hline 1 & 2 \\\hline 3 & 5 \\\hline 4 & 6 \\\hline\end{array} \quad \begin{array}{|c|c|}\hline 1 & 3 \\\hline 2 & 5 \\\hline 4 & 6 \\\hline\end{array} \quad \begin{array}{|c|c|}\hline 1 & 2 \\\hline 3 & 4 \\\hline 5 & 6 \\\hline\end{array} \tag{8}$$

$$y(T) \in D_2^3 : \text{abcabc } \text{ababcc } \text{aabcbc } \text{abacbc } \text{aabbcc}$$

$$y(T') \in D_3^2 : \text{aaabbb } \text{aabbab } \text{abaabb } \text{aababb } \text{ababab}$$

4 Dyck Words as Iterated Shuffles

Can we give an algorithm that generates the words of D_n^d in a unique way, and that provides the combinatorial *raison d'être* for the hook length formula for rectangular tableaux?

We approach this question in two steps. First we model the interleaving of n copies of the word $a_1 \cdots a_d$ as an iterated *shuffle* [6], where for strings w_1, w_2 the shuffle operation gives all possibilities of interleaving them in such a way that the order of the letters in w_1 and in w_2 is preserved. The operation takes its name from the familiar way of shuffling a deck of cards: the deck is cut in two packets, which are then merged by releasing the cards with the thumb of the left and of the right hand.

When shuffling a deck of cards, all cards are distinct. When interleaving copies of a Dyck word $a_1 \cdots a_d$, there are multiple occurrences of the letters, hence duplicate outcomes of their interleaving.

As a second step, then, we break the symmetry, and introduce a biased form of shuffling that generates the words of D_n^d without duplicates. We give a redundancy-free representation of these unique outputs in the form of a deterministic acyclic word automaton; the size (number of states) of this automaton directly correlates with the cardinality of D_n^d as given by the hook length formula (7).

Formally, shuffling is an operation $\odot : \Sigma^* \times \Sigma^* \to \mathcal{P}(\Sigma^*)$ defined as in (9) below (for $\alpha_1, \alpha_2 \in \Sigma$, and $u, u_1, u_2 \in \Sigma^*$):

$$
\begin{aligned}
(i) \quad & u \odot \epsilon = \epsilon \odot u = \{u\} \\
(ii) \quad & \alpha_1 u_1 \odot \alpha_2 u_2 = \{\alpha_1 w \mid w \in u_1 \odot \alpha_2 u_2\} \cup \{\alpha_2 w \mid w \in \alpha_1 u_1 \odot u_2\}
\end{aligned}
\tag{9}
$$

The $\odot$ operation can be lifted to the level of languages in the usual way (we use the same symbol for the lifted operation): for languages L_1, L_2,

$$
L_1 \odot L_2 = \bigcup_{\substack{w_1 \in L_1 \\ w_2 \in L_2}} w_1 \odot w_2
$$

Analogous to the Kleene star, we have the notion of the *shuffle closure* of a language, $L^{(*)}$, defined in terms of shuffle iteration $L^{(n)}$:

$$
L^{(0)} = \{\epsilon\}, \qquad L^{(n)} = L \odot L^{(n-1)}, \qquad L^{(*)} = \bigcup_{0 \leq i} L^{(i)}
$$

With these definitions in place, we see that D_n^d is the n-th shuffle interation of the language that has the concatenation of the d alphabet symbols in their lexicographic order as its single word. D^d is the shuffle closure of that language.

$$
D_n^d = \{a_1 \cdots a_d\}^{(n)}, \qquad D^d = \{a_1 \cdots a_d\}^{(*)}
\tag{10}
$$

As noted, (10) correctly describes D_n^d as a *set* of words. But if, in the definition of shuffling in (9), we would substitute *lists* for sets and list concatenation for set union, we unfortunately would not obtain the desired procedure to produce the outcomes of shuffling in a repetition-free way.

To generate D_n^d without repetitions, we build a deterministic acyclic finite state automaton (DAFSA) representing the words of D_n^d in a redundancy-free way. The construction adapts a technique used by Warmuth and Haussler to prove that for any regular language R, its shuffle closure $R^{(*)}$ can be recognized in deterministic polynomial time [26, Theorem 5.1]. Informally, the idea is to decorate the states of a finite automaton with 'pebbles'; these pebbles control when and how often a transition for a given alphabet symbol can be made.

The construction of DAFSA(D_n^d) proceeds as follows. Let (11) below be the $(d + 1)$-state finite automaton for $\{a_1 \cdots a_d\}$, i.e. the singleton language that has the concatenation of the alphabet symbols in their lexicographic order as its single word.

$$\tag{11}$$

Then $\text{DAFSA}(D_n^d) = (Q, \Sigma, q_{start}, \delta, \{q_f\})$ is the following machine.

- ALPHABET $\Sigma = \{a_1, \ldots, a_d\}$
- STATES Q: all $(d+1)$-tuples of non-negative integers $(p_0, \ldots, p_d)$ such that $p_0 + \cdots + p_d = n$.
- INITIAL STATE q_{start}: the tuple that has $p_0 = n$ and $p_i = 0$ for $0 < i \leq d$
- FINAL STATE q_f: the tuple that has $p_d = n$ and $p_i = 0$ for $0 \leq i < d$
- TRANSITION FUNCTION δ: for states $q = (p_0, \ldots, p_d)$ and $q' = (p'_0, \ldots, p'_d)$ and alphabet symbols a_i $(1 \leq i \leq d)$, we have a transition $\delta(q, a_i) = q'$ whenever $p_{i-1} > 0$, $p'_{i-1} = p_{i-1} - 1$, $p'_i = p_i + 1$, and $p'_j = p_j$ for the other elements of the state tuples.

Note that the *size* (number of states) of the constructed $\text{DAFSA}(D_n^d)$ is given by the number of possible distributions of n pebbles over the $(d+1)$ states of (11), i.e. the number of weak compositions of n in $d+1$ parts which is $\binom{n+d}{d}$.

Allowing the case of $n = 0$, there is one state, $q_{start} = q_f = (0, \ldots, 0)$, and no transitions: this automaton accepts only the empty word.

The graph in (12) illustrates the construction with D_2^3, Dyck words over a three-letter alphabet with letter multiplicity 2. The diagram as a whole represents *all* ways of shuffling two copies of the word abc. An upward step releases a card from the left hand packet, a step downward from the right hand packet. There are twenty walks through this graph, enumerating the five words of D_2^3 in a non-unique way.

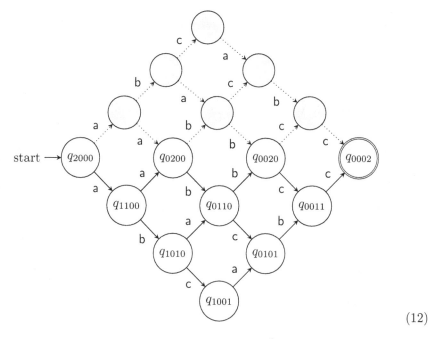

$$(12)$$

The state diagram for the constructed $\text{DAFSA}(D_2^3)$ restricts the graph to the lower half, with the ten named states for the weak compositions of 2 in 4 parts.

One can picture this as a *biased* form of shuffling: the dealer releases a card/letter from the packet in the left hand (move upwards) only if in the lexicographic order it is strictly smaller than the competing card/letter in the right hand packet.

Under the above interpretation, a walk through the graph spells the Yamanouchi word of a rectangular tableau. The $\text{DAFSA}(D_n^d)$ word graph allows another interpretation, discussed in [19], where a walk through the graph now provides a recipe for constructing a tableau (rather than its Yamanouchi word).

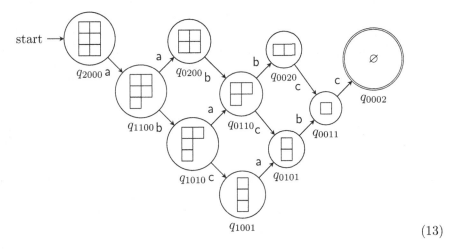

$$(13)$$

States, under this second interpretation, are Young diagrams for all possible partitions of integers 1 to $d \times n$ into maximally d parts of maximal size n. There is a directed edge from partition λ to λ' if λ' is the result of breaking off a *corner* of λ. A corner is a box that appears at the end of a row and a column. We read off the tableau from a walk in this graph as follows. Start at the d-part partition $\lambda_{start} = (n, \ldots, n)$ and finish at the empty partition $\lambda_\varnothing$. At each step, insert the highest integer from $d \times n$ not yet inserted into the corner that is deleted at that edge.

In (13), the Young diagrams corresponding to the states of $\text{DAFSA}(D_2^3)$ have been added. As an example, take the path for the word $\mathsf{abacbc} \in D_2^3$. The corresponding tableau is obtained by successively inserting 6 in the corner of the bottom row, 5 in the corner of the middle row, 4 in the corner of the first column, etc, with as final result the tableau

$$\begin{array}{|c|c|} \hline 1 & 3 \\ \hline 2 & 5 \\ \hline 4 & 6 \\ \hline \end{array}$$

We have the ingredients in place now for a combinatorial interpretation of the hook-length formula (7) counting the rectangular standard Young tableaux of shape $d \times n$ (n positive) or equivalently the words of D_n^d they are in bijection with. It is convenient to model the situation with an n-handed card dealer. We can then represent the shuffle of n copies of the word $a_1 \cdots a_d$ as an n-dimensional diagram with sides of length d; steps forward in each direction are labeled with the alphabet symbols $a_1, \ldots, a_d$ in order, as in the two-dimensional case depicted in (12) above. Now consider the formula in (14) below.

$$\prod_{k=1}^{n} \left[\binom{kd}{d} \middle/ \binom{k+d-1}{d} \right] \tag{14}$$

The product of the numerators counts all outcomes of shuffling, including duplicates: in other words, all monotone paths from the origin $(0, \ldots, 0)$ to $(n, \ldots, n)$. The product of the denominators divides this by the size (number of states) of the successive DAFSA$(D^d_{(k-1)})$ word graphs. The result of the division counts the *distinct* outcomes of shuffling, i.e. gives the cardinality of D^d_n.

It is not hard to see that (14) is equivalent to (7). Indeed, expressing the binomial coefficients $\binom{n}{m}$ in factorial form $\frac{n^{\underline{m}}}{m!}$, with $n^{\underline{m}}$ for the falling factorial powers $n(n-1)\cdots(n-m+1)$, and noticing that $(k+d-1)^{\underline{d}} = k^{\overline{d}}$, we have

$$(14) \quad = \quad \prod_{k=1}^{n} \frac{kd^{\underline{d}}}{(k+d-1)^{\underline{d}}} \quad = \quad \prod_{k=1}^{n} \frac{kd^{\underline{d}}}{k^{\overline{d}}} \quad = \quad \frac{dn!}{\prod_{k=1}^{n} k^{\overline{d}}} \quad = \quad (7) \quad (15)$$

The table below illustrates with the first iterations of the shuffle of the word abc. For increasing k, the first row gives the number of possible positions for the letters of the kth copy of abc among $3k$ available positions. The second row gives the size of the word automaton for $(k-1)$ iterations (pebbles). In the bottom row, the result of the divisions: $(1 \times 20)/4 = 5$, $(5 \times 84)/10 = 42$, etc, i.e. *OEIS* sequence *A005789* of Table 1.

$$\begin{array}{l|llll}
k: & 1\ 2 & 3 & 4 & \ldots \\
\hline
\binom{3k}{3} & 1\ 20 & 84 & 220 & \ldots \\
|\text{DAFSA}^3_{(k-1)}| & 1\ 4 & 10 & 20 & \ldots \\
\hline
& 1\ 5 & 42 & 462 & \ldots
\end{array} \tag{16}$$

The diagram for the third iteration is given in (17). There is a general recipe for obtaining these growing diagrams: draw the paths for $(a_1 \cdots a_d)^n$ (no discontinuity) and $a_1^n \cdots a_d^n$ (maximal discontinuity) and fill in the missing connections.

$$\tag{17}$$

The word graph construction discussed here gives a non-redundant representation of the *finite* sets of elements of D^d_n, given symbol multiplicity n. For the *infinite* languages D^d, the transition function of the DAFSA word graph suggests a natural automaton model in the form of Greibach's [8] partially blind one-way multicounter machines. In these automata (see [9] for discussion) the stack holds a representation of a natural number, which can be incremented, decremented, or tested for zero. A counter is *blind* if it cannot be tested for zero except at the end of the computation as part of the acceptance condition. A *partially* blind counter cannot hold a negative value: the machine blocks on decrementing zero.

D^d then is accepted by a partially blind $(d-1)$-counter machine: on reading symbol a_1, the first counter is incremented; on reading symbol a_d, the last counter is decremented; on reading intermediate symbols a_i ($1 < i < d$), the i-th counter is incremented and the $(i-1)$-th counter decremented. Failure on decrementing zero captures the prefix constraint of Equation (1) on Dyck words.

5   Dyck Words and Irreducible Webs for $\mathfrak{sl}_3$

In §3, we discussed the correspondence between multidimentional Dyck languages and rectangular standard Young tableaux for arbitrary dimension (number of rows) d. In the three-dimensional case, there is a further bijection between $3 \times n$ tableaux (hence: D_n^3 Dyck words) and a class of combinatorial graphs ('webs') underlying Khovanov and Kuperberg's work [13] on combinatorial *spiders*, planar categories describing the invariant space of a tensor product of irreducible representations of a Lie algebra of given rank. The correspondence between webs for rank 3 and D_n^3 Dyck words throws a surprising light on the nature of the discontinuity involved: the $\mathfrak{sl}_3$ webs are in fact *planar* graphs.

Our presentation essentially follows [25] except that where Tymoczko formulates the bijection in terms of tableaux, we rely on §3 and phrase the map directly in terms of the 3D Dyck words corresponding to these tableaux.

A web for the $\mathfrak{sl}_3$ spider category is a planar directed graph embedded in a disk where (i) internal vertices are of degree 3, boundary vertices are of degree 1; (ii) each vertex is either a source (all edges outgoing) or a sink (all edges incoming). A web is irreducible (or non-elliptic) if all internal faces have at least 6 sides. For the webs considered here all boundary vertices are sources.

Let w be a word of D_n^3, i.e. the Yamanouchi word of a $3 \times n$ tableau. To build the web corresponding to w, take the following steps.

(i) Place the letters of $w = w_1, \ldots, w_{3n}$ counterclockwise on the boundary of a disk.

(ii) Create n internal sink vertices; each of these has incoming edges from boundary vertices labeled a, b and c. To determine how these triples are grouped, read w from left to right; for each b, find the a that precedes it most closely and that is not already linked, and direct the outgoing edges for this a and b pair to the same internal sink; then match c's and b's in the same way: for each c, find the b that precedes it most closely and that is not already co-linked with a c and connect them to the same internal sink.

(iii) Step (ii) may lead to crossing edges. If not, we are done. If there are crossing edges, restore planarity by replacing the degree 4 vertex v at their intersection by two degree 3 vertices: a sink v_1 and a source v_2; v_1 receives the two incoming arrows that meet at the intersection point v, v_2 emits the two outgoing edges and has an arrow to v_1.

We refer the reader to [25] for proofs that the outcome of removing crossings is uniquely determined, and independent of the order of their removal, in the case of multiple crossings. Note that the grouping algorithm of step (ii) easily generalizes to alphabets of cardinality greater than three, a point we will return to in §6.

We illustrate with the word abacbc $\in D_2^3$. To avoid a clutter of arrows, we represent sources with dark, sinks with open circles. For step (a), start reading at the one o'clock vertex. Step (ii) here leads to the crossing shown in (b): the edge that connects the first c to the internal vertex that has an incoming edge from the closest b preceding it intersects an edge that was created in linking the second a and b to the same internal sink. Resolving the crossing with the recipe in (iii) leads to the planar web in (c).

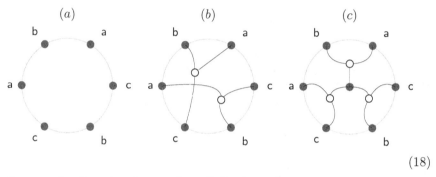

$$(18)$$

For the opposite direction, from web to 3D Dyck word or tableau, we cut the disk, creating an unbounded face f_∞. All faces f are associated with a *depth*, $d(f)$, defined as the minimal number of edges one has to cross to reach the unbounded face, which then has $d(f_\infty) = 0$. Visit the boundary vertices counterclockwise and calculate $d(f_l) - d(f_r)$, the difference between the depth of the face to its left and to its right, looking in the direction of the outgoing edge. These differences have $\{1, 0, -1\}$ as possible outcomes. Interpret these as the three alphabet symbols of D_n^3: read 1 as a, 0 as b and -1 as c.

$$(19)$$

In (19), we illustrate with the web we obtained in (18)(c). We have cut the disk between the one o'clock and the three o'clock vertices, so that the one o'clock vertex is the starting point for the computation of the corresponding

Dyck word, as it was for (18)(c). Faces are annotated with the depth values resulting from that cut. Visiting the vertices counterclockwise, starting from one o'clock, and computing $d(f_l) - d(f_r)$, gives rise to the sequence $1, 0, 1, -1, 0, -1$, which translates as abacbc.

In general, for a word of length dn, there are dn possibilities of cutting the disk, with a maximum of dn distinct words to be read off from these cuts. In (20a) below, the cut is between the vertices at 11 o'clock and one o'clock, making the 11 o'clock vertex the starting point for the computation of the associated Dyck word. Another way of picturing this is as a counterclockwise *rotation* of the boundary vertices. The calculation of $d(f_l) - d(f_r)$ for the boundary vertices starting from 11 o'clock now gives rise to the sequence $1, 1, 0, 0, -1, -1$, which translates as the word aabbcc. We leave it to the reader to verify that the six possible cuts of the web of (19)/(20a) give rise to the two words abacbc, aabbcc. In the case of (20b), a web that doesn't involve any crossing dependencies, the six possible cuts give rise to three words: abcabc, ababcc, aabcbc. The five words of D_2^3, then, are read off from the rotations of the two webs in (20).

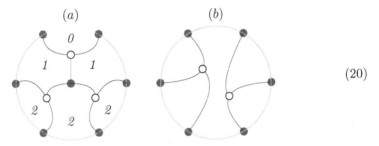

$$(a) \qquad\qquad (b) \tag{20}$$

In [25,20] it is shown that web rotation corresponds to Schützenberger's [23] *jeu de taquin* promotion operation on tableaux. The promotion operation (web rotation in the 3D case) induces an equivalence relation on multidimensional Dyck words that further reduces the apparent diversity of surface forms.

6 Grammars

After this exploration of the combinatorics of multidimensional Dyck languages, let us turn now to their analysis in terms of categorial grammar.

As we saw in the Introduction, D^2, the Dyck language over a two-symbol alphabet, is the language consisting of balanced strings of parentheses, one of the archetypical examples of a context-free language. A grammar with start symbol S and rules $S \longrightarrow a\,S\,b\,S$, $S \longrightarrow \epsilon$ generates D^2 with the empty word included in the language. A *lexicalized* version of this grammar excludes the empty word; each rule is required to introduce a single terminal element, the lexical anchor. The grammar with alphabet $\{a, b\}$, non-terminals $\{S, B\}$, start symbol S and the set of productions P of (21)

$$P: \qquad S \longrightarrow a\,B \mid a\,S\,B \mid a\,B\,S \mid a\,S\,B\,S \qquad ; \qquad B \longrightarrow b \tag{21}$$

is in Greibach normal form (GNF), with the lexical anchor as the leftmost element in a rule's right-hand side. As is well-known, a grammar in GNF format can be directly transformed in a categorial type lexicon associating each terminal symbol with a finite number of types. For the resulting categorial grammar, atomic types are the non-terminals of the rewrite grammar; the only type constructor is '/' which we read left-associatively to economize on parentheses; the rules in P are turned into first-order type assignments for the lexical anchors according to the recipe in (22).

$$\text{LEX}(a) = \{A_0/A_n/\cdots/A_1 \mid (A_0 \longrightarrow aA_1\ldots A_n) \in P\} \tag{22}$$

(or $A_0\, A_n^l\, \cdots\, A_1^l$, in the pregroup type format). In (23) one finds the lexicon for D^2 that results from the transformation of the rules in (21) into categorial types.

$$\text{LEX}(\mathsf{a}) = \{S/B,\ S/B/S,\ S/S/B,\ S/S/B/S\} \qquad \text{LEX}(\mathsf{b}) = \{B\} \tag{23}$$

For Dyck languages D^d with dimensionality $d > 2$, context-free expressivity is no longer sufficient, so one has to turn to generalizations of Lambek's Syntactic Calculus or its simplified pregroup incarnation. There is a choice of options here. The *multimodal* typelogical grammars of the Barcelona School (see Morrill and Valentín, this volume) combine residuated families of concatenation operations $(/, \otimes, \backslash)$ with *wrapping* operations $(\uparrow, \odot, \downarrow)$ for the combination of discontinuous expressions consisting of detached parts. Multimodal typelogical grammar, Utrecht style, extends the language of the Syntactic Calculus with *control features* $\Diamond, \Box$: a residuated pair of *unary* multiplicative operations licensing or blocking structural rules of inference. Bi-Lambek (or Lambek-Grishin) calculus (see [16]) adds duals for the $/, \otimes, \backslash$ operations: a multiplicative *sum* $\oplus$ together with left and right *difference* operations $\oslash, \obslash$; discontinuous dependencies then arise out of the *linear distributivity* principles relating the product and sum operations. Finally, one can move beyond the propositional language of the Syntactic Calculus, and handle the composition of discontinuous expressions in terms of structure variables and first-order quantification over them (as in [17] and Moot, this volume).

The recipient of this volume has shown a certain reluctance to consider any of these extended categorial formalisms (indeed, the title of [15] 'Should Pregroup Grammars be Adorned with Additional Operations?' can be taken as a *rhetorical* question), except for extensions of the formula language with meet and/or join operations, and for grammars based on products of pregroups (see for example Béchet, this volume). So let us here explore these strategies in the case of generalized Dyck languages of higher dimension.

In §5 we discussed a method for grouping the (a,b,c) triples that are interleaved to make up the words of D^3 — a method that generalizes to alphabets of greater cardinality. The crucial observation about these d-symbol alphabets is that for each pair of successive symbols a_i, a_{i+1} $(1 \leq i < d)$, the projections $\pi_{a_i,a_{i+1}}(D^d)$ constitute D^2 Dyck languages over the two-letter alphabets $\{a_i, a_{i+1}\}$. As an example, take the word $w = \mathsf{aabacbbcc} \in D^3$. The projection $\pi_{\mathsf{a,b}}(w)$ removes all occurrences of c, producing aababb $\in D^2$. Similarly, $\pi_{b,c}(w) = $ bcbbcc removes all occurrences of a, resulting in a D^2 Dyck word over the alphabet $\{b, c\}$.

$$(24)$$

The D^d $(d > 2)$ languages, then, can be obtained through the *intersection* of $d - 1$ context-free languages, where each of the constituting languages respects well-nestedness for a pair of successive alphabet symbols and ignores intervening symbols from the rest of the alphabet.

We work this out for $d = 3$; the generalization to alphabets of greater size is straightforward. Consider $G = (\{a,b,c\}, \{S, B, C\}, S, P)$ with the productions P in (25). It recognizes the same a,b patterns as the D^2 grammar with productions (21), but allows each occurrence of the letters a and b to be followed by an arbitrary number of c's.

$$
P: \quad
\begin{aligned}
S &\longrightarrow \mathsf{a}\,B \mid \mathsf{a}\,S\,B \mid \mathsf{a}\,B\,S \mid \mathsf{a}\,S\,B\,S \\
&\mid \mathsf{a}\,C\,B \mid \mathsf{a}\,C\,S\,B \mid \mathsf{a}\,C\,B\,S \mid \mathsf{a}\,C\,S\,B\,S \\
B &\longrightarrow \mathsf{b} \mid \mathsf{b}\,C \\
C &\longrightarrow \mathsf{c} \mid \mathsf{c}\,C
\end{aligned}
\qquad (25)
$$

Similarly, $G' = (\{a,b,c\}, \{S', B', C'\}, S', P')$ with the productions P' in (26) recognizes balanced subsequences of a bracket pair $\{b,c\}$, but allows occurrences of the letters b and c to be preceded by an arbitrary number of a's.

$$
P': \quad
\begin{aligned}
S' &\longrightarrow \mathsf{b}\,C' \mid \mathsf{b}\,C'\,S' \mid \mathsf{b}\,S'\,C' \mid \mathsf{b}\,S'\,C'\,S' \\
&\mid \mathsf{a}\,B'\,C' \mid \mathsf{a}\,B'\,C'\,S' \mid \mathsf{a}\,B'\,S'\,C' \mid \mathsf{a}\,B'\,S'\,C'\,S' \\
B' &\longrightarrow \mathsf{b} \mid \mathsf{a}\,B' \\
C' &\longrightarrow \mathsf{c} \mid \mathsf{a}\,C'
\end{aligned}
\qquad (26)
$$

We now have $D^3 = L(G) \cap L(G')$: the words of $L(G)$ have a and b with equal multiplicity, and at least as many a's as b's in every prefix; the words of $L(G')$ have an equal number of occurrences of b and c, and every prefix counts at least as many b's as c's.

The grammars G and G' are in Greibach normal form, so again, the transformation to categorial lexica is immediate. Intersection of context-free languages can be expressed categorially in a number of ways. One option, investigated in [11], is to enrich the type language with an intersective conjunction $\sqcap$, as originally proposed in [14]. Sequent rules for $\sqcap$ are given below.

$$
\frac{\Gamma, A, \Gamma' \Rightarrow C}{\Gamma, A \sqcap B, \Gamma' \Rightarrow C} \, \sqcap L
\qquad
\frac{\Gamma, A, \Gamma' \Rightarrow C}{\Gamma, A \sqcap B, \Gamma' \Rightarrow C} \, \sqcap L
\qquad
\frac{\Gamma \Rightarrow A \quad \Gamma \Rightarrow B}{\Gamma \Rightarrow A \sqcap B} \, \sqcap R
$$

Under this approach, the lexicon for D^3 assigns type $A \sqcap B$ to an alphabet symbol a iff the lexicon obtained from G by the rule-to-type map (22) assigns A to a

and the lexicon obtained from G' assigns B; the distinguished symbol for the D^3 categorial grammar then becomes $S \cap S'$.

Alternatively, one can work in a product pregroup, where alphabet symbols are associated with a *pair* of pregroup types $\langle p, q \rangle$, with p obtained from G and q from G', and distinguished symbol $\langle S, S' \rangle$. Below in (27) is the computation that accepts aabacbbcc as a word of D^3.

$$
\begin{array}{ccccccccc}
\text{a} & \text{a} & \text{b} & \text{a} & \text{c} & \text{b} & \text{b} & \text{c} & \text{c}
\end{array}
$$

$$
\begin{pmatrix} SB^lS^l \\ S'S''C''B'' \end{pmatrix} \begin{pmatrix} SS^lB^l \\ B'B'' \end{pmatrix} \begin{pmatrix} B \\ B' \end{pmatrix} \begin{pmatrix} SB^lC^l \\ C'C'' \end{pmatrix} \begin{pmatrix} C \\ C' \end{pmatrix} \begin{pmatrix} B \\ S'C''S'' \end{pmatrix} \begin{pmatrix} BC^l \\ S'C'' \end{pmatrix} \begin{pmatrix} CC^l \\ C' \end{pmatrix} \begin{pmatrix} C \\ C' \end{pmatrix} \quad (27)
$$

$$
= \begin{pmatrix} SB^lS^lSS^lB^lBSB^lC^lCBBC^lCC^lC \\ S'S''C''B''B'B''B'C'C''C'S'C''S''S'C''C'C' \end{pmatrix} \leq \begin{pmatrix} S \\ S' \end{pmatrix}
$$

An even simpler pregroup construction obtains if we use the *multiplicative identity* to handle the letters that have to be ignored in checking balance for pairs of alphabet symbols. The type-assignment schema in (28) is the pregroup version of (23), handling successive pairs of alphabet symbols $\{a_i, a_{i+1}\}$ $(1 \leq i < d)$. The non-terminals B_i here introduce the 'closing bracket' a_{i+1} matching the 'opening bracket' a_i.

$$
\text{LEX}_i(a_i) = \{S_i B_i^l, \ S_i B_i^l S_i^l, \ S_i S_i^l B_i^l, \ S_i S_i^l B_i^l S_i^l\} \qquad \text{LEX}_i(a_{i+1}) = \{B_i\} \quad (28)
$$

The construction for the D^3 lexicon $\text{LEX}_{1,2}$ now associates alphabet symbols with pairs of types. The first coordinate checks balance for the $\pi_{\text{a,b}}$ projection; the letter that is to be ignored for this check (c) has the multiplicative identity in this position. The second coordinate, similarly, checks balance for the $\pi_{\text{b,c}}$ projection; the letter a can be ignored in this case.

$$
\begin{aligned}
\text{LEX}_{1,2}(\text{a}) &= \{\langle p, 1 \rangle \mid p \in \text{LEX}_1(\text{a})\} \\
\text{LEX}_{1,2}(\text{b}) &= \{\langle p, q \rangle \mid p \in \text{LEX}_1(\text{b}) \wedge q \in \text{LEX}_2(\text{b})\} \\
\text{LEX}_{1,2}(\text{c}) &= \{\langle 1, q \rangle \mid q \in \text{LEX}_2(\text{c})\}
\end{aligned} \quad (29)
$$

The computation for aabacbbcc $\in D^3$ now takes the form of (30) (distinguished type: $\langle S_1, S_2 \rangle$).

$$
\begin{array}{ccccccccc}
\text{a} & \text{a} & \text{b} & \text{a} & \text{c} & \text{b} & \text{b} & \text{c} & \text{c}
\end{array}
$$

$$
\begin{pmatrix} S_1 B_1^l S_1^l \\ 1 \end{pmatrix} \begin{pmatrix} S_1 S_1^l B_1^l \\ 1 \end{pmatrix} \begin{pmatrix} B_1 \\ S_2 S_2^l B_2^l \end{pmatrix} \begin{pmatrix} S_1 B_1^l \\ 1 \end{pmatrix} \begin{pmatrix} 1 \\ B_2 \end{pmatrix} \begin{pmatrix} B_1 \\ S_2 B_2^l S_2^l \end{pmatrix} \begin{pmatrix} B_1 \\ S_2 B_2^l \end{pmatrix} \begin{pmatrix} 1 \\ B_2 \end{pmatrix} \begin{pmatrix} 1 \\ B_2 \end{pmatrix} \quad (30)
$$

Again, the generalization to alphabets $\{a_1, \ldots, a_d\}$ of size $d > 3$ will be clear. Type assignments are $d - 1$ tuples. The symbol a_1 has 1 in all coordinates except the first, which has the LEX_1 values for a_1; the symbol a_d has 1 in all coordinates except the last, which has the LEX_{d-1} value for a_d; intermediate letters a_j $(1 < j < d)$ serve simultaneously as closing bracket for a_{j-1} and as opening bracket for a_{j+1}: they have LEX_{j-1} and LEX_j values in the relevant coordinates, and 1 elsewhere.

Notice that with this construction, the first and the last alphabet symbols commute. In the case of the LEX$_{1,2}$ type assignments, for example, we have $\langle p, 1 \rangle \circ \langle 1, q \rangle = \langle p, q \rangle = \langle 1, q \rangle \circ \langle p, 1 \rangle$, and indeed substrings ac of D^3 words commute: $u\,\mathrm{ac}\,v$ is in D^3 iff $u\,\mathrm{ca}\,v$ is.

7 Discussion, Conclusion

For discontinuous dependencies of the interleaving type discussed here, Bach [1,2] proposed action-at-a-distance function types $A \mathbin{/\!\!/} B$ ($B \mathbin{\backslash\!\!\backslash} A$): functions producing a result of type A when provided with a B argument *somewhere* to their right (left), ignoring material that might intervene. The constructions of the previous section implement this idea of 'delayed' function application in a rather direct way. They apply to multidimensional Dyck languages with alphabets of arbitrary size, handling d-symbol alphabets by means of type tuples of size $(d - 1)$ (or $(d-1)$ counter machines, if one prefers the automaton model discussed at the end of §4).

In §5 we briefly discussed web rotations and the *jeu de taquin* promotion operation on the $3 \times n$ tableaux corresponding to these webs. The promotion operation is not restricted to the three-dimensional case; it applies to tableaux of arbitrary dimension. This opens up the possibility of viewing Dyck words related by *jeu de taquin* promotion as different surface realisations that can be read off from a more abstract underlying representation. We leave this as a subject for further research.

References

1. Bach, E.: Discontinuous constituents in generalized categorial grammars. In: Proceedings of the 11th Annual Meeting of the North Eastern Linguistics Society, New York, pp. 1–12 (1981)
2. Bach, E.: Some generalizations of categorial grammar. In: Landman, F., Veltman, F. (eds.) Varieties of Formal Semantics: Proceedings of the Fourth Amsterdam Colloquium, September 1982. Groningen-Amsterdam Studies in Semantics, pp. 1–23. Walter de Gruyter (1984)
3. Berglund, M., Björklund, H., Högberg, J.: Recognizing shuffled languages. In: Dediu, A.-H., Inenaga, S., Martín-Vide, C. (eds.) LATA 2011. LNCS, vol. 6638, pp. 142–154. Springer, Heidelberg (2011)
4. Frame, J.S., de, G., Robinson, B., Thrall, R.M.: The hook graphs of the symmetric group. Canad. J. Math 6, 316–325 (1954)
5. Fulton, W.: Young Tableaux: With Applications to Representation Theory and Geometry. London Mathematical Society Student Texts, vol. 35. Cambridge University Press (1997)
6. Gischer, J.: Shuffle languages, petri nets, and context-sensitive grammars. Communications of the ACM 24(9), 597–605 (1981)
7. Gorska, K., Penson, K.A.: Multidimensional Catalan and related numbers as Hausdorff moments. ArXiv e-prints (April 2013)
8. Greibach, S.A.: Remarks on blind and partially blind one-way multicounter machines. Theoretical Computer Science 7(3), 311–324 (1978)

9. Hoogeboom, H., Engelfriet, J.: Pushdown automata. In: Martín-Vide, C., Mitrana, V., Păun, G. (eds.) Formal Languages and Applications. STUDFUZZ, vol. 148, pp. 117–138. Springer, Heidelberg (2004)

10. Joshi, A., Schabes, Y.: Tree-adjoining grammars. In: Rozenberg, G., Salomaa, A. (eds.) Handbook of Formal Languages, pp. 69–123. Springer, Heidelberg (1997)

11. Kanazawa, M.: The Lambek calculus enriched with additional connectives. Journal of Logic, Language and Information 1(2), 141–171 (1992)

12. Kanazawa, M., Salvati, S.: MIX is not a tree-adjoining language. In: Proceedings of the 50th Annual Meeting of the Association for Computational Linguistics, pp. 666–674. Association for Computational Linguistics (2012)

13. Khovanov, M., Kuperberg, G.: Web bases for $\mathfrak{sl}_3$ are not dual canonical. Pacific J. Math 188(1), 129–153 (1999)

14. Lambek, J.: On the calculus of syntactic types. In: Jakobson, R. (ed.) Structure of Language and its Mathematical Aspects, Proceedings of the Symposia in Applied Mathematics XII, pp. 166–178. American Mathematical Society (1961)

15. Lambek, J.: Should pregroup grammars be adorned with additional operations? Studia Logica 87(2-3), 343–358 (2007)

16. Moortgat, M.: Symmetric categorial grammar. Journal of Philosophical Logic 38(6), 681–710 (2009)

17. Moot, R., Piazza, M.: Linguistic applications of first order intuitionistic linear logic. Journal of Logic, Language and Information 10(2), 211–232 (2001)

18. Nederhof, M.-J., Satta, G., Shieber, S.M.: Partially ordered multiset context-free grammars and ID/LP parsing. In: Proceedings of the Eighth International Workshop on Parsing Technologies, Nancy, France, pp. 171–182 (April 2003)

19. Nijenhuis, A., Wilf, H.S.: Combinatorial algorithms for computers and calculators. Computer science and applied mathematics. Academic Press, New York (1978); First ed. published in 1975 under title: Combinatorial algorithms

20. Petersen, T.K., Pylyavskyy, P., Rhoades, B.: Promotion and cyclic sieving via webs. Journal of Algebraic Combinatorics 30(1), 19–41 (2009)

21. Reape, M.: A Logical Treatment of Semi-free Word Order and Bounded Discontinuous Constituency. In: Proceedings of the Fourth Conference on European Chapter of the Association for Computational Linguistics, EACL 1989, Manchester, England, pp. 103–110. Association for Computational Linguistics, Stroudsburg (1989)

22. Salvati, S.: MIX is a 2-MCFL and the word problem in $\mathbb{Z}^2$ is solved by a third-order collapsible pushdown automaton. Rapport de recherche (February 2011)

23. Schützenberger, M.P.: Promotion des morphismes d'ensembles ordonnés. Discrete Mathematics 2(1), 73–94 (1972)

24. Seki, H., Matsumura, T., Fujii, M., Kasami, T.: On multiple context-free grammars. Theoretical Computer Science 88(2), 191–229 (1991)

25. Tymoczko, J.: A simple bijection between standard $3 \times n$ tableaux and irreducible webs for $\mathfrak{sl}_3$. Journal of Algebraic Combinatorics 35(4), 611–632 (2012)

26. Warmuth, M.K., Haussler, D.: On the complexity of iterated shuffle. Journal of Computer and System Sciences 28(3), 345–358 (1984)

Extended Lambek Calculi
and First-Order Linear Logic

Richard Moot

CNRS, LaBRI, University of Bordeaux
Richard.Moot@labri.fr

1 Introduction

The Syntactic Calculus [27] — often simply called the Lambek calculus, L, — is a beautiful system in many ways: Lambek grammars give a satisfactory syntactic analysis for the (context-free) core of natural language and, in addition, it provides a simple and elegant syntax-semantics interface.

However, since Lambek grammars generate only context-free languages [49], there are some well-know linguistic phenomena (Dutch verb clusters, at least if we want to get the semantics right [21], Swiss-German verb clusters [54], etc.) which cannot be treated by Lambek grammars.

In addition, though the syntax-semantics interface works for many of the standard examples, the Lambek calculus does not allow a non-peripheral quantifier to take wide scope (as we would need for sentence (1) below if we want the existential quantifier to have wide scope, the so-called "de re" reading) or non-peripheral extraction (as illustrated by sentence (2) below); see [34, Section 2.3] for discussion.

(1)   John believes someone left.

(2)   John will pick up the package which Mary left here yesterday.

To deal with these problems, several extensions of the Lambek calculus have been proposed. Though this is not the time and place to review them — I recommend [33,34] and the references cited therein for an up-to-date overview of the most prominent extensions; they include multimodal categorial grammar (MMCG,[31]), the Lambek-Grishin calculus (LG,[32]) and the Displacement calculus (D,[46]) — I will begin by listing a number of properties which I consider desirable for such an extension. In essence, these desiderata are all ways of keeping as many of good points of the Lambek calculus as possible while at the same time dealing with the inadequacies sketched above.[1]

---

[1] To the reader who is justifiably skeptical of any author who writes down a list of desiderata, followed by an argument by this same author arguing how well he scores on his own list, I say only that, in my opinion, this list is uncontroversial and at least implicitly shared by most of the work on extensions of the Lambek calculus and that the list still allows for a considerable debate as to *how well* each extension responds to each desideratum as well as discussion about the relative importance of the different items.

C. Casadio et al. (Eds.): Lambek Festschrift, LNCS 8222, pp. 297–330, 2014.
© Springer-Verlag Berlin Heidelberg 2014

1. The logic should have a simple proof theory,
2. generate the mildly context-sensitive languages,
3. have a simple syntax-semantics interface giving a correct and simple account of medial scope for quantifiers and of medial extraction,
4. have a reasonable computational complexity.

None of these desiderata is absolute: there are matters of degree for each of them. First of all, it is often hard to distinguish familiarity from simplicity, but I think that having multiple equivalent proof systems for a single calculus is a sign that the calculus is a natural one: the Lambek calculus has a sequent calculus, natural deduction, proof nets, etc. and we would like its extensions to have as many of these as possible, each formulated in the simplest possible way.

The mildly context-sensitive languages [22] are a family of languages which extend the context-free language in a limited way, and opinions vary as to which of the members of this family is the most appropriate for the description of natural language. Throughout this article, I will only make the (rather conservative and uncontroversial) claim that any extension of the Lambek calculus should at least generate the tree adjoining languages, the multiple context-free languages [53] (the well-nested 2-MCFLs [24] are weakly equivalent to the tree adjoining languages) or the simple, positive range concatenation grammars (sRCG, weakly equivalent to MCFG, [7]).

With respect to the semantics, it generally takes the form of a simple homomorphism from proofs in the source logic to proofs in the Lambek-van Benthem calculus LP (which is multiplicative intuitionistic linear logic, MILL, for the linear logicians), though somewhat more elaborate continuation-based mappings [5,35] have been used as well.

Finally, what counts as reasonable computational complexity is open to discussion as well: since theorem-proving for the Lambek calculus is NP complete [50], I will consider NP-complete to be "reasonable", though polynomial parsing is generally considered a requirement for mildly context-sensitive formalisms [22]. Since the complexity of the logic used corresponds to the *universal* recognition problem in formal language theory, NP completeness is not as bad as it may seem, as it corresponds to the complexity of the universal recognition problem for multiple context-free grammars, which is a prototypical mildly context-sensitive formalism (NP completeness holds when we fix the maximum number of string tuples a non-terminal is allowed to have, non-deleting MCFGs without this restriction are PSPACE complete [23]). In addition, many NP hard grammar formalisms such as LFG and HPSG have very efficient parsers [8,30]. Little is known about fragments of extended Lambek calculi with better complexity (though some partial results can be found in [37,39]). Parsing the Lambek calculus itself is known to be polynomial when we fix the order of formulas [51].

Table 1 gives an overview of the Lambek calculus as well as several of its prominent extensions with respect to the complexity of the universal recognition problem, the class of languages generated and the facilities in the formalism for

Table 1. The Lambek calculus and several of its variants/extensions, together with the complexity of the universal recognition problem, classes of languages generated and the appropriateness of the formalism for handling medial quantifier scope and medial extraction

Calculus	Complexity	Languages	Scope	Extraction
L	NP complete	CFL	−	−
MMCG	PSPACE complete	CSL	+	+
LG	NP complete	$\supseteq$ MCFL	+	−
D	NP complete	$\supseteq$ MCFL	+	+
MILL1	NP complete	$\supseteq$ MCFL	+	+

handling medial quantifier scope and medial extraction. Note that few exact upper bounds for language classes are known.

In this paper, I will present an alternative extension of the Lambek calculus: first-order multiplicative intuitionistic linear logic (MILL1) [15,40]. It generates the right class of languages (MCFG are a subset of the Horn clause fragment, as shown in Section 3.3), and embeds the Displacement calculus (D, as shown in Section 4 and 5). As can be seen in Table 1, it has the lowest complexity class among the different extensions, generates (at least) the right class of languages, but also handles medial scope and medial extraction in a very simple way (as shown already in [40]). In addition, as we will see in Section 2, MILL1 has a very simple proof theory, essentially a resource-conscious version of first-order logic, with a proof net calculus which is a simple extension of the proof nets of multiplicative linear logic [10,15]. Finally, the homomorphism from MILL1 to MILL for semantics consists simply of dropping the first-order quantifiers.

I will also look at the (deterministic, unit-free) Displacement calculus from the perspective of MILL1 and give a translation of D into MILL1, indirectly solving two open problems from [43] by providing a proof net calculus for D and showing that D is NP complete. In addition it is also worth mentioning briefly that the simpler proof theory of MILL1 (ie. proof nets) greatly simplifies the cut elimination proofs of D [46,57]: as for the multiplicative case, cut elimination for MILL1 consists of simple, local conversions with only three distinct cases to verify (axiom, tensor/par and existential/universal).

The remainder of this paper is structured as follows. In the next section, I will briefly introduce MILL1 and its proof theory, including a novel correctness condition for first-order proof nets, which is a simple extension of the contraction criterion from Danos [9]. Section 3 will introduce the Displacement calculus, D, using a presentation of the calculus from [46] which emphasizes the operations on string tuples and, equivalently, on string positions. Section 4 will present a translation from D to MILL1, with a correctness proof in Section 5. Section 6 will briefly mention some other possible applications of MILL1, which include agreement, non-associativity and island constraints, and quantifier scope. Finally, I will reflect on the implications of the results in this paper and give some interesting open problems.

2   MILL1

First-order multiplicative intuitionistic linear logic (MILL1) extends (multiplicative) intuitionistic linear logic with the first-order quantifiers $\exists$ and $\forall$. The first-order multiplicative fragment shares many of the good properties of the propositional fragment: the decision problem is NP complete [28] and it has a simple proof net calculus which is an extension of the proof net calculus for multiplicative linear logic.

Table 2 presents the natural deduction calculus for MILL1, which is without surprises, though readers familiar with intuitionistic logic should note that the $\otimes E$, $\multimap I$ and $\exists E$ rule discharge exactly one occurrence of each of the hypotheses with which it is coindexed.

Table 2. Natural deduction rules for MILL1

$$
\begin{array}{c}
[A]^i[B]^i \\
\vdots \\
\dfrac{A \otimes B \qquad C}{C} \; \otimes E_i \qquad \dfrac{A \quad B}{A \otimes B} \; \otimes I
\end{array}
$$

$$
\begin{array}{c}
[A]^i \\
\vdots \\
\dfrac{A \quad A \multimap B}{B} \; \multimap E \qquad \dfrac{B}{A \multimap B} \; \multimap I
\end{array}
$$

$$
\begin{array}{c}
[A]^i \\
\vdots \\
\dfrac{\exists x.A \quad C}{C} \; \exists E_i^* \qquad \dfrac{A[x := t]}{\exists x.A} \; \exists I
\end{array}
$$

$$
\dfrac{\forall x.A}{A[x := t]} \; \forall E \qquad \dfrac{A}{\forall x.A} \; \forall I^*
$$

* no free occurrences of x in any of the free hypotheses

The presentation of proof nets is (out of necessity) somewhat terse. A more gentle introduction to proof nets can be found, for example in [16,42]. I will present the proof net calculus in three steps, which also form a basic proof search procedure: for a given statement $\Gamma \vdash C$ (with C a formula and Γ a multiset of formulas) we form a *proof frame* by unfolding the formulas according to the logical links shown in the bottom two rows of Table 3, using the negative unfolding for the formulas in Γ and the positive unfolding for the formula C. We then connect the atomic formulas using the axiom link (shown on the top left of the table) until we have found a complete matching of the atomic formulas, forming a *proof structure*. Finally, we check if the resulting proof structure is a *proof net* (ie. we verify if $\Gamma \vdash C$ is derivable) by verifying it satisfies a correctness condition.

As is usual, I will use the following conventions, which will make formulating the proof net calculus simpler.

- dotted binary links are called *par* links, solid binary links are called *tensor* links,
- dotted unary links are called *universal* links, solid unary links are called *existential* links, the bound variables of these links are called universally bound and existentially bound respectively.
- each occurrence of a quantifier link uses a distinct bound variable,
- the variable of a positive $\forall$ and a negative $\exists$ link (ie. the universal links and universally quantified variables) are called its *eigenvariable*,
- following [4], I require eigenvariables of existential links to be used *strictly*, meaning that replacing the eigenvariable throughout a proof with a special, unused constant will *not* result in a proof (in other words, we never unnecessarily instantiate an existentially quantified variable with the eigenvariable of a universal link).

The fact that both par links and universal links are drawn with dotted lines is not a notational accident: one of the fundamental insights of focusing proofs and ludics [2,17] is that these two types of links naturally group together, as do the existential and tensor links, both drawn with solid lines. This property is also what makes the correctness proof of Section 5 work. When it is convenient to refer to the par and universal links together, I will call them *asynchronous* links, similarly I will refer to the existential and tensor links as *synchronous* links (following Andreoli [2]).

In Table 3, the formulas drawn below the link are its conclusions (the axiom link, on the top left of the table, is the only multiple conclusion link, the cut link, on the top right, does not have a conclusion, all logical links have a single conclusion), the formulas drawn above the link are its premisses.

Definition 1. *A* proof structure *is a set of polarized formulas connected by instances of the links shown in Table 3 such that each formula is at most once the premiss of a link and exactly once the conclusion of a link. Formulas which are not the premiss of any link are called the* conclusions *of the proof structure. We say a proof structure with negative conclusions Γ and positive conclusions Δ is a proof structure of the statement $\Gamma \vdash \Delta$.*

Definition 2. *Given a proof structure Π a switching is*
- *for each of the par links a choice of one of its two premisses,*
- *for each of the universal links a choice either of a formula containing the eigenvariable of the link or of the premiss of the link.*

Definition 3. *Given a proof structure Π and a switching s we obtain a correction graph G by*
- *replacing each par link by an edge connecting the conclusion of the link to the premiss selected by s*
- *replacing each universal link by an edge connecting the conclusion of the link to the formula selected by s*

Table 3. Logical links for MILL1 proof structures

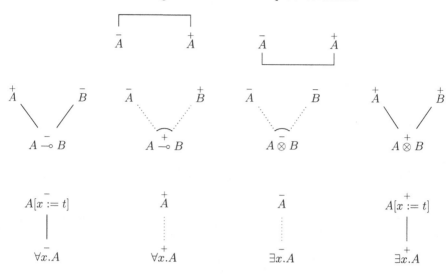

Whereas a proof structure is a graph with some additional structure (paired edges, draw as connected dotted lines for the par links, and "universal" edges, draw as dotted lines) a correction graph is a plain graph as used in graph theory: both types of special edges are replaced by normal edges according to the switching s.

Definition 4. *A proof structure is a proof net iff for all switchings s the corresponding correction graph G is acyclic and connected.*

Remarkably, the proof nets correspond exactly to the provable statements in MILL1 [15].

The basic idea of [40] is very simple: instead of using the well-known translation of Lambek calculus formulas into first-order logic (used for model-theory, see e.g. [11]), we use this same translation to obtain formulas of first-order multiplicative *linear* logic. In this paper, I extend this result to the discontinuous Lambek calculus D, while at the same time sketching some novel applications of the system which correspond more closely to analyses in multimodal categorial grammars.

2.1   A Danos-Style Correctness Condition

Though the correctness condition is conceptually simple, a proof structure has a number of correction graphs which is exponential in the number of asynchronous links, making the correctness condition hard to verify directly (though linear-time algorithms for checking the correctness condition exist in the quantifier-free case, eg. [20,47]).

Here, I present an extension of the correctness condition of [9] to the first-order case, which avoids this exponential complexity. Let G be a proof structure, where each vertex of the proof structure is a assigned the set of eigenvariables which occur in the corresponding formula. Then we have the following contractions.

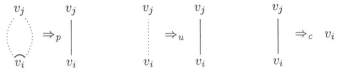

There is one contraction for the par links (p), one contraction for the universal links (u) and a final contraction which contracts components (connected subgraphs consisting only of synchronous, axiom and cut links) to a single vertex (c). The u contraction has the condition that there are no occurrences of the eigenvariable of the universal variable corresponding to the link outside of v_j. The c contraction has as condition that $i \neq j$; it contracts the vertex connecting i and j and the set of eigenvariables of v_i on the right hand side of the contraction corresponds to the set union of the eigenvariables of v_i and v_j on the left hand side of the contraction.

The following proposition is easy to prove using induction on the number of asynchronous links in the proof structure, using a variant of the "splitting par" sequentialization proof of Danos [9]:

Proposition 1. *A proof structure is a proof net iff it contracts to a single vertex using the contractions p, u and c.*

It is also easy to verify that the contractions are confluent, and can therefore be applied in any desired order.

To give an idea of how these contractions are applied, Figure 1 shows (on the left) a proof structure for the underivable statement $\forall x \exists y. f(x,y) \vdash \exists v \forall w. f(w,v)$. In the middle of the figure, we see the proof structure with each formula replaced by the set of its free variables and before any contractions, with the eigenvariables shown next to their universal links. On the right, we see the structure after all c contractions have been applied. It is clear that we cannot apply the u contraction for y, since y occurs at a vertex other than the top vertex. Similarly, we cannot apply the u contraction for w either, meaning the proof structure is not contractible and therefore not a proof net.

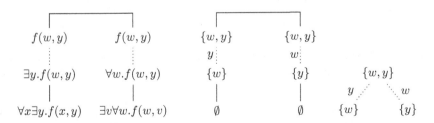

Fig. 1. Proof structure and partial contraction sequence for the underivable statement $\forall x \exists y. f(x,y) \vdash \exists v \forall w. f(w,v)$

Figure 2 shows the proof structure and part of the contraction sequence for the derivable statement $\exists x \forall y. f(x, y) \vdash \forall v \exists w. f(w, v)$. In this case, the structure on the right *does* allow us to perform the u contractions (in any order), producing a single vertex and thereby showing the proof structure is a proof net.

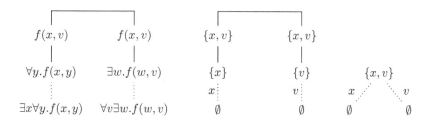

Fig. 2. Proof net and partial contraction sequence for the derivable statement $\exists x \forall y. f(x, y) \vdash \forall v \exists w. f(w, v)$

2.2 Eager Application of the Contractions

Though the contraction condition can be efficiently implemented, when verifying whether or not a given statement is a proof net it is often possible to disqualify partial proof structures (that is, proof structures where only some of the axiom links have been performed). Since the number of potential axiom links is enormous ($n!$ in the worst case), efficient methods for limiting the combinatorial explosion as much as possible are a prerequisite for performing proof search on realistic examples.

The contractions allow us to give a compact representation of the search space by reducing the partial proof structure produced so far. When each vertex is assigned a multiset of literals (in addition to the set of eigenvariables already required for the contractions), the axiom rule corresponds to selecting, if necessary, while unifying the existentially quantified variables, two conjugate literals $+A$ and $-A$ from two different vertices (since the axiom rule corresponds to an application of the c contraction), identifying the two vertices and taking the multiset union of the remaining literals from the two vertices, in addition to taking the set union of the eigenvariables of the vertices. When the input consists of (curried) Horn clauses, each vertex will correspond to a Horn clause; therefore this partial proof structure approach generalizes resolution theorem proving. However, it allows for a lot of freedom in the strategy of literal selection, so we can apply "smart backtracking" strategies such as selecting the literal which has the smallest number of conjugates [36,38]. The contraction condition immediately suggest the following.

– never connect a literal to a descendant or an ancestor (generalizes "formulas from different vertices" for the Horn clause case); failure to respect this constraint will result in a cyclic proof structure,

– if the premiss of an asynchronous link is a leaf with the empty set of literals, then we must be able to contract it immediately ; failure to respect this constraint will result in a disconnected proof structure.
– similarly, if an isolated vertex which is not the only vertex in the graph has the empty set of literals, then the proof structure is disconnected.

3 The Displacement Calculus

The Displacement calculus [46] is an extension of the Lambek calculus using tuples of strings as their basic units. Unless otherwise noted, I will restrict myself Displacement calculus without the identity elements, the synthetic connectives (though see the discussion in Section 4.2 on how some of the synthetic connectives can be included) or the non-deterministic connectives.

3.1 String Tuples

Whereas the Lambek calculus is the logic of strings, several formalisms are using *tuples* of strings as their basic units (eg. MCFGs, RCGs).

In what follows I use s, $s_0, s_1, \ldots, s', s'', \ldots$ to refer to *simple* strings (ie. the 1-tuples) with the constant ϵ for the empty string. The letters t, u, v etc. refer to i-tuples of strings for $i \geq 1$. I will write a i-tuple of strings as $s_1, \ldots, s_i$, but also (if $i \geq 2$) as s_1, t or t', s_i where t is understood to be the string tuple $s_2, \ldots, s_i$ and t' the string tuple $s_1, \ldots, s_{i-1}$, both $(i-1)$-tuples.

The basic operation for simple strings is concatenation. How does this operation extend to string tuples? For our current purposes, the natural extension of concatenation to string tuples is the following

$$(s_1, \ldots, s_m) \circ (s'_1, \ldots, s'_n) = s_1, \ldots, s_m s'_1, \ldots, s'_n$$

where $s_m s'_1$ is the string concatenation of the two simple strings s_m and s'_1. In other words, the result of concatenating an m-tuple t and an n-tuple u is the $n + m - 1$ tuple obtained by first taking the first $m - 1$ elements of t, then the simple string concatenation of the last element of t with the first element of u and finally the last $n - 1$ elements of u. When both t and u are simple strings, then their concatenation is the string concatenation of their single element.[2] In what follows, I will simply write tu for the concatenation of two string tuples t and u and $u[t]$ to abbreviate $u_1 t u_2$.

3.2 Position Pairs

As is common in computational linguistics, it is sometimes more convenient to represent a simple string as a pair of string positions, the first element of the pair representing the leftmost string position and the second element its rightmost

---

[2] Another natural way to define concatenation is as point-wise concatenation of the different elements of two (equal-sized) tuples, as done by Stabler [56].

position. These positions are commonly represented as integers (to make the implicit linear precedence relation more easily visible). Likewise, we can represent an n-tuple of strings as a $2n$ tuple of string positions. This representation has the advantage that it makes string concatenation trivial: if x_0, x_1 is a string starting at position x_0 and ending at position x_1 and x_1, x_2 is a string starting at position x_1 and ending at position x_2 then the concatenation of these two strings is simply x_0, x_2 (this is the familiar difference list concatenation from Prolog [52]).

Definition 5. *We say a grammar is* simple in the input string *if for each input string $w_1, \ldots, w_n$ we have that w_i spans positions $i, i + 1$.*

Much of the work in parsing presupposes grammars are simple in the input string [48], since it makes the definition of the standard parsing algorithms much neater. However, the original construction of Bar-Hillel et al. [3] on which it is based is much more general: it computes the intersection of a context-free grammar and a finite-state automaton (FSA), where each non-terminal is assigned an input state and an output state of the FSA. For grammars which are not simple in the input string, this FSA can have self-loops and complex cycles, whereas the input string for a simple grammar is an FSA with a simple, deterministic linear path as shown in the example below. With the exception of Section 4.2, where I discusses the possibility of abandoning this constraint, the grammars I use will be simple in the input string.

$$1 \xrightarrow{\text{Jan}} 2 \xrightarrow{\text{Henk}} 3 \xrightarrow{\text{Cecilia}} 4 \xrightarrow{\text{de}} 5 \xrightarrow{\text{nijlpaarden}} 6 \xrightarrow{\text{zag}} 7 \xrightarrow{\text{helpen}} 8 \xrightarrow{\text{voeren}} 9$$

Suppose "nijlpaarden" (hippos) above is assigned the category n, for noun. Incorporating its string positions produces $n(5, 6)$. It gets more interesting with the determiner "de" (the): we assign it the formula $\forall x.n(5, x) \multimap np(4, x)$, which says that whenever it finds an n to its immediate right (starting at position 5 and ending at any position x) it will return an np from position 4 to this same x (this is the MILL1 translation of np/n at position $4, 5$). In a chart parser, we would indicate this by adding an np arc from 4 to 6. There is an important difference with a standard chart parser though: since we are operating in a resource-conscious logic, we know that in a correct proof each rule is used exactly once (though their order is only partially determined).

Figure 3 shows the three elementary string operations of the Displacement calculus both in the form of operations of string tuples and in the form of operations of string position pairs.

Concatenation takes an i-tuple t (shown in the top of the figure as the white blocks, with corresponding string positions $x_0, \ldots, x_n$ for $n = 2i - 1$) and a j-tuple u (shown in the top of the figure as the gray blocks, with corresponding string positions $x_n, \ldots, x_{n+m}$ for $m = 2j - 1$) and the resulting concatenation tu (with the last element of t concatenated to the first element of u, indicated as the gray-white block x_{n-1}, x_{n+1}; x_n is not a string position in the resulting $i + j - 1$-tuple tu, which consists of the string positions $x_0, \ldots, x_{n-1}, x_{n+1}, \ldots, x_{n+m}$.

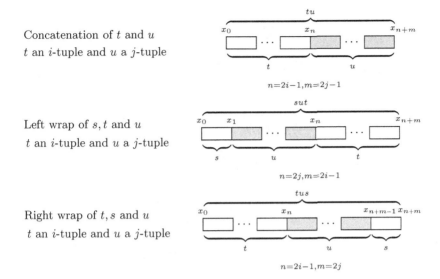

Concatenation of t and u
t an i-tuple and u a j-tuple

Left wrap of s, t and u
t an i-tuple and u a j-tuple

Right wrap of t, s and u
t an i-tuple and u a j-tuple

Fig. 3. String operations and their equivalent string positions operations

Left wrap takes an $i + 1$-tuple s, t (with s a simple string and t an i-tuple, with string positions $x_0, x_1, x_n, \ldots, x_{n+m}$) and a j-tuple u (with string positions $x_1, \ldots, x_n$) and wraps s, t around u producing and $i + j - 1$-tuple sut with string positions $x_0, x_2, \ldots, x_{n-1}, x_{n+1}, \ldots, x_{n+m}$, with positions x_1 and x_n removed because of the two concatenations.

Symmetrically, *right wrap* takes an $i + 1$-tuple t, s (with s a simple string and t an i-tuple) and a j-tuple u and wraps t, s around u producing tus.

Given these operations, the proof rules for D are simple to state. I give a notational variant of the natural deduction calculus of [46]. As usual, natural deduction proofs start with a hypothesis $t : A$ (for $t : A$ an entry the lexicon of the grammar, in which case t is a lexical constant, or for a hypothesis discharged by the product elimination and implication introduction rules, in which case t is an appropriate n-tuple). In each case the string tuple t is unique in the proof.

For a given (sub-)proof, the *active* hypotheses are all hypotheses which have not been discharged by a product elimination of implication introduction rule in this (sub-)proof.

For the logical rules, we can see that the different families of connectives correspond to the three basic string tuple operations: with concatenation for $/$, $\bullet$ and $\backslash$ (the rules are shown in Figure 4 and Figure 7 with the corresponding string tuples), left wrap for $\uparrow_>$, $\odot_>$ and $\downarrow_>$ (shown in Figure 5 and Figure 8) and right wrap for $\uparrow_<$, $\odot_<$ and $\downarrow_<$ (shown in Figure 6 and Figure 9).

In the discontinuous Lambek calculus, we define the *sort* of a formula F, written $s(F)$ as the number of items in its string tuple minus 1. Given sorts for the atomic formulas, we compute the sort of a complex formula as shown

$$\frac{t:A \quad u:A \backslash C}{tu:C} \backslash E \qquad \begin{array}{c} [t:A]^i \\ \vdots \\ \dfrac{tu:C}{u:A \backslash C} \backslash I_i \end{array}$$

$$\frac{t:C/B \quad u:B}{tu:C} /E \qquad \begin{array}{c} [u:B]^i \\ \vdots \\ \dfrac{tu:C}{t:C/B} /I_i \end{array}$$

$$\frac{t:A \bullet B \qquad u[t_1 t_2]:C}{u[t]:C} \bullet E_i \qquad \frac{t:A \quad u:B}{tu:A \bullet B} \bullet I$$

with hypotheses $[t_1:A]^i \quad [t_2:B]^i$

Fig. 4. Proof rules – Lambek calculus

$$\frac{s,t:A \quad u:A \downarrow_{>} C}{sut:C} \downarrow_{>} E \qquad \begin{array}{c} [s,t:A]^i \\ \vdots \\ \dfrac{sut:C}{u:A \downarrow_{>} C} \downarrow_{>} I_i \end{array}$$

$$\frac{s,t:C \uparrow_{>} B \quad u:B}{sut:C} \uparrow_{>} E \qquad \begin{array}{c} [u:B]^i \\ \vdots \\ \dfrac{sut:C}{s,t:C \uparrow_{>} B} \uparrow_{>} I_i \end{array}$$

$$\frac{t:A \odot_{>} B \qquad u[st_1 t_2]:C}{u[t]:C} \odot_{>} E_i \qquad \frac{s,t:A \quad u:B}{sut:A \odot_{>} B} \odot_{>} I$$

with hypotheses $[s,t_2:A]^i \quad [t_1:B]^i$

Fig. 5. Proof rules — leftmost infixation,extraction

in Table 4 (the distinction between the left wrap and right wrap connectives is irrelevant for the sorts).

3.3   MILL1 and Multiple Context-Free Grammars

It is fairly easy to see that MILL1 generates (at least) the multiple context-free languages (or equivalently, the languages generated by simple, positive range concatenation grammars [7]) by using a lexicalized form of the grammars as defined below.

Definition 6. *A grammar is* lexicalized *if each grammar rule uses exactly one non-terminal symbol.*

Table 4. Computing the sort of a complex formula given the sort of its immediate subformulas

$$s(A \bullet B) = s(A) + s(B)$$
$$s(A \setminus C) = s(C) - s(A)$$
$$s(C \mathbin{/} B) = s(C) - s(B)$$

$$s(A \odot B) = s(A) + s(B) - 1$$
$$s(A \downarrow C) = s(C) + 1 - s(A)$$
$$s(C \uparrow B) = s(C) + 1 - s(B)$$

$$
\frac{t,s : A \quad u : A \downarrow_< C}{tus : C} \ \downarrow_< E
\qquad
\frac{\begin{array}{c}[t,s : A]^i \\ \vdots \\ tus : C\end{array}}{u : A \downarrow_< C} \ \downarrow_< I_i
$$

$$
\frac{t,s : C \uparrow_< B \quad u : B}{tus : C} \ \uparrow_< E
\qquad
\frac{\begin{array}{c}[u : B]^i \\ \vdots \\ tus : C\end{array}}{t,s : C \uparrow_< B} \ \uparrow_< I_i
$$

$$
\frac{t : A \odot_< B \quad u[t_1 t_2 s] : C}{u[t] : C} \ \odot_< E_i
\qquad
\frac{t,s : A \quad u : B}{tus : A \odot_< B} \ \odot_< I
$$

$$\begin{array}{cc}[t_1, s : A]^i & [t_2 : B]^i\end{array}$$

Fig. 6. Proof rules — rightmost infixation,extraction

Lexicalization is one of the principal differences between traditional phrase structure grammars and categorial grammars: categorial grammars generally require a form of lexicalization, whereas phrase structure grammars do not. The most well-known lexicalized form is the Greibach normal form for context-free grammars [19]. Wijnholds [58] shows that any (ϵ-free) simple, positive range concatenation grammar has a lexicalized grammar generating the same language (see also [55])[3]. Since ranges are simply pairs of non-negative integers (see Section 3.2 and [7]) these translate directly to Horn clauses in MILL1. The following rules are therefore both a notational variant of a (lexicalized) MCFG/sRCG and an MILL1 lexicon (corresponding to the verbs of the example on page 306).

---

[3] Wijnholds [58] and Sorokin [55] also given translations between MCFG/sRCG and D, to which we will return in Section 4.2.

$$\forall x_0 x_1 x_2 x_3 . np(x_0, x_1) \otimes np(x_1, x_2) \otimes \mathit{inf}(x_2, 6, 7, x_3) \multimap s(x_0, x_3) \qquad \mathit{zag}$$

$$\forall x_0 x_1 x_2 x_3 . np(x_0, x_1) \otimes \mathit{inf}(x_1, x_2, 8, x_3) \multimap \mathit{inf}(x_0, x_2, 7, x_3) \quad \mathit{helpen}$$

$$\forall x_0 x_1 . np(x_0, x_1) \multimap \mathit{inf}(x_0, x_1, 8, 9) \qquad \mathit{voeren}$$

In Section 4.2, we will see how we can obtain (the Curried versions of) these formulas via a translation of D and provide the corresponding MILL1 proof net.

4   Translations

The proof rules and the corresponding string tuple operations (shown in Figures 7, 8 and 9) suggest the translation shown in Table 5 of D formulas into MILL1 formulas. It is an extension of the translation of Lambek calculus formulas of [40], while at the same time extending the translation of [44,12] for the simple displacement calculus (1-D, where all formulas are of sort ≤ 1). Fadda [12] also gives a proof net calculus for the simple displacement calculus, which is a special case of the one presented here.

The reader intimidated by the number variable indices in Table 5 is invited to look at Figures 7, 8 and 9 for the correspondence between the string position numbers and the string components of the different formulas in the translation. Section 4.1 will illustrate the translation using some examples, whereas Section 5

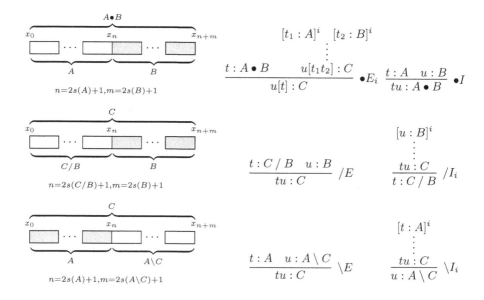

Fig. 7. String positions – Lambek calculus

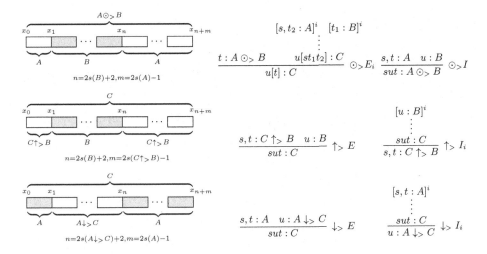

Fig. 8. String positions – leftmost infix/extraction

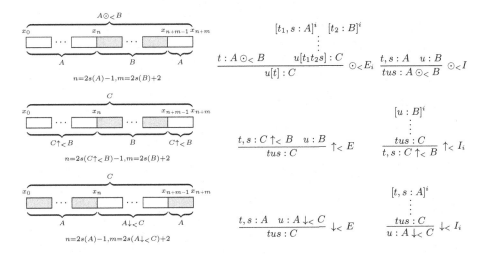

Fig. 9. String positions – rightmost infix/extraction

will make the correspondence with the rules from the Displacement calculus more precise.

Note: the sequence $x_i, \ldots, x_i$ is of course simply the unit sequence x_i whereas the sequence $x_i, \ldots, x_{i-1}$ is the empty sequence.

If there are at most two string tuples, both $C \uparrow_> B$ (Equation 5 with $n = 2$, $m = 1$, remembering that $x_2, \ldots, x_{n-1} \equiv x_2, \ldots, x_1$ which is equivalent to the empty sequence of string positions and the empty sequence of quantifier prefixes,

Table 5. Translation of D formulas to MILL1 formulas

(1)
$$\|A \bullet B\|^{x_0,\dots,x_{n-1},x_{n+1},\dots,x_{n+m}} =$$
$$\exists x_n \|A\|^{x_0,\dots,x_n} \otimes \|B\|^{x_n,\dots,x_{n+m}} \left.\right\} n=2s(A)+1, m=2s(B)+1$$

(2)
$$\|C \mathbin{/} B\|^{x_0,\dots,x_n} =$$
$$\forall x_{n+1},\dots,x_{n+m} \|B\|^{x_n,\dots,x_{n+m}} \multimap$$
$$\|C\|^{x_0,\dots,x_{n-1},x_{n+1},\dots,x_{n+m}} \left.\right\} n=2s(C/B)+1, m=2s(B)+1$$

(3)
$$\|A \backslash C\|^{x_n,\dots,x_{n+m}} =$$
$$\forall x_0,\dots,x_{n-1} \|A\|^{x_0,\dots,x_n} \multimap$$
$$\|C\|^{x_0,\dots,x_{n-1},x_{n+1},\dots,x_{n+m}} \left.\right\} n=2s(A)+1, m=2s(A\backslash C)+1$$

(4)
$$\|A \odot_> B\|^{x_0,x_2,\dots,x_{n-1},x_{n+1},\dots,x_{n+m}} =$$
$$\exists x_1,x_n \|A\|^{x_0,x_1,x_n,\dots,x_{n+m}} \otimes \|B\|^{x_1,\dots,x_n} \left.\right\} n=2s(B)+2, m=2s(A)-1$$

(5)
$$\|C \uparrow_> B\|^{x_0,x_1,x_n,\dots,x_{n+m}} =$$
$$\forall x_2,\dots,x_{n-1} \|B\|^{x_1,\dots,x_n} \multimap$$
$$\|C\|^{x_0,x_2,\dots,x_{n-1},x_{n+1},\dots,x_{n+m}} \left.\right\} n=2s(B)+2, m=2s(C\uparrow_> B)-1$$

(6)
$$\|A \downarrow_> C\|^{x_1,\dots,x_n} =$$
$$\forall x_0,x_{n+1},\dots,x_{n+m} \|A\|^{x_0,x_1,x_n,\dots,x_{n+m}} \multimap$$
$$\|C\|^{x_0,x_2,\dots,x_{n-1},x_{n+1},\dots,x_{n+m}} \left.\right\} n=2s(A\downarrow_> C)+2, m=2s(A)-1$$

(7)

$$\|A \odot_< B\|^{x_0,\dots,x_{n-1},x_{n+1},\dots,x_{n+m-2},x_{n+m}} =$$
$$\exists x_n,x_{n+m-1} \|A\|^{x_0,\dots,x_n,x_{n+m-1},x_{n+m}} \otimes \|B\|^{x_n,\dots,x_{n+m-1}} \left.\right\} n=2s(A)-1, m=2s(B)+2$$

(8)
$$\|C \uparrow_< B\|^{x_0,\dots,x_n,x_{n+m-1},x_{n+m}} =$$
$$\forall x_{n+1},\dots,x_{n+m-2} \|B\|^{x_n,\dots,x_{n+m-1}} \multimap$$
$$\|C\|^{x_0,\dots,x_{n-1},x_{n+1},\dots,x_{n+m-2},x_{n+m}} \left.\right\} n=2s(C\uparrow_> B)-1, m=2s(B)+2$$

(9)
$$\|A \downarrow_< C\|^{x_n,\dots,x_{n+m-1}} =$$
$$\forall x_0,\dots,x_{n-1},x_{n+m} \|A\|^{x_0,\dots,x_n,x_{n+m-1},x_{n+m}} \multimap$$
$$\|C\|^{x_0,\dots,x_{n-1},x_{n+1},\dots,x_{n+m-2},x_{n+m}} \left.\right\} n=2s(A)-1, m=2s(A\downarrow_> C)+2$$

and that $x_{n+1}, \ldots, x_{n+m} \equiv x_3, \ldots, x_3 \equiv x_3$) and $C \uparrow_< B$ (Equation 8 with $n = 1$, $m = 2$) translate to the following

$$\|C \uparrow B\|^{x_0, x_1, x_2, x_3} = \|B\|^{x_1, x_2} \multimap \|C\|^{x_0, x_3}$$

Similarly, it is easy to verify that both $A \downarrow_> C$ (Equation 6 with $n = 2$, $m = 1$, remember that $x_2, \ldots x_{n-1} \equiv x_2, \ldots, x_1$ and therefore equal to the empty sequence and that $x_{n+1}, \ldots, x_{n+m} \equiv x_3, \ldots, x_3 \equiv x_3$) and $A \downarrow_< C$ (Equation 9 with $n = 1$, $m = 2$) produce the following translation for D formulas with at most two string tuples.

$$\|A \downarrow C\|^{x_1, x_2} = \forall x_0, x_3 \|A\|^{x_0, x_1, x_2, x_3} \multimap \|C\|^{x_0, x_3}$$

In the Lambek calculus, all sorts are zero, therefore instantiating Equation 3 with n=1, m=1 produces the following

$$\|A \backslash C\|^{x_1, x_2} = \forall x_0 \|A\|^{x_0, x_1} \multimap \|C\|^{x_0, x_2}$$

and therefore has the translation of [40] as a special case.

4.1 Examples

As an illustration, let's look at the formula unfolding of $((vp \uparrow vp)/vp)\backslash(vp \uparrow vp)$, which is the formula for "did" assigned to sentences like

(3)   John slept before Mary did.

by [46]. This lexical entry for "did" is of sort 0 and therefore has two string positions (I use 4 and 5) to start off its translation. However, since both direct subformulas are of sort 1, these subformulas have four position variables each. Applying the translation for $\backslash$ shown in Equation 3 with $n = 3$ ($= 2s((vp \uparrow vp)/vp) + 1$), $m = 1$ (the sort of the complete formula being 0) gives us the following partial translation.

$$\forall x_0 x_1 x_2 \|(vp \uparrow vp)/vp\|^{x_0, x_1, x_2, 4} \multimap \|vp \uparrow vp\|^{x_0, x_1, x_2, 5}$$

I first translate the leftmost subformula, which is of sort 1, and apply the / rule (Equation 2) with $n = 3$ ($= 2s((vp \uparrow vp)/vp) + 1$) and $m = 1$ ($= 2s(vp) + 1$) giving the following partial translation.

$$\forall x_0 x_1 x_2 [\forall x_3 [\|vp\|^{4, x_3} \multimap \|vp \uparrow vp\|^{x_0, x_1, x_2, x_3}] \multimap \|vp \uparrow vp\|^{x_0, x_1, x_2, 5}]$$

Applying the translation rule for $C \uparrow B$ (Equation 5) twice produces.

$$\forall x_0 x_1 x_2 [\forall x_3 [\|vp\|^{4, x_3} \multimap \|vp\|^{x_1, x_2} \multimap \|vp\|^{x_0, x_3}] \multimap \|vp\|^{x_1, x_2} \multimap \|vp\|^{x_0, 5}]$$

Figure 10 on the facing page shows a proof net for the complete sentence — slightly abbreviated, in that not all vp's have been expanded and that the existential links and the corresponding substitutions have not been included in the figure.

The intelligent backtracking solution of Section 2.2 (and [38]) guarantees that at each step of the computation we can make a deterministic choice for literal selection, though the reader is invited to try and find a proof by hand to convince himself that this is by no means a trivial example!

As a slightly more complicated example translation, which depends on the distinction between left wrap and right wrap, Morrill et al. [46] give the following formula for an object reflexive:

$$((vp \uparrow_> np) \uparrow_< np) \downarrow_< (vp \uparrow_> np)$$

Translating the $\downarrow_<$ connective, with input positions 3 and 4 using Equation 9 with $n = 3$ (since $s((vp \uparrow_> np) \uparrow_< np) = 2$) and $m = 2$ gives the following partial translation.

$$\forall x_0, x_1, x_2, x_5 \| (vp \uparrow_> np) \uparrow_< np \|^{x_0, x_1, x_2, 3, 4, x_5} \multimap \| vp \uparrow_> np \|^{x_0, x_1, x_2, x_5}$$

Translating the $\uparrow_<$ connective using Equation 8 with $n = 3$ and $m = 2$ gives.

$$\forall x_0, x_1, x_2, x_5 [\| np \|^{3,4} \multimap \| vp \uparrow_> np \|^{x_0, x_1, x_2, x_5}] \multimap \| vp \uparrow_> np \|^{x_0, x_1, x_2, x_5}$$

Finally, unfolding the two $\uparrow_>$ connectives (using Equation 5) gives.

$$\forall x_0, x_1, x_2, x_5 [np(3, 4) \multimap np(x_1, x_2) \multimap \| vp \|^{x_0, x_5}] \multimap np(x_1, x_2) \multimap \| vp \|^{x_0, x_5}$$

Indicating that an object reflexive described by the given formula takes a ditransitive verb (with a first object argument spanning the input positions $3 - 4$ of the reflexive and a second without constraints on the position) to produce a transitive verb, a vp still missing an np spanning the positions of the second np of the ditransitive verb, which corresponds to the intuitive meaning of the lexical entry.

4.2 Synthetic Connectives

Morrill et al. [46] introduce the synthetic connectives[4] for the simple displacement calculus 1-D (with formulas of sort ≤ 1), whereas Valentín [57] presents

---

[4] Note that from the point of view of MILL1, *all* D-connectives are synthetic MILL1-connectives, that is, combinations of a series of quantifiers and a binary connective, as can already be seen from the translation in Table 5 on page 312; we will return to this point in Section 5. The synthetic connectives of D are combinations of a binary connective and an identity element. The idea for both is essentially the same: to treat a combination of rules as a single rule, which can be added to the logic as a conservative extension.

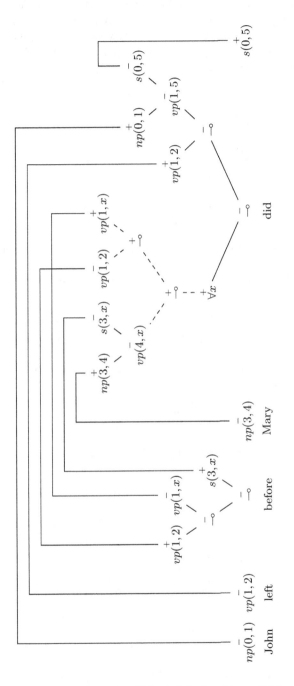

Fig. 10. Proof net for "John left before Mary did."

their natural extension to D (as well as non-deterministic versions of these connectives, which will not be treated here). The synthetic connectives can be seen as abbreviations of combinations of a connective and an identity element (I denoting the empty string ϵ and J denoting the pair of empty strings ϵ, ϵ) as shown in the list below.

$$\begin{aligned}
{}^{\vee}A &=_{def} A \uparrow I & &\text{Split} \\
{}^{\wedge}A &=_{def} A \odot I & &\text{Bridge} \\
\rhd^{-1}A &=_{def} J \setminus A & &\text{Right projection} \\
\rhd A &=_{def} J \bullet A & &\text{Right injection} \\
\lhd^{-1}A &=_{def} A / J & &\text{Left projection} \\
\lhd A &=_{def} A \bullet J & &\text{Left injection}
\end{aligned}$$

Figures 11 and 12 show the proof rules for leftmost bridge/split and right projection/injection (the proof rules for left projection and injection as well as the proof rules for rightmost bridge and split are symmetric).

$$\frac{s,t : {}^{\vee}A}{st : A} \ {}^{\vee}E \qquad \frac{st : A}{s,t : {}^{\vee}A} \ {}^{\vee}I$$

$$\begin{array}{c} s,t' : A \\ \vdots \\ \dfrac{t : {}^{\wedge}A \quad u[st'] : C}{u[t] : C} \ {}^{\wedge}E \qquad \dfrac{s,t : A}{st : {}^{\wedge}A} \ {}^{\wedge}I \end{array}$$

Fig. 11. Proof rules — leftmost split, bridge

$$\frac{t : \rhd^{-1}A}{\epsilon, t : A} \ \rhd^{-1}E \qquad \frac{t : A}{\epsilon, t : \rhd^{-1}A} \ \rhd^{-1}I$$

$$\begin{array}{c} t : A \\ \vdots \\ \dfrac{v : \rhd A \quad u, tu' : C}{uvu' : C} \ \rhd E \qquad \dfrac{t : A}{\epsilon, t : \rhd A} \ \rhd I \end{array}$$

Fig. 12. Proof rules — right projection, injection

The synthetic connectives are translated as follows (only the leftmost split and wedge are shown, the rightmost versions are symmetric in the variables):

$$(10) \qquad \|\check{}A\|^{x_0,x_1,x_1,x_2,\ldots,x_n} = \quad \|A\|^{x_0,x_2,\ldots,x_n}$$

$$(11) \qquad \|\hat{}A\|^{x_0,x_2,\ldots,x_n} = \exists x_1.\|A\|^{x_0,x_1,x_1,x_2,\ldots,x_n}$$

$$(12) \qquad \|\rhd A\|^{x_0,x_0,x_1,\ldots,x_n} = \quad \|A\|^{x_1,\ldots,x_n}$$

$$(13) \qquad \|\rhd^{-1} A\|^{x_1,\ldots,x_n} = \forall x_0.\|A\|^{x_0,x_0,x_1,\ldots,x_n}$$

$$(14) \qquad \|\lhd A\|^{x_0,\ldots,x_n,x_{n+1},x_{n+1}} = \quad \|A\|^{x_0,\ldots,x_n}$$

$$(15) \qquad \|\lhd^{-1} A\|^{x_0,\ldots,x_n} = \forall x_{n+1}.\|A\|^{x_0,\ldots,x_n,x_{n+1},x_{n+1}}$$

In [46], the bridge connective appears exclusively in (positive) contexts $\hat{}(A \uparrow B)$ where it translates as.

$$\|\hat{}(A \uparrow B)\|^{x_0,x_2} = \exists x_1.\|A \uparrow B\|^{x_0,x_1,x_1,x_2}$$
$$= \exists x_1.[\|B\|^{x_1,x_1} \multimap \|A\|^{x_0,x_2}]$$

The resulting formula indicates that it takes a B argument spanning the empty string (anywhere) to produce an A covering the original string position x_0 and x_2. Intuitively, this formalizes (in positive contexts) an A constituent with a B trace. The final translation is positive subformula of the extraction type used in [40].

The split connective ($\check{}$, but also $\lhd$ and $\rhd$) is more delicate, since it identifies string position variables. This can force the identification of variables, which means that direct application of the translation above can produce formulas which have "vacuous" quantifications, though this is not harmful (and these are easily removed in a post-processing step if desired). However, this identification of variables means that the grammars are no longer necessarily simple in the input string as discussed in Section 3.2. As an example, unfolding the formula below (which is patterned after the formula for "unfortunately" from [46]) with input variables x_i and x_j forces us to identify x_i and x_j as shown below, hence producing a self-loop in the input FSA.

$$\|\check{}A \downarrow B\|^{x_i,x_i} = \forall x_0, x_2.\|\check{}A\|^{x_0,x_i,x_i,x_2} \multimap \|B\|^{x_0,x_2}$$
$$= \forall x_0, x_2.\|A\|^{x_0,x_2} \multimap \|B\|^{x_0,x_2}$$

Intuitively, this translation indicates that a formula of the form $\check{}A \downarrow B$ takes its A argument at any span of the string and produces a B at the same position, with the complete formula spanning the empty string. It is, in essence, a translation of the (commutative) linear logic or LP implication into our current context. The MIX language can easily be generated using this property [45].

It is easy to find complex formulas which, together with their arguments, produce a complex cycle. The following formula spans the empty string after it combines with its C argument.

$$\|(\check{}A \downarrow B)/C\|^{x_i,x_j} = \forall x_1 \|C\|^{x_j,x_1} \multimap \|\check{}A \downarrow B\|^{x_i,x_1}$$
$$= \forall x_1 [\|C\|^{x_j,x_1} \multimap \forall x_0, x_2. [\|\check{}A\|^{x_0,x_i,x_1,x_2} \multimap \|B\|^{x_0,x_2}]]$$
$$= \forall x_1 [\|C\|^{x_j,x_i} \multimap \forall x_0, x_2. [\|A\|^{x_0,x_2} \multimap \|B\|^{x_0,x_2}]]$$
$$= \|C\|^{x_j,x_i} \multimap \forall x_0, x_2. [\|A\|^{x_0,x_2} \multimap \|B\|^{x_0,x_2}]$$

The final line in the equation simply removes the x_1 quantifier. Since there are no longer any occurrences of the x_1 variable in the rest of the formula, this produces the equivalent formula shown. The translation specifies that the formula, which spans positions x_i to x_j takes an np argument spanning positions x_j to x_i, ie. the rightmost position of the np argument is the leftmost position of the complete formula.

If we want a displacement grammar to be simple in the input string, we can restrict the synthetic connectives used for its lexical entries to $\check{}$, $\triangleright^{-1}$ and $\triangleleft^{-1}$; in addition, no formulas contain the units I and J except where these occurrences are instances of the allowed synthetic connectives.[5] The only lexical entries proposed for D which are not simple in this sense are those of the MIX grammar and the type for "supposedly" discussed above.

The $\triangleright^{-1}$ and $\triangleleft^{-1}$ connectives, together with atomic formulas of sort greater than 0, allow us to encode MCFG-style analyses, as we have seen them in Section 3.3, into D [6]. As an example, let's look at the unfolding of "lezen" which is assigned formula $\triangleright^{-1}np \setminus (np \setminus si)$ and assume "lezen" occupies the string position 4,5.

$$\forall x_2 \|np \setminus (np \setminus si)\|^{x_2,x_2,4,5}$$

Given that $s(np) = 0$ this reduces further to.

$$\forall x_2 \forall x_1 \|np\|^{x_1,x_2} \multimap \|np \setminus si\|^{x_1,x_2,4,5}$$

If we combine this entry with "boeken" from positions 1,2 (ie. the formula $np(1,2)$, instantiating x_1 to 1 and x_2 to 2, this gives the following partial translation for "boeken lezen"

$$\|np \setminus si\|^{1,2,4,5}$$

Similarly, "kunnen" with formula $\triangleright^{-1}(np \setminus si) \downarrow (np \setminus si)$ reduces as follows when occupying string position 3,4.

$$\forall x_2 \|(np \setminus si) \downarrow (np \setminus si)\|^{x_2,x_2,3,4}$$

---

[5] Alternatively, we can allow the $\check{}$, $\triangleright$ and $\triangleleft$ connectives but restrict them to cases where there is strict identity of the two string positions (disallowing instantiation of variables to obtain identity). Note that this means that formulas of the form $\check{}A \downarrow B$ are valid only in contexts spanning the empty string.

[6] Wijnholds [58] and Sorokin [55] show that D of order 1 (ie. containing only the synchronous rules, but also the bridge and projection connectives) generates the well-nested multiple context-free languages.

Which unfolds further as.

$$\forall x_2 \forall x_0 \forall x_5 \|np \setminus si\|^{x_0,x_2,4,x_5} \multimap \|np \setminus si\|^{x_0,x_2,3,x_5}$$

This combines with the previous translation of "boeken lezen", instantiating x_0 to 1, x_2 to 2 and x_5 to 5, giving "boeken kunnen lezen" with translation $\|np \setminus si\|^{1,2,3,5}$.

Finally, the tensed verb "wil" with formula $(np \setminus si) \downarrow (np \setminus s)$ unfolds at position 2,3 as.

$$\forall x_0 \forall x_3 \|np \setminus si\|^{x_0,2,3,x_3} \multimap \|np \setminus s\|^{x_0,x_3}$$

Instantiating x_0 to 1 and x_3 to 5 and combining this with the previously computed translation of "boeken kunnen lezen" produces $\|np \setminus s\|^{1,5}$ for "boeken wil kunnen lezen". Figure 13 on the next page shows a proof net derivation of the slightly more complex "(dat) Jan Henk Cecilia de nijlpaarden zag helpen voeren". Note that the axiom linkings are again fully deterministic.

5   Correctness of the Translation

The basic idea of the correctness proof is again very simple: we use the property of focused proof search and of ludics that combinations of synchronous connectives can always be seen as instances of a synthetic synchronous connective, whereas the same holds for the asynchronous connectives. Since the translations either use a combination of $\exists$ and $\otimes$ (both synchronous) or a combination of $\forall$ and $\multimap$ (both asynchronous), it follows immediately that we can treat these combinations as synthetic connectives, giving a rule to (synthetic) rule translation.

We only prove the case for the binary continuous and discontinuous connectives. As noted in Section 4.2, the extension to the bridge and projection connectives is simple, whereas the split and injection are more complicated and will be left for future research. In addition, none of the non-deterministic connectives of [46,57] are considered: their obvious translation would use the additive connectives from linear logic, which would complicate the proof nets and increase the computational complexity [28,29][7].

Lemma 1. *For every proof of $t_1 : A_1, \ldots, t_n : A_n \vdash t : C$ in D, there is a proof of its translation in MILL1.*

Proof. Refer back to Figure 8 to see the correspondence between pairs of string positions and tuples of strings more clearly. The rules are simply the translation of the natural deduction rules of D, where the string tuples have been replaced by pairs of string positions.

For the case of $\setminus E$ we are in the following situation (let $i = \frac{1}{2}(n-1)$, $j = \frac{1}{2}(m-1)$, then $x_0, \ldots, x_n$ corresponds to a i-tuple t, $x_n, \ldots, x_{n+m}$ to a j-tuple u and $x_0, \ldots, x_{n-1}, x_{n+1}, \ldots, x_{n+m}$ to their concatenation tu). The translations

---

[7] Though if the non-deterministic connectives occur only in negative contexts, we can treat them by simply multiplying the lexical entries.

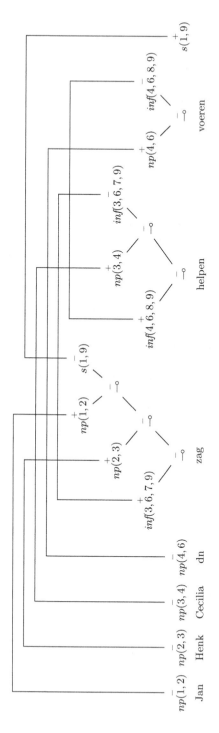

Fig. 13. Proof net for "(dat) Jan Henk Cecilia de nijlpaarden zag helpen voeren"

of A and $A \setminus C$ share point x_n and we can instantiate the universally quantified variables of the other points of A (x_0 to x_{n-1}) applying the $\forall E$ rule n times (/ is symmetric).

$$\cfrac{\cfrac{\cfrac{\|A \setminus C\|^{x_n,\ldots,x_{n+m}}}{\forall y_0,\ldots,y_{n-1}\|A\|^{y_0,\ldots,y_{n-1},x_n} \multimap \|C\|^{y_0,\ldots,y_{n-1},x_{n+1},\ldots,x_{n+m}}} =_{def}}{\|A\|^{x_0,\ldots,x_n} \multimap \|C\|^{x_0,\ldots,x_{n-1},x_{n+1},\ldots,x_{n+m}}} \forall E \ (n \text{ times}) \quad \|A\|^{x_0,\ldots,x_n}}{\|C\|^{x_0,\ldots,x_{n-1},x_{n+1},\ldots,x_{n+m}}} \multimap E$$

For the introduction rule, we again set i to $\frac{1}{2}(n-1)$ and j to $\frac{1}{2}(m-1)$, making $x_0,\ldots,x_n$ corresponds to a i-tuple t, $x_n,\ldots,x_{n+m}$ to a j-tuple u and $x_0,\ldots,x_{n-1},x_{n+1},\ldots,x_{n+m}$ to their concatenation tu. In this case, induction hypothesis gives us a MILL1 proof corresponding to $\Gamma, t : A \vdash tu : C$. To extend this proof to a MILL1 proof corresponding to $\Gamma \vdash u : A \setminus C$ (/ is again symmetric), we can continue the translated proof of $\Gamma, t : A \vdash tu : C$ as follows.

$$\cfrac{\cfrac{\cfrac{\cfrac{[\|A\|^{x_0,\ldots,x_n}]^i \ \ldots \ \Gamma}{\vdots \atop \|C\|^{x_0,\ldots,x_{n-1},x_{n+1},\ldots,x_{n+m}}}}{\|A\|^{x_0,\ldots,x_n} \multimap \|C\|^{x_0,\ldots,x_{n-1},x_{n+1},\ldots,x_{n+m}}} \multimap I_i}{\forall x_0,\ldots,x_{n-1}\|A\|^{x_0,\ldots,x_n} \multimap \|C\|^{x_0,\ldots,x_{n-1},x_{n+1},\ldots,x_{n+m}}} \forall I \ (n \text{ times})}{\|A \setminus C\|^{x_n,\ldots,x_{x+m}}} =_{def}$$

The cases for $\uparrow_>$ are shown below ($\downarrow_>$ is easily verified).

$$\cfrac{\cfrac{\cfrac{\|C \uparrow_> B\|^{x_0,x_1,x_n,\ldots,x_{n+m}}}{\forall y_2,\ldots,y_{n-1}\|B\|^{x_1,y_2,\ldots,y_{n-1},x_n} \multimap \|C\|^{x_0,y_2,\ldots,y_{n-1},x_{n+1},\ldots,x_{n+m}}} =_{def}}{\|B\|^{x_1,\ldots,x_n} \multimap \|C\|^{x_0,x_2,\ldots,x_{n-1},x_{n+1},\ldots,x_{n+m}}} \forall E \ (n-2 \text{ times}) \quad \|B\|^{x_1,\ldots,x_n}}{\|C\|^{x_0,x_2,\ldots,x_{n-1},x_{n+1},\ldots,x_{n+m}}} \multimap E$$

$$\cfrac{\cfrac{\cfrac{\cfrac{[\|B\|^{x_1,\ldots,x_n}]^i \ \ldots \ \Gamma}{\vdots \atop \|C\|^{x_0,x_2,\ldots,x_{n-1},x_{n+1},\ldots,x_{n+m}}}}{\|B\|^{x_1,\ldots,x_n} \multimap \|C\|^{x_0,x_2,\ldots,x_{n-1},x_{n+1},\ldots,x_{n+m}}} \multimap I_i}{\forall x_2,\ldots,x_{n-1}\|B\|^{x_1,\ldots,x_n} \multimap \|C\|^{x_0,x_2,\ldots,x_{n-1},x_{n+1},\ldots,x_{n+m}}} \forall I \ (n-2 \text{ times})}{\|C \uparrow_> B\|^{x_0,x_2,\ldots,x_{n-1},x_{n+1},\ldots,x_{x+m}}} =_{def}$$

Finally, the cases for $\odot_>$ are as follows.

$$\cfrac{\cfrac{\|A \odot_> B\|^{x_0,x_2,\ldots,x_{n-1},x_{n+1},\ldots,x_{n+m}}}{\exists x_1 \exists x_n \|A\|^{x_0,\ldots,x_n} \otimes \|B\|^{x_n,\ldots,x_{n+m}}} =_{def}}{C} \cfrac{\cfrac{[\|A\|^{x_0,x_1,x_n,\ldots,x_{n+m}}]^i \otimes \|B\|^{x_1,\ldots,x_n}}{C} \quad \cfrac{[\|A\|^{x_0,x_1,x_n,\ldots,x_{n+m}}]^j [\|B\|^{x_1,\ldots,x_n}]^j}{\vdots \atop C}}{C} \otimes E_j}{C} \exists E_i \text{ twice}$$

$$\cfrac{\cfrac{\cfrac{\cfrac{\|A\|^{x_0,x_1,x_n,\ldots,x_{n+m}} \quad \|B\|^{x_1,\ldots,x_n}}{\|A\|^{x_0,x_1,x_n,\ldots,x_{n+m}} \otimes \|B\|^{x_1,\ldots,x_n}} \otimes I}{\exists x_n \|A\|^{x_0,x_1,x_n\ldots,x_{n+m}} \otimes \|B\|^{x_1,\ldots,x_n}} \exists I}{\exists x_1 \exists x_n \|A\|^{x_0,x_1,x_n\ldots,x_{n+m}} \otimes \|B\|^{x_1,\ldots,x_n}} \exists I}{\|A \odot_> B\|^{x_0,x_2,\ldots,x_{n-1},x_{n+1},\ldots,x_{n+m}}} =_{def}$$

$\square$

Lemma 2. *If the translation of a D sequent $t_1 : A_1,\ldots,t_n : A_n \vdash t : C$ is provable, then there is a D proof of $t_1 : A_1,\ldots,t_n : A_n \vdash t : C$.*

Proof. This is most easily shown using proof nets, using induction on the number of links while removing them in groups of synchronous or asynchronous links corresponding to a D connective.

If there are terminal asynchronous links, then we proceed by case analysis knowing that we are dealing the result of the translation of D formulas.

The case for $C \uparrow_> B$ looks as follows.

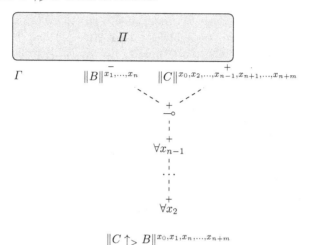

$$\|C \uparrow_> B\|^{x_0, x_1, x_n, \ldots, x_{n+m}}$$

Given that removing the portrayed links produces a proof net Π of $\Gamma, B \vdash C$, we can apply the induction hypothesis, which gives a proof δ of $\Gamma, u : B \vdash sut : C$, which we can extend as follows.

$$\frac{\begin{array}{cc} \Gamma & u : B \\ & \vdots \ \delta \\ C : sut \end{array}}{s, t : C \uparrow_> B} \uparrow_> I$$

Similarly, the par case for $\odot_>$ looks as follows.

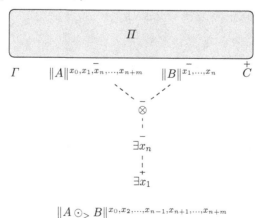

$$\|A \odot_> B\|^{x_0, x_2, \ldots, x_{n-1}, x_{n+1}, \ldots, x_{n+m}}$$

Again, we know by induction hypothesis that there is a proof δ of $\Gamma, s, t : A, u : B \vdash v[sut] : C$ and we need to show that there is a proof of $\Gamma, sut : A \odot_> B \vdash v[sut] : C$, which we do as follows.

$$
\begin{array}{c}
\Gamma \quad [s, t : A]^i \quad [u : B]^i \\
\vdots \; \delta \\
\underline{sut : A \odot_> B \qquad v[sut] : C} \\
v[sut] : C
\end{array} \odot_> E^i
$$

Suppose there are no terminal asynchronous links, then we know there must be a group of splitting synchronous links corresponding to a D connective (a series of universal links ended by a tensor link which splits the proof net into two subnets, though the synthetic connectives of Section 4.2 allow for a single universal link, which is splitting by definition, since after removal of the link, all premisses of the link are the conclusion of disjoint subnets), using the standard splitting tensor argument [14,10,4].

Suppose this group of splitting links is the translation of $\uparrow_>$, then the proof net is of the following form. Note that the translation of B corresponds to the string tuple u (with $i = \frac{1}{2}n$ components), the translation of $C \uparrow_> B$ to the string tuple sut and the translation of C to the string tuple s, t.

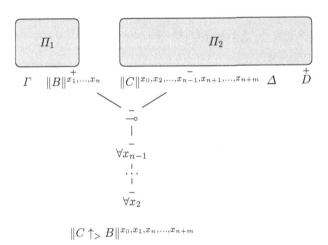

Therefore, we know by induction hypothesis that there is a proof δ_1 of $\Gamma \vdash u : B$ and a proof δ_2 of $\Delta, sut : C \vdash D$. We need to show that there is a proof $\Gamma, \Delta, s, t : C \uparrow_> B \vdash D$, which we can do as follows[8].

---

[8] The observant reader has surely noted that this step is just the standard sequentialization step followed by the translation of sequent calculus into natural deduction, see for example [18,42].

$$
\Delta \quad \cfrac{u:B \quad s,t:C \uparrow_> B}{\underset{\vdots\ \delta_2}{\overset{\underset{\vdots\ \delta_1}{\overset{\Gamma}{}}}{sut:C}}} \uparrow_> E
$$

$$
D
$$

In case the splitting tensor link and associated existential quantifiers are the translation of a $\odot_>$ formula, we are in the following case.

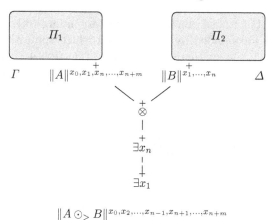

$$
\|A \odot_> B\|^{x_0,x_2,\ldots,x_{n-1},x_{n+1},\ldots,x_{n+m}}
$$

Induction hypothesis gives us a proof δ_1 of $\Gamma \vdash s,t : A$ and a proof δ_2 of $\Delta \vdash u : B$, which we combine as follows.

$$
\cfrac{\underset{s,t:A}{\overset{\underset{\vdots\ \delta_1}{\Gamma}}{}} \quad \underset{u:B}{\overset{\underset{\vdots\ \delta_2}{\Delta}}{}}}{sut:A \odot_> B} \odot_> I
$$

Theorem 1. *Derivability of a statement in D and derivability of the translation of this statement into MILL1 coincide.*

Proof. Immediate from Lemma 1 and 2.

The main theorem gives a simple solution to two of the main open problems from [43].

Corollary 1. *D is NP-complete.*

Proof. We have that the derivability of L, D and MILL1 are related as follows (given the translations of L and D into MILL1) $L \subset D \subset MILL1$. Therefore NP-completeness of L and MILL1 gives us NP-completeness of D.

Corollary 2. *MILL1 provides a proof net calculus for D.*

Proof. This is immediate from the fact that the D connectives correspond to synthetic MILL1 connectives. Therefore, adding these synthetic connectives to MILL1 provides a conservative extension of MILL1, which contains a proof net calculus of D. For the synchronous links, this possibility is already implicit in the proof nets of Figures 10 and 13, where the existential links are not portrayed; the combination of the asynchronous $\forall$ and the $\multimap$ link in Figure 10 can be similarly replaced by a single link and a switch to either one of the premisses of the $\multimap$ link or to one of the formulas containing the variable x.[9]

Corollary 3. *D satisfies cut elimination.*

Cut elimination for D, including the non-deterministic connectives and the units, is already proved directly in [46,57]. However, using the translation into MILL1 gives us a very easy cut elimination proof.

6 Agreement, Non-associativity and Scope Restrictions

Though I have focused only on using the first-order terms for representing string positions, I will sketch a number of other applications of the first-order terms which are orthogonal to their use for string positions, for which other extension of the Lambek calculus have introduced additional connectives and logical rules, such as the unary modalities of multimodal categorial grammar [26].

The most obvious of these applications is for the use of linguistic features, allowing us, for example, to distinguish between nominative and accusative noun phrases $np(nom)$ and $np(acc)$ but also allowing a lexical entry to fill either role by assigning it the formula $\forall x.np(x)$.

Until now, we have only seen variables and constants as arguments of predicate symbols. When we allow more complex terms, things get more interesting. Let's only consider complex terms of the form $s(T)$ — the well-known successor term from unary arithmetic not to be confused with the predicate symbol s for sentence — where T is itself a term (complex, a variable or a constant). These complex terms allow us to implement non-associativity when we need it, using the following translation (remember that the string positions are orthogonal and can be included if needed).

$$\|A \bullet B\|^x = \|A\|^{s(x)} \otimes \|B\|^{s(x)}$$
$$\|C/B\|^{s(x)} = \|B\|^{s(x)} \multimap \|C\|^x$$
$$\|A\backslash C\|^{s(x)} = \|A\|^{s(x)} \multimap \|C\|^x$$

The translation is parametric in a single variable x unique to the formula, which can get partially instantiated during the translation, producing a formula with a single free variable which is universally quantified to complete the translation. For example, a prototypical statement whose derivability presupposes associativity

---

[9] Another proof net calculus for D would be a direct adaptation of the results from Section 7 of [41]. However, this footnote is too small to contain it.

$$a/b, b/c \vdash a/c$$

translates as

$$\forall x[b(s(x)) \multimap a(x)], \forall y[c(s(y)) \multimap b(y)] \vdash \forall z[c(s(z)) \multimap a(z)]$$

which the reader can easily verify to be underivable. This translation generalizes both the translation of NL to MILL1 and the implementation of island constraints of [40].

In addition, we can handle scope restrictions in the same spirit as [6], by translating s_1 as $\forall x.s(x)$, s_2 as $\forall x.s(s(x))$ and s_3 as $\forall x.s(s(s(x)))$, which are easily verified to satisfy $s_i \vdash s_j$ for $i \le j$ and $s_i \nvdash s_j$ for $i > j$.

Scope restrictions and island constraints are some of the iconic applications of the unary modalities of multimodal categorial grammars and I consider it an attractive feature of MILL1 they permit a transparent translation of these applications.

The use of complex terms moves us rather close to the indexed grammars [1], where complex unary term symbols play the role of a stack of indices. The linear indexed grammars [13] would then correspond to the restriction of quantified variables to two occurrences of opposite polarity[10] (or a single occurrence of any polarity; for the string position variables, they occur twice: either once as a left (resp. right) position of a positive atomic formula and once as a left (resp. right) position of a negative atomic formula or once as a left position and once as a right position of atomic formulas of the same polarity). If we restrict variables to at most two occurrences of each variable, without any restriction on the polarities, we are closer to an extension of linear indexed grammars proposed by [25], which they call partially linear PATR, and thereby closer to unification-based grammars such as LFG and HPSG. This restriction on quantified variables seems very interesting and naturally encompasses the restriction on string position variables.

These are of course only suggestions, which need to be studied in more detail in future work.

7 Conclusions and Open Questions

First-order multiplicative intuitionistic linear logic includes several interesting subsystems: multiple context-free grammars, the Lambek calculus and the Displacement calculus. In spite of this, the computational complexity of MILL1 is the same as the complexity of the universal recognition problem for each of these individual systems. In addition, it gives a natural implementation of several additional linguistic phenomena, which would require further machinery in each of the other calculi.

MILL1 satisfies all conditions of extended Lambek calculi: it has a simple proof theory, which includes a proof net calculus, it generates the mildly context-free languages, it is NP-complete and the homomorphism for semantics consists

---

[10] The encoding of non-associativity above is a clear violation of this constraint, since the quantified variable will occur in all atomic subformulas.

of simply dropping the quantifiers to obtain an MILL proof — though it is conceivable to use the first-order quantifiers for *semantic* features which would have a reflection in the homomorphism.

Many important questions have been left open. Do MILL1 grammars without complex terms (simple in the input string or not) generate exactly the MCFLs or strictly more? Do MILL1 grammars *with* complex terms generate exactly the indexed languages and can we get interesting subclasses (eg. partially linear PATR) by restricting the variables to occur at most twice? Are there interesting fragments of MILL1 grammars which have a polynomial recognition problem? I hope these questions will receive definite answers in the future.

References

1. Aho, A.: Indexed grammars: An extension of context-free grammars. Journal of the ACM 15(4), 647–671 (1968)
2. Andreoli, J.-M.: Logic programming with focussing proofs in linear logic. Journal of Logic and Computation 2(3) (1992)
3. Bar-Hillel, Y., Perles, M., Shamir, E.: On formal properties of simple phrase structure grammars. In: Bar-Hillel, Y. (ed.) Language and Information. Selected Essays on their Theory and Application, pp. 116–150. Addison-Wesley, New York (1964)
4. Bellin, G., van de Wiele, J.: Empires and kingdoms in MLL. In: Girard, J.-Y., Lafont, Y., Regnier, L. (eds.) Advances in Linear Logic, pp. 249–270. Cambridge University Press (1995)
5. Bernardi, R., Moortgat, M.: Continuation semantics for symmetric categorial grammar. In: Leivant, D., de Queiroz, R. (eds.) WoLLIC 2007. LNCS, vol. 4576, pp. 53–71. Springer, Heidelberg (2007)
6. Bernardi, R., Moot, R.: Generalized quantifiers in declarative and interrogative sentences. Logic Journal of the IGPL 11(4), 419–434 (2003)
7. Boullier, P.: Proposal for a natural language processing syntactic backbone. Technical Report 3342, INRIA, Rocquencourt (1998)
8. Boullier, P., Sagot, B.: Efficient and robust LFG parsing: SxLfg. In: International Workshop on Parsing Technologies (2005)
9. Danos, V.: La Logique Linéaire Appliquée à l'étude de Divers Processus de Normalisation (Principalement du λ-Calcul). PhD thesis, University of Paris VII (June 1990)
10. Danos, V., Regnier, L.: The structure of multiplicatives. Archive for Mathematical Logic 28, 181–203 (1989)
11. Došen, K.: A brief survey of frames for the Lambek calculus. Zeitschrift für Mathematische Logic und Grundlagen der Mathematik 38, 179–187 (1992)
12. Fadda, M.: Geometry of Grammar: Exercises in Lambek Style. PhD thesis, Universitat Politècnica de Catalunya (2010)
13. Gazdar, G.: Applicability of indexed grammars to natural languages. In: Reyle, U., Rohrer, C. (eds.) Natural Language Parsing and Linguistic Theories, pp. 69–94. D. Reidel, Dordrecht (1988)
14. Girard, J.-Y.: Linear logic. Theoretical Computer Science 50, 1–102 (1987)
15. Girard, J.-Y.: Quantifiers in linear logic II. In: Corsi, G., Sambin, G. (eds.) Nuovi Problemi Della Logica e Della Filosofia Della Scienza, Bologna, Italy, vol. II. CLUEB (1991). Proceedings of the conference with the same name, Viareggio, Italy (January 1990)

16. Girard, J.-Y.: Linear logic: Its syntax and semantics. In: Girard, J.-Y., Lafont, Y., Regnier, L. (eds.) Advances in Linear Logic, pp. 1–42. Cambridge University Press (1995)
17. Girard, J.-Y.: Locus solum: From the rules of logic to the logic of rules. Mathematical Structures in Computer Science 11, 301–506 (2001)
18. Girard, J.-Y., Lafont, Y., Taylor, P.: Proofs and Types. Cambridge Tracts in Theoretical Computer Science 7. Cambridge University Press (1988)
19. Greibach, S.A.: A new normal-form theorem for context-free phrase structure grammars. Journal of the ACM 12(1), 42–52 (1965)
20. Guerrini, S.: Correctness of multiplicative proof nets is linear. In: Fourteenth Annual IEEE Symposium on Logic in Computer Science, pp. 454–263. IEEE Computer Science Society (1999)
21. Huybregts, R.: The weak inadequacy of context-free phrase structure grammars. In: de Haan, G., Trommelen, M., Zonneveld, W. (eds.) Van Periferie naar Kern. Foris, Dordrecht (1984)
22. Joshi, A.: Tree-adjoining grammars: How much context sensitivity is required to provide reasonable structural descriptions. In: Dowty, D., Karttunen, L., Zwicky, A. (eds.) Natural Language Processing: Theoretical, Computational, and Psychological Perspectives. Cambridge University Press (1985)
23. Kaji, Y., Nakanishi, R., Seki, H., Kasami, T.: The computational complexity of the universal recognition problem for parallel multiple context-free grammars. Computational Intelligence 10(4), 440–452 (1994)
24. Kanazawa, M.: The pumping lemma for well-nested multiple context-free languages. In: Diekert, V., Nowotka, D. (eds.) DLT 2009. LNCS, vol. 5583, pp. 312–325. Springer, Heidelberg (2009)
25. Keller, B., Weir, D.: A tractable extension of linear indexed grammars. In: Proceedings of the Seventh Meeting of the European Chapter of the Association for Computational Linguistics, pp. 75–82 (1995)
26. Kurtonina, N., Moortgat, M.: Structural control. In: Blackburn, P., de Rijke, M. (eds.) Specifying Syntactic Structures, pp. 75–113. CSLI, Stanford (1997)
27. Lambek, J.: The mathematics of sentence structure. American Mathematical Monthly 65, 154–170 (1958)
28. Lincoln, P.: Deciding provability of linear logic formulas. In: Girard, Y., Lafont, Y., Regnier, L. (eds.) Advances in Linear Logic, pp. 109–122. Cambridge University Press (1995)
29. Lincoln, P., Scedrov, A.: First order linear logic without modalities is NEXPTIME-hard. Theoretical Computer Science 135(1), 139–154 (1994)
30. Matsuzaki, T., Miyao, Y., Tsujii, J.: Efficient HPSG parsing with supertagging and CFG-filtering. In: Proceedings of the 20th International Joint Conference on Artifical Intelligence, pp. 1671–1676 (2007)
31. Moortgat, M.: Multimodal linguistic inference. Journal of Logic, Language and Information 5(3-4), 349–385 (1996)
32. Moortgat, M.: Symmetries in natural language syntax and semantics: The Lambek-Grishin calculus. In: Leivant, D., de Queiroz, R. (eds.) WoLLIC 2007. LNCS, vol. 4576, pp. 264–284. Springer, Heidelberg (2007)
33. Moortgat, M.: Typelogical grammar. Stanford Encyclopedia of Philosophy Website (2010), http://plato.stanford.edu/entries/typelogical-grammar/

34. Moortgat, M.: Categorial type logics. In: van Benthem, J., ter Meulen, A. (eds.) Handbook of Logic and Language, ch. 2, pp. 95–179. Elsevier/MIT Press (2011)
35. Moortgat, M., Moot, R.: Proof nets for the lambek-grishin calculus. In: Grefenstette, E., Heunen, C., Sadrzadeh, M. (eds.) Compositional Methods in Physics and Linguistics, pp. 283–320. Oxford University Press (2013)
36. Moot, R.: Proof nets and labeling for categorial grammar logics. Master's thesis, Utrecht University, Utrecht (1996)
37. Moot, R.: Proof Nets for Linguistic Analysis. PhD thesis, Utrecht Institute of Linguistics OTS, Utrecht University (2002)
38. Moot, R.: Filtering axiom links for proof nets. In: Kallmeyer, L., Monachesi, P., Penn, G., Satta, G. (eds.) Proceedings of Formal Grammar 2007 (2007) (to appear with CSLI)
39. Moot, R.: Lambek grammars, tree adjoining grammars and hyperedge replacement grammars. In: Gardent, C., Sarkar, A. (eds.) Proceedings of TAG+9, The Ninth International Workshop on Tree Adjoining Grammars and Related Formalisms, pp. 65–72 (2008)
40. Moot, R., Piazza, M.: Linguistic applications of first order multiplicative linear logic. Journal of Logic, Language and Information 10(2), 211–232 (2001)
41. Moot, R., Puite, Q.: Proof nets for the multimodal Lambek calculus. Studia Logica 71(3), 415–442 (2002)
42. Moot, R., Retoré, C.: The Logic of Categorial Grammars. LNCS, vol. 6850. Springer, Heidelberg (2012)
43. Morrill, G.: Categorial Grammar: Logical Syntax, Semantics, and Processing. Oxford University Press (2011)
44. Morrill, G., Fadda, M.: Proof nets for basic discontinuous Lambek calculus. Journal of Logic and Computation 18(2), 239–256 (2008)
45. Morrill, G., Valentín, O.: On calculus of displacement. In: Proceedings of TAG+Related Formalisms. University of Yale (2010)
46. Morrill, G., Valentín, O., Fadda, M.: The displacement calculus. Journal of Logic, Language and Information 20(1), 1–48 (2011)
47. Murawski, A.S., Ong, C.-H.L.: Dominator trees and fast verification of proof nets. In: Logic in Computer Science, pp. 181–191 (2000)
48. Nederhof, M.-J., Satta, G.: Theory of parsing. In: Clark, A., Fox, C., Lappin, S. (eds.) The Handbook of Computational Linguistics and Natural Language Processing, pp. 105–130. Wiley-Blackwell (2010)
49. Pentus, M.: Product-free Lambek calculus and context-free grammars. Journal of Symbolic Logic 62, 648–660 (1997)
50. Pentus, M.: Lambek calculus is NP-complete. Theoretical Computer Science 357(1), 186–201 (2006)
51. Pentus, M.: A polynomial-time algorithm for Lambek grammars of bounded order. Linguistic Analysis 36(1-4), 441–471 (2010)
52. Pereira, F., Shieber, S.: Prolog and Natural Language Analysis. CSLI, Stanford (1987)
53. Seki, H., Matsumura, T., Fujii, M., Kasami, T.: On multiple context-free grammars. Theoretical Computer Science 88, 191–229 (1991)
54. Shieber, S.: Evidence against the context-freeness of natural language. Linguistics & Philosophy 8, 333–343 (1985)
55. Sorokin, A.: Normal forms for multiple context-free languages and displacement Lambek grammars. In: Artemov, S., Nerode, A. (eds.) LFCS 2013. LNCS, vol. 7734, pp. 319–334. Springer, Heidelberg (2013)

56. Stabler, E.: Tupled pregroup grammars. Technical report, University of California, Los Angeles (2003)
57. Valentín, O.: Theory of Discontinuous Lambek Calculus. PhD thesis, Universitat Autònoma de Catalunya (2012)
58. Wijnholds, G.: Investigations into categorial grammar: Symmetric pregroup grammar and displacement calculus, Bachelor thesis, Utrecht University (2011)

A Categorial Type Logic⋆

Glyn Morrill

Universitat Politècnica de Catalunya
morrill@lsi.upc.edu
http://www.lsi.upc.edu/~morrill

Abstract. In logical categorial grammar [23,11] syntactic structures are
categorial proofs and semantic structures are intuitionistic proofs, and
the syntax-semantics interface comprises a homomorphism from syntactic proofs to semantic proofs. Thereby, logical categorial grammar
embodies in a pure logical form the principles of compositionality, lex-
icalism, and parsing as deduction. Interest has focused on multimodal
versions but the advent of the (dis)placement calculus of Morrill, Valentín and Fadda [21] suggests that the role of structural rules can be
reduced, and this facilitates computational implementation. In this paper we specify a comprehensive formalism of (dis)placement logic for the
parser/theorem prover CatLog integrating categorial logic connectives
proposed to date and illustrate with a cover grammar of the Montague
fragment.

1 Introduction

According to the principle of compositionality of Frege the meaning of an expression is a function of the meanings of its parts and their mode of composition. This
is refined in Montague grammar where the syntax-semantics interface comprises
a homomorphism from a syntactic algebra to a semantic algebra. In logical categorial grammar [23,11] both syntactic structures and semantic structures are
proofs and the Montagovian rendering of Fregean compositionality is further
refined to a homomorphism from syntactic (categorial) proofs to semantic (intuitionistic) proofs. Thus we see successive refinements of Frege's principle in
theories of the syntax-semantics interface which are expressed first as algebra
and then further as algebraic logic. The present paper gathers together and inte-
grates categorial connectives proposed to date to specify a particular formalism
according to this design, one implemented in the parser/theorem-prover CatLog
[16,15] and illustrates with a cover grammar of the Montague fragment.

Multimodal categorial grammar [25,9,6,22,7,8,24] constitutes a methodology
rather than a particular categorial calculus, admitting an open class of residuated connective families for multiple modes of composition related by structural
rules of interaction and inclusion. On the one hand, since no particular system is
identified, the problem of computational implementation is an open-ended one;

⋆ This research was partially supported by an ICREA Acadèmia 2012, by BASMATI
MICINN project (TIN2011-27479-C04-03) and by SGR2009-1428 (LARCA).

C. Casadio et al. (Eds.): Lambek Festschrift, LNCS 8222, pp. 331–352, 2014.

and on the other hand, the structural rules add to the proof search-space. Moot [10] and Moot and Retoré ([11], Ch. 7) provide a general-purpose implementation Grail. It supports the so-called Weak Sahlqvist structural inclusions and is based on proof-net contraction criteria, with certain contractions according to the structural rules. This seems to constitute the computational scope of the potential of the multimodal framework.

The displacement calculus of Morrill et al. [21] creates another option. This calculus provides a solution to the problem of discontinuous connectives in categorial grammar initiated in [1,2]. The calculus addresses a wide range of empirical phenomena, and it does so without the use of structural rules since the rules effecting displacement are *defined*. This opens the possibility of categorial calculus in which the role of structural rules is reduced. To accommodate discontinuity of resources the calculus invokes sorting of types according to their syntactical datatype (number of points of discontinuity), and this requires a novel kind of sequent calculus which we call a hypersequent calculus. In this paper we consider how displacement calculus and existing categorial logic can be integrated in a uniform hypersequent displacement logic, which we call simply placement logic.[1] We observe that this admits a relatively straightforward implementation which we use to illustrate a Montague fragment and we define as a program the goal of implementing increasing fragments of this logic with proof nets.

In the course of the present paper we shall specify the formalism and its calculus. This incorporates connectives introduced over many years addressing numerous linguistic phenomena, but the whole enterprise is characterized by the features of the placement calculus which is extended: sorting for the types and hypersequents for the calculus. In Section 2 we define the semantic representation language; in Section 3 we define the types; in Section 4 we define the calculus. In Section 5 we give a cover grammar of the Montague fragment of Dowty, Wall and Peters ([4], Ch. 7). In Section 6 we give analyses of the examples from the second half of that Chapter. We conclude in Section 7.

2   Semantic Representation Language

Recall the following operations on sets:

(1)   a. Functional exponentiation: X^Y = the set of all total functions from Y to X
      b. Cartesian product: $X \times Y = \{\langle x, y \rangle \mid x \in X \;\&\; y \in Y\}$
      c. Disjoint union: $X \uplus Y = (\{1\} \times X) \cup (\{2\} \times Y)$
      d. i-th Cross product, $i \geq 0$: $X^0 = \{0\}$
$$X^{1+i} = X \times (X^i)$$

The set $\mathcal{T}$ of *semantic types* of the semantic representation language is defined on the basis of a set δ of *basic semantic types* as follows:

---

[1] The prefix 'dis-' is dropped since reversing the line of reasoning which *dis*places items *places* items.

(2) $\mathcal{T} ::= \delta \mid \top \mid \bot \mid \mathcal{T} + \mathcal{T} \mid \mathcal{T}\&\mathcal{T} \mid \mathcal{T} \to \mathcal{T} \mid \mathbf{L}\mathcal{T} \mid \mathcal{T}^+$

A *semantic frame* comprises a family $\{D_\tau\}_{\tau \in \delta}$ of non-empty *basic type domains* and a non-empty set W of worlds. This induces a *type domain* D_τ for each type τ as follows:

(3)
$$
\begin{aligned}
D_\top &= \{\emptyset\} \\
D_\bot &= \{\} \\
D_{\tau_1 + \tau_2} &= D_{\tau_2 \& \tau_1} \\
D_{\tau_1 \& \tau_2} &= D_{\tau_1 \& \tau_2} \\
D_{\tau_1 \to \tau_2} &= D_{\tau_2}^{D_{\tau_1}} \\
D_{\mathbf{L}\tau} &= D_\tau^W \\
D_{\tau^+} &= \bigcup_{i>0}(D_\tau)^i
\end{aligned}
$$

The sets Φ_τ of *terms* of type τ for each type τ are defined on the basis of sets C_τ of constants of type τ and enumerably infinite sets V_τ of variables of type τ for each type τ as follows:

(4)

$$
\begin{aligned}
\Phi_\tau &::= C_\tau & \text{constants} \\
\Phi_\tau &::= V_\tau & \text{variables} \\
\Phi_\tau &::= \Phi_{\tau_1 + \tau_2} \to V_{\tau_1}.\Phi_\tau;\ V_{\tau_2}.\Phi_\tau & \text{case statement} \\
\Phi_{\tau + \tau'} &::= \iota_1\Phi_\tau & \text{first injection} \\
\Phi_{\tau' + \tau} &::= \iota_2\Phi_\tau & \text{second injection} \\
\Phi_\tau &::= \pi_1\Phi_{\tau \& \tau'} & \text{first projection} \\
\Phi_\tau &::= \pi_2\Phi_{\tau' \& \tau} & \text{second projection} \\
\Phi_{\tau \& \tau'} &::= (\Phi_\tau, \Phi_{\tau'}) & \text{ordered pair formation} \\
\Phi_\tau &::= (\Phi_{\tau' \to \tau}\ \Phi_{\tau'}) & \text{functional application} \\
\Phi_{\tau \to \tau'} &::= \lambda V_\tau \Phi_{\tau'} & \text{functional abstraction} \\
\Phi_\tau &::= {}^\vee\Phi_{\mathbf{L}\tau} & \text{extensionalization} \\
\Phi_{\mathbf{L}\tau} &::= {}^\wedge\Phi_\tau & \text{intensionalization} \\
\Phi_{\tau^+} &::= [\Phi_\tau] \mid [\Phi_\tau | \Phi_{\tau^+}] & \text{non-empty list construction}
\end{aligned}
$$

Given a semantic frame, a *valuation* f mapping each constant of type τ into an element of D_τ, an assignment g mapping each variable of type τ into an element of D_τ, and a world $i \in W$, each term ϕ of type τ receives an interpretation $[\phi]^{g,i} \in D_\tau$ as shown in Figure 1.

An occurrence of a variable x in a term is called *free* if and only if it does not fall within any part of the term of the form $x.\cdot$ or $\lambda x\cdot$; otherwise it is *bound* (by the closest $x.$ or λx within the scope of which it falls). The result $\phi\{\psi_1/x_1, \ldots, \psi_n/x_n\}$ of substituting terms $\psi_1, \ldots, \psi_n$ (of types $\tau_1, \ldots, \tau_n$) for variables $x_1, \ldots, x_n$ (of types $\tau_1, \ldots, \tau_n$) respectively in a term ϕ is the result of simultaneously replacing by ψ_i every free occurrence of x_i in ϕ. We say that ψ is *free for x in ϕ* if and only if no variable in ψ becomes bound in $\phi\{\psi/x\}$. We say that a term is *modally closed* if and only if every occurrence of $^\vee$ occurs within the scope of an $^\wedge$. A modally closed term is denotationally invariant across worlds. We say that a term ψ is *modally free for x in ϕ* if and only if either

$$[a]^{g,i} = f(a) \text{ for constant } a \in C_\tau$$
$$[x]^{g,i} = g(x) \text{ for variable } x \in V_\tau$$
$$[\phi \rightarrow x.\psi; y.\chi]^{g,i} = \begin{cases} [\psi]^{(g-\{(x,g(x))\})\cup\{(x,\mathbf{snd}([\phi]^{g,i}))\},i} & \text{if } \mathbf{fst}([\phi]^{g,i}) = 1 \\ [\chi]^{(g-\{(y,g(y))\})\cup\{(y,\mathbf{snd}([\phi]^{g,i}))\},i} & \text{if } \mathbf{fst}([\phi]^{g,i}) = 2 \end{cases}$$
$$[\iota_1\phi]^{g,i} = \langle 1, [\phi]^{g,i}\rangle$$
$$[\iota_2\phi]^{g,i} = \langle 2, [\phi]^{g,i}\rangle$$
$$[\pi_1\phi]^{g,i} = \mathbf{fst}([\phi]^{g,i})$$
$$[\pi_2\phi]^{g,i} = \mathbf{snd}([\phi]^{g,i})$$
$$[(\phi, \psi)]^{g,i} = \langle [\phi]^{g,i}, [\psi]^{g,i}\rangle$$
$$[(\phi\ \psi)]^{g,i} = [\phi]^{g,i}([\psi]^{g,i})$$
$$[\lambda x\phi]^{g,i} = d \mapsto [\phi]^{(g-\{(x,g(x))\})\cup\{(x,d)\},i}$$
$$[^\vee\phi]^{g,i} = [\phi]^{g,i}(i)$$
$$[^\wedge\phi]^{g,i} = j \mapsto [\phi]^{g,j}$$
$$[[\phi]]^{g,i} = \langle [\phi]^{g,i}, 0\rangle$$
$$[[\phi|\psi]]^{g,i} = \langle [\phi]^{g,i}, [\psi]^{g,i}\rangle$$

Fig. 1. Interpretation of the semantic representation language

ψ is modally closed, or no free occurrence of x in ϕ is within the scope of an $^\wedge$. The laws of conversion in Figure 2 obtain; we omit the so-called commuting conversions for the case statement.

3   Syntactic Types

The types in (dis)placement calculus and placement logic which extends it are sorted according to the number of points of discontinuity (placeholders) their expressions contain. Each *type predicate letter* will have a sort and an arity which are naturals, and a corresponding semantic type. Assuming ordinary terms to be already given, where P is a type predicate letter of sort i and arity n and $t_1, \ldots, t_n$ are terms, $Pt_1 \ldots t_n$ is an (atomic) type of sort i of the corresponding semantic type. Compound types are formed by connectives given in the following subsections, and the homomorphic semantic type map T associates these with semantic types. In Subsection 3.1 we give relevant details of the multiplicative (dis)placement calculus basis and in Subsection 3.2 we define types for all connectives.

3.1   The Placement Calculus Connectives

Let a *vocabulary* V be a set which includes a distinguished placeholder symbol 1 called the *separator*. For $i \in \mathcal{N}$ we define L_i as the set of strings over V containing i separators:

(5) $L_i = \{s \in V^* | \ |s|_1 = i\}$

V induces the *placement algebra*

$$(\{L_i\}_{i\in\mathcal{N}}, +, \{\times_k\}_{k\in\mathbb{Z}^\pm}, 0, 1)$$

$$\phi \rightarrow y.\psi; z.\chi = \phi \rightarrow x.(\psi\{x/y\}); z.\chi$$

if x is not free in ψ and is free for y in ψ

$$\phi \rightarrow y.\psi; z.\chi = \phi \rightarrow y.\psi; x.(\chi\{x/z\})$$

if x is not free in χ and is free for z in χ

$$\lambda y\phi = \lambda x(\phi\{x/y\})$$

if x is not free in ϕ and is free for y in ϕ

α-conversion

$$\iota_1\phi \rightarrow y.\psi; z.\chi = \psi\{\phi/y\}$$

if ϕ is free for y in ψ and modally free for y in ψ

$$\iota_2\phi \rightarrow y.\psi; z.\chi = \chi\{\phi/z\}$$

if ϕ is free for z in χ and modally free for z in χ

$$\pi_1(\phi, \psi) = \phi$$
$$\pi_2(\phi, \psi) = \psi$$
$$(\lambda x\phi \ \psi) = \phi\{\psi/x\}$$

if ψ is free for x in ϕ, and modally free for x in ϕ

$$^{\vee\wedge}\phi = \phi$$

β-conversion

$$(\pi_1\phi, \pi_2\phi) = \phi$$
$$\lambda x(\phi \ x) = \phi$$

if x is not free in ϕ

$$^{\wedge\vee}\phi = \phi$$

if ϕ is modally closed

η-conversion

Fig. 2. Semantic conversion laws

where $+ : L_i, L_j \rightarrow L_{i+j}$ is concatenation, and k-th wrapping $\times_k : L_{i+|k|}, L_j \rightarrow L_{i+|k|-1+j}$ is defined as replacing by its second operand the $|k|$-th separator in its first operand, counting from the left for positive k and from the right for negative k.[2] 0 is the empty string. Note that 0 is a left and right identity element for $+$ and that 1 is a left and right identity element for $\times$:

(6)

$$0+s = s \quad s = s+0$$
$$1\times s = s \quad s = s\times 1$$

Sorted types $\mathcal{F}_i, i \in \mathcal{N}$, are defined and interpreted sort-wise as shown in Figure 3. Where A is a type, let sA denotes its sort. The sorting discipline ensures that $[A] \subseteq L_{sA}$. Note that $\{\backslash, \bullet, /\}$ and $\{\downarrow_k, \odot_k, \uparrow_k\}$ are residuated triples with parents $\bullet$ and $\odot_k$, and that as the canonical extensions of the operations of the placement algebra, I is a left and right identity for $\bullet$ and J is a left and right identity for $\odot_k$.

---

[2] In the version of Morrill and Valentín [18] wrapping is only counted from the left, and in the "edge" version of Morrill et al. [21] there is only leftmost and rightmost wrapping, hence these can be seen as subinstances of the general case given in this paper where $k > 0$ and $k \in \{+1, -1\}$ respectively.

$$
\begin{array}{lll}
\mathcal{F}_j ::= \mathcal{F}_i \backslash \mathcal{F}_{i+j} & [A\backslash C] = \{s_2 | \ \forall s_1 \in [A], s_1 + s_2 \in [C]\} & \text{under} \\
\mathcal{F}_i ::= \mathcal{F}_{i+j}/\mathcal{F}_j & [C/B] = \{s_1 | \ \forall s_2 \in [B], s_1 + s_2 \in [C]\} & \text{over} \\
\mathcal{F}_{i+j} ::= \mathcal{F}_i \bullet \mathcal{F}_j & [A\bullet B] = \{s_1 + s_2 | \ s_1 \in [A] \ \& \ s_2 \in [B]\} & \text{product} \\
\mathcal{F}_0 ::= I & [I] = \{0\} & \text{product unit} \\
\mathcal{F}_j ::= \mathcal{F}_{i+1} \downarrow_k \mathcal{F}_{i+j} & [A\downarrow_k C] = \{s_2 | \ \forall s_1 \in [A], s_1 \times_k s_2 \in [C]\} & \text{infix} \\
\mathcal{F}_{i+1} ::= \mathcal{F}_{i+j} \uparrow_k \mathcal{F}_j & [C\uparrow_k B] = \{s_1 | \ \forall s_2 \in [B], s_1 \times_k s_2 \in [C]\} & \text{circumfix} \\
\mathcal{F}_{i+j} ::= \mathcal{F}_{i+1} \odot_k \mathcal{F}_j & [A\odot_k B] = \{s_1 \times_k s_2 | \ s_1 \in [A] \ \& \ s_2 \in [B]\} & \text{wrap} \\
\mathcal{F}_1 ::= J & [J] = \{1\} & \text{wrap unit}
\end{array}
$$

Fig. 3. Types of the placement calculus **D** and their interpretation

3.2   All Connectives

We consider type-logical connectives in the context of the placement sorting discipline. The connectives in types may surface as main connectives in either the antecedent or the succedent of sequents and some connectives are restricted with respect to which of these may occur. Hence we define sorted types of each of two polarities: input ($\bullet$) or antecedent and output ($\circ$) or succedent; where p is a polarity, $\bar{p}$ is the opposite polarity. The types formed by primitive connectives together with the type map T are defined as shown in Figure 4. The structural

$$
\begin{array}{lll}
\mathcal{F}_j^p ::= \mathcal{F}_i^{\bar{p}} \backslash \mathcal{F}_{i+j}^p & T(A\backslash C) = T(A) \to T(C) & \\
\mathcal{F}_i^p ::= \mathcal{F}_{i+j}^p / \mathcal{F}_j^{\bar{p}} & T(C/B) = T(B) \to T(C) & \\
\mathcal{F}_{i+j}^p ::= \mathcal{F}_i^p \bullet \mathcal{F}_j^p & T(A\bullet B) = T(A) \& T(B) & \\
\mathcal{F}_0^{\bar{p}} ::= I & T(I) = \top & \\
\mathcal{F}_j^p ::= \mathcal{F}_{i+1}^{\bar{p}} \downarrow_k \mathcal{F}_{i+j}^p & T(A\downarrow_k C) = T(A) \to T(C) & \\
\mathcal{F}_{i+1}^p ::= \mathcal{F}_{i+j}^p \uparrow_k \mathcal{F}_j^{\bar{p}} & T(C\uparrow_k B) = T(B) \to T(C) & \\
\mathcal{F}_{i+j}^p ::= \mathcal{F}_{i+1}^p \odot_k \mathcal{F}_j^p & T(A\odot_k B) = T(A) \& T(B) & \\
\mathcal{F}_1^{\bar{p}} ::= J & T(J) = \top & \\
\mathcal{F}_i^p ::= \mathcal{F}_i^p \& \mathcal{F}_i^p & T(A\&B) = T(A) \& T(B) & \text{additive conjunction [5,12]} \\
\mathcal{F}_i^p ::= \mathcal{F}_i^p \oplus \mathcal{F}_i^p & T(A\oplus B) = T(A) + T(B) & \text{additive disjunction [5,12]} \\
\mathcal{F}_i^p ::= \mathcal{F}_i^p \sqcap \mathcal{F}_i^p & T(A\sqcap B) = T(A) = T(B) & \text{sem. inert additive conjunction [22]} \\
\mathcal{F}_i^p ::= \mathcal{F}_i^p \sqcup \mathcal{F}_i^p & T(A\sqcup B) = T(A) = T(B) & \text{sem. inert additive disjunction [22]} \\
\mathcal{F}_i^p ::= \Box \mathcal{F}_i^p & T(\Box A) = \mathbf{L}T(A) & \text{modality [13]} \\
\mathcal{F}_i^p ::= \blacksquare \mathcal{F}_i^p & T(\blacksquare A) = T(A) & \text{rigid designator modality} \\
\mathcal{F}_0^p ::= \,! \mathcal{F}_0^p & T(!A) = T(A) & \text{structural modality [3]} \\
\mathcal{F}_i^p ::= \langle\,\rangle \mathcal{F}_i^p & T(\langle\,\rangle A) = T(A) & \text{exist. bracket modality [14,7]} \\
\mathcal{F}_i^p ::= [\,]^{-1} \mathcal{F}_i^p & T([\,]^{-1} A) = T(A) & \text{univ. bracket modality [14,7]} \\
\mathcal{F}_i^p ::= \forall X \mathcal{F}_i^p & T(\forall x A) = T(A) & \text{1st order univ. qu. [22]} \\
\mathcal{F}_i^p ::= \exists X \mathcal{F}_i & T(\exists x A) = T(A) & \text{1st order exist. qu. [22]} \\
\mathcal{F}_0^\circ ::= \mathcal{F}_0^\circ{}^+ & T(A^+) = list(T(A)) & \text{Kleene plus [22]} \\
\mathcal{F}_i^\circ ::= \neg \mathcal{F}_i^\circ & T(\neg A) = \bot & \text{negation-as-failure [19]}
\end{array}
$$

Fig. 4. Primitive connectives

$$\triangleright^{-1}A =_{df} J\backslash A \quad \{s|\ 1{+}s \in A\} \quad T(\triangleright^{-1}A) = T(A) \text{ right projection [20]}$$
$$\triangleleft^{-1}A =_{df} A/J \quad \{s|s{+}1 \in A\} \quad T(\triangleleft^{-1}A) = T(A) \text{ left projection [20]}$$
$$\triangleright A =_{df} J{\bullet}A \quad \{1{+}s|\ s \in A\} \quad T(\triangleright A) = T(A) \text{ right injection [20]}$$
$$\triangleleft A =_{df} A{\bullet}J \quad \{s{+}1|\ s \in A\} \quad T(\triangleleft A) = T(A) \text{ left injection [20]}$$
$$\check{}^{k}A =_{df} A{\uparrow}_{k}I \quad \{s|\ s{\times}_{k}0 \in A\} \quad T(\check{}^{k}A) = T(A) \text{ split [17]}$$
$$\hat{}^{k}A =_{df} A{\odot}_{k}I \quad \{s{\times}_{k}0|\ s \in A\} \quad T(\hat{}^{k}A) = T(A) \text{ bridge [17]}$$

Fig. 5. Unary derived connectives

modality and Kleene plus are limited to types of sort 0 because structural operations of contraction and expansion would not preserve other sorts. The Kleene plus and negation-as-failure are restricted to succedent polarity occurrences.

In addition to the primitive connectives we may define derived connectives which do not extend expressivity, but which permit abbreviations. Unary derived connectives are given in Figure 5. Continuous and discontinuous nondeterministic binary derived connectives are given in Figure 6, where $+(s_1, s_2, s_3)$ if and only if $s_3 = s_1{+}s_2$ or $s_3 = s_2{+}s_1$, and $\times(s_1, s_2, s_3)$ if and only if $s_3 = s_1 \times_1 s_2$ or ... or $s_3 = s_1 \times_n s_2$ where s_1 is of sort n.

$$\frac{B}{A} \qquad (A\backslash B) \sqcap (B/A) \qquad \{s|\ \forall s' \in A, s_3, +(s, s', s_3) \Rightarrow s_3 \in B\}$$
$$A \otimes B \qquad (A{\bullet}B) \sqcup (B{\bullet}A) \qquad \{s_3|\ \exists s_1 \in A, s_2 \in B, +(s_1, s_2, s_3)\}$$
$$A{\Downarrow}C \qquad (A{\downarrow}_1 C) \sqcap \cdots \sqcap (A{\downarrow}_{\sigma A}C) \qquad \{s_2|\ \forall s_1 \in A, s_3, \times(s_1, s_2, s_3) \Rightarrow s_3 \in C\}$$
$$C{\Uparrow}B \qquad (C{\uparrow}_1 B) \sqcap \cdots \sqcap (C{\uparrow}_{\sigma C}B) \qquad \{s_1|\ \forall s_2 \in B, s_3, \times(s_1, s_2, s_3) \Rightarrow s_3 \in C\}$$
$$A{\circledcirc}B \qquad (A{\odot}_1 B) \sqcup \cdots \sqcup (A{\odot}_{\sigma A}B) \qquad \{s_3|\ \exists s_1 \in A, s_2 \in B, \times(s_1, s_2, s_3)\}$$

$$T(\tfrac{B}{A}) = T(A) \to T(B) \qquad \text{nondet. division}$$
$$T(A \otimes B) = T(A)\&T(B) \qquad \text{nondet. product}$$
$$T(A{\Downarrow}C) = T(A) \to T(C) \qquad \text{nondet. infix}$$
$$T(C{\Uparrow}B) = T(B) \to T(C) \qquad \text{nondet. circumfix}$$
$$T(A{\circledcirc}B) = T(A)\&T(B) \qquad \text{nondet. wrap}$$

Fig. 6. Binary nondeterministic derived connectives

4   Calculus

The set $\mathcal{O}$ of *configurations* of hypersequent calculus for our categorial logic is defined as follows, where Λ is the empty string and * is the metalinguistic separator or *hole*:

$$(7)\ \mathcal{O} ::= \Lambda\ |\ *\ |\ \mathcal{F}_0\ |\ \mathcal{F}_{i+1}\{\underbrace{\mathcal{O}:\ldots:\mathcal{O}}_{i+1\ \mathcal{O}\text{'s}}\}\ |\ \mathcal{O},\mathcal{O}\ |\ [\mathcal{O}]$$

The sort of a configuration Γ is the number of holes it contains: $|\Gamma|_*$. Where Δ is a configuration of sort $k{+}i, k > 0$ and Γ is a configuration, $\Delta|_{+k}\Gamma$ ($\Delta|_{-k}\Gamma$)

is the configuration resulting from replacing by Γ the k-th hole from the left (right) in Δ. The *figure* $\overrightarrow{A}$ of a type A is defined by:

$$(8) \quad \overrightarrow{A} = \begin{cases} A & \text{if } sA = 0 \\ A\{\underbrace{* : \ldots : *}_{sA \ *\text{'s}}\} & \text{if } sA > 0 \end{cases}$$

The usual configuration distinguished occurrence notation $\Delta(\Gamma)$ signifies a configuration Δ with a distinguished subconfiguration Γ, i.e. a configuration occurrence Γ with (external) context Δ. In the hypersequent calculus the distinguished *hyperoccurrence* notation $\Delta\langle\Gamma\rangle$ signifies a configuration *hyper*occurrence Γ with external *and* internal context Δ as follows: where Γ is a configuration of sort i and $\Delta_1, \ldots, \Delta_i$ are configurations, the *fold* $\Gamma \otimes \langle\Delta_1, \ldots, \Delta_i\rangle$ is the result of replacing the successive holes in Γ by $\Delta_1, \ldots, \Delta_i$ respectively; the *distinguished hyperoccurrence* notation $\Delta\langle\Gamma\rangle$ represents $\Delta_0(\Gamma \otimes \langle\Delta_1, \ldots, \Delta_i\rangle)$.

A *sequent* $\Gamma \Rightarrow A$ comprises an antecedent configuration Γ of sort i and a succedent type A of sort i. The types which are allowed to enter into the antecedent are the input ($^\bullet$) types and the types which are allowed to enter into the succedent are the output ($^\circ$) types. The hypersequent calculus for the placement categorial logic defined in the previous section has the following identity axiom:

$$(9) \quad \overline{\overrightarrow{A} \Rightarrow A} \ id$$

The logical rules for primitive multiplicatives, additives, exponentials,[3] modalities and quantifiers are given in Figures 7, 8, 9, 10 and 11 respectively.

The rules for the unary and binary derived connectives are shown in Figures 12 and 13.

5 Grammar

We give a grammar for the Montague fragment of Dowty, Wall and Peters ([4], Ch. 7). We structure atomic types N for name or (referring) nominal and CN for common noun or count noun with feature terms for gender for which there are feature constants m (masculine), f (feminine) and n (neuter) and a denumerably infinit supply of feature variables. Feature variables are understood as being

---

[3] As given, the contraction rules, which are for parastic gaps ([23], Ch. 5), can be applied only a finite number of times in backward-chaining proof search since they are conditioned on brackets. Alternatively, the contraction rules may be given the form:

$$\frac{\Delta\langle !A, [!A, \Gamma]\rangle \Rightarrow B}{\Delta\langle !A, \Gamma\rangle \Rightarrow B} \ !C \qquad \frac{\Delta\langle [\Gamma, !A], !A\rangle \Rightarrow B}{\Delta\langle \Gamma, !A, \rangle \Rightarrow B} \ !C$$

We think there would still be decidability if there were a bound on the number of brackets it would be appropriate to introduce applying the rules from conclusion to premise, but this needs to be examined in detail.

$$\frac{\Gamma \Rightarrow A \qquad \Delta\langle \overrightarrow{C}\rangle \Rightarrow D}{\Delta\langle \Gamma, \overrightarrow{A\backslash C}\rangle \Rightarrow D} \backslash L \qquad \frac{\overrightarrow{A}, \Gamma \Rightarrow C}{\Gamma \Rightarrow A\backslash C} \backslash R$$

$$\frac{\Gamma \Rightarrow B \qquad \Delta\langle \overrightarrow{C}\rangle \Rightarrow D}{\Delta\langle \overrightarrow{C/B}, \Gamma\rangle \Rightarrow D} /L \qquad \frac{\Gamma, \overrightarrow{B} \Rightarrow C}{\Gamma \Rightarrow C/B} /R$$

$$\frac{\Delta\langle \overrightarrow{A}, \overrightarrow{B}\rangle \Rightarrow D}{\Delta\langle \overrightarrow{A \bullet B}\rangle \Rightarrow D} \bullet L \qquad \frac{\Gamma_1 \Rightarrow A \qquad \Gamma_2 \Rightarrow B}{\Gamma_1, \Gamma_2 \Rightarrow A \bullet B} \bullet R$$

$$\frac{\Delta\langle \Lambda\rangle \Rightarrow A}{\Delta\langle \overrightarrow{I}\rangle \Rightarrow A} IL \qquad \frac{}{\Lambda \Rightarrow I} IR$$

$$\frac{\Gamma \Rightarrow A \qquad \Delta\langle \overrightarrow{C}\rangle \Rightarrow D}{\Delta\langle \Gamma|_k \overrightarrow{A\downarrow_k C}\rangle \Rightarrow D} \downarrow_k L \qquad \frac{\overrightarrow{A}|_k \Gamma \Rightarrow C}{\Gamma \Rightarrow A\downarrow_k C} \downarrow_k R$$

$$\frac{\Gamma \Rightarrow B \qquad \Delta\langle \overrightarrow{C}\rangle \Rightarrow D}{\Delta\langle \overrightarrow{C\uparrow_k B}|_k \Gamma\rangle \Rightarrow D} \uparrow_k L \qquad \frac{\Gamma|_k \overrightarrow{B} \Rightarrow C}{\Gamma \Rightarrow C\uparrow_k B} \uparrow_k R$$

$$\frac{\Delta\langle \overrightarrow{A}|_k \overrightarrow{B}\rangle \Rightarrow D}{\Delta\langle \overrightarrow{A\odot_k B}\rangle \Rightarrow D} \odot_k L \qquad \frac{\Gamma_1 \Rightarrow A \qquad \Gamma_2 \Rightarrow B}{\Gamma_1|_k \Gamma_2 \Rightarrow A\odot_k B} \odot_k R$$

$$\frac{\Delta\langle *\rangle \Rightarrow A}{\Delta\langle \overrightarrow{J}\rangle \Rightarrow A} JL \qquad \frac{}{* \Rightarrow J} JR$$

Fig. 7. Multiplicative rules

universally quantified outermost in types and thus undergo unification in the usual way. Other atomic types are S for statement or (declarative) sentence and CP for complementizer phrase. All these atomic types are of sort 0. Our lexicon for the Montague fragment is as shown in Figure 14; henceforth we omit the subscript $(+)1$ for first wrap on connectives and abbreviate as $-$ the subscript -1 for last wrap.

6   Analyses

We analyse the examples from the second half of Chapter 7 of Dowty, Wall and Peters [4] — DWP; the example numbers of that source are included within displays. The first examples involve the copula of identity. Minimally:

(10) (7-73) **john+is+bill** : S

$$\frac{\Gamma\langle\overrightarrow{A}\rangle \Rightarrow C}{\Gamma\langle\overrightarrow{A\&B}\rangle \Rightarrow C}\&L_1 \qquad \frac{\Gamma\langle\overrightarrow{B}\rangle \Rightarrow C}{\Gamma\langle\overrightarrow{A\&B}\rangle \Rightarrow C}\&L_2$$

$$\frac{\Gamma \Rightarrow A \qquad \Gamma \Rightarrow B}{\Gamma \Rightarrow A\&B}\&R$$

$$\frac{\Gamma\langle\overrightarrow{A}\rangle \Rightarrow C \qquad \Gamma\langle\overrightarrow{B}\rangle \Rightarrow C}{\Gamma\langle\overrightarrow{A\oplus B}\rangle \Rightarrow C}\oplus L$$

$$\frac{\Gamma \Rightarrow A}{\Gamma \Rightarrow A\oplus B}\oplus L_1 \qquad \frac{\Gamma \Rightarrow B}{\Gamma \Rightarrow A\oplus B}\oplus L_2$$

$$\frac{\Gamma\langle\overrightarrow{A}\rangle \Rightarrow C}{\Gamma\langle\overrightarrow{A\sqcap B}\rangle \Rightarrow C}\sqcap L_1 \qquad \frac{\Gamma\langle\overrightarrow{B}\rangle \Rightarrow C}{\Gamma\langle\overrightarrow{A\sqcap B}\rangle \Rightarrow C}\sqcap L_2$$

$$\frac{\Gamma \Rightarrow A \qquad \Gamma \Rightarrow B}{\Gamma \Rightarrow A\sqcap B}\sqcap R$$

$$\frac{\Gamma\langle\overrightarrow{A}\rangle \Rightarrow C \qquad \Gamma\langle\overrightarrow{B}\rangle \Rightarrow C}{\Gamma\langle\overrightarrow{A\sqcup B}\rangle \Rightarrow C}\sqcup L$$

$$\frac{\Gamma \Rightarrow A}{\Gamma \Rightarrow A\sqcup B}\sqcup L_1 \qquad \frac{\Gamma \Rightarrow B}{\Gamma \Rightarrow A\sqcup B}\sqcup L_2$$

Fig. 8. Additive rules

For this there is the semantically labelled sequent:

(11) $\Box Nm : \hat{\ }j, \Box((NA\backslash S)/NB) : \hat{\ }\lambda C\lambda D[D = C], \Box Nm : \hat{\ }b \Rightarrow S$

This has the derivation given in Figure 15. It delivers semantics:

(12) $[j = b]$

More subtly:

(13) (7-76) **john+is+a+man** : S

Inserting the same lexical entry for the copula, lexical lookup yields the semantically annotated sequent:

(14) $\Box Nm : \hat{\ }j, \Box((NA\backslash S)/NB) : \hat{\ }\lambda C\lambda D[D = C], \Box(((S\uparrow\Box NE)\downarrow S)/CNE) :$
$\hat{\ }\lambda F\lambda G\exists H[(F\ H) \wedge (G\ \hat{\ }H)], \Box CNm : man \Rightarrow S$

This has the derivation given in Figure 16. The derivation delivers the semantics:

(15) $\exists C[(\check{\ }man\ C) \wedge [j = C]]$

$$\frac{\Gamma(A) \Rightarrow B}{\Gamma(!A) \Rightarrow B}\,!L \qquad \frac{!A_1,\ldots,!A_n \Rightarrow A}{!A_1,\ldots,!A_n \Rightarrow !A}\,!R$$

$$\frac{\Delta\langle !A,\Gamma\rangle \Rightarrow B}{\Delta\langle\Gamma,!A\rangle \Rightarrow B}\,!P \qquad \frac{\Delta\langle\Gamma,!A\rangle \Rightarrow B}{\Delta\langle !A,\Gamma\rangle \Rightarrow B}\,!P$$

$$\frac{\Delta\langle !A,[!A,\Gamma]\rangle \Rightarrow B}{\Delta\langle !A,[[\Gamma]]\rangle \Rightarrow B}\,!C \qquad \frac{\Delta\langle[\Gamma,!A],!A\rangle \Rightarrow B}{\Delta\langle[[\Gamma]],!A,\rangle \Rightarrow B}\,!C$$

$$\frac{\Gamma \Rightarrow A}{\Gamma \Rightarrow A^+}\,{}^+R \qquad \frac{\Gamma \Rightarrow A \quad \Delta \Rightarrow A^+}{\Gamma,\Delta \Rightarrow A^+}\,{}^+R$$

Fig. 9. Exponential rules

$$\frac{\Gamma\langle\overrightarrow{A}\rangle \Rightarrow B}{\Gamma\langle\overrightarrow{\Box A}\rangle \Rightarrow B}\,\Box L \qquad \frac{\Box/\blacksquare\Gamma \Rightarrow A}{\Box/\blacksquare\Gamma \Rightarrow \Box A}\,\Box R$$

$$\frac{\Gamma\langle\overrightarrow{A}\rangle \Rightarrow B}{\Gamma\langle\overrightarrow{\blacksquare A}\rangle \Rightarrow B}\,\blacksquare L \qquad \frac{\Box/\blacksquare\Gamma \Rightarrow A}{\Box/\blacksquare\Gamma \Rightarrow \blacksquare A}\,\blacksquare R$$

$$\frac{\Delta\langle\overrightarrow{A}\rangle \Rightarrow B}{\Delta\langle\overrightarrow{[[]^{-1}A]}\rangle \Rightarrow B}\,[]^{-1}L \qquad \frac{[\Gamma] \Rightarrow A}{\Gamma \Rightarrow []^{-1}A}\,[]^{-1}R$$

$$\frac{\Delta\langle\overrightarrow{[A]}\rangle \Rightarrow B}{\Delta\langle\overrightarrow{\langle\rangle A}\rangle \Rightarrow B}\,\langle\rangle L \qquad \frac{\Gamma \Rightarrow A}{[\Gamma] \Rightarrow \langle\rangle A}\,\langle\rangle R$$

Fig. 10. Normal (semantic) and bracket (syntactic) modality rules, where $\Box/\blacksquare\Gamma$ signifies a configuration all the types of which have main connective $\Box$ or $\blacksquare$

$$\frac{\Gamma\langle\overrightarrow{A[t/x]}\rangle \Rightarrow B}{\Gamma\langle\overrightarrow{\forall x A}\rangle \Rightarrow B}\,\forall L \qquad \frac{\Gamma \Rightarrow A[a/x]}{\Gamma \Rightarrow \forall x A}\,\forall R^\dagger$$

$$\frac{\Gamma\langle\overrightarrow{A[a/x]}\rangle \Rightarrow B}{\Gamma\langle\overrightarrow{\exists x A}\rangle \Rightarrow B}\,\exists L^\dagger \qquad \frac{\Gamma \Rightarrow A[t/x]}{\Gamma \Rightarrow \exists x A}\,\exists R$$

Fig. 11. Quantifier rules, where † indicates that there is no a in the conclusion

$$\frac{\Gamma\langle\overrightarrow{A}\rangle \Rightarrow B}{\Gamma\langle\overrightarrow{\triangleleft^{-1}A},*\rangle \Rightarrow B}\triangleleft^{-1}L \qquad \frac{\Gamma,* \Rightarrow A}{\Gamma \Rightarrow \triangleleft^{-1}A}\triangleleft^{-1}R$$

$$\frac{\Gamma\langle\overrightarrow{A},*\rangle \Rightarrow B}{\Gamma\langle\overrightarrow{\triangleleft A}\rangle \Rightarrow B}\triangleleft L \qquad \frac{\Gamma \Rightarrow A}{\Gamma,* \Rightarrow \triangleleft A}\triangleleft R$$

$$\frac{\Gamma\langle\overrightarrow{A}\rangle \Rightarrow B}{\Gamma\langle *,\overrightarrow{\triangleright^{-1}A}\rangle \Rightarrow B}\triangleright^{-1}L \qquad \frac{*,\Gamma \Rightarrow A}{\Gamma \Rightarrow \triangleright^{-1}A}\triangleright^{-1}R$$

$$\frac{\Gamma\langle *,\overrightarrow{A}\rangle \Rightarrow B}{\Gamma\langle\overrightarrow{\triangleright A}\rangle \Rightarrow B}\triangleright L \qquad \frac{\Gamma \Rightarrow A}{*,\Gamma \Rightarrow \triangleright A}\triangleright R$$

$$\frac{\Delta\langle\overrightarrow{B}\rangle \Rightarrow C}{\Delta\langle\overrightarrow{{}^{\smile k}B}|_k\Lambda\rangle \Rightarrow C}{}^{\smile k}L \qquad \frac{\Delta|_k\Lambda \Rightarrow B}{\Delta \Rightarrow {}^{\smile k}B}{}^{\smile k}R$$

$$\frac{\Delta\langle\overrightarrow{B}|_k\Lambda\rangle \Rightarrow C}{\Delta\langle\overrightarrow{{}^{\wedge k}B}\rangle \Rightarrow C}{}^{\wedge k}L \qquad \frac{\Delta \Rightarrow B}{\Delta|_k\Lambda \Rightarrow {}^{\wedge k}B}{}^{\wedge k}R$$

Fig. 12. Unary derived connective rules

This is logically equivalent to $({}^{\vee}man\ j)$, as required. This correct interaction of the copula of identity with an indefinitely quantified complement is a nice prediction of Montague grammar, conserved in type logical grammar, and simplified by the lower type of the copula.

The next example involves an intensional adsentential modifier:

(16) (7-83) **necessarily+john+walks** : S

Lexical lookup yields the following semantically labelled sequent:

(17) $\Box(S/\Box S) : {}^{\wedge}nec, \Box Nm : {}^{\wedge}j, \Box(NA\backslash S) : walk \Rightarrow S$

This has the derivation given in Figure 17. The derivation delivers semantics:

(18) $(nec\ {}^{\wedge}({}^{\smile}walk\ j))$

The following example involves an adverb:

(19) (7-86) **john+walks+slowly** : S

This is also assumed to create an intensional context. Lexical lookup yields:

(20) $\Box Nm : {}^{\wedge}j, \Box(NA\backslash S) : walk, \Box(\Box(NB\backslash S)\backslash(NB\backslash S)) : slowly \Rightarrow S$

This has the derivation given in Figure 18, which delivers semantics (in η-long form):

$$\frac{\Gamma \Rightarrow A \quad \Delta\langle \vec{C}\rangle \Rightarrow D}{\Delta\langle \Gamma, \dfrac{\vec{C}}{A}\rangle \Rightarrow D}\,\text{-}L_1 \qquad\qquad \frac{\Gamma \Rightarrow A \quad \Delta\langle \vec{C}\rangle \Rightarrow D}{\Delta\langle \dfrac{\vec{C}}{A}, \Gamma\rangle \Rightarrow D}\,\text{-}L_2$$

$$\frac{\vec{A}, \Gamma \Rightarrow C \quad \Gamma, \vec{A} \Rightarrow C}{\Gamma \Rightarrow \dfrac{C}{A}}\,\text{-}R$$

$$\frac{\Delta\langle \vec{A}, \vec{B}\rangle \Rightarrow D \quad \Delta\langle \vec{B}, \vec{A}\rangle \Rightarrow D}{\Delta\langle \overrightarrow{A \otimes B}\rangle \Rightarrow D}\,\otimes L$$

$$\frac{\Gamma_1 \Rightarrow A \quad \Gamma_2 \Rightarrow B}{\Gamma_1, \Gamma_2 \Rightarrow A \otimes B}\,\otimes R_1 \qquad\qquad \frac{\Gamma_1 \Rightarrow B \quad \Gamma_2 \Rightarrow A}{\Gamma_1, \Gamma_2 \Rightarrow A \otimes B}\,\otimes R_2$$

$$\frac{\Gamma \Rightarrow A \quad \Delta\langle \vec{C}\rangle \Rightarrow D}{\Delta\langle \Gamma|_k \overrightarrow{A{\Downarrow}C}\rangle \Rightarrow D}\,{\Downarrow}L \qquad\qquad \frac{\vec{A}|_1\Gamma \Rightarrow C \quad \cdots \quad \vec{A}|_{\sigma A}\Gamma \Rightarrow C}{\Gamma \Rightarrow A{\Downarrow}C}\,{\Downarrow}R$$

$$\frac{\Gamma \Rightarrow B \quad \Delta\langle \vec{C}\rangle \Rightarrow D}{\Delta\langle \overrightarrow{C{\Uparrow}B}|_k\Gamma\rangle \Rightarrow D}\,{\Uparrow}L \qquad\qquad \frac{\Gamma|_1\vec{B} \Rightarrow C \quad \cdots \quad \Gamma|_{\sigma C}\vec{B} \Rightarrow C}{\Gamma \Rightarrow C{\Uparrow}B}\,{\Uparrow}R$$

$$\frac{\Delta\langle \vec{A}|_1\vec{B}\rangle \Rightarrow D \quad \cdots \quad \Delta\langle \vec{A}|_{\sigma A}\vec{B}\rangle \Rightarrow D}{\Delta\langle \overrightarrow{A{\odot}B}\rangle \Rightarrow D}\,{\odot}L \qquad \frac{\Gamma_1 \Rightarrow A \quad \Gamma_2 \Rightarrow B}{\Gamma_1|_k\Gamma_2 \Rightarrow A{\odot}B}\,{\odot}R$$

Fig. 13. Binary derived connective rules

(21) $((\check{\ }slowly\ \hat{\ }\lambda A(\check{\ }walk\ A))\ j)$

The next example involves an equi control verb:

(22) (7-91) **john+tries+to+walk** : S

We lexically analyse the equi semantics as a relation of trying between the subject and a proposition of which the subject is agent (something Montague did not do). Lexical lookup yields:

(23) $\Box Nm : \hat{\ }j, \Box((NA\backslash S)/\Box(NA\backslash S)) : \hat{\ }\lambda B\lambda C((\check{\ }tries\ \hat{\ }(\check{\ }B\ C))\ C),$
$\Box((ND\backslash S)/(ND\backslash S)) : \hat{\ }\lambda EE, \Box(NF\backslash S) : walk \Rightarrow S$

This has the derivation given in Figure 19, which delivers the semantics:

(24) $((\check{\ }tries\ \hat{\ }(\check{\ }walk\ j))\ j)$

I.e. that John tries to bring about the state of affairs that he (John) walks.

The next example involves control, quantification, coordination and also anaphora:

a : $\Box(((S{\uparrow}\Box NA){\downarrow}S)/CNA)$: ˆ$\lambda B\lambda C\exists D[(B\ D)\wedge(C\ {\char`^}D)]$
and : $\Box((S\backslash S)/S)$: ˆ$\lambda A\lambda B[B\wedge A]$
and : $\Box(((NA\backslash S)\backslash(NA\backslash S))/(NA\backslash S))$: ˆ$\lambda B\lambda C\lambda D[(C\ D)\wedge(B\ D)]$
believes : $\Box((NA\backslash S)/CP)$: $believe$
bill : $\Box Nm$: ˆb
catch : $\Box((NA\backslash S)/NB)$: $catch$
doesnt : $\Box((NA\backslash S)/(NA\backslash S))$: ˆ$\lambda B\lambda C\neg(B\ C)$
eat : $\Box((NA\backslash S)/NB)$: eat
every : $\Box(((S{\uparrow}NA){\downarrow}S)/CNA)$: ˆ$\lambda B\lambda C\forall D[(B\ D)\to(C\ D)]$
finds : $\Box((NA\backslash S)/NB)$: $finds$
fish : $\Box CNn$: $fish$
he : $\Box((\Box S|Nm)/\Box(Nm\backslash S))$: ˆ$\lambda A\lambda B{\char`^}(\check{}\ A\ B)$
her : $\Box(\Box((S{\uparrow}Nf)-(J\bullet(Nf\backslash S))){\downarrow}(\Box S|Nf))$: ˆ$\lambda A\lambda B{\char`^}(\check{}\ A\ B)$
her : $\Box(((((S{\uparrow}Nf)-(J\bullet(Nf\backslash S))){\uparrow}\Box Nf)-(J\bullet((Nf\backslash S){\uparrow}Nf))){\downarrow}_{-}(S{\uparrow}\Box Nf))$: ˆ$\lambda A\lambda B((A\ B)\ \check{}\ B)$
in : $\Box(((NA\backslash S)\backslash(NA\backslash S))/NB)$: ˆ$\lambda C\lambda D\lambda E((\check{}\ in\ C)\ (D\ E))$
is : $\Box((NA\backslash S)/NB)$: ˆ$\lambda C\lambda D[D=C]$
it : $\Box(\Box(S{\uparrow}Nn){\downarrow}(\Box S|Nn))$: ˆ$\lambda A\lambda B{\char`^}(\check{}\ A\ B)$
it : $\Box((((((S{\uparrow}Nn)-(J\bullet(Nn\backslash S))){\uparrow}\Box Nn)-(J\bullet((Nn\backslash S){\uparrow}Nn))){\downarrow}_{-}(S{\uparrow}\Box Nn))$: ˆ$\lambda A\lambda B((A\ B)\ \check{}\ B)$
john : $\Box Nm$: ˆj
loses : $\Box((NA\backslash S)/NB)$: $loses$
loves : $\Box((NA\backslash S)/NB)$: $loves$
man : $\Box CNm$: man
necessarily : $\Box(S/\Box S)$: ˆnec
or : $\Box((S\backslash S)/S)$: ˆ$\lambda A\lambda B[B\vee A]$
or : $\Box(((NA\backslash S)\backslash(NA\backslash S))/(NA\backslash S))$: ˆ$\lambda B\lambda C\lambda D[(C\ D)\vee(B\ D)]$
park : $\Box CNn$: $park$
seeks : $\Box((NA\backslash S)/\Box(((NB\backslash S)/NC)\backslash(NB\backslash S)))$: ˆ$\lambda D\lambda E((tries\ {\char`^}((\check{}\ D\ find)\ E))\ E)$
she : $\Box((\Box S|Nf)/\Box(Nf\backslash S))$: ˆ$\lambda A\lambda B{\char`^}(\check{}\ A\ B)$
slowly : $\Box(\Box(NA\backslash S)\backslash(NA\backslash S))$: $slowly$
such+that : $\Box((CNA\backslash CNA)/(S|NA))$: ˆ$\lambda B\lambda C\lambda D[(C\ D)\wedge(B\ D)]$
talks : $\Box(NA\backslash S)$: $talk$
that : $\Box(CP/\Box S)$: ˆλAA
the : $\Box(NA/CNA)$: the
to : $\Box((NA\backslash S)/(NA\backslash S))$: ˆλBB
tries : $\Box((NA\backslash S)/\Box(NA\backslash S))$: ˆ$\lambda B\lambda C((\check{}\ tries\ {\char`^}(\check{}\ B\ C))\ C)$
unicorn : $\Box CNn$: $unicorn$
walk : $\Box(NA\backslash S)$: $walk$
walks : $\Box(NA\backslash S)$: $walk$
woman : $\Box CNf$: $woman$

Fig. 14. The Montague fragment

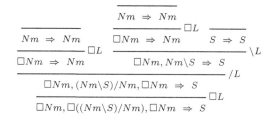

Fig. 15. Derivation for *John is Bill*

$$
\cfrac{
  \cfrac{
    \cfrac{Nm \Rightarrow Nm}{\Box Nm \Rightarrow Nm}\,\Box L
    \qquad
    \cfrac{
      \cfrac{
        \cfrac{
          \cfrac{
            \cfrac{Nm \Rightarrow Nm}{\Box Nm \Rightarrow Nm}\,\Box L \qquad S \Rightarrow S
          }{\Box Nm, Nm\backslash S \Rightarrow S}\,\backslash L
        }{\Box Nm, (Nm\backslash S)/Nm, \Box Nm \Rightarrow S}\,/L
      }{\Box Nm, \Box((Nm\backslash S)/Nm), \Box Nm \Rightarrow S}\,\Box L
      }{\Box Nm, \Box((Nm\backslash S)/Nm), 1 \Rightarrow S{\uparrow}\Box Nm}\,{\uparrow}R
      \qquad S \Rightarrow S
    }{\Box Nm, \Box((Nm\backslash S)/Nm), (S{\uparrow}\Box Nm){\downarrow}S \Rightarrow S}\,{\downarrow}L
  }{}\,
}{}
$$

$$
\cfrac{
  \cfrac{CNm \Rightarrow CNm}{\Box CNm \Rightarrow CNm}\,\Box L
  \qquad
  \Box Nm, \Box((Nm\backslash S)/Nm), (S{\uparrow}\Box Nm){\downarrow}S \Rightarrow S
}{
  \cfrac{\Box Nm, \Box((Nm\backslash S)/Nm), ((S{\uparrow}\Box Nm){\downarrow}S)/CNm, \Box CNm \Rightarrow S}{\Box Nm, \Box((Nm\backslash S)/Nm), \Box(((S{\uparrow}\Box Nm){\downarrow}S)/CNm), \Box CNm \Rightarrow S}\,\Box L
}\,/L
$$

Fig. 16. Derivation for *John is a man*

$$
\cfrac{
  \cfrac{
    \cfrac{
      \cfrac{
        \cfrac{
          \cfrac{\cfrac{Nm \Rightarrow Nm}{\Box Nm \Rightarrow Nm}\,\Box L \qquad S \Rightarrow S}{\Box Nm, Nm\backslash S \Rightarrow S}\,\backslash L
        }{\Box Nm, \Box(Nm\backslash S) \Rightarrow S}\,\Box L
      }{\Box Nm, \Box(Nm\backslash S) \Rightarrow \Box S}\,\Box R
      \qquad S \Rightarrow S
    }{S/\Box S, \Box Nm, \Box(Nm\backslash S) \Rightarrow S}\,/L
  }{\Box(S/\Box S), \Box Nm, \Box(Nm\backslash S) \Rightarrow S}\,\Box L
}{}
$$

Fig. 17. Derivation for *Necessarily John walks*

$$
\cfrac{
  \cfrac{
    \cfrac{
      \cfrac{\cfrac{Nm \Rightarrow Nm \qquad S \Rightarrow S}{Nm, Nm\backslash S \Rightarrow S}\,\backslash L}{Nm, \Box(Nm\backslash S) \Rightarrow S}\,\Box L
    }{\Box(Nm\backslash S) \Rightarrow Nm\backslash S}\,\backslash R
  }{\Box(Nm\backslash S) \Rightarrow \Box(Nm\backslash S)}\,\Box R
  \qquad
  \cfrac{
    \cfrac{\cfrac{Nm \Rightarrow Nm}{\Box Nm \Rightarrow Nm}\,\Box L \qquad S \Rightarrow S}{\Box Nm, Nm\backslash S \Rightarrow S}\,\backslash L
  }{}
}{
  \cfrac{\Box Nm, \Box(Nm\backslash S), \Box(Nm\backslash S)\backslash(Nm\backslash S) \Rightarrow S}{\Box Nm, \Box(Nm\backslash S), \Box(\Box(Nm\backslash S)\backslash(Nm\backslash S)) \Rightarrow S}\,\Box L
}\,\backslash L
$$

Fig. 18. Derivation for *John walks slowly*

$$\cfrac{\cfrac{\cfrac{\cfrac{\cfrac{\cfrac{Nm \Rightarrow Nm \quad S \Rightarrow S}{Nm, Nm\backslash S \Rightarrow S}\backslash L}{Nm, \Box(Nm\backslash S) \Rightarrow S}\Box L}{\Box(Nm\backslash S) \Rightarrow Nm\backslash S}\backslash R \quad \cfrac{Nm \Rightarrow Nm \quad S \Rightarrow S}{Nm, Nm\backslash S \Rightarrow S}\backslash L}{Nm, (Nm\backslash S)/(Nm\backslash S), \Box(Nm\backslash S) \Rightarrow S}/L}{\cfrac{\cfrac{Nm, \Box((Nm\backslash S)/(Nm\backslash S)), \Box(Nm\backslash S) \Rightarrow S}{\Box((Nm\backslash S)/(Nm\backslash S)), \Box(Nm\backslash S) \Rightarrow Nm\backslash S}\backslash R}{\Box((Nm\backslash S)/(Nm\backslash S)), \Box(Nm\backslash S) \Rightarrow \Box(Nm\backslash S)}\Box R \quad \cfrac{\cfrac{Nm \Rightarrow Nm}{\Box Nm \Rightarrow Nm}\Box L \quad S \Rightarrow S}{\Box Nm, Nm\backslash S \Rightarrow S}\backslash L}{\cfrac{\Box Nm, (Nm\backslash S)/\Box(Nm\backslash S), \Box((Nm\backslash S)/(Nm\backslash S)), \Box(Nm\backslash S) \Rightarrow S}{\Box Nm, \Box((Nm\backslash S)/\Box(Nm\backslash S)), \Box((Nm\backslash S)/(Nm\backslash S)), \Box(Nm\backslash S) \Rightarrow S}\Box L}$$/L ... □L

Fig. 19. Derivation for *John tries to walk*

(25) (7-94) **john+tries+to+catch+a+fish+and+eat+it** : S

The sentence is ambiguous as to whether *a fish* is wide scope (with existential commitment) or narrow scope (without existential commitment) with respect to *tries*, but in both cases it must be the antecedent of *it*. Lexical lookup inserting a sentential coordinator or the (clause) external anaphora pronoun assignment has no derivation. Lexical lookup inserting the verb phrase coordinator and the internal (clause local) anaphora pronoun assignment yields the semantically labelled sequent:

(26) $\Box Nm : \hat{\ } j, \Box((NA\backslash S)/\Box(NA\backslash S)) : \hat{\ }\lambda B\lambda C((\check{\ }tries \hat{\ }(\check{\ }B\ C))\ C),$
   $\Box((ND\backslash S)/(ND\backslash S)) : \hat{\ }\lambda EE, \Box((NF\backslash S)/NG) : catch,$
   $\Box(((S\!\uparrow\!\Box NH)\!\downarrow\!S)/CNH) : \hat{\ }\lambda I\lambda J\exists K[(I\ K) \wedge (J\ \hat{\ }K)], \Box CNn : fish,$
   $\Box(((NL\backslash S)\backslash(NL\backslash S))/(NL\backslash S)) : \hat{\ }\lambda M\lambda N\lambda O[(N\ O) \wedge (M\ O)],$
   $\Box((NP\backslash S)/NQ) : eat,$
   $\Box((((((S\!\uparrow\!Nn) - (J\bullet(Nn\backslash S)))\!\uparrow\!\Box Nn) - (J\bullet((Nn\backslash S)\!\uparrow\!Nn)))\!\downarrow_<(S\!\uparrow\!\Box Nn)) :$
   $\hat{\ }\lambda R\lambda S((R\ S)\ \check{\ }S) \Rightarrow S$

Because we do not have verb form features on S this has one derivation on the pattern *[tries to catch a fish]* and *[eat it]* in which a finite verb phrase coordinates with a base form verb phrase. This would be excluded as required by adding the features. A wide scope existential derivation delivers semantics with existential commitment as follows; the derivation is too large to fit on a page.

(27) $\exists C[(\check{\ }fish\ C) \wedge ((\check{\ }tries\ \hat{\ }[((\check{\ }catch\ C)\ j) \wedge ((\check{\ }eat\ C)\ j)])\ j)]$

Also because of the absence of verb form features, there is an existential narrow scope derivation on the pattern of *[to catch a fish]* and *[eat it]* in which an infinitive verb phrase coordinates with a base form verb phrase. This would also be straightforwardly ruled out by including the relevant features on S. An appropriate existential narrow scope derivation, which is too large to fit on the page, delivers the semantics without existential commitment:

(28) $((\check{~}tries\ \hat{~}\exists H[(\check{~}fish\ H) \wedge [((\check{~}catch\ H)\ j) \wedge ((\check{~}eat\ H)\ j)]])\ j)$

The next example involves an extensional transitive verb:

(29) (7-98) **john+finds+a+unicorn** : S

This sentence cannot be true unless a unicorn exists. Our treatment of this is simpler than Montague's because while Montague had to raise the type of extensional verbs to accommodate intensional verbs ("raising to the worst case"), and then use meaning postulates to capture the existential commitment, type logical grammar allows assignment of the lower types which capture it automatically. Lexical lookup yields:

(30) $\Box Nm : \hat{~}j, \Box((NA\backslash S)/NB) : finds, \Box(((S{\uparrow}\Box NC){\downarrow}S)/CN\,C) :$
$\hat{~}\lambda D\lambda E\exists F[(D\ F) \wedge (E\ \hat{~}F)], \Box CNn : unicorn\ \Rightarrow\ S$

$$\dfrac{\dfrac{}{Nm \Rightarrow Nm}}{\Box Nm \Rightarrow Nm}\,\Box L \qquad \dfrac{}{S \Rightarrow S}$$

$$\dfrac{\dfrac{}{Nn \Rightarrow Nn}}{\Box Nn \Rightarrow Nn}\,\Box L \qquad \dfrac{\Box Nm \Rightarrow Nm \quad S \Rightarrow S}{\Box Nm, Nm\backslash S \Rightarrow S}\,\backslash L$$

$$\dfrac{\Box Nn \Rightarrow Nn \quad \Box Nm, Nm\backslash S \Rightarrow S}{\Box Nm, (Nm\backslash S)/Nn, \Box Nn \Rightarrow S}\,/L$$

$$\dfrac{\Box Nm, (Nm\backslash S)/Nn, \Box Nn \Rightarrow S}{\Box Nm, \Box((Nm\backslash S)/Nn), \Box Nn \Rightarrow S}\,\Box L$$

$$\dfrac{\Box Nm, \Box((Nm\backslash S)/Nn), \Box Nn \Rightarrow S}{\Box Nm, \Box((Nm\backslash S)/Nn), 1 \Rightarrow S{\uparrow}\Box Nn}\,{\uparrow}R$$

$$\dfrac{\dfrac{}{CNn \Rightarrow CNn}}{\Box CNn \Rightarrow CNn}\,\Box L \qquad \dfrac{\Box Nm, \Box((Nm\backslash S)/Nn), 1 \Rightarrow S{\uparrow}\Box Nn \quad \dfrac{}{S \Rightarrow S}}{\Box Nm, \Box((Nm\backslash S)/Nn), (S{\uparrow}\Box Nn){\downarrow}S \Rightarrow S}\,{\downarrow}L$$

$$\dfrac{\Box CNn \Rightarrow CNn \quad \Box Nm, \Box((Nm\backslash S)/Nn), (S{\uparrow}\Box Nn){\downarrow}S \Rightarrow S}{\Box Nm, \Box((Nm\backslash S)/Nn), ((S{\uparrow}\Box Nn){\downarrow}S)/CNn, \Box CNn \Rightarrow S}\,/L$$

$$\dfrac{\Box Nm, \Box((Nm\backslash S)/Nn), ((S{\uparrow}\Box Nn){\downarrow}S)/CNn, \Box CNn \Rightarrow S}{\Box Nm, \Box((Nm\backslash S)/Nn), \Box(((S{\uparrow}\Box Nn){\downarrow}S)/CNn), \Box CNn \Rightarrow S}\,\Box L$$

Fig. 20. Derivation for *John finds a unicorn*

This has the derivation given in Figure 20, which yields the semantics with existential commitment:

(31) $\exists C[(\check{~}unicorn\ C) \wedge ((\check{~}finds\ C)\ j)]$

DWP continue with a donkey sentence, for which of course Montague grammar and our cover grammar make the wrong prediction:

(32) (7-105) **every+man+such+that+he+loves+a+woman+loses+her** : S

There is a dominant reading in which *a woman* which is the donkey anaphora antecedent is understood universally, but Montague semantics obtains only an at best subordinate reading in which *a woman* is quantified existentially at the matrix level. Lexical lookup inserting the external anaphora assignment to *her* yields no derivation. Lexical insertion of the internal anaphora assignment yields:

(33) $\Box(((S{\uparrow}NA){\downarrow}S)/CNA) : {}^{\wedge}\lambda B\lambda C\forall D[(B\ D) \rightarrow (C\ D)], \Box CNm : man,$
$\Box((CNE\backslash CNE)/(S|NE)) : {}^{\wedge}\lambda F\lambda G\lambda H[(G\ H) \wedge (F\ H)],$
$\Box((\Box S|Nm)/\Box(Nm\backslash S)) : {}^{\wedge}\lambda I\lambda J{}^{\wedge}({}^{\vee}I\ J), \Box((NK\backslash S)/NL) : loves,$
$\Box(((S{\uparrow}\Box NM){\downarrow}S)/CNM) : {}^{\wedge}\lambda N\lambda O\exists P[(N\ P) \wedge (O\ {}^{\wedge}P)], \Box CNf : woman,$
$\Box((NQ\backslash S)/NR) : loses,$
$\Box(((((S{\uparrow}Nf) - (J{\bullet}(Nf\backslash S))){\uparrow}\Box Nf) - (J{\bullet}((Nf\backslash S){\uparrow}Nf))){\downarrow}_{-}(S{\uparrow}\Box Nf)) :$
${}^{\wedge}\lambda S\lambda T((S\ T)\ {}^{\vee}T) \Rightarrow S$

The derivation of this is too large for the page, but it delivers semantics:

(34) $\exists C[({}^{\vee}woman\ C) \wedge \forall K[[({}^{\vee}man\ K) \wedge (({}^{\vee}loves\ C)\ K)] \rightarrow (({}^{\vee}loses\ C)\ K)]]$

The assignment of lowest types in type logical grammar also means that existential commitment of a preposition comes without the need for devices such as meaning postulates in Montague grammar:

(35) (7-110) **john+walks+in+a+park** : S

Lexical lookup for this example yields the semantically labelled sequent:

(36) $\Box Nm : {}^{\wedge}j, \Box(NA\backslash S) : walk, \Box(((NB\backslash S)\backslash(NB\backslash S))/NC) :$
${}^{\wedge}\lambda D\lambda E\lambda F(({}^{\vee}in\ D)\ (E\ F)), \Box(((S{\uparrow}\Box NG){\downarrow}S)/CNG) :$
${}^{\wedge}\lambda H\lambda I\exists J[(H\ J) \wedge (I\ {}^{\wedge}J)], \Box CNn : park \Rightarrow S$

This sequent has the proof given in Figure 21, which delivers the semantics (with existential commitment):

(37) $\exists C[({}^{\vee}park\ C) \wedge (({}^{\vee}in\ C)\ ({}^{\vee}walk\ j))]$

Finally, DWP analyse the ambiguous example:

(38) (7-116, 7-118) **every+man+doesnt+walk** : S

This has a dominant reading in which the universal has narrow scope with respect to the negation, and a subordinate reading in which the universal has wide scope with respect to the negation. Our grammar generates only the subordinate reading. Lexical lookup yields:

(39) $\Box(((S{\uparrow}NA){\downarrow}S)/CNA) : {}^{\wedge}\lambda B\lambda C\forall D[(B\ D) \rightarrow (C\ D)], \Box CNm : man,$
$\Box((NE\backslash S)/(NE\backslash S)) : {}^{\wedge}\lambda F\lambda G\neg(F\ G), \Box(NH\backslash S) : walk \Rightarrow S$

This has the derivation given in Figure 22, which delivers semantics:

(40) $\forall C[({}^{\vee}man\ C) \rightarrow \neg({}^{\vee}walk\ C)]$

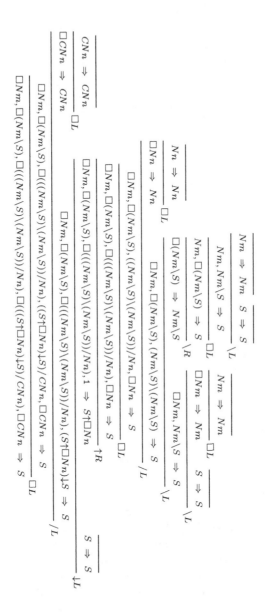

Fig. 21. Derivation for *John walks in a park*

$$
\dfrac{
\dfrac{
\dfrac{
\dfrac{
\dfrac{
\dfrac{
\dfrac{
\dfrac{
\dfrac{Nm \Rightarrow Nm \quad S \Rightarrow S}{Nm, Nm\backslash S \Rightarrow S}\backslash L}
{Nm, \square(Nm\backslash S) \Rightarrow S}\square L
\;\;
\dfrac{Nm \Rightarrow Nm \quad S \Rightarrow S}{Nm, Nm\backslash S \Rightarrow S}\backslash L}
{\;}
}{\;}
}{\;}
}{\;}
}{\;}
}{\;}
$$

$$
\begin{array}{c}
\dfrac{Nm \Rightarrow Nm \quad S \Rightarrow S}{Nm, Nm\backslash S \Rightarrow S}\backslash L \\[4pt]
\dfrac{\;}{Nm, \square(Nm\backslash S) \Rightarrow S}\square L
\end{array}
\qquad
\dfrac{\dfrac{Nm \Rightarrow Nm \quad S \Rightarrow S}{Nm, Nm\backslash S \Rightarrow S}\backslash L}{\square(Nm\backslash S) \Rightarrow Nm\backslash S}\backslash R
$$

$$
\cfrac{
\cfrac{
\cfrac{
\cfrac{
\cfrac{
\cfrac{\square(Nm\backslash S) \Rightarrow Nm\backslash S \qquad \cfrac{Nm \Rightarrow Nm \quad S \Rightarrow S}{Nm, Nm\backslash S \Rightarrow S}\backslash L}
{Nm, (Nm\backslash S)/(Nm\backslash S), \square(Nm\backslash S) \Rightarrow S}\;/L}
{Nm, \square((Nm\backslash S)/(Nm\backslash S)), \square(Nm\backslash S) \Rightarrow S}\;\square L}
{1, \square((Nm\backslash S)/(Nm\backslash S)), \square(Nm\backslash S) \Rightarrow S{\uparrow}Nm \qquad S \Rightarrow S}\;{\uparrow}R}
{(S{\uparrow}Nm){\downarrow}S, \square((Nm\backslash S)/(Nm\backslash S)), \square(Nm\backslash S) \Rightarrow S}\;{\downarrow}L}
{
\begin{array}{l}
\cfrac{CNm \Rightarrow CNm}{\square CNm \Rightarrow CNm}\square L
\end{array}
\quad ((S{\uparrow}Nm){\downarrow}S)/CNm, \square CNm, \square((Nm\backslash S)/(Nm\backslash S)), \square(Nm\backslash S) \Rightarrow S}/L}
{\square(((S{\uparrow}Nm){\downarrow}S)/CNm), \square CNm, \square((Nm\backslash S)/(Nm\backslash S)), \square(Nm\backslash S) \Rightarrow S}\;\square L
$$

Fig. 22. Derivation for *Every man doesn't walk*

7   Conclusion

The negation-as-failure rule is as follows:

$$(41) \quad \dfrac{\not\vdash \Gamma \Rightarrow A}{\Gamma \Rightarrow \neg A}\;\neg R$$

The calculus is presented without the Cut rule:

$$(42) \quad \dfrac{\Gamma \Rightarrow A \qquad \Delta\langle \overrightarrow{A}\rangle \Rightarrow B}{\Delta\langle \Gamma\rangle \Rightarrow B}\;Cut$$

This is because transitivity of inference is unsuitable in the presence of the negation-as-failure [19]. We believe that the remaining rules enjoy Cut-elimination. Thus, Morrill et al. [21] appendix proves Cut-elimination for the displacement calculus **D**; Moortgat [7] proves Cut-elimination for the bracket modalities in ordinary sequent calculus, and the other rules follow patterns in standard logic or linear logic for which there is Cut-elimination. Cut-free backward chaining hypersequent proof search operates in a finite space and so constitutes a terminating procedure for parsing/theorem-proving. Cut-free categorial sequent proof search still suffers from (finite) spurious ambiguity, but this can be treated by normalisation [16]. This is the basis of the implementation of the placement logic used for this paper: the parser/theorem prover CatLog of Morrill [15]. Apart from the shorter-term objective of refining the CatLog implementation of the current type formalism in hypersequent calculus, we define as a longer times goal the implementation of the same logic in proof nets.

References

1. Bach, E.: Discontinuous constituents in generalized categorial grammars. In: Burke, V.A., Pustejovsky, J. (eds.) Proceedings of the 11th Annual Meeting of the North Eastern Linguistics Society, pp. 1–12. GLSA Publications, Department of Linguistics, University of Massachussets at Amherst, Amherst (1981)
2. Bach, E.: Some Generalizations of Categorial Grammars. In: Landman, F., Veltman, F. (eds.) Varieties of Formal Semantics: Proceedings of the Fourth Amsterdam Colloquium, Foris, Dordrecht, pp. 1–23 (1984); Reprinted in Savitch, W.J., Bach, E., Marsh, W., Safran-Naveh, G. (eds.) The Formal Complexity of Natural Language, pp. 251–279. D. Reidel, Dordrecht (1987)
3. Barry, G., Hepple, M., Leslie, N., Morrill, G.: Proof Figures and Structural Operators for Categorial Grammar. In: Proceedings of the Fifth Conference of the European Chapter of the Association for Computational Linguistics, Berlin (1991)
4. Dowty, D.R., Wall, R.E., Peters, S.: Introduction to Montague Semantics. Synthese Language Library, vol. 11. D. Reidel, Dordrecht (1981)
5. Lambek, J.: On the Calculus of Syntactic Types. In: Jakobson, R. (ed.) Structure of Language and its Mathematical Aspects, Proceedings of the Symposia in Applied Mathematics XII, pp. 166–178. American Mathematical Society, Providence (1961)
6. Moortgat, M., Oehrle, R.T.: Adjacency, dependency and order. In: Dekker, P., Stokhof, M. (eds.) Proceedings of the Ninth Amsterdam Colloquim, pp. 447–466. ILLC, Amsterdam (1994)
7. Moortgat, M.: Multimodal linguistic inference. Journal of Logic, Language and Information 5(3,4), 349–385 (1996); Also in Bulletin of the IGPL 3(2,3), 371–401 (1995)
8. Moortgat, M.: Categorial Type Logics. In: van Benthem, J., ter Meulen, A. (eds.) Handbook of Logic and Language, pp. 93–177. Elsevier Science B.V. and The MIT Press, Amsterdam and Cambridge (1997)
9. Moortgat, M., Morrill, G.: Heads and phrases: Type calculus for dependency and constituent structure. Manuscript, Universiteit Utrecht (1991)
10. Moot, R.: Grail: An automated proof assistant for categorial grammar logics. In: Backhouse, R.C. (ed.) Proceedings of the 1998 User Interfaces or Theorem Provers Conference (1998)
11. Moot, R., Retoré, C.: The Logic of Categorial Grammars: A Deductive Account of Natural Language Syntax and Semantics. Springer, Heidelberg (2012)
12. Morrill, G.: Grammar and Logical Types. In: Stokhof, M., Torenvelt, L. (eds.) Proceedings of the 1989 Seventh Amsterdam Colloquium, pp. 429–450 (1989)
13. Morrill, G.: Intensionality and Boundedness. Linguistics and Philosophy 13(6), 699–726 (1990)
14. Morrill, G.: Categorial Formalisation of Relativisation: Pied Piping, Islands, and Extraction Sites. Technical Report LSI-92-23-R, Departament de Llenguatges i Sistemes Informàtics, Universitat Politècnica de Catalunya (1992)
15. Morrill, G.: CatLog: A Categorial Parser/Theorem-Prover. In: LACL 2012 System Demonstrations, Logical Aspects of Computational Linguistics 2012, pp. 13–16 (2012)
16. Morrill, G.: Logic Programming of the Displacement Calculus. In: Pogodalla, S., Prost, J.-P. (eds.) LACL 2011. LNCS (LNAI), vol. 6736, pp. 175–189. Springer, Heidelberg (2011)
17. Morrill, G., Merenciano, J.-M.: Generalising discontinuity. Traitement automatique des langues 37(2), 119–143 (1996)

18. Morrill, G., Valentín, O.: Displacement Calculus. Linguistic Analysis 36(1-4), 167–192 (2010); Special issue Festschrift for Joachim Lambek, http://arxiv.org/abs/1004.4181
19. Morrill, G., Valentín, O.: On Anaphora and the Binding Principles in Categorial Grammar. In: Dawar, A., de Queiroz, R. (eds.) WoLLIC 2010. LNCS (LNAI), vol. 6188, pp. 176–190. Springer, Heidelberg (2010)
20. Morrill, G., Valentín, O., Fadda, M.: Dutch Grammar and Processing: A Case Study in TLG. In: Bosch, P., Gabelaia, D., Lang, J. (eds.) TbiLLC 2007. LNCS (LNAI), vol. 5422, pp. 272–286. Springer, Heidelberg (2009)
21. Morrill, G., Valentín, O., Fadda, M.: The Displacement Calculus. Journal of Logic, Language and Information 20(1), 1–48 (2011), doi:10.1007/s10849-010-9129-2
22. Morrill, G.V.: Type Logical Grammar: Categorial Logic of Signs. Kluwer Academic Publishers, Dordrecht (1994)
23. Morrill, G.V.: Categorial Grammar: Logical Syntax, Semantics, and Processing. Oxford University Press, New York (2011)
24. Oehrle, R.T.: Multi-Modal Type-Logical Grammar. In: Borsley, R.D., Börjars, K. (eds.) Non-transformational Syntax: Formal and Explicit Models of Grammar. Wiley-Blackwell, Oxford (2011), doi:10.1002/9781444395037.ch6
25. Oehrle, R.T., Zhang, S.: Lambek calculus and preposing of embedded subjects. In: Chicago Linguistics Society, Chicago, vol. 25 (1989)

Chasing Diagrams in Cryptography

Dusko Pavlovic*

Royal Holloway, University of London
dusko.pavlovic@rhul.ac.uk

Abstract. Cryptography is a theory of secret functions. Category theory is a general theory of functions. Cryptography has reached a stage
where its structures often take several pages to define, and its formulas
sometime run from page to page. Category theory has some complicated
definitions as well, but one of its specialties is taming the flood of structure. Cryptography seems to be in need of high level methods, whereas
category theory always needs concrete applications. So why is there no
categorical cryptography? One reason may be that the foundations of
modern cryptography are built from probabilistic polynomial-time Turing machines, and category theory does not have a good handle on such
things. On the other hand, such foundational problems might be the
very reason why cryptographic constructions often resemble low level
machine programming. I present some preliminary explorations towards
categorical cryptography. It turns out that some of the main security
concepts are easily characterized through *diagram chasing*, going back
to Lambek's seminal *'Lecture Notes on Rings and Modules'*.

To Jim Lambek for his 90th birthday.

1   Introduction

Idea

For a long time, mathematics was subdivided into geometry and arithmetic, later
algebra. The obvious difference between the two was that the geometric reasoning
was supported by pictures and diagrams, whereas the algebraic reasoning relied
upon the equations and abstract text. For various reasons, the textual reasoning
seemed dominant in XX century mathematics: there were relatively few pictures
in the mathematical publications, and even the formal systems for geometry
were presented as lists of formulas. But as the algebraic constructions grew
more complex, the task to stratify and organize them grew into a mathematical
problem on its own. Category theory was proposed as a solution for this problem.
The earliest categorical diagrams expanded the textual reasoning from exact
sequences to matrices of exact exact sequences [22,7]. The technique of diagram
chasing seems to have emerged around the time of Lambek's classic "Lectures on
Rings and Modules" [19], where it was used not just as a convenient visualization

---

* Recent primary affiliation: University of Hawaii at Menoa. dusko@hawaii.edu

C. Casadio et al. (Eds.): Lambek Festschrift, LNCS 8222, pp. 353–367, 2014.

of lists of equations, but also as a geometric view of universal constructions. This unassuming idea then proceeded to form a germ of geometric reasoning in category theory, uncovering the geometric patterns behind abstract logical structures [23]. Other forms of geometric reasoning emerged in various forms of categorical research [14,12,6, to mention just a few], providing some of the most abstract algebraic structures with some of the most concrete geometric tools.

The present paper reports about the beginnings of an exploration towards applying categorical diagrams in a young and exciting area of mathematics: modern cryptography. Initiated in the late 1970s [3] by introducing algorithmic hardness as a tool of security, modern cryptography developed a rich conceptual and technical apparatus in a relatively short period of time. The increasing complexity of its proofs and constructions, usually presented in a textual, "command line" mode, akin to low-level programming, occasionally engendered doubts that its formalisms may sometimes conceal as many errors as they prevent [16,15,17]. Would a high level categorical view help?

Background

Modern cryptography is a theory of effectively computable, randomized boolean functions. A boolean function is a mapping over bitstrings, i.e. in the form $f : 2^M \longrightarrow 2^N$, where $2 = \{0,1\}$ denotes the set of two elements, and M, N are finite sets. So 2^M denotes the set of M-tuples of 0 and 1; or equivalently of the subsets of M. Which view of 2^M is more convenietn depends on the application. Formally, the algebraic structure of 2^M is induced by the algebraic structure of 2, which is usually viewed as

- Boolean algebra $(2, \wedge, \vee, \neg, 0, 1)$
- Boolean ring $(\mathbb{Z}_2, \oplus, \cdot, 0, 1)$
- submonoid $\{1, -1\} \subseteq (\mathbb{Z}_3, \cdot)$

A boolean function f is *effectively computable*, or *feasible*, and denoted by $f : 2^M \xrightarrow{\mathcal{F}} 2^N$, when it is implemented by a boolean circuit, a Turing machine with suitable time and space bounds, or in some other model of computation. Computations in general are, of course, generally expressed as effective boolean functions over the representations of mathematical structures by bitstrings, all the way up to the continuum [27].

A *randomized* boolean function $g : 2^M \xrightarrow{\mathcal{R}} 2^N$ is in fact a boolean function of two arguments, say $g : 2^R \times 2^M \longrightarrow 2^N$, where the first argument is interpreted as a *random seed*. The output of a randomized function is viewed as a random variable. The probability that a randomized boolean function g, given an input x produces an output y is estimated by counting for how many values of the random seed ρ it takes that value, i.e.

$$\Pr(y \leftarrow gx) = \frac{\#\{\rho \in 2^R \mid y = g(\rho, x)\}}{2^R}$$

where $\#S$ denotes the number of elements of the set S, and R is the length of the random seeds ρ.

An *effective, randomized* boolean function $h : 2^M \xrightarrow{\mathcal{RF}} 2^N$ is thus an effectively computable boolean function $h : 2^R \times 2^M \xrightarrow{\mathcal{F}} 2^N$. It is usually realized by a Deterministic Polynomial-time Turing (DPT) machine, i.e. as $h : 2^R \times 2^M \xrightarrow{DPT} 2^N$. A DPT with two input tapes, one of which is interpreted as providing the random seeds, is called a Probabilistic Polynomial-time Turing (PPT) machine. So for the same function h we would write $h : 2^M \xrightarrow{PPT} 2^N$, leaving the random seeds implicit. This is what cryptographers talk about in their formal proofs, although they seldom specify any actual PPTs. Building a PPT is tedious work, in fact an abstract form of low level machine programming. For a high level view of cryptographic programming, an abstract theory of feasible functions is needed.

Before we proceed in that direction, let us quickly summarize what cryptographers actually build from effective randomized boolean functions and PPTs.

A *crypto system* is a structure given over three finite sets

- $\mathcal{M}$ of *plaintexts*
- $\mathcal{C}$ of *cyphertexts*
- $\mathcal{K}$ of *keys*

plus a set of random seeds, that we leave implicit. They are all given with their bitstring representations. The structure of the crypto-system consists of three feasible functions

- key generation $\langle k, \overline{k} \rangle : 1 \xrightarrow{PPT} \mathcal{K} \times \mathcal{K}$,
- encryption $\mathsf{E} : \mathcal{K} \times \mathcal{M} \xrightarrow{PPT} \mathcal{C}$, and
- decryption $\mathsf{D} : \mathcal{K} \times \mathcal{C} \xrightarrow{DPT} \mathcal{M}$,

that together provide

- unique decryption: $\mathsf{D}(\overline{k}, \mathsf{E}(r, k, m)) = m$,
- and secrecy.

This secrecy is in fact what cryptography is all about. Even defining it took a while.

The earliest formal definition of secrecy is due to Shannon [29]. His idea was to require that the ciphertext discloses nothing about the plaintext. He viewed the attacker as a statistician, equipped with the precise frequency distribution of the language of the meaningful expressions in $\mathcal{M}$, i.e. knowing exactly the values of $\Pr(m \leftarrow \mathcal{M})$, the probability that a randomly sampled string from $\mathcal{M}$ is the plaintext m. Shannon's requirement was that knowing the encryption $c = \mathsf{E}(r, k, m)$ should not make it any easier for this attacker to guess m, or formally

$$\Pr\left(m \leftarrow \mathcal{M} \mid \exists rk.\ c = \mathsf{E}(r, k, m)\right) = \Pr\left(m \leftarrow \mathcal{M}\right) \tag{1}$$

Shannon wrote this in a different, but equivalent form[1], and called it *perfect security*.

---

[1] Except that the encryption was not randomized at the time.

When the age of modern cryptography broke out, the concept of secrecy got refined by considering the feasibility of the encryption and decryption operations, and moreover strengthened by requiring that the attacker is unlikely to guess not only the plaintext m, but even a single bit from it. Otherwise, the concept of secrecy would miss the possibility that the plaintext is hard to guess as a whole, but that it may be easy to guess bit by bit. The original formalization of this requirement is due to Goldwasser and Micali [9,10] under the name *semantic security*, but it was later somewhat simplified to the form of *chosen plaintext indistinguishability* (IND-CPA), which looks something like this:

$$\Pr\left(b \leftarrow \mathsf{A}_1(m_0, m_1, c, s) \mid c \leftarrow \mathsf{E}(k, m_b),\ b \leftarrow 2,\ m_0, m_1, s \leftarrow \mathsf{A}_0\right) \sim \frac{1}{2} \quad (2)$$

The attacker consists of two PPTs, A_0 and A_1, which communicate through a tape. He tests the crypto system as follows. First A_0 chooses and announces two plaintexts m_0 and m_1. She may also convey to A_1 a part of her state, by writing s on their shared tape. Then the crypto system tosses a fair coin b, computes the encryption $c \leftarrow \mathsf{E}(k, m_b)$ of one of the chosen plaintexts, and gives it to the attacker. The attacker A_1 is now supposed to guess which of the two plaintexts was encrypted. The system is secure if knowing c does not give him any advantage in this, i.e. if his chance to guess b is indistinguishable from $\Pr(b \leftarrow 2) = \frac{1}{2}$.

The point that I am trying to make is that this is mouthful of a definition. Especially when we are defining secrecy, which is one of the most basic concepts of cryptography. The upshot is that the most basic cryptographic proofs need to show that some crypto system satisfies the above property.

It is, of course, not unheard of that the fundamental concepts tend to be subtle, and require complicated formal definitions. In cryptography, however, this phenomenon seems to be escalating. First of all, the above definition of secrecy as chosen plaintext indistinguishability turns out to be too weak, and too simple. In reality, the attacker can usually access a decryption oracle, which she can consult before she chooses any plaintexts, and also after she receives back the encryption of one of them, but before she attempts to guess which one it is. So the attacker actually consists of four PPTs, A_0, A_1, A_2 and A_3, where A_0 begins with choosing some cyphertexts, which it submits to the decryption oracle, etc. A reader who is not a cryptographer may enjoy deciphering the interactions between the crypto system and the attacker from the formula below, describing the *chosen cyphertext indistinguishability* (IND-CCA2), due to Rackoff and Simon [28]. The PPTs again share a tape, which they can use to pass each other a part of the state, denoted s_0, s_1 etc.

$$\Pr\left(b \leftarrow \mathsf{A}_3(c_0, m, m_0, m_1, c, c_1, \widetilde{m}, s_2) \left| \begin{array}{l} m = \mathsf{D}(\overline{k}, c_0),\ c_0, s_0 \leftarrow \mathsf{A}_0, \\ c \leftarrow \mathsf{E}(k, m_b),\ b \leftarrow 2,\ m_0, m_1, s_1 \leftarrow \mathsf{A}_1(c_0, m, s_0) \\ \widetilde{m} = \mathsf{D}(\overline{k}, c_1),\ c_1, s_2 \leftarrow \mathsf{A}_2(c_0, m, m_0, m_1, c^{\neq}, s_1) \end{array} \right.\right) \sim \frac{1}{2}$$

$$(3)$$

This formula is nowadays one of the centerpieces of cryptography. As verbose as it may look, and as prohibitive as its requirements may be[2], it came to be a solid and useful concept. The problem is, however, that the story does not end with it, and that the concepts of ever greater complexity and verbosity rapidly proliferate. This makes cryptographic proofs fragile, with some errors surviving extensive examination [30]. The argument that mandatory formal proofs, if they are too complex, may decrease, rather than increase, the reliability of the proven statements, by decreasing the expert scrutiny over the proven statements, while concealing subtle errors, has been raised from within the cryptographic community [1,2,15,16,17]. At the same time, the efforts towards the formalization have ostensibly consolidated the field and clarified some of its conceptual foundations [8,13]. Maybe we have good reasons and enough insight to start looking for better notations?

Outline of the Paper

Section 2 presents a symbolic model of a crypto system, and a very crude symbolic definition of secrecy. These definitions can be stated in any relational calculus, and thus also in the category of relations. Section 3 presents an information theoretic model of a crypto system. The symbolic definition of secrecy refines here to Shannon's familiar definition of perfect security. We formalize it all in the category of sets and stochastic operators between them. And finally, Section 4 introduces a category where the modern cryptographic concepts can be formalized, such as (IND-CPA) and (IND-CCA2). The upshot of this development is to show how the incremental approach, refining the crude abstract concepts, while enriching the categorical structures, motivates the conceptual development and provides technical tools. Section 5 invites for further work.

2   Symbolic Cryptography

In [5,4], Dolev, Yao, Even and Karp describe public key cryptosystems using an algebraic theory — roughly what mathematicians would call *bicyclic semigroups* [11].

2.1   Dolev-Yao Crypto Systems

Definition 1. *A* message algebra $\mathcal{A}$ *consists of three operations:*

- *encryption* $\mathsf{E} : \mathcal{A} \times \mathcal{A} \longrightarrow \mathcal{A}$,
- *decryption* $\mathsf{D} : \mathcal{A} \times \mathcal{A} \longrightarrow \mathcal{A}$, *and*

---

[2] The attacker may submit, e.g. two very large plaintexts, say video blocks, as m_0 and m_1. After she receives the encryption c of one of them, she can then flip just one bit of it, and make that into c_1, which is submitted back for decryption. Although c and c_1 differ in a single bit, the decryption of c_1 should not disclose even a single bit of information about c_0.

– *key pairing* $\overline{(-)} : \mathcal{A} \longrightarrow \mathcal{A}$,

and one equation:

$$\mathsf{D}\left(\overline{k}, \mathsf{E}(k, m)\right) = m$$

called decryption condition. *By convention, the first arguments of* E *and* D *are called* keys, *the second arguments* messages. *A message that occurs in* E *is a* plaintext; *a message that occurs in* D *is a* cyphertext.

Definition 2. *A Dolev-Yao crypto system is given by*

– *a message algebra*
– *a set* $M \subseteq \mathcal{A}$ *of* well-formed *plaintexts;*
– *the* hiding condition: *"knowing* $E(k, m)$ *does not reveal anything about* m*"*

Remarks. The above definitions are close in spirit to Dolev and Yao's definitions, but deviate in details from their presentation. First of all, Dolev and Yao do not present the encryption and decryption operations as binary operations, but as families of unary operations indexed by the keys. More importantly, their results also require the *encryption equation*

$$\mathsf{E}\left(k, \mathsf{D}(\overline{k}, c)\right) = c$$

that should hold for all keys k and all ciphertexts c. Nowadays even toy crypto systems do not satisfy this, so we allow that $\mathsf{E}(k, -)$ may not be surjective. Restricted to its image, of course, the decryption equation implies the encryption equation; but not generally. Finally, Dolev and Yao do not take $M \subseteq \mathcal{A}$ as a part of the structure. Intuitively, if $\mathcal{A}$ is the set of character strings, then M may be construed as the set of the meaningful words meaningful of some language. For a cryptographer, the capability to distinguish the meaningful words, and recognize a decryption when he finds it, is often critical. The set $M \subseteq \mathcal{A}$ is thus a first, very crude step towards the concepts of source redundancy and frequency distribution, which are of course crucial for cryptanalysis.

The main challenge left behind Dolev and Yao's analysis is that the hiding condition, which is clearly the heart of the matter, is left completely informal. At the first sight, there seem to be many ways to make it precise. We present one in the next section. Its conceptual analogy with the more familiar information theoretic and computational notions of secrecy are clear, but its technical utility seems limited.

2.2   Algebraic Perfect Security

An attacker sees a ciphertext c and wants to know the plaintext m, such that $\mathsf{E}(k, m) = c$. But since she does not know the key k, she can only form the set of possible[3] plaintexts m that may correspond to c

$$c\widetilde{D} = \{m \in M \mid \exists k.\ \mathsf{E}(k, m) = c\} \tag{4}$$

---

[3] I.e., this is the only thing that she can do in the *possibilistic* world of mere relations. In the *probabilistic* world of stochastic relations, she can of course do more, and that will be discussed in the next section.

One way to formalize the hiding condition is to require that any well-formed message m must be a candidate for a decryption of c, and thus lie in $c\widetilde{D}$.

Definition 3. *A Dolev-Yao crypto system $\mathcal{A}$ is algebraically perfectly secure if every ciphertext can be an encryption of any well-formed message, i.e. if for all $c, m \in \mathcal{A}$ holds*

$$m \in M \ \wedge \ \exists k \in \mathcal{A}. \ \mathsf{E}(k, m) = c \iff m \in M \tag{5}$$

The following lemma says that this captures the intended requirement that the set $c\widetilde{D}$ does not tell anything about m.

Lemma 1. *A Dolev-Yao crypto system $\mathcal{A}$ is algebraically perfectly secure if and only if for all $c, m \in \mathcal{A}$ and the binary relation $\widetilde{D}$ from (4) holds*

$$c\widetilde{D}m \iff m \in M \tag{6}$$

A convenient framework to work with algebraic security is the category Rel of sets and binary relations

$$|\mathsf{Rel}| = |\mathsf{Set}|$$
$$\mathsf{Rel}(A, B) = \{0, 1\}^{A \times B}$$

with the usual relational composition of $A \xrightarrow{R} B$ and $B \xrightarrow{S} C$

$$a(R; S)c \iff \exists b \in B. \ aRb \wedge bSc$$

and the equality $aIb \iff a = b$ as the identity relation $A \xrightarrow{I} A$. Note that any subset, say $M \subseteq \mathcal{A}$, can be viewed as a relation $M \in \{0, 1\}^{1 \times \mathcal{A}}$, where $1 = \{0\}$, and thus as an arrow $1 \xrightarrow{M} \mathcal{A}$ in Rel with $0Mx \iff x \in M$.

Proposition 4. *A Dolev-Yao crypto system $\mathcal{A}$ is algebraically perfectly secure if and only if the following diagram commutes in the category of relations Rel*

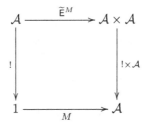

where

– *$\mathcal{A} \xrightarrow{!} 1$ denotes the total relation, i.e. $x!0$ holds for all x and $1 = \{0\}$, and*
– *$\widetilde{\mathsf{E}}^M$ is by definition $c \, \widetilde{\mathsf{E}}^M(k, m) \iff m \in M \wedge \mathsf{E}(k, m) = c$.*

3   Information Theoretic Cryptography

Shannon [29] brought cryptography to the solid ground of information theory, recognizing the fact that an attacker has access not just to the set $M \subseteq \mathcal{A}$ of possible plaintexts, but also to their probabilities $\mu : \mathcal{A} \longrightarrow [0,1]$. And just like we viewed the former one in the form $M \in \{0,1\}^{1 \times \mathcal{A}}$ as an arrow $1 \xrightarrow{M} \mathcal{A}$ in Rel, we shall now view the latter, in the form $\mu \in [0,1]^{1 \times \mathcal{A}}$ as an arrow $1 \xrightarrow{\mu} \mathcal{A}$ in the category Sto of stochastic matrices.

3.1   Shannon Crypto Systems

To begin, Shannon introduced into analysis *mixed* crypto systems, in the form $R = pS + (1 - p)T$ where S and T can be thought of as two Dolev-Yao crypto systems, and $p \in [0,1]$. The idea is that the system R behaves like S with probability p, and like T with probability $1 - p$. In summary, Shannon considered message algebras $\mathcal{A}$

(a) given with a probability distribution $\mu : \mathcal{A} \longrightarrow [0,1]$ that assigns to each plaintext m a *frequency*, $\mu(m)$, and moreover
(b) convex closed, in the sense that for any $p \in [0,1]$

$$\mathsf{E}\left(pk + (1-p)h, m\right) = p\mathsf{E}(k,m) + (1-p)\mathsf{E}(h,m)$$
$$\mathsf{D}\left(pk + (1-p)h, m\right) = p\mathsf{D}(k,m) + (1-p)\mathsf{D}(h,m)$$

But (b) makes it convenient to draw the keys from the convex hull of $\mathcal{A}$

$$\Delta\mathcal{A} = \left\{ \kappa : \mathcal{A} \longrightarrow [0,1] \mid \#\varsigma\kappa < \infty \wedge \sum_{x \in \varsigma\kappa} \kappa(x) = 1 \right\}$$

where $\varsigma\kappa = \{x \in \mathcal{A} \mid \kappa(x) > 0\}$ is the support. As a consequence, the encryption and decryption maps are not functions any more, but stochastic matrices E^κ and D^κ with the entries

$$\mathsf{E}^\kappa_{cm} = \Pr_\kappa(c|m) = \sum_{\substack{x \in \varsigma\kappa \\ \mathsf{E}(x,m)=c}} \kappa(x)$$

$$\mathsf{D}^\kappa_{mc} = \Pr_\kappa(m|c) = \sum_{\substack{x \in \varsigma\kappa \\ \mathsf{D}(\overline{x},c)=m}} \kappa(x)$$

Condition (a) similarly suggests that a plaintext, or the available partial information about it, should also be viewed as a stochastic vector $\mu \in \Delta\mathcal{A}$. A crypto system is now an algebra in the category of sets and stochastic operators

$$|\mathsf{Sto}| = |\mathsf{Set}|$$
$$\mathsf{Sto}(M,N) = \left\{ \Phi \in [0,1]^{M \times N} \mid \#\varsigma\Phi < \infty \wedge \sum_{i \in \varsigma\Phi} \Phi_{ij} = 1 \right\}$$

Indeed, the encryption and the decryption operations are now stochastic operators $\mathsf{E}^\kappa, \mathsf{D}^\kappa \in \mathsf{Sto}(\mathcal{A}, \mathcal{A})$; whereas the mixed plaintexts are the points $\mu \in \mathsf{Sto}(1, \mathcal{A})$.

Definition 5. *A* Shannon crypto system *is given by*

- *a message algebra in the category* Sto, *i.e. stochastic operators for*
  - *encryption* $\mathsf{E} : \mathcal{A} \times \mathcal{A} \longrightarrow \mathcal{A}$,
  - *decryption* $\mathsf{D} : \mathcal{A} \times \mathcal{A} \longrightarrow \mathcal{A}$, *and*
  - *key pairing* $\overline{(-)} : \mathcal{A} \longrightarrow \mathcal{A}$,
- *a frequency distribution of the plaintexts* $\mu : \mathcal{A} \longrightarrow [0, 1]$, *and*
- *the hiding condition.*

This time, the formal definition of the hiding condition available, and well known.

3.2 Perfect Security

Shannon [29] considers an attacker who makes a *probabilistic* model of the observed crypto system. More precisely, when she observes a cyphtertext c, instead of forming the set $c\widetilde{D} \subseteq M$ of *possible* decryptions, like in Sec. 2.2, she now tries to compute the conditional distribution $\Pr(m|c) \geq \Pr(m)$ of the *probable* decryptions of c.

But now the cyphertext c is a random variable $\gamma = \Pr(c)$, which can be viewed as an arrow $1 \xrightarrow{\Pr(c)} \mathcal{A}$ in Sto. An observation of a cyphertext thus provides knowledge about the distribution of $\Pr(c)$. We assume that the attacker knows the distribution $\kappa = \Pr(k)$ of the keys, and the frequency distribution $\mu = \Pr(m)$.

Definition 6. *A* Shannon crypto system *is* perfectly secure *if the plaintexts are statistically independent on the cyphertexts, i.e. if for all* $c, m \in \mathcal{A}$ *holds*

$$\Pr\left(m \leftarrow \mu \mid \exists k.\ c = \mathsf{E}(k, m)\right) = \Pr(m \leftarrow \mu) \tag{7}$$

where the conditional probability on the left stands for

$$\Pr\left(m \leftarrow \mu \mid \exists k.\ c = \mathsf{E}(k, m)\right) = \sum_{x \in \varsigma\kappa} \Pr\left(m \leftarrow \mu \mid c = \mathsf{E}(x, m)\right) \cdot \kappa(x)$$

Aligning definitions 3 and 6 shows that algebraic perfect security is an algebraic approximation of Shannon's probabilistic perfect security [29, II.10]. The following proposition shows that the connection extends to the categorical characterizations.

Proposition 7. *A* Shannon crypto system $\mathcal{A}$ *with finite support is perfectly secure if and only if the following diagram commutes in the category of stochastic operators* Sto

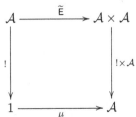

where

- $\mathcal{A} \xrightarrow{!} 1$ *is the row vector of* $\frac{1}{\#_\varsigma \mathcal{A}}$,
- $1 \xrightarrow{\mu} \mathcal{A}$ *is the distribution* μ *viewed as a column vector,*
- $\mathcal{A} \times \mathcal{A} \xrightarrow{! \times \mathcal{A}} \mathcal{A}$ *is the stochastic matrix with the entries*

$$(! \times \mathcal{A})_{i(jk)} = \begin{cases} \frac{1}{\#_\varsigma \mathcal{A}} & \text{if } i = k \\ 0 & \text{otheriwise} \end{cases}$$

- $\widetilde{\mathsf{E}}$ *is the stochastic matrix with the entries*

$$\widetilde{\mathsf{E}}_{c(km)} = \begin{cases} \kappa(k) \cdot \mu(m) & \text{if } c = \mathsf{E}(k, m) \\ 0 & \text{otherwise} \end{cases}$$

4   Computational Cryptography

Modern cryptography arose from the idea to use computational complexity as a tool, and attacker's computational limitations as the persistent assumptions upon which the cryptographer can built the desired security guarantees. To represent modern crypto system, we need to lift the preceding considerations beyond the mere frequency distributions and randomness, captured in the category Sto, to a category suitable to represent randomized feasible computations, graded by a security parameter.

4.1   Category of Effective Stochastic Ensembles up to Indistinguishability

The category suitable to present cryptographic constructions will be build by incremental refinement of the category of sets and functions, in three steps: we first make functions feasible, then randomize them, and finally capture the security parameter.

Effective Functions. Suppose that every set is given with an encoding: e.g., each element is encoded as a bitstring. A function between encoded sets can then be considered feasible if it is realized by a feasible boolean function on the codes.

Let us begin with a crude realization of this idea, just to get a feeling for it. Let $R = (2^*)^{2^*}$ be the monoid of boolean functions and $F \subseteq R$ a submonoid of functions that we call *feasible*. For concreteness, we could assume that the functions from F are just those realized by some suitable family of boolean circuits or Turing-machines. The category Set_F of F-computable functions is then defined

$$|\mathsf{Set}_F| = |\mathsf{Set}/2^*| = \sum_{A \in |\mathsf{Set}|} \{[\![-]\!]_A : A \longrightarrow 2^*\}$$

$$\mathsf{Set}_F(A, B) = \{f \in \mathsf{Set}(A, B) \mid \exists \varphi \in F \; \forall a \in A. \; [\![f(a)]\!]_B = \varphi[\![a]\!]_A\}$$

$$A \xrightarrow{f} B$$

$$[\![-]\!]_A \downarrow \qquad \downarrow [\![-]\!]_B$$

$$2^* \xrightarrow{\varphi} 2^*$$

Effective Substochastic Operators. Now we want to refine the category Set_F effective functions to a category of *randomized* effective functions. The step is analogous to the step from Set to Sto. So randomized effective functions will actually be effective stochastic operators. But since feasible functions may not be total, we will actually work with effective *sub*stochastic operators.

The first task is to define the monoid of randomized boolean functions that will operate on the codes. Consider the set of partial functions

$$\mathcal{R} = \{\gamma : 2^* \times 2^* \rightharpoonup 2^* \mid \forall x \forall \rho_1 \forall \rho_2.\ \gamma(\rho_1, x)\downarrow\ \wedge\ \gamma(\rho_2, y)\downarrow\ \wedge\ |x| = |y|$$
$$\implies |\rho_1| = |\rho_2|\ \wedge\ |\gamma(\rho_1, x)| = |\gamma(\rho_2, y)|\}$$

where $f(x)\downarrow$ asserts that the partial function f is defined at x, and $|\xi|$ denotes the length of the bitstring ξ. The set $\mathcal{R}$ forms a monoid $(\mathcal{R}, \circ, \iota)$ where

$$\gamma \circ \beta(\rho_2 :: \rho_1, x) = \gamma(\rho_2, \beta(\rho_1, x)) \tag{8}$$

and $\iota(\langle\rangle, x) = x$, where $\langle\rangle$ denotes the empty string. This monoid was previously used in [26]. Let $\mathcal{F} \subseteq \mathcal{R}$ be a submonoid of functions that we consider feasible. An example are the functions realized by DPT machines. The category $\mathsf{Sto}_\mathcal{F}$ of effective substochastic operators is now defined as follows

$$|\mathsf{Sto}_\mathcal{F}| = |\mathsf{Set}/2^*| = \sum_{A \in |\mathsf{Set}|} \{[\![-]\!]_A : A \longrightarrow 2^*\}$$

$$\mathsf{Sto}_\mathcal{F}(A, B) = \{\Phi \in [0, 1]^{A \times B} \mid \exists \varphi \in \mathcal{F}\ \forall a \in A\ \forall b \in B.\ \Phi_{ab} = \Pr\left([\![b]\!]_B \leftarrow \varphi[\![a]\!]_A\right)\}$$

Ensembles. In order to capture security parameters, we must expand randomized functions to ensembles. A *feasible ensemble* is a sequence of feasible functions

$$\psi = \left\{\psi_\ell : 2^{r(\ell)} \times 2^{s(\ell)} \xrightarrow{\mathcal{F}} 2^{t(\ell)} \mid \ell \in \omega\right\}$$

where $\omega = \{0, 1, 2, \ldots\}$, and such that

$$k < \ell \implies \psi_k = \psi_\ell \!\restriction_{(2^{r(k)} \times 2^{s(k)})}\ \wedge\ r(k) < r(\ell)\ \wedge\ s(k) < s(\ell)\ \wedge\ t(k) < t(\ell)$$

Write $\mathcal{F}^\omega$ for the set of feasible ensembles. A typical example of an ensemble is the extensional (i.e. input-output) view of a PPT machine, which can consume longer inputs, and then it produces longer outputs.

The monoid structure on $\mathcal{F}^\omega$ is induced by the monoid structure of $\mathcal{F}$. The composite $\vartheta \circ \psi$ of

$$\vartheta = \left\{\vartheta_k : 2^{u(\ell)} \times 2^{v(\ell)} \xrightarrow{\mathcal{F}} 2^{w(\ell)} \mid \ell \in \omega\right\}$$

and

$$\psi = \left\{ \psi_\ell : 2^{r(\ell)} \times 2^{s(\ell)} \xrightarrow{\mathcal{F}} 2^{t(\ell)} \mid \ell \in \omega \right\}$$

consists of the components

$$(\vartheta \circ \psi)_\ell = \overline{\vartheta}_{\overline{\ell}} \circ \psi_\ell : 2^{\overline{u}(\ell)} \times 2^{\overline{v}(\ell)} \xrightarrow{\mathcal{F}} 2^{t(\ell)}$$

where

- $\overline{\ell}$ is the smallest number such that $w(\overline{\ell}) \geq s(\ell)$,
- $\overline{\vartheta}_{\overline{\ell}} = \vartheta_{\overline{\ell}} \upharpoonright_{2^{s(\ell)}}$,
- $\overline{\vartheta}_{\overline{\ell}} \circ \psi_\ell$ is defined by (8),
- $\overline{u}(\ell) = u(\overline{\ell})$ and $\overline{v}(\ell) = v(\overline{\ell})$.

The category $\mathsf{Sto}_{\mathcal{F}}^{\omega}$ of effective substochastic ensembles is now defined as follows

$$|\mathsf{Ens}_{\mathcal{F}}| = |\mathsf{Set}/2^\omega| = \sum_{A \in |\mathsf{Set}|} \{[\![-]\!]_A : A \longrightarrow 2^\omega\}$$

$\mathsf{Ens}_{\mathcal{F}}(A, B) =$

$$\left\{ \Psi \in [0,1]^{\omega \times A \times B} \mid \exists \psi \in \mathcal{F}^\omega \ \forall \ell \in \omega \ \forall a \in A \ \forall b \in B. \ \Psi_{ab}^\ell = \Pr\left([\![b]\!] \leftarrow \psi_\ell[\![a]\!]\right) \right\}$$

where

$$\Pr\left([\![b]\!] \leftarrow \psi_\ell([\![a]\!])\right) = \frac{\#\left\{\rho \in 2^{r(\ell)} \mid [\![b]\!]_{t(\ell)} = \psi_\ell\left(\rho, [\![a]\!]_{s(\ell)}\right)\right\}}{2^{r(\ell)}}$$

In the special case when $\mathcal{F}^\omega$ consists of the actions of PPT machines, we get the category $\mathsf{Ens}_{\mathrm{PPT}}$, where the morphisms are the extensional views of PPTs. More precisely, a morphism is a sequence of substochastic matrices $\Psi = \{\Psi^\ell\}_{\ell \in \omega}$ such that there is a PPT Π and the ab-entry of Ψ^ℓ is $\Psi_{ab}^\ell = \Pr(b \leftarrow \Pi_\ell a)$, where ℓ is the security parameter.

So $\mathsf{Ens}_{\mathcal{F}}$ comes close to providing an abstract view of the universe in which the cryptographers work. The view is abstract in the sense that $\mathcal{F}$ does not have to be realized by PPTs, but can be any submonoid of $\mathcal{R}$. By taking $\mathcal{F}$ to be the PPT realized stochastic operations we get the usual probabilistic algorithms — *except* that those that are indistinguishable, because their difference is a negligible function still correspond to different morphisms in $\mathsf{Ens}_{\mathrm{PPT}}$.

Indistinguishability. Note, first of all, that $[0,1]$ is not only a monoid, but an ordered semiring[4]. The semiring structure lifts to $[0,1]^\omega$. A *semi-ideal* in an ordered semiring is a lower closed subset closed under addition and multiplication. Since it is lower closed, it contains 0, but generally not 1.

Let $\Upsilon \subseteq [0,1]^\omega$ be a semi-ideal. The canonical example is the semi-ideal of *negligible functions* [8]. A function $\nu : \omega \longrightarrow [0,1]$ is called negligible if $\nu(x) <$

---

[4] A *semiring* is a structure $(R, +, \cdot, 0, 1)$ such that $(R, +, 0)$ and $(R, \cdot, 1)$ are commutative monoids such that $a(b+c) = ab + ac$ and $a0 = 0$.

$\frac{1}{q(x)}$ holds eventually, for every positive polynomial q. Any semi-ideal Υ induces on $[0,1]^\omega$ the equivalence relation

$$\sigma \underset{\Upsilon}{\sim} \tau \iff \exists \nu \in \Upsilon. \, |\sigma_\ell - \tau_\ell| < \nu(\ell)$$

and we define $\mathsf{Ens}_{\mathcal{F}}^{\Upsilon}$ to be the category with the same objects as $\mathsf{Ens}_{\mathcal{F}}$, but

$$\mathsf{Ens}_{\mathcal{F}}^{\Upsilon}(A,B) = \mathsf{Ens}_{\mathcal{F}}(A,B) / \underset{\Upsilon}{\sim}$$

Unfolding this definition over the semiring $\mathcal{J}_\Upsilon = [0,1]^\omega / \underset{\Upsilon}{\sim}$, we have $\mathsf{Ens}_{\mathcal{F}}^{\Upsilon}(A,B) =$

$$\left\{ \Psi \in \mathcal{J}_\Upsilon^{A \times B} \mid \exists \psi \in \mathcal{F}^\omega \, \forall \ell \in \omega \, \forall a \in A \, \forall b \in B. \, \Psi_{ab}^\ell = \Pr\left(\llbracket b \rrbracket \leftarrow \psi_\ell \llbracket a \rrbracket\right) \right\}$$

4.2 Characterizing Semantic Security

The usual definition of a crypto system from the Introduction can now be stated abstractly, in a categorical form. While the definition follows the pattern of Def. 2 and Def. 5, this time we revert to the usual *multi-sorted* specification, where the plaintexts, the cyphertexts and the keys are drawn from different sets.

Definition 8. *An* abstract crypto system, *relative to a monoid* $\mathcal{F}$ *of feasible functions, and a semi-ideal* Υ *of negligible functions is given by*

- *a multi-sorted message algebra in the category* $\mathsf{Ens}_{\mathcal{F}}^{\Upsilon}$, *such that*
  - *encryption* $\mathsf{E} : \mathcal{K} \times \mathcal{M} \longrightarrow \mathcal{C}$, *is a stochastic ensemble, whereas*
  - *decryption* $\mathsf{D} : \mathcal{K} \times \mathcal{C} \longrightarrow \mathcal{M}$, *and*
  - *key pairing* $\overline{(-)} : \mathcal{K} \longrightarrow \mathcal{K}$ *are deterministic functions*[5].
- *a frequency distribution of the plaintexts* $\mu : \mathcal{M} \longrightarrow [0,1]$, *and*
- *the hiding condition.*

The upshot of it all. The abstract versions of the hiding conditions, such as (IND-CPA) and (IND-CCA2), described in the Introduction, boil down to commutative diagrams in $\mathsf{Ens}_{\mathcal{F}}^{\Upsilon}$. We illustrate this fact for (IND-CPA).

Proposition 9. *Let* $\mathsf{Ens}_{\mathrm{PPT}}^{\mathcal{V}}$ *be the category of ensembles of PPT-realized boolean functions modulo negligible functions. A crypto system in the usual sense (as described in the Introduction) is equivalent to an abstract crypto system in this category. Such a crypto system is semantically secure, i.e. it satisfies (IND-CPA), as defined by (2), if and only if the following diagram commutes for all arrows* A_0 *and* A_1 *in* $\mathsf{Ens}_{\mathrm{PPT}}^{\mathcal{V}}$.

$$
\begin{array}{ccccc}
\mathcal{K} & \xrightarrow{\langle \mathrm{id}_{\mathcal{K}}, A_0 \rangle} \mathcal{K} \times \mathcal{M}^2 \times \mathcal{S} \xrightarrow{\langle \pi_{\mathcal{K}}, \pi_b \pi_{\mathcal{M}^2}, \pi_{\mathcal{M}^2}, \pi_{\mathcal{S}} \rangle} \mathcal{K} \times \mathcal{M} \times \mathcal{M}^2 \times \mathcal{S} & \xrightarrow{\mathsf{E} \times \mathcal{M}^2 \times \mathcal{S}} & \mathcal{C} \times \mathcal{M}^2 \times \mathcal{S} \\
\downarrow{\scriptstyle !} & & & \downarrow{\scriptstyle A_1} \\
1 & \xrightarrow{\hspace{5cm} b \hspace{5cm}} & & 2
\end{array}
$$

A similar proposition holds for (IND-CCA2).

---

[5] Deterministic functions can be characterized intrinsically in $\mathsf{Sto}_{\mathcal{F}}$, $\mathsf{Ens}_{\mathcal{F}}$ and $\mathsf{Ens}_{\mathcal{F}}^{\Upsilon}$.

5   Further Work

While the various notions of secrecy can thus be characterized by commutative diagrams in suitable categories, the notions of one-way function and pseudorandom generator correspond to the requirements that some diagrams *do not* commute. This leads to interesting categorical structures, which seem to be best expressed in terms of enriched categories, and the suitable convolution operations. This observation led to an different approach, through *monoidal computer* [24,25], lifting the ideas from another strand of Lambek's work, leading from infinite abacus as an intensional model of computation [18], to the extensional models [20], elaborated in the book with P.J. Scott [21].

But how useful might our categorical models of computation be for cryptography? Can the categorical tools, developed for high level program semantics, really be used to stratify cryptographic constructions? The preliminary evidence, some of which was presented here, suggests that certain types of cryptographic proofs and constructions can be significantly simplified by using categorical tools to 'hide the implementation details'. The price to be paid, though, is that this hiding requires some preliminary work. For instance, we have seen that the secrecy conditions can be captured by simple diagrams, albeit in randomized categories. This approach echoes the well established programming methodologies, where complex structures are encapsulate into components that hide the irrelevant implementation details, and only the fragments that need to be manipulated are displayed at the interface. The categorical approach developed in Lambek's work has made such strategies available across a broad gamut of sciences.

References

1. Choo, K.-K.R., Boyd, C., Hitchcock, Y.: Errors in computational complexity proofs for protocols. In: Roy, B. (ed.) ASIACRYPT 2005. LNCS, vol. 3788, pp. 624–643. Springer, Heidelberg (2005)
2. Dent, A.W.: Fundamental problems in provable security and cryptography. Philosophical Transactions of the Royal Society A: Mathematical, Physical and Engineering Sciences 364(1849), 3215–3230 (2006)
3. Diffie, W., Hellman, M.E.: New directions in cryptography. IEEE Transactions on Information Theory IT-22(6), 644–654 (1976)
4. Dolev, D., Even, S., Karp, R.M.: On the security of ping-pong protocols. In: CRYPTO, pp. 177–186 (1982)
5. Dolev, D., Yao, A.C.: On the security of public key protocols. IEEE Transactions on Information Theory 29(2), 198–208 (1983)
6. Pavlovic, D.: Geometry of abstraction in quantum computation. Proceedings of Symposia in Applied Mathematics 71, 233–267 (2012) arxiv.org:1006.1010
7. Freyd, P.: Abelian Categories: An Introduction to the Theory of Functors. Harper and Row (1964)
8. Goldreich, O.: Foundations of Cryptography. Cambridge University Press (2000)
9. Goldwasser, S., Micali, S.: Probabilistic encryption & how to play mental poker keeping secret all partial information. In: STOC 1982: Proceedings of the Fourteenth Annual ACM Symposium on Theory of Computing, pp. 365–377. ACM Press, New York (1982)

10. Goldwasser, S., Micali, S.: Probabilistic encryption. J. Comput. Syst. Sci. 28(2), 270–299 (1984)
11. Grillet, P.A.: Semigroups: An introduction to the structure theory. Marcel Dekker, Inc. (1995)
12. Joyal, A., Street, R.: The geometry of tensor calculus I. Adv. in Math. 88, 55–113 (1991)
13. Katz, J., Lindell, Y.: Introduction to Modern Cryptography. Chapman & Hall/CRC Series in Cryptography and Network Security. Chapman & Hall/CRC (2007)
14. Kelly, G.M.: On clubs and doctrines. In: Kelly, G.M. (ed.) Category Seminar. Sydney 1972/73, pp. 181–256. Springer, Berlin (1974)
15. Koblitz, N., Menezes, A.: Another look at "Provable Security". II. In: Barua, R., Lange, T. (eds.) INDOCRYPT 2006. LNCS, vol. 4329, pp. 148–175. Springer, Heidelberg (2006)
16. Koblitz, N., Menezes, A.: Another look at "Provable Security". J. Cryptology 20(1), 3–37 (2007)
17. Koblitz, N., Menezes, A.: The brave new world of bodacious assumptions in cryptography. Notices of the American Mathematical Society 57(3), 357–365 (2010)
18. Lambek, J.: How to program an infinite abacus. Canad. Math. Bull. 4(3), 295–302 (1961)
19. Lambek, J.: Lectures on Rings and Modules. Blaisdell Publishing Co. (1966)
20. Lambek, J.: From types to sets. Adv. in Math. 36, 113–164 (1980)
21. Lambek, J., Scott, P.J.: Introduction to higher order categorical logic. Cambridge Stud. Adv. Math., vol. 7. Cambridge University Press, New York (1986)
22. Lane, S.M.: Homology. Springer (1963)
23. Pavlovic, D.: Maps II: Chasing diagrams in categorical proof theory. J. of the IGPL 4(2), 1–36 (1996)
24. Pavlovic, D.: Categorical logic of names and abstraction in action calculus. Math. Structures in Comp. Sci. 7, 619–637 (1997)
25. Pavlovic, D.: Monoidal computer I: Basic computability by string diagrams. Information and Computation (2013) (to appear) arxiv:1208.5205
26. Pavlovic, D., Meadows, C.: Bayesian authentication: Quantifying security of the Hancke-Kuhn protocol. E. Notes in Theor. Comp. Sci. 265, 97–122 (2010)
27. Pavlovic, D., Pratt, V.: The continuum as a final coalgebra. Theor. Comp. Sci. 280(1-2), 105–122 (2002)
28. Rackoff, C., Simon, D.R.: Non-interactive zero-knowledge proof of knowledge and chosen ciphertext attack. In: Feigenbaum, J. (ed.) CRYPTO 1991. LNCS, vol. 576, pp. 433–444. Springer, Heidelberg (1992)
29. Shannon, C.E.: Communication theory of secrecy systems. Bell Systems Technical Journal 28, 656–715 (1949)
30. Shoup, V.: OAEP reconsidered. In: Kilian, J. (ed.) CRYPTO 2001. LNCS, vol. 2139, pp. 239–259. Springer, Heidelberg (2001)

The Monotone Lambek Calculus
Is NP-Complete

Mati Pentus

Moscow State University, Moscow, Russia
http://lpcs.math.msu.su/~pentus/

Abstract. We consider the Lambek calculus with the additional structural rule of monotonicity (weakening). We show that the derivability problem for this calculus is NP-complete (both for the full calculus and for the product-free fragment). The same holds for the variant that allows empty antecedents. To prove NP-hardness of the product-free fragment, we provide a mapping reduction from the classical satisfiability problem *SAT*. This reduction is similar to the one used by Yury Savateev in 2008 to prove NP-hardness (and hence NP-completeness) of the product-free Lambek calculus.

Keywords: Lambek calculus, complexity, structural rule.

Introduction

The Lambek syntactic calculus L (introduced in [4]) is one of the logical calculi used in the paradigm of categorial grammar for deriving reduction laws of syntactic types (also called *categories*) in natural and formal languages. In categorial grammars based on the Lambek calculus (or its variants) an expression is assigned to category $B\,/\,A$ (respectively, $A\setminus B$) if and only if the expression produces an expression of category B whenever it is followed (respectively, preceded) by an expression of category A. An expression is assigned to category $A\cdot B$ if and only if the expression can be obtained by concatenation of an expression of category A and an expression of category B. The reduction laws derivable in this calculus are of the form $A\to B$ (meaning "every expression of category A is also assigned to category B"). In the sequent form of the calculus, a law of the form $A_1\dots A_n\to B$ means the same as the law $A_1\cdot\dots\cdot A_n\to B$.

In this paper, we consider the Lambek calculus with the added rule of monotonicity (also called *weakening*) and denote it by LM. It is known that LM is sound and complete with respect to semiring models [6], where types are interpreted as two-sided ideals (given two ideals I and J, the product $I\cdot J$ is the minimal ideal that contains all pairwise products of elements from I and J, the right division is defined as $I/J=\{c\mid c\cdot J\subseteq I\}$ and the left division is defined as $J\setminus I=\{c\mid J\cdot c\subseteq I\}$). The calculus LM is obviously also sound with respect to ring models where types are interpreted as two-sided ideals (the division operations of ideals were considered, e.g., in [5]), but the completeness with respect to ring models seems to be an open question.

C. Casadio et al. (Eds.): Lambek Festschrift, LNCS 8222, pp. 368–380, 2014.
© Springer-Verlag Berlin Heidelberg 2014

There is a natural modification of the original Lambek calculus, which we call *the Lambek calculus allowing empty premises* and denote L* (see [11, p. 44]). Intuitively, the modified calculus assigns the empty expression to some categories. A similar modification can also be considered in the presence of the monotonicity (weakening) rule. Thus we obtain the calculus LM*, which is sound with respect to models on rings with unit.

In all these calculi, the cut rule can be eliminated and cut-free proofs are of polynomial size. Thus, the derivability problem for these calculi is in NP.

We show that the classical satisfiability problem *SAT* is polynomial time reducible to the derivability problem for the product-free fragment of LM and thus this fragment and the full calculus LM are NP-complete. The same reduction from *SAT* works also for the product-free fragment of LM*.

In [7], it was proved that the fragment of LM* with only one connective, left division, is decidable in polynomial time. The same holds for the fragment with only right division and for the fragment with only product, as well as for the one-connective fragments of LM. Thus, for all these fragments the complexity of deciding derivability is in the same class as in the absence of the monotonicity rule. A survey of complexity results for fragments of L and L* can be found in [9].

This paper is organized as follows. The first section contains definitions of the calculi LM and LM*. In Section 2, we give the main construction that reduces *SAT* to the derivability problem for the product-free fragment of LM (and also to the derivability problem for the product-free fragment of LM*). The correctness of this reduction is proved in Section 3.

1   The Monotone Lambek Calculus

First we define *the Lambek calculus allowing empty premises* (denoted by L*).

Assume that an enumerable set of *variables* Var is given. The *types* of L* are built of variables (also called *primitive types*) and three binary connectives $\cdot$, $/$, and $\backslash$. The set of all types is denoted by Tp. The letters p, q, r, ... range over the set Var, capital letters A, B, ... range over types, and capital Greek letters range over finite (possibly empty) sequences of types. For notational convenience, we assume that the product operator $\cdot$ associates to the left. We say that in B / A and $A \backslash B$ the *numerator* is B and the *denominator* is A.

The *sequents* of L* are of the form $\Gamma \to A$ (note that Γ can be the empty sequence). Here Γ is called the *antecedent*, and A is the *succedent*. The calculus L* has the following axioms and rules of inference:

$$A \to A,$$

$$\frac{\Phi \to B \quad \Gamma B \Delta \to A}{\Gamma \Phi \Delta \to A} \ (\text{cut}),$$

$$\frac{\Pi A \to B}{\Pi \to B / A} \ (\to/),$$

$$\frac{\Phi \to A \quad \Gamma B \Delta \to C}{\Gamma (B / A) \Phi \Delta \to C} \ (/\to),$$

$$\frac{A \Pi \to B}{\Pi \to A \backslash B} \ (\to\backslash),$$

$$\frac{\Phi \to A \quad \Gamma B \Delta \to C}{\Gamma \Phi (A \backslash B) \Delta \to C} \ (\backslash\to),$$

$$\frac{\Gamma \to A \quad \Delta \to B}{\Gamma\Delta \to A \cdot B} \ (\to \cdot), \qquad\qquad \frac{\Gamma AB\Delta \to C}{\Gamma(A \cdot B)\Delta \to C} \ (\cdot \to).$$

As usual, we write $L^* \vdash \Gamma \to A$ to indicate that the sequent $\Gamma \to A$ is derivable in L^*.

The calculus L has the same axioms and rules with the only exception that in the rules $(\to\backslash)$ and $(\to/)$ we require Π to be nonempty. The calculus L is the original syntactic calculus introduced in [4]. In L, every derivable sequent has non-empty antecedent. Evidently, if $L \vdash \Gamma \to A$, then $L^* \vdash \Gamma \to A$.

It is known that dropping the rule (cut) from the calculus L (or L^*) does not change the set of derivable sequents. The standard way to prove this is to consider a derivation with only one instance of (cut) and proceed by induction on the total size of its two premises (see [4]). Here the size of a sequent is defined as the total number of primitive type occurrences and operation symbol occurrences.

Example 1. The derivation

$$\frac{\dfrac{p_1 \to p_1}{\to p_1 \backslash p_1} \ (\to\backslash) \qquad p_2 \to p_2}{(p_1 \backslash p_1) \backslash p_2 \to p_2} \ (\backslash\to)$$

demonstrates that $L^* \vdash (p_1 \backslash p_1) \backslash p_2 \to p_2$.

Note that $L \nvdash (p_1 \backslash p_1) \backslash p_2 \to p_2$ (evidently, this sequent can have no cut-free derivation in L).

The monotone Lambek calculus (we denote it by LM) is obtained from the calculus L by adding the rule

$$\frac{\Gamma\Delta \to B}{\Gamma A\Delta \to B} \ (\text{M}).$$

Sometimes the rule (M) is called the *monotonicity rule* (see [11, p. 47]) or the *weakening rule* (see [2, p. 359]). Similarly, the monotone Lambek calculus allowing empty premises (we denote it by LM^*) is obtained from the calculus L^* by adding the rule (M).

Example 2. The derivation

$$\frac{\dfrac{p_2 \to p_2}{(p_1 \backslash p_1)\, p_2 \to p_2} \ (\text{M})}{p_2 \to (p_1 \backslash p_1) \backslash p_2} \ (\to\backslash)$$

demonstrates that $LM \vdash p_2 \to (p_1 \backslash p_1) \backslash p_2$.

Example 3. The derivation

$$\frac{\dfrac{\dfrac{\dfrac{p_1 \to p_1}{p_1\, p_3 \to p_1} \ (\text{M})}{p_3 \to p_1 \backslash p_1} \ (\to\backslash) \qquad p_2 \to p_2}{p_3 \,((p_1 \backslash p_1) \backslash p_2) \to p_2} \ (\backslash\to)}{(p_1 \backslash p_1) \backslash p_2 \to p_3 \backslash p_2} \ (\to\backslash)$$

demonstrates that $\text{LM} \vdash (p_1 \setminus p_1) \setminus p_2 \to p_3 \setminus p_2$.

Note that $\text{LM} \nvdash (p_1 \setminus p_1) \setminus p_2 \to p_2$ (this follows from the cut-elimination theorem below).

Theorem 1 (the cut-elimination theorem). *Dropping the rule* (cut) *from the calculus* LM *does not change the set of derivable sequents. The same holds for* LM^*.

This theorem can be proved similarly to the cut-elimination theorem for the original Lambek calculus. The claim for LM^* is also a particular case of Corollary 4.8 from [3].

Corollary 1 (the subtype property). *If a sequent is derivable in* LM, *then there exists a derivation where each type is a subtype of a type in the final sequent. The same holds for* LM^*.

Corollary 2. *For the calculi* LM *and* LM^* *the derivability problem is decidable in nondeterministic polynomial time.*

Proof. In a bottom-up derivation search (for each rule, from the conclusion to the premises), the size of sequents decreases provided we do not use the cut rule. Here the size of a sequent means the total number of primitive type occurrences and operation symbol occurrences.

Therefore, for a given sequent we can calculate an upper bound for the derivation size (which is defined as the total size of the sequents in the derivation). It is easy to see that the derivation size is less than $\frac{n^2+n}{2}$, where n is the size of the given sequent. □

For the calculi L, L^*, LM, and LM^* we denote their product-free fragments by $\text{L}(\setminus, /)$, $\text{L}^*(\setminus, /)$, $\text{LM}(\setminus, /)$, and $\text{LM}^*(\setminus, /)$, respectively. In view of the subformula property, the rules $(\to \cdot)$ and $(\cdot \to)$ are not needed in these fragments.

2   Reduction from *SAT*

Let $c_1 \wedge \ldots \wedge c_m$ be a Boolean formula in conjunctive normal form with clauses $c_1, \ldots, c_m$ and variables $x_1, \ldots, x_n$. The reduction maps the formula to a sequent, which is derivable in LM^* (and in LM) if and only if the formula $c_1 \wedge \ldots \wedge c_m$ is satisfiable.

For any Boolean variable x_i let $\neg_0 x_i$ stand for the literal $\neg x_i$ and $\neg_1 x_i$ stand for the literal x_i. Note that $\langle t_1, \ldots, t_n \rangle \in \{0,1\}^n$ is a satisfying assignment for the Boolean formula $c_1 \wedge \ldots \wedge c_m$ if and only if for every index $j \leq m$ there exists an index $i \leq n$ such that the literal $\neg_{t_i} x_i$ appears in the clause c_j (as usual, 1 stands for "true" and 0 stands for "false").

Here we follow the notation from [10].

Let p_i^j (where $0 \leq i \leq n$ and $0 \leq j \leq m$) and q_i^j (where $0 \leq i \leq n$ and $1 \leq j \leq m$) be distinct primitive types from Var.

We define the following types:

$$G^0 = p_0^0 \setminus p_n^0,$$
$$G^j = (q_n^j / ((q_0^j \setminus p_0^j) \setminus G^{j-1})) \setminus p_n^j \quad \text{for } 1 \leq j \leq m,$$
$$G = G^m$$
$$E_i^0(t) = p_{i-1}^0 \quad \text{for } 1 \leq i \leq n \text{ and } t \in \{0,1\},$$
$$E_i^j(t) = \begin{cases} q_i^j / (((q_{i-1}^j / E_i^{j-1}(t)) \setminus p_{i-1}^j) \setminus p_i^{j-1}), & \text{if } \neg_t x_i \text{ occurs in } c_j, \\ (q_{i-1}^j / (q_i^j / (E_i^{j-1}(t) \setminus p_i^{j-1}))) \setminus p_{i-1}^j, & \text{otherwise,} \end{cases}$$
$$\text{for } 1 \leq i \leq n, \ 1 \leq j \leq m, \text{ and } t \in \{0,1\},$$
$$F_i(t) = E_i^m(t) \setminus p_i^m \quad \text{for } 1 \leq i \leq n \text{ and } t \in \{0,1\}.$$

Let Θ_i denote the sequence of types $(F_i(1) / F_i(0)) F_i(0)$ (for each i, the sequence Θ_i consists of two types).

Our aim is to prove that the sequent $\Theta_1 \ldots \Theta_n \to G$ is derivable in $\text{LM}(\backslash, /)$ (and in $\text{LM}^*(\backslash, /)$) if and only if the formula $c_1 \wedge \ldots \wedge c_m$ is satisfiable.

Example 4. Consider the Boolean formula $x_1 \wedge \neg x_1$. Here $n = 1$, $m = 2$, $c_1 = x_1$, and $c_2 = \neg x_1$. By construction,

$$G = ((q_1^2 / ((q_0^2 \setminus p_0^2) \setminus ((q_1^1 / ((q_0^1 \setminus p_0^1) \setminus (p_0^0 \setminus p_0^0))) \setminus p_1^1))) \setminus p_1^2),$$
$$F_1(0) = ((q_1^2 / (((q_0^2 / ((q_0^1 / (q_1^1 / (p_0^0 \setminus p_1^0))) \setminus p_0^1)) \setminus p_0^2) \setminus p_1^1)) \setminus p_1^2),$$
$$F_1(1) = (((q_0^2 / (q_1^2 / ((q_1^1 / (((q_0^1 / p_0^0) \setminus p_0^1) \setminus p_1^0)) \setminus p_1^1)) \setminus p_0^2) \setminus p_1^2).$$

The reduction described in this section maps $x_1 \wedge \neg x_1$ to the sequent

$$(F_1(1) / F_1(0)) F_1(0) \to G.$$

The Boolean formula $x_1 \wedge \neg x_1$ is not satisfiable, and indeed

$$\text{LM}^* \nvdash (F_1(1) / F_1(0)) F_1(0) \to G.$$

3   Correctness of the Reduction

The following lemma is proved in [10].

Lemma 1. *Let $\langle t_1, \ldots, t_n \rangle \in \{0,1\}^n$. The following conditions are equivalent:*

1. *$\langle t_1, \ldots, t_n \rangle$ is a satisfying assignment for the Boolean formula $c_1 \wedge \ldots \wedge c_m$,*
2. *$\text{L}(\backslash, /) \vdash F_1(t_1) \ldots F_n(t_n) \to G$,*
3. *$\text{L}^*(\backslash, /) \vdash F_1(t_1) \ldots F_n(t_n) \to G$.*

Lemma 2. *If $c_1 \wedge \ldots \wedge c_m$ is satisfiable, then $\text{LM}(\backslash, /) \vdash \Theta_1 \ldots \Theta_n \to G$.*

Proof. First, we note that $\text{LM}(\backslash, /) \vdash \Theta_i \to F_i(0)$ and $\text{LM}(\backslash, /) \vdash \Theta_i \to F_i(1)$ for all i. Let $\langle t_1, \ldots, t_n \rangle$ be a satisfying assignment for the Boolean formula $c_1 \wedge \ldots \wedge c_m$. In view of Lemma 1, we have $\text{L}(\backslash, /) \vdash F_1(t_1) \ldots F_n(t_n) \to G$, whence $\text{LM}(\backslash, /) \vdash F_1(t_1) \ldots F_n(t_n) \to G$. Applying the cut rule m times, we obtain $\text{LM}(\backslash, /) \vdash \Theta_1 \ldots \Theta_n \to G$. $\qquad\square$

It remains to prove that $c_1 \wedge \ldots \wedge c_m$ is satisfiable whenever

$$\text{LM}^*(\backslash, /) \vdash \Theta_1 \ldots \Theta_n \to G$$

(this is slightly stronger than the converse of Lemma 2, because we use $\text{LM}^*(\backslash, /)$ instead of $\text{LM}(\backslash, /)$). We start with some auxiliary notions and results.

For every $r \in \text{Var}$ we define a function $\#_r$ that maps types to integers as follows:

$$\#_r(s) = \begin{cases} 1 & \text{if } s = r, \\ 0 & \text{if } s \in \text{Var and } s \neq r, \end{cases}$$
$$\#_r(A \cdot B) = \#_r(A) + \#_r(B),$$
$$\#_r(A \backslash B) = \#_r(B) - \#_r(A),$$
$$\#_r(B / A) = \#_r(B) - \#_r(A).$$

This definition is extended to sequences of types as follows:

$$\#_r(A_1 \ldots A_n) = \#_r(A_1) + \ldots + \#_r(A_n).$$

We also define a function $\#$ that maps types to integers as follows:

$$\#(A) = \sum_{r \in \text{Var}} \#_r(A).$$

Example 5. Let $A = (q_0^1 \backslash p_0^1) \backslash q_0^1$. Then $\#_{p_0^1}(A) = -1$, $\#_{q_0^1}(A) = 2$, and $\#(A) = 1$.

Lemma 3. *If* $\text{L}^* \vdash \Pi \to C$, *then* $\#(\Pi) = \#(C)$ *and* $\#_r(\Pi) = \#_r(C)$ *for every* $r \in \text{Var}$.

Proof. Straightforward induction on derivations. $\qquad\qquad\qquad\qquad\qquad$ □

Lemma 4. *Let* $\text{LM}^* \vdash \Pi \to C$ *and* $r \in \text{Var}$. *Then the sum of* $\#_r(A)$ *over all instances of the rule*

$$\frac{\Gamma \Delta \to B}{\Gamma A \Delta \to B} \ (\text{M})$$

in a given derivation of $\Pi \to C$ *equals* $\#_r(\Pi) - \#_r(C)$. *Similarly, the sum of* $\#(A)$ *over all instances of the rule* (M) *in a given derivation of* $\Pi \to C$ *equals* $\#(\Pi) - \#(C)$.

Proof. Induction on derivations. $\qquad\qquad\qquad\qquad\qquad\qquad\qquad\qquad$ □

Example 6. Let us consider the derivation

$$\cfrac{p \to p \qquad \cfrac{\cfrac{\cfrac{\cfrac{s \to s}{q\,s \to s}\ (\text{M})}{q\,q\,s \to s}\ (\text{M})}{q\,s \to q \backslash s}\ (\to \backslash)}{q\,s \to q \backslash s}}{p\,(p \backslash q)\,s \to q \backslash s}\ (\backslash \to)$$

in LM*. For $r = q$ Lemma 4 claims that $\#_q(q) + \#_q(q) = \#_q(p\,(p \setminus q)\,s) -$ $\#_q(q \setminus s)$, i.e., $1 + 1 = 1 - (-1)$. For $r = s$ Lemma 4 claims that $\#_s(q) + \#_s(q) =$ $\#_s(p\,(p \setminus q)\,s) - \#_s(q \setminus s)$, i.e., $0 + 0 = 1 - 1$.

Lemma 5. *The type G contains only primitive types p_0^j, q_0^j, p_n^j, and q_n^j. For any i and t the type $F_i(t)$ contains only primitive types p_{i-1}^j, q_{i-1}^j, p_i^j, and q_i^j. For any i, j, k, and t satisfying $1 \le i \le n$, $0 \le j \le m$, $1 \le k \le m$, and $t \in \{0,1\}$ we have $\#_{p_0^j}(G) = -1$, $\#_{q_0^k}(G) = 1$, $\#_{p_n^j}(G) = 1$, $\#_{q_n^k}(G) = -1$, $\#_{p_{i-1}^j}(F_i(t)) = -1$, $\#_{q_{i-1}^k}(F_i(t)) = 1$, $\#_{p_i^j}(F_i(t)) = 1$, and $\#_{q_i^k}(F_i(t)) = -1$.*

Lemma 6. *Let $1 \le i \le n$. For any $r \in \mathrm{Var}$ we have $\#_r(F_i(1) \,/\, F_i(0)) = 0$. Hence, $\#(F_i(1) \,/\, F_i(0)) = 0$.*

Proof. Immediate from Lemma 5. □

Lemma 7. *For any $r \in \mathrm{Var}$ we have $\#_r(\Theta_1 \ldots \Theta_n) = \#_r(G)$. As a corollary, we see that $\#(\Theta_1 \ldots \Theta_n) = \#(G)$.*

Proof. Immediate from Lemma 5. □

Following [1], we denote the set of all positive occurrences of subtypes of C by $\mathrm{Sub}_1(C)$ and the set of all negative occurrences of subtypes of C by $\mathrm{Sub}_2(C)$. Formally,

$$\mathrm{Sub}_1(s) = \{s\} \text{ if } s \in \mathrm{Var},$$
$$\mathrm{Sub}_2(s) = \emptyset \text{ if } s \in \mathrm{Var},$$
$$\mathrm{Sub}_1(A \cdot B) = \mathrm{Sub}_1(A) \cup \mathrm{Sub}_1(B) \cup \{A \cdot B\},$$
$$\mathrm{Sub}_2(A \cdot B) = \mathrm{Sub}_2(A) \cup \mathrm{Sub}_2(B),$$
$$\mathrm{Sub}_1(A \setminus B) = \mathrm{Sub}_2(A) \cup \mathrm{Sub}_1(B) \cup \{A \setminus B\},$$
$$\mathrm{Sub}_2(A \setminus B) = \mathrm{Sub}_1(A) \cup \mathrm{Sub}_2(B),$$
$$\mathrm{Sub}_1(A \,/\, B) = \mathrm{Sub}_1(A) \cup \mathrm{Sub}_2(B) \cup \{B \,/\, A\},$$
$$\mathrm{Sub}_2(A \,/\, B) = \mathrm{Sub}_2(A) \cup \mathrm{Sub}_1(B).$$

We extend Sub_1 and Sub_2 to sequents as follows:

$$\mathrm{Sub}_1(D_1 \ldots D_k \to C) = \mathrm{Sub}_2(D_1) \cup \ldots \cup \mathrm{Sub}_2(D_k) \cup \mathrm{Sub}_1(C),$$
$$\mathrm{Sub}_2(D_1 \ldots D_k \to C) = \mathrm{Sub}_1(D_1) \cup \ldots \cup \mathrm{Sub}_1(D_k) \cup \mathrm{Sub}_2(C).$$

Given a derivation, we identify subtype occurrences of the conclusion of a rule instance with the corresponding subtype occurrences in the premises.

In the sequel, when there is no danger of confusion, we denote an occurrence of a type and the type itself by the same letter.

Lemma 8. *Let $\mathrm{LM}^*(\setminus, /) \vdash \Pi \to C$. Then there exists a derivation of $\Pi \to C$ where each subtype occurrence A introduced in an instance of the rule (M) is either an element of Π (at the top level) or the denominator of a positive occurrence of a subtype in the sequent $\Pi \to C$ (i.e., $D \,/\, A \in \mathrm{Sub}_1(\Pi \to C)$ or $A \setminus D \in \mathrm{Sub}_1(\Pi \to C)$ for some D).*

Proof. We fix a derivation of $\Pi \to C$ with minimal total size of axiom instances. Let us consider a subtype occurrence A introduced in an instance of the rule (M). Evidently, the occurrence in the sequent $\Pi \to C$ identified with A is a negative occurrence (i.e., $A \in \mathrm{Sub}_2(\Pi \to C)$). It remains to prove that A can not be the numerator of a subtype occurrence in the sequent $\Pi \to C$. Suppose to the contrary that we have a rule instance

$$\frac{\Phi \to B \quad \Gamma A \Delta \to E}{\Gamma \Phi (B \backslash A) \Delta \to E} \; (\backslash \to),$$

where the subtype occurrence A in $\Gamma A \Delta \to E$ was introduced in an instance of the rule (M). Then we can move this instance of (M) down (without changing the total size of axiom instances) and obtain

$$\frac{\Phi \to B \quad \dfrac{\Gamma \Delta \to E}{\Gamma A \Delta \to E} \; (M)}{\Gamma \Phi (B \backslash A) \Delta \to E} \; (\backslash \to).$$

This fragment can be replaced by

$$\frac{\dfrac{\dfrac{\Gamma \Delta \to E}{\Gamma (B \backslash A) \Delta \to E} \; (M)}{\vdots}}{\Gamma \Phi (B \backslash A) \Delta \to E} \; (M),$$

which reduces the total size of axiom instances. This contradicts the choice of the original derivation. The case of $(/\to)$ is similar. $\qquad\square$

Example 7. The derivation from Example 6 does not have the property formulated in Lemma 8. However, if we follow the proof of the lemma, then we obtain the following simpler derivation of the same sequent:

$$\frac{\dfrac{\dfrac{\dfrac{s \to s}{q\, s \to s} \; (M)}{s \to q \backslash s} \; (\to\backslash)}{(p \backslash q)\, s \to q \backslash s} \; (M)}{p\,(p \backslash q)\, s \to q \backslash s} \; (M).$$

Note that the derivations from Examples 2 and 3 do have the property formulated in Lemma 8.

Lemma 9. *If* $\#_{p_i^j}(E_{i'}^{j'}(t')) < 0$, *then* $E_{i'}^{j'}(t')$ *contains an occurrence of* $E_i^j(t)$ *for some* t. *If* $\#_{q_i^j}(E_{i'}^{j'}(t')) < 0$, *then* $E_{i'}^{j'}(t')$ *contains an occurrence of* $E_{i+1}^{j-1}(t)$ *for some* t.

Proof. Straightforward induction on j'. $\qquad\square$

Lemma 10. *Let* $\mathrm{LM}^*(\backslash, /) \vdash \Theta_1 \ldots \Theta_n \to G$. *Then there is a sequence* Ψ *such that* $\mathrm{L}^*(\backslash, /) \vdash \Psi \to G$ *and* Ψ *is a subsequence of* $\Theta_1 \ldots \Theta_n$ *(i.e., the sequent* $\Theta_1 \ldots \Theta_n \to G$ *can be derived from* $\Psi \to G$ *using the rule* (M) *only).*

Proof. We consider a derivation of $\Theta_1 \ldots \Theta_n \to G$ where each subtype occurrence A introduced in an instance of the rule (M) is either an element of $\Theta_1 \ldots \Theta_n$ (at the top level) or the denominator of a positive occurrence of a subtype in $\Theta_1 \ldots \Theta_n \to G$ (such a derivation exists according to Lemma 8). We denote by $\mathcal{M}$ the set of occurrences of subtypes in $\Theta_1 \ldots \Theta_n \to G$ that were introduced by the rule (M).

We denote by $\mathcal{N}$ the subset of $\mathcal{M}$ where we exclude all types of the form $F_i(1) / F_i(0)$. If $\mathcal{N} = \emptyset$, then we can move all instances of the rule (M) to the end of the derivation and the "middle" sequent immediately above all the instances of the rule (M) is the required sequent $\Psi \to G$, where Ψ is a subsequence of $\Theta_1 \ldots \Theta_n$.

It remains to consider the case where $\mathcal{N} \neq \emptyset$.

From the construction of G we see that denominators of positive occurrences of subtypes of G can only be of the form p_0^0, $q_n^j / ((q_0^j \setminus p_0^j) \setminus G^{j-1})$, or $q_0^j \setminus p_0^j$. Denominators of negative occurrences of subtypes of $F_i(0)$ and $F_i(1) / F_i(0)$ can only be of the form $E_i^j(t)$, $((q_{i-1}^j / E_i^{j-1}(t)) \setminus p_{i-1}^j) \setminus p_i^{j-1}$, $(q_{i-1}^j / E_i^{j-1}(t)) \setminus p_{i-1}^j$, $q_{i-1}^j / E_i^{j-1}(t)$, $q_{i-1}^j / (q_i^j / (E_i^{j-1}(t) \setminus p_i^{j-1}))$, $q_i^j / (E_i^{j-1}(t) \setminus p_i^{j-1})$, or $E_i^{j-1}(t) \setminus p_i^{j-1}$. In addition to the above-listed types, $\mathcal{M}$ can only contain occurrences of types of the form $F_i(0)$ and $F_i(1) / F_i(0)$. Note that for all types listed in this paragraph the value of $\#$ is either 0 or 1.

In view of Lemma 4,

$$\sum_{A \in \mathcal{M}} \#(A) = \#(\Theta_1 \ldots \Theta_n) - \#(G).$$

Using Lemma 7, we obtain

$$\sum_{A \in \mathcal{M}} \#(A) = 0.$$

We have seen that all addends in this sum are nonnegative. Thus, they are all zero. This means that the set $\mathcal{N}$ can only contain occurrences of $q_0^j \setminus p_0^j$, $((q_{i-1}^j / E_i^{j-1}(t)) \setminus p_{i-1}^j) \setminus p_i^{j-1}$, $q_{i-1}^j / E_i^{j-1}(t)$, $q_{i-1}^j / (q_i^j / (E_i^{j-1}(t) \setminus p_i^{j-1}))$, $E_i^{j-1}(t) \setminus p_i^{j-1}$, and $F_i(0)$.

We define an auxiliary function

$$f: \{p_i^j \mid 0 \le i \le n \text{ and } 0 \le j \le m\} \cup \{q_i^j \mid 0 \le i \le n \text{ and } 1 \le j \le m\} \to \mathbb{N}$$

as follows:

$$f(p_i^j) = 2i + 2j,$$
$$f(q_i^j) = 2i + 2j - 1.$$

Let s be the primitive type with the least value $f(s)$ among all primitive types that occur in elements of $\mathcal{N}$.

No element of $\mathcal{N}$ can contain more than one occurrence of s. Thus, there is $A \in \mathcal{N}$ such that $\#_s(A) \neq 0$. In view of Lemmas 4 and 7, the sum of $\#_s(A)$ over $A \in \mathcal{M}$ equals 0. According to Lemma 6, the sum of $\#_s(A)$ over $A \in \mathcal{N}$ equals 0. Thus, there is $A \in \mathcal{N}$ such that $\#_s(A) > 0$. We fix such a type A and consider three cases.

Case 1. Assume that $s = p_0^j$. Then $A = q_0^j \setminus p_0^j$. Obviously, $f(q_0^j) < f(p_0^j)$, which contradicts the choice of s.

Case 2. Assume that $s = p_i^j$, where $i > 0$. It is easy to see that then $E_i^j(t)$ is a subtype of A for some $t \in \{0,1\}$ (here we use the list of possible forms of A and Lemma 9). Note that $E_i^j(t)$ contains p_{i-1}^j. Obviously, $f(p_{i-1}^j) < f(p_i^j)$, which contradicts the choice of s.

Case 3. Assume that $s = q_i^j$. It is easy to see that then $j > 0$ and $E_{i+1}^{j-1}(t)$ is a subtype of A for some $t \in \{0,1\}$ (here we use the list of possible forms of A and Lemma 9). Note that $E_{i+1}^{j-1}(t)$ contains p_i^{j-1}. Obviously, $f(p_i^{j-1}) < f(q_i^j)$, which contradicts the choice of s.

Thus, we see that the case $\mathcal{N} \neq \emptyset$ is impossible.   □

Lemma 11. *Let $s \in \mathrm{Var}$ and $\mathrm{L}^*(\backslash, /) \vdash \Theta (E_1 \setminus s) \Xi \to s$. Let the sequent $\Theta (E_1 \setminus s) \Xi \to s$ contain only one negative occurrence of s (the occurrence explicitly shown in $E_1 \setminus s$). Then Ξ is empty.*

Proof. Induction on cut-free derivations.   □

Lemma 12. *Let $s \in \mathrm{Var}$. Let the sequent $\Theta ((E_1 \setminus s) / (E_2 \setminus s)) \Xi \to D$ contain only one negative occurrence of s (the occurrence explicitly shown in $E_1 \setminus s$). Then $\mathrm{L}^*(\backslash, /) \nvdash \Theta ((E_1 \setminus s) / (E_2 \setminus s)) \Xi \to D$.*

Proof. Induction on cut-free derivations.   □

Lemma 13. *Let $s \in \mathrm{Var}$ and $\mathrm{L}^*(\backslash, /) \vdash \Theta ((E_1 \setminus s) / (E_2 \setminus s)) (E_3 \setminus s) \Xi \to D$. Let the sequent $\Theta ((E_1 \setminus s) / (E_2 \setminus s)) (E_3 \setminus s) \Xi \to D$ contain only two negative occurrences of s (the occurrences explicitly shown in $E_1 \setminus s$ and $E_3 \setminus s$). Then $\mathrm{L}^*(\backslash, /) \vdash \Theta (E_1 \setminus s) \Xi \to D$.*

Proof. We generalize the claim by considering also similar sequents without E_3, i.e., sequents of the form $\Theta ((E_1 \setminus s) / (E_2 \setminus s)) s \Xi \to D$, and proceed by induction on cut-free derivations. Most cases in the induction step are straightforward. We consider the nontrivial cases.

Case 1. Assume that the last rule is

$$\frac{\to (E_2 \setminus s) \qquad \Theta (E_1 \setminus s) H \Xi \to D}{\Theta ((E_1 \setminus s) / (E_2 \setminus s)) H \Xi \to D} \ (/\to),$$

where H is either s or $(E_3 \setminus s)$. Since the sequent $\to (E_2 \setminus s)$ contains no negative occurrences of s it is not derivable in $\mathrm{L}^*(\backslash, /)$, a contradiction.

Case 2. Assume that the last rule is

$$\frac{H \Xi_1 \to (E_2 \setminus s) \qquad \Theta (E_1 \setminus s) \Xi_2 \to D}{\Theta ((E_1 \setminus s) / (E_2 \setminus s)) H \Xi_1 \Xi_2 \to D} \ (/\to),$$

where H is either s or $(E_3 \setminus s)$ and $\Xi_1 \Xi_2 = \Xi$. It is well-known that the rule $(\to\setminus)$ is reversible in L^* (the converse rule is easy to derive with the help of the cut rule). Thus, from $L^*(\setminus, /) \vdash H \, \Xi_1 \to (E_2 \setminus s)$ we obtain $L^*(\setminus, /) \vdash E_2 H \, \Xi_1 \to s$. Lemma 11 yields that Ξ_1 is empty, whence $\Xi_2 = \Xi$. Thus, the right premise is the required derivable sequent.

Case 3. Assume that the last rule is

$$\frac{\to E_3 \qquad \Theta\left((E_1 \setminus s) / (E_2 \setminus s)\right) s \, \Xi \to D}{\Theta\left((E_1 \setminus s) / (E_2 \setminus s)\right) (E_3 \setminus s) \, \Xi \to D} \; (\setminus\to).$$

Here we apply induction hypothesis to the right premise.

Case 4. Assume that the last rule is

$$\frac{\Theta_2\left((E_1 \setminus s) / (E_2 \setminus s)\right) \to E_3 \qquad \Theta_1 \, s \, \Xi \to D}{\Theta_1 \Theta_2\left((E_1 \setminus s) / (E_2 \setminus s)\right) (E_3 \setminus s) \, \Xi \to D} \; (\setminus\to),$$

where $\Theta_1 \Theta_2 = \Theta$. Here we apply Lemma 12 to the left premise. □

Remark 1. Lemma 13 can also be proved using the proof nets for $L^*(\setminus, /)$ defined in [10] or the proof nets for the multiplicative noncommutative linear logic introduced in [8]. In these proof nets, primitive type occurrences are divided into pairs by connecting them to each other by axiom links. These axiom links show which primitive type occurrences come from the same axiom in a derivation. Below, we sketch an argument based on proof nets. In this argument, we do not need Lemmas 11 and 12.

Using the derivability criteria associated with the above-mentioned proof nets, it can be shown that the two occurrences of s in $(E_1 \setminus s) / (E_2 \setminus s)$ can not be connected to each other by an axiom link. Thus, the occurrence of s in $E_2 \setminus s$ must be connected to the occurrence of s in $E_3 \setminus s$. Due to the planarity condition in the proof net criteria, we see that all primitive type occurrences in $E_2 \setminus s$ and $E_3 \setminus s$ are connected to each other. This yields that we can remove $E_2 \setminus s$ and $E_3 \setminus s$ from the sequent and obtain a proof net for the sequent $\Theta\left(E_1 \setminus s\right) \Xi \to D$, which means that this sequent is derivable in $L^*(\setminus, /)$.

Lemma 14. *Let Ψ be a subsequence of $\Theta_1 \ldots \Theta_n$ and $L^*(\setminus, /) \vdash \Psi \to G$. Then $L^*(\setminus, /) \vdash F_1(t_1) \ldots F_n(t_n) \to G$ for some $\langle t_1, \ldots, t_n \rangle \in \{0, 1\}^n$.*

Proof. According to the construction,

$$\Theta_1 \ldots \Theta_n = (F_1(1) / F_1(0)) \, F_1(0) \, (F_2(1) / F_2(0)) \, F_2(0) \; \ldots \; (F_n(1) / F_n(0)) \, F_n(0).$$

In view of Lemma 3, $\#_r(\Psi) = \#_r(G)$ for every $r \in \mathrm{Var}$. Evidently, $\Psi = \Psi_1 \ldots \Psi_n$, where Ψ_i is a subsequence of Θ_i for each i. By induction on i one can prove that either $\Psi_i = \Theta_i$ or $\Psi_i = F_i(0)$ (this follows from $\#_{p_{i-1}^0}(\Psi) = \#_{p_{i-1}^0}(G)$ in view of Lemma 5).

By induction on i we prove that

$$L^*(\setminus, /) \vdash F_1(t_1) \ldots F_i(t_i) \, \Psi_{i+1} \ldots \Psi_n \to G$$

for some $\langle t_1, \ldots, t_i \rangle \in \{0,1\}^i$. In the induction step, we assume that

$$\mathrm{L}^*(\backslash, /) \vdash F_1(t_1) \ldots F_{i-1}(t_{i-1}) \, \Psi_i \, \Psi_{i+1} \ldots \Psi_n \to G.$$

If $\Psi_i = F_i(0)$, then we put $t_i = 0$. It remains to consider the case $\Psi_i = (F_i(1)/F_i(0)) \, F_i(0)$. In this case, we put $t_i = 1$ and apply Lemma 13 with $s = p_i^m$, $E_1 = E_i^m(1)$, $E_2 = E_3 = E_i^m(0)$, $\Theta = F_1(t_1) \ldots F_{i-1}(t_{i-1})$, $\Xi = \Psi_{i+1} \ldots \Psi_n$, and $D = G$. $\square$

Lemma 15. *If* $\mathrm{LM}^*(\backslash, /) \vdash \Theta_1 \ldots \Theta_n \to G$, *then the formula* $c_1 \wedge \ldots \wedge c_m$ *is satisfiable.*

Proof. Immediate from Lemmas 10, 14, and 1. $\square$

Theorem 2. *The derivability problem for* $\mathrm{LM}(\backslash, /)$ *is NP-complete. The derivability problem for* $\mathrm{LM}^*(\backslash, /)$ *is NP-complete.*

Proof. In Corollary 2, it was shown that these problems are in NP. To prove their NP-hardness, we use the construction in Section 2. According to Lemmas 2 and 15, this construction provides a mapping reduction from the classical satisfiability problem SAT to the derivability problem for $\mathrm{LM}(\backslash, /)$ and also to the derivability problem for $\mathrm{LM}^*(\backslash, /)$. $\square$

Corollary 3. *The derivability problem for* LM *is NP-complete. The derivability problem for* LM^* *is NP-complete.*

This research was partially supported by the Russian Foundation for Basic Research (grants 14-01-00127-a, 11-01-00958-a, 12-01-00888-a, NSh-5593.2012.1) and by the Scientific and Technological Cooperation Programme Switzerland–Russia (STCP-CH-RU, project "Computational Proof Theory").

References

1. Buszkowski, W.: Completeness results for Lambek syntactic calculus. Z. Math. Logik Grundlag. Math. 32(1), 13–28 (1986)
2. Buszkowski, W.: Type logics in grammar. In: Hendricks, V.F., Malinowski, J. (eds.) Trends in Logic: 50 Years of Studia Logica, pp. 337–382. Kluwer Academic, Dordrecht (2003)
3. Galatos, N., Ono, H.: Cut elimination and strong separation for substructural logics: An algebraic approach. Ann. Pure Appl. Logic 161(9), 1097–1133 (2010)
4. Lambek, J.: The mathematics of sentence structure. Am. Math. Mon. 65(3), 154–170 (1958)
5. Lambek, J.: Lectures on Rings and Modules. Blaisdell, London (1966)
6. Pentus, A.E., Pentus, M.R.: The atomic theory of multiplication and division of semiring ideals. J. Math. Sci. 167(6), 841–856 (2010)
7. Pentus, A.E., Pentus, M.R.: The atomic theory of left division of two-sided ideals of semirings with unit. J. Math. Sci. 193(4), 566–579 (2013)

8. Pentus, M.: Free monoid completeness of the Lambek calculus allowing empty premises. In: Larrazabal, J.M., Lascar, D., Mints, G. (eds.) Logic Colloquium 1996. Lecture Notes in Logic, vol. 12, pp. 171–209. Springer, Berlin (1998)
9. Pentus, M.: Complexity of the Lambek calculus and its fragments. In: Beklemishev, L.D., Goranko, V., Shehtman, V. (eds.) Advances in Modal Logic 2010, pp. 310–329. College Publications, London (2010)
10. Savateev, Y.: Product-free Lambek calculus is NP-complete. In: Artemov, S., Nerode, A. (eds.) LFCS 2009. LNCS, vol. 5407, pp. 380–394. Springer, Heidelberg (2008)
11. van Benthem, J.: Language in Action. Categories, Lambdas and Dynamic Logic. North-Holland, Amsterdam (1991)

A Mathematical Analysis of Masaccio's *Trinity*

Gonzalo E. Reyes

Université de Montréal
gonzalo@reyes-reyes.com

The florentine Tommaso Cassai (1401-c.1427)[1] better known as Masaccio, has been hailed as the first great painter of the Italian Renaissance and his fresco *Trinity* (c.1425, Santa Maria Novella, Florence)[2] as the first work of Western art that used full perspective. It appears that the author was inspired and actually helped by his friend Filippo Brunelleschi (1377-1446), the celebrated architect of the cupola of Il Duomo di Firenzi (the cathedral of S. Maria del Fiore in Florence). It is less well-known that Brunelleschi was a pioneer in perspective and that he devised a method for representing objects in depth on a flat surface by using a single vanishing point.

The aim of this note is to study several questions of a mathematical nature suggested by this fresco:

(1) How accurate is the perspective of the fresco?
(2) What are the dimensions of the chapel?
(3) What are the dimensions of the coffers of the vaulted ceiling of the chapel?
(4) Where is the point of view situated with respect to the fresco?
(5) Where are the different characters situated inside the chapel?
(6) What are the "real" heights of the characters portrayed?

Questions (1)-(4) admit answers that may be computed starting from the data of the fresco, by using some rules of perspective and simple mathematical facts. This is not true for the others. Nevertheless, we will show that under some reasonable hypotheses estimates may be made.

The mathematical methods used are elementary and were known to Euclid. So they were accessible to Masaccio. To make the text more readable, mathematical developments are relegated to the Appendix. All the figures are given in cm.

1   Checking the Data

We will check whether the data provided by [1] (page 100) fits the theoretical criterion developed in the Appendix (Section 5.1 (3).) We let b_n ($n < 10$) be the distance from the line of the reclining figures to the intersection of the nth circle (starting from the top) with the line of symmetry, i.e., the vertical line in the painting going through the vanishing point.

To understand what is going on, we look at the simpler problem of the representation of a plane, rather than cylindric vault.

---

[1] Actually born near Florence, in San Giovanni Valdarno.
[2] See http://www.kfki.hu/~arthp/html/m/masaccio/trinity/trinity.html

C. Casadio et al. (Eds.): Lambek Festschrift, LNCS 8222, pp. 381–392, 2014.

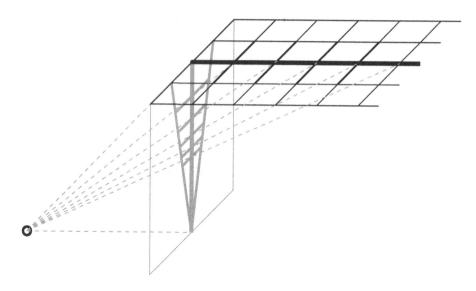

From this diagram, we just keep "its dorsal spine" (i.e. the bold line), which is all what the data in [1] (page 100) is about and which is the same for both vaults (plane and cylindric)

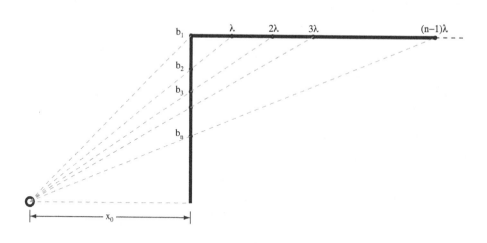

The following data ("empirical b_n") is either given explicitly in the above reference or follows from additions and subtractions from the data therein. We will refer to these data as "robust" and use the term "fragile" to our own measures on magnified copies of Xerox reproductions of pictures in books or WEB.

The values obtained from the criterion ("theoretical b_n") are computed in the Appendix (Section 5.1.)

n	empirical b_n	theoretical b_n	error
1	416.80	416.80	0.00%
2	393.06	393.06	0.00%
3	371.95	371.88	0.02%
4	352.46	352.86	0.11%
5	335.67	335.70	0.01%
6	319.63	320.12	0.15%
7	305.95	305.93	0.01%
8	292.72	292.94	0.08%
9	280.92	281.01	0.03%

The fit is excellent. In all cases the error is less than two parts in a hundred. This shows that Masaccio constructed a projection from a point on the line situated on the horizontal plane of the reclining figures, perpendicular to the horizon and going through the point of intersection of the horizon and the line of symmetry in the fresco.

On the other hand the magnified copy [3] would lead us to think that the "rays" of the vault do not converge to this point, but rather to a point below. This is certainly wrong, as shown by the fit of the data with the theoretical analysis. It is possible that the damage caused to the fresco by removing it from the original place, putting it back again, covering by an altar (by Vasari) and renovating it in a period when no sound scientific basis for this operation was known, may account for this and other mismatches that we shall point out later on.

In the Appendix (Section 5.1) it is shown that the distance from the point of view to the entrance of the chapel is

$$x_0 = 16.56\lambda$$

where λ is the length of an individual coffer. Similarly, the length of the interior of the chapel, i.e., the space under the coffers, may be expressed in terms of λ: since there are 7 rows of coffers, each of length λ,

$$\text{length of the interior of chapel} = 7\lambda$$

The other dimensions are given in the data [1]:

$$\text{height of the chapel} = 416.8^3$$

$$\text{width of the chapel} = 211.6$$

This solves the problem of the dimension of the chapel *provided that* λ *may be computed.*

---

[3] Measured from the horizon rather than from the floor of the chapel.

2   Dimensions of the Chapel

As we pointed out in the remarks, we need only to compute the length of an individual coffer. But first, we tackle the question of the width. The width of an individual coffer may be computed quite easily. In fact, since the radius of the "virtual" cylinder is given in the data of [1], namely 105.8, the total length of the frontal arch is $L = \pi \times 105.8 = 332.38$. Since there are 8 rows of coffers, the width of each is $w = 323.38/8 = 41.55$. On the other hand, the length is not so straightforward, The trouble is that although we can measure heights and widths, we cannot measure depths. However, there is one object whose depth may be computed on the base of the given data: the square abacus on top of the columns. This gives the missing clue to compute depths.

To compute the length λ of an individual coffer we use both front and back columns and make the following

Assumption 1. *The abacus on top of the four columns under the arches (the two in front and the two in the back) are squares. More precisely, the top of the capital of each of the four columns is a square.*

This supposition is natural, since the columns are cylindric. Now the idea is to take the square abacus on top of a column as a "patron" or unit of measure for the depth of the chapel. The (real) length of this abacus can be measured directly from the fresco or rather inferred, since part of the horizontal side of the abacus (the one that can be measured) is hidden. The trouble is that the apparent length of the patron decreases as we take it along the "diagonal" between the top of the front column to the top of the corresponding back column in the fresco. But we can take averages. In details: if we imagine identical abacus between the columns, their number n is

(app. distance between columns/average app. length of abacus) $+ (a - \lambda)/a$.

(The term $(a-\lambda)/a$ is due to the fact that the fraction $1/2((a-\lambda)/a)$ of the first abacus and the same fraction of the last abacus are inside the chapel). Since the apparent lengths of these abacus are known, the distance between the columns is roughly $n \times$ real length of abacus. On the other hand, this distance is $7 \times \lambda$ and this allows us to compute λ. In what follows a_f is the apparent length of the front top abacus and a_b the apparent length of the back front abacus.

	app. distance	a_f	a_b	average	a	λ
Left	5.5	1.2	0.9	1.05	42.16	32.87
Right	5.3	1.3	1.0	1.15	41.39	29.02

Without taking averages, we have the following inequalities corresponding to the apparent lengths of the front abacus (a_f) and the back abacus (a_b) for left and right columns, respectively

$$\begin{cases} 29.42 < \lambda < 37.48 \\ 26.27 < \lambda < 32.60 \end{cases}$$

We notice that we have a "robust" lower bound, in contrast to our "fragile" ones for λ, namely

$$\lambda > 23.74$$

To explain where this value comes from, look at the Appendix (Section 5.3). Unfortunately, we don't have a "robust" upper bound for λ.

It seems likely that nothing more precise may come out of these measures and that the value of λ is between 26.27 and 37.48. Correspondingly, the length of the interior of the chapel is between 183.89 and 262.36 and the distance x_0 from the viewpoint to the entrance of the chapel is between 435.03 and 620.67.

For definiteness sake we take $\lambda = 31$ (roughly the average of the values given by the above table) as the length of an individual coffer. As a consequence, we obtain

$$\begin{cases} \text{distance from the viewpoint to the chapel} = 513.36 \\ \qquad\text{length of the interior of the chapel} \quad = \quad 217 \end{cases}$$

From a "practical" point of view, the exact value of λ does not matter too much. If λ were 33, for instance, the length of the interior of the chapel would be 231, rather than 217, a difference of 14 cm. Now, we turn to problems (5) and (6).

3 Position of Characters on the Ground

The problem of finding the positions of the characters of the fresco on the ground of the chapel can not be solved by studying the fresco only and the measures therein and some external clues as well as some tinkering is needed to proceed. The reason is that, grosso modo, figures of different heights and situated in different places may have the same projection.

We have an historical clue: the height of Christ. According to J.A. Aiken, "...four columns formerly in the *aula del Concilio* of the lateran in Rome were believed in Masaccio's time to establish the height of Christ at approximately 1.78 m." ([1]). Thus, we make the following

Assumption 2. *Masaccio took the real height of Christ to be 178 cm*

This assumption allows us to find the position of Christ (and the cross) inside the chapel. In fact, as shown in the Appendix (Section 5.2), the depth of Christ (i.e., the distance from the entrance of the chapel) is

$$d = 3.36\lambda.$$

For $\lambda = 31$, we obtain that Christ depth is 104.16. Thus, all the scene of the crucifixion with the Virgin and St. John as witnesses takes place in this rather reduced space. (Recall that the figures are *inside* the chapel and this leaves a space of 104.16-31=73.16 as the available space for the whole scene).

Had we taken the height of Christ to be three "braccia" (approximately 172), as the canon of what an ideal man should measure in the Renaissance, its depth would be 83.25, a figure that seems too small as a theater of the scene. We keep the first figure, the one given by the historical clue.

We next tackle the Father's depth. Although the apparent height may be measured directly on the fresco (155), some tinkering seems necessary to proceed as the notion of real height does not make sense. We shall concentrate on the distance between the Father and Christ. Notice that the Father is holding the cross with his arms extended in such a way that his hands are approximately 95 apart and this suggests a distance between the two not far from 10 or 15 from the cross. In fact, this seems to be a comfortable position to hold the cross. At any rate, we present a table for his depth and real height with different choices of separation between Christ and the Father around 10 or 15. Furthermore, we tabulate the distance between his head and the vault and the length of the support of the Father. Notice, however, that the height of the chapel is 416.8-26.44=390.37, since the first figure is the height measured from the horizon, i.e. the level of the kneeling figures, rather than the floor of the chapel. The step to go from that level to the floor is 26.44 from the magnified copy. Details of these calculations are in Appendix (Section 5.2).

separation	d	d/λ	real height	head/vault	length support
0	104.16	3.36	186.46	52.38	174.84
5	109.16	3.52	187.97	49.65	169.84
10	114.16	3.68	189.48	46.91	164.84
15	119.16	3.84	190.99	44.18	159.84
20	124.16	4.01	192.50	41.44	154.84
25	129.16	4.17	194.01	38.71	149.84
30	134.16	4.33	195.52	35.98	144.84

For the separation 5, the Father would be directly in the middle of the chapel; for the separation 15, his height would be three and a third bracia (florentine measure=57.33cm). The majority of people I showed the picture chose a point in the vault between the third and the fourth coffers and rather closer to the fourth as lying directly above the Father. This choice corresponds to a separation between 5 and 20 and closer to 20.

Now we tackle the depths of the Virgin and St. John. First, notice that they seem to be approximately the same and both are standing just before Christ, their separation to the cross being not far from 15, say. The following is a a table for their depth and real heights around this figure

separation	depth	real height Virgin	real height St. John
0	104.16	166.40	162.74
5	99.16	165.06	161.42
10	94.16	163.71	160.10
15	89.16	162.37	158.78
20	84.16	161.02	157.47
25	79.16	159.68	156.15
30	74.16	158.34	154.83

Unfortunately, we don't seem to have records of their believed heights, contrary to the case of Christ. On the other hand, the average height of a man in the period was about 160.

4   Conclusions

The questions raised could be divided roughly into three types: those that can be answered from measures performed on the fresco itself; those that can be answered using historical clues and finally, those whose answers can only be guessed from common sense. The question of the accuracy of the perspective in so far as the apparent decreasing distances between the vault ribs may be answered quite precisely and we have done so. The error between theory and practice is less than two parts in one hundred. The question of the dimensions of the chapel may be answered in principle from measures performed on the fresco. The only problem is the accuracy of the measures. We should notice, however, that there seem to be discrepancies between measures that in theory should be the same. Such is the case with the heights of the front columns and the size of the abacus on top of them. As we said before, it is likely that the fresco has been badly damaged after the different transformations that it suffered: changes of place in the church and renovations in a period when there was no sound scientific base for this operation. (For the saga of the fresco, see [2]). Finally, the question of the position of the characters inside the chapel require some external clues, such as the historical clue on the height of Christ and some tinkering about the relation between Christ and the rest of the figures.

We sum up our conclusions in the form of a diagram with 20 as the distance between Christ (C) and the Father (F) who stands on the support (s), and as the difference between the depths of Christ and the Virgin (V).

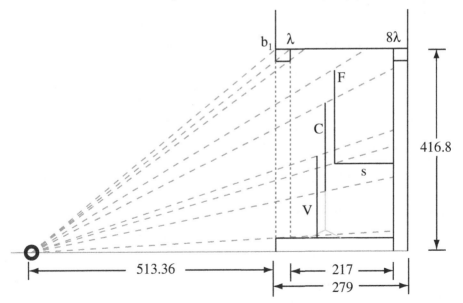

One mystery remains: what is the length of the Father's support? From the assumption that the distance between Christ and the Father is about 20, it follows that the length of the support *inside the chapel* is about 124 and the whole support is 154.84, a figure that seems excessive. Most people I have asked the question gave answers of the order of 80 for the whole support and, consequently, about 40 for the part of the support inside the chapel, although one person, a painter, suggested 150 for the length of the whole support. Given the precision of the painting, we are inclined to think that Masaccio did this on purpose, as if he would like to leave the mystery of the Trinity untouched at this visual level.

5   Mathematical Appendix

5.1   A Theoretical Criterion

To solve the first question, we first formulate a purely mathematical criterion for the existence of a projection centered on the line situated on the horizontal plane of the reclining figures, perpendicular to the horizon and going through the point of intersection of the horizon and the line of symmetry in the fresco

Proposition 1. *Let $b_1, b_2, b_3, \ldots, b_n, \ldots$ be a strictly decreasing sequence of positive real numbers. Then the following are equivalent:*

(1) For every $\lambda > 0$, there is a unique point $(x_0, 0)$ on the x-axis such that the projection of the y-axis on the line $y = b_1$ from this point projects $(0, b_n)$ into $(-(n-1)\lambda, b_1)$. In symbols: $(0, b_n)\overline{\wedge}(-(n-1)\lambda, b_1)$.
(2) For every natural number $n \geq 1$

$$b_n = b_1/[1 + (n-1)\omega]$$

where $\omega = (b_1 - b_2)/b_2$

Proof:

$(1) \Rightarrow (2)$: By similarity of triangles in the diagram below

$$\begin{cases} b_n/x_0 = (b_1 - b_n)/[(n-1)\lambda] \\ b_2/x_0 = \quad (b_1 - b_2)/\lambda \end{cases}$$

Dividing the first equation by the second and isolating b_n we obtain the desired formula.

$(2) \Rightarrow (1)$: Let λ be an arbitrary positive real number. We define

$$x_0 = \lambda/\omega.$$

We have to show that

$$(0, b_n) \overline{\wedge} (-(n-1)\alpha, b_1)$$

In other words, we have to show that the intersection of the line l_n joining $(x_0, 0)$ and $(0, b_n)$ with the line $y = b_1$ is the point $(-(n-1)\alpha, b_1)$. First notice that the equation of l_n is $y = -(b_n/x_0)x + b_n$. A simple computation shows that the lines in question meet at the point $((b_n - b_1)x_0/b_n, b_1)$. But it follows from (2) that $(b_1 - b_n)/b_n = (n-1)\omega$. Replacing x_0 by λ/ω, we obtain the desired result. Uniqueness of x_0 is obvious.

Notice that (2) implies that all the b_n's are known once that we know the first two of them.

Remark. Although we don't need them, we may add without proof the following equivalent conditions to (1) and (2):

(3) If n_1, n_2, n_3, n_4 are natural numbers such that $n_1 \neq n_4$ and $n_2 \neq n_3$, the cross-ratio of the corresponding $b's$ is given by

$$(b_{n_1} b_{n_2} b_{n_3} b_{n_4}) = (n_1 - n_3)/(n_1 - n_4) \times (n_2 - n_4)/(n_2 - n_3)$$

Furthermore, $\lim_{n \to \infty} b_n = 0$

(4) For every m, n

$$(b_n - b_m)/b_n b_m = (m - n)(b_1 - b_2)/b_1 b_2$$

5.2 Apparent vs. Real Lengths

Let us call the depth of A the distance from the front of the chapel to A. From the figure

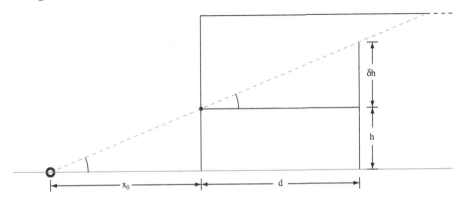

we deduce (from similarity of triangles) the formula for the depth of a vertical segment A

$$d = x_0 \delta h / h = \frac{16.56 \lambda \delta h}{h}$$

where h the apparent height (measured on the painting) of A and δh the difference between the real height an the apparent height of A.

Since the apparent height of Christ is $h = 148$, $\delta h = 30$ and it follows from this formula that the depth of Christ is 3.36λ. For $\lambda = 25.68$, we obtain that Christ depth is 86.28.

5.3   A Robust Lower Bound

To fully explain this value and provide further information that may be useful, we assume that the arches of the vault of the fresco are represented by the nine circles $C_1, C_2, C_3, \ldots, C_9$ shown in [1], page 100. The fact that the real arches that we assume are circular are represented by circles follows from the fact that the visual circular cone of the painter, whose directrix is the frontal arch, is cut by the vertical fresco in a circle, an elementary fact known to Apolonius. To compute the radius of these circles, let $(0, k_n)$ be the center of the circle C_n. By similarity of triangles, $r_n / k_n = r_1 / k_1$. Fortunately, $r_1 = 105.8$ is given as data and k_1 is easily calculated from the data (by additions and subtractions) in [1] (page 100). In fact, $k_1 = 311.2$. Thus, $r_n / k_n = 0.34$. We may also notice that $b_n - k_n = r_n$. Thus, $r_n = b_n - r_n/0.34$ and this formula may be rewritten as

$$r_n = 0.34 b_n / 1.34$$

Using this formula, $r_1 = 0.34 \times 416.8/1.34 = 105.76$ which is very near the value 105.8 given in the data (error: 0.04%).

The apparent width of a coffer on the nth rib is $w_n = r_n \times \pi/8$ and its apparent length is $\lambda_n = b_n - b_{n+1}$. The ratio between the apparent width and the apparent length of a coffer on the same rib is $\phi_n = w_n/l_n$. We organize these values in the following table

n	r_n	λ_n	w_n	ϕ_n
1	105.76	23.74	41.55	1.75
2	99.73	21.11	39.16	1.86
3	94.38	19.49	37.06	1.90
4	89.43	16.79	35.12	2.09
5	85.17	16.04	33.45	2.09
6	81.10	13.68	31.85	2.33
7	77.63	13.23	30.49	2.31
8	74.27	11.80	29.17	2.47
9	71.28	-	28.00	-

Clearly, $\lambda > \lambda_1 = 23.74$, the lower bound mentioned in the text. Furthermore, $\phi < \phi_1 = 1.75$.

6 A Comparison with the Results of the crs4 Group

After finishing this paper, my attention was called to a WEB site ([5]) where a reconstruction of Masaccio's chapel is attempted. Unfortunately, there are no details, no statement on their assumptions and the phrase "From the geometry it is actually possible to work backwards to reconstruct the full volume in measured accuracy of the 3-dimensional space Masaccio depicts" gives the erroneous impression that no further assumptions are needed for the reconstruction. The final result is given by means of a drawing.[4] The measures below are based exclusively on this drawing and hence are very rough. I computed the scale using the measures in [1]. In the column "this paper" I took 20 as the distance between Christ and the Father and as the difference between the depths of Christ and the Virgin and approximate to .5.

	crs4 Group	this paper
λ	32	31
x_0	$544 = 17\lambda$	$513.5 = 16.56\lambda$
length chapel	$224 = 7\lambda$	$217 = 7\lambda$
height Christ	179	178
height Father	192	192.5
head/vault	32	41.5
height Virgin	-	164
height St. John	166.5	160.5
length support Father	147	155
depth Christ	$134.5 = 4.2\lambda$	$104 = 3.36\lambda$
depth Father	$160 = 5\lambda$	$124 = 4\lambda$

There is considerable overall agreement, although with some important differences. It is interesting to note the coincidence on the real height of Christ and I suspect that they took this figure, just as we did, as an assumption for their reconstruction. Furthermore, there is no difference about λ since only a range of values around 30 was determined by the fresco and I took 31 for definiteness sake.

The main discrepancy is about x_0, the distance from the viewpoint to the chapel. This distance can be *proved* to be 16.56λ. (as we did in the Mathematical Appendix), rather than 17λ as they suggest. This discrepancy accounts for the difference of depths of Christ and the other figures. At the depth they suggest, the height of Christ would be 185.51. A more precise comparison can only be made when the details of their work will be available.

Acknowledgments. First, I would like to thank Prof. Kirsti Andersen for her meticulous critical reading of this paper. She corrected several architectural terms and pointed out wrong or incomplete formulations in the text. She also

---

[4] See `http://www.crs4.it/Ars/arsgifs/zmasacciodiagram.gif`

indicated some relevant literature. I'm very grateful to her for helping me to improve the paper. I would also like to thank Nicolas Fournier and Geneviève Habel for reading a first version of this paper. They suggested some improvements and/or referred me to other studies on the *Trinity*. I had discussions on these matters with Trish Campbell, Marie La Palme Reyes, François Magnan, Luis-Emilio Reyes and Houman Zolfaghari. Special thanks are due to Nicolas Fournier who made the diagrams, making the paper easier to understand. Finally, I would like to thank Phil Scott for translating my latex files into the latex2e files which were used for the publication of this paper.

References

1. Aiken, J.A.: The perspective construction of Masccio's *Trinity* fresco and medieval astronomical graphics. In: Goffen, R. (ed.) Masaccio's Trinity. Masterpieces of Western Painting, pp. 90–107. Cambridge University Press (1998)
2. Casazza, O.: Masaccio's fresco technique and problems of conservation. In: Goffen, R. (ed.) Masaccio's Trinity. Masterpieces of Western Painting, pp. 65–89. Cambridge University Press (1998)
3. Zuffi, S.: The Renaissance. Barnes & Noble Books, New York (2003)
4. Web Gallery of Art. Trinity by MASACCIO,
   http://www.kfki.hu/arthp/html/m/masaccio/trinity/trinity.html
5. The csr4 Group: part 2 of Renaissance Art and mathematical perspective. In: The Art of Renaissance Science, http://www.crs4.it/Ars/arshtml/arstoc.html, http://www.crs4.it/Ars/arsgifs/zmasacciodiagram.gif

Conjoinability in 1-Discontinuous Lambek Calculus

Alexey Sorokin

Moscow State University, Faculty of Mechanics and Mathematics,
Moscow Institute of Physics and Technology

Abstract. In the present work we prove a conjoinability criterion for 1-discontinuous Lambek calculus. It turns out that types of this calculus are conjoinable if and only if they have the same sort and the same interpretation in the free abelian group generated by the primitive types.

1 Introduction

Lambek calculus was introduced by Joachim Lambek in 1958 for modelling the syntactic structure of natural languages. In 1994 M. Pentus proved that Lambek grammars generate exactly context-free languages ([6]). Since context-free languages are well-known to be too weak for adequate representation of natural languages, G. Morrill introduced a generalization of Lambek calculus, the so-called discontinuous Lambek calculus ([3], [4]). In our work we consider 1-discontinuous Lambek calculus, which is sufficient for most linguistic applications (this calculus was thoroughly studied in [7]).

Let A and B be types of a particular categorial calculus. A type C is called a join for A and B (in this calculus) if both the sequents $A \to C$ and $B \to C$ are derivable. In this case the types A and B are called conjoinable. The conjoinability problem is interesting both from the linguistic ([1]) and from the algebraic point of view. For example, two types are conjoinable in Lambek calculus iff they have the same interpretation in the free group generated by the primitive types (this criterion was proved in [5]). If we replace the free group by the free abelian group, then we obtain the criterion of conjoinability in commutative Lambek calculus. It is worth noting that the criterion of conjoinability in Lambek-Grishin calculus also uses the interpretation in a free abelian group ([2]), though this calculus lacks commutativity (and even associativity). In our paper we prove that the same conjoinability criterion holds for 1-discontinuous Lambek calculus. The result is rather surprising because this calculus is not commutative either.

2 Discontinuous Lambek Calculus

Let Pr be a countable ranked set of primitive types and $rk \colon \mathrm{Pr} \to \mathbb{N}$ be a rank function. Then the set Tp of 1-types is the smallest ranked set satisfying the following conditions (s is a sort function which extends the rank function to the set of types):

C. Casadio et al. (Eds.): Lambek Festschrift, LNCS 8222, pp. 393–401, 2014.

1. $\Pr \subset \mathrm{Tp}$, $\forall A \in \Pr s(A) = rk(A)$, $I \in \mathrm{Tp}$, $s(I) = 0$, $J \in \mathrm{Tp}$, $s(J) = 1$.
2. $\forall A, B \in \mathrm{Tp}$ $((s(A) \geq s(B)) \Rightarrow (A/B), (B\backslash A) \in \mathrm{Tp}$, $s(A/B) = s(B\backslash A) = s(A) - s(B))$.
3. $\forall A, B \in \mathrm{Tp}$ $(A \cdot B) \in \mathrm{Tp}$, $s(A \cdot B) = s(A) + s(B)$.
4. $\forall A, B \in \mathrm{Tp}((s(A) \geq s(B) - 1) \Rightarrow (B \downarrow A) \in \mathrm{Tp}$, $s(B \downarrow A) = s(A) - s(B) + 1)$.
5. $\forall A, B \in \mathrm{Tp}$ $((s(A) \geq s(B)) \Rightarrow (A \uparrow B \in \mathrm{Tp})$, $s(A \uparrow B) = s(A) - s(B) + 1)$.
6. $\forall A, B \in \mathrm{Tp}$ $((s(A) \geq 1) \Rightarrow A \odot B \in \mathrm{Tp}$, $s(A \odot B) = s(A) + s(B) - 1)$.

Below $\mathcal{F}_i$ denotes the set of all 1-types having the sort i, Λ denotes the empty string, and $[]$ is a metalinguistic separator. The set of 1-hyperconfigurations is generated by the following grammar:

$$\mathcal{O} ::== \Lambda | [] | \mathcal{F}_0 | \mathcal{F}_i \underbrace{\{\mathcal{O} : \ldots : \mathcal{O}\}}_{i \text{ times}} | \mathcal{O}, \mathcal{O}.$$

External brackets are often omitted when writing a type. We use capital Latin letters $A, B, C, \ldots$ for types and capital Greek letters $\Gamma, \Delta, \Pi, \ldots$ for hyperconfigurations. A sort of a hyperconfiguration $s(\Gamma)$ is defined inductively and corresponds to the number of separators in Γ: $s(\Lambda) = 0$, $s([]) = 1$; $s(A) = i, A \in \mathcal{F}_i$; $s(A\{\Gamma_1, \ldots, \Gamma_{s(A)}\} = s(\Gamma_1) + \ldots + s(\Gamma_{s(A)})$; $s(\Gamma, \Delta) = s(\Gamma) + s(\Delta)$. The sequents of 1-discontinuous Lambek calculus have the form $\Gamma \to A$, where $s(\Gamma) = s(A)$. For every 1-type A we define its vector representation $\overrightarrow{A}$, which equals A in the case $sA = 0$ and $A\underbrace{\{[] : \ldots : []\}}_{sA \text{ times}}$ otherwise.

Let Γ, Δ be hyperconfigurations and $k \leq s(\Gamma)$, then we denote by $\Gamma|_k \Delta$ the result of substituting Δ for the k-th separator in Γ. If Γ is of sort i, then $\Gamma \otimes \langle \Delta_1, \ldots, \Delta_i \rangle$ denotes the result of simultaneous replacement of all the separators in Γ by the hyperconfigurations $\Delta_1, \ldots, \Delta_i$. If $\Delta, \Delta_1, \ldots, \Delta_i$ are hyperconfigurations and Γ is a hyperconfiguration of sort i, then $\Delta\langle \Gamma \rangle$ denotes the hyperconfiguration $\Delta(\Gamma \otimes \rangle \Delta_1, \ldots, \Delta_i))$ with the distinguished occurrence of Γ. The standard occurrence notation $\Delta[\Gamma]$ refers to a hyperconfiguration Γ in the external context Δ, the notation $\Delta\langle \Gamma \rangle$ refers to hyperconfiguration Γ with the external context Δ and the internal context $\Delta_1, \ldots, \Delta_i$. For a more detailed presentation see [4]. Now we formulate the discontinuous calculus **D**:

$$\frac{}{A \to A}(ax) \qquad \frac{\Gamma \to A \quad \Delta\langle \overrightarrow{A} \rangle \to B}{\Delta\langle \Gamma \rangle \to B}(cut)$$

$$\frac{\overrightarrow{A}, \Gamma \to C}{\Gamma \to A\backslash C}(\to \backslash) \qquad \frac{\Gamma \to A \quad \Delta\langle \overrightarrow{C} \rangle \to D}{\Delta\langle \Gamma, \overline{A\backslash C} \rangle \to D}(\backslash \to)$$

$$\frac{\Gamma, \overrightarrow{A} \to C}{\Gamma \to C/A}(\to /) \qquad \frac{\Gamma \to A \quad \Delta\langle \overrightarrow{C} \rangle \to D}{\Delta\langle \overline{C/A}, \Gamma \rangle \to D}(/ \to)$$

$$\frac{\Gamma \to A \quad \Delta \to B}{\Gamma, \Delta \to A \cdot B}(\to \cdot) \qquad \frac{\Delta\langle \overrightarrow{A}, \overrightarrow{B} \rangle \to D}{\Delta\langle \overrightarrow{A \cdot B} \rangle \to D}(\cdot \to)$$

$$\frac{}{\Lambda \to I}(\to I) \qquad \frac{\Delta\langle \Lambda \rangle \to A}{\Delta\langle I \rangle \to A}(I \to)$$

$$\frac{\overrightarrow{A}|_k \Gamma \to C}{\Gamma \to A \downarrow_k C}(\to\downarrow) \qquad \frac{\Gamma \to A \quad \Delta\langle \overrightarrow{C} \rangle \to D}{\Delta\langle \Gamma|_k \overrightarrow{A} \downarrow_k \overrightarrow{C} \rangle \to D}(\downarrow\to)$$

$$\frac{\Gamma|_k \vec{A} \to C}{\Gamma \to C \uparrow_k A}(\to\uparrow)$$

$$\frac{\Gamma \to A \quad \Delta\langle\vec{C}\rangle \to D}{\Delta\langle\overrightarrow{C \uparrow_k A}|_k\Gamma\rangle \to D}(\uparrow\to)$$

$$\frac{\Gamma \to A \quad \Delta \to B}{\Gamma|_k\Delta \to A \odot_k B}(\to\odot)$$

$$\frac{\Delta\langle\overrightarrow{A}|_k\vec{B}\rangle \to D}{\Delta\langle\overrightarrow{A \odot_k B}\rangle \to D}(\odot\to)$$

$$\frac{}{[] \to J}(\to J)$$

$$\frac{\Delta\langle[]\rangle \to A}{\Delta\langle J\rangle \to A}(I\to)$$

Note that the continuous product operation is explicitly associative, so we omit the parentheses in the expressions $A \cdot B \cdot C$, $B\backslash A/C$. Let Tp_1 be the subset of Tp satysfying the following condition: a type A belongs to Tp_1 iff it is of sort 1 or less and all its subtypes are also of sort 1 or less. Analogously $\mathcal{O}_1$ denotes the set of hyperconfigurations which do not involve subhyperconfigurations of sort greater than 1. Then the calculus $\mathbf{D}_1$ under consideration has the same rules as $\mathbf{D}$ but admits only hyperconfigurations from $\mathcal{O}_1$ and types from Tp_1. We will omit the subscript of the operations $\downarrow_1, \uparrow_1, \odot_1$ writing down the sequents of $\mathbf{D}_1$.

3   Conjoinability in Discontinuous Lambek Calculus

In this section we study the conjoinability relation in the calculus $\mathbf{D}_1$. In what follows we omit the vector sign in the sequents of the form $\vec{A} \to B$ simply writing $A \to B$. We will write $\mathbf{D}_1 \vdash A \to B$ if the sequent $A \to B$ is derivable in $\mathbf{D}_1$.

Definition 1. *The two types $A, B \in \mathrm{Tp}_1$ are called conjoinable if there exists a type C such that the sequents $A \to C$ and $B \to C$ are both derivable in the calculus $\mathbf{D}_1$.*

Lemma 1. *The following conditions are equivalent:*
1)$\exists C\,(\mathbf{D}_1 \vdash A \to C \wedge \mathbf{D}_1 \vdash B \to C)$ 2)$\exists D\,(\mathbf{D}_1 \vdash D \to A \wedge \mathbf{D}_1 \vdash D \to B)$.

Proof. This proof is due to [5]. 1) $\to$ 2) We can take $D = (A/C) \cdot C \cdot (C\backslash B)$. 2) $\to$ 1) We can take $C = (A\backslash D)\backslash D/(D\backslash B)$. $\square$

This lemma implies that the conjoinability relation $\sim$ is an equivalence. To formulate the conjoinability criterion we need some auxiliary notions. Note that only types of the same sort can be conjoinable.

Let Pr_1 be the set of primitive types of sort not greater than 1 and $\alpha \notin \mathrm{Pr}_1$. Then FAG is a free abelian group generated by the set $\mathrm{Pr}_1 \cup \{\alpha\}$. For every type $A \in \mathrm{Tp}_1$ we define its interpretation in the group $FAG = \langle FAG, \circ \rangle$ and denote this interpretation by $[A]$: $[p_i] = p_i$, $[I] = \varepsilon$, $[J] = \alpha$, $[A/B] = [B\backslash A] = [A] \circ [B]^{-1}$ $[B \downarrow A] = [A \uparrow B] = [A] \circ \alpha \circ [B]^{-1}$, $[A \cdot B] = [A] \circ [B]$, $[A \odot B] = [A] \circ \alpha^{-1} \circ [B]$

The introduction of an additional element α is a standard technique in the calculi with multiple residual families of product operations; the same method was used in [2] for characterizing the conjoinability relation in Lambek-Grishin calculus.

Lemma 2. *The condition* $[A] = [B]$ *is necessary for the types A and B to be conjoinable.*

Proof. Immediately follows from the fact that for every derivable sequent $A \to B$ it holds that $[A] = [B]$, which is proved by induction on derivation length in $\mathbf{D}_1$. □

We want to prove that this condition is also sufficient so it is a criterion. We will use the following notation: $|A|$ stands for the number of primitive type occurrences in A and $|A|_p$ for the number of occurrences of a particular primitive type p. $|A|_p^+$ and $|A|_p^-$ denote the number of positive and negative occurrences of p in A where the positive and negative occurrences are defined below (in this definition I and J are also considered to be primitive):

1. A primitive type p occurs positively in itself.
2. If p occurs positively/negatively in A, then it occurs positively/negatively in $A/B, B\backslash A, A \uparrow B, B \downarrow A, A \cdot B, B \cdot A, A \odot B$ and $B \odot A$.
3. If p occurs positively/negatively in B, then it occurs negatively/positively in $A/B, B\backslash A, A \uparrow B$ and $B \downarrow A$.

So the primitive type changes its polarity when it is placed under the continuous or discontinuous division operation. The polarity of the connective occurrence is defined in the same way. Let $* \in \{/, \backslash, \uparrow, \downarrow, \odot, \cdot\}$, then the following conditions hold:

1. The main connective of the type $A * B$ is positive in this type.
2. If an occurrence of the connective $*$ in A is positive/negative, then it is positive/negative in the types $A/B, B\backslash A, A \uparrow B, B \downarrow A, A \cdot B, B \cdot A, A \odot B$, $B \odot A$ for any type B.
3. If an occurrence of the connective $*$ in B is positive/negative, then it is negative/positive in the types $A/B, B\backslash A, A \uparrow B, B \downarrow A$.

The notation $|A|_*, |A|_*^+, |A|_*^-$, where $*$ is a binary connective, has the same meaning as in the case of primitive types. Further we denote $[A]_p = |A|_p^+ - |A|_p^-$, $[A]_* = |A|_*^+ - |A|_*^-$. The next lemma is proved by induction on type structure.

Lemma 3. *For any type $A \in \mathrm{Tp}_1$ it holds that* $[A] \vDash \prod_{p \in \mathrm{Pr}} p^{[A]_p} \circ (\alpha^{[A]_J + [A]_\uparrow + [A]_\downarrow - [A]_\odot})$.

In the proof of the conjoinability criterion we will use without mention the following statements (we suppose that both sides of every statement belong to Tp_1): $A \sim (A/B) \cdot B \sim B \cdot (B\backslash A)$, $A \sim (A \cdot B)/B \sim B\backslash(B \cdot A)$, $A \sim B \odot (B \downarrow A) \sim (A \uparrow B) \odot B, A \cdot J \sim (A \cdot B) \uparrow B, J \cdot A \sim B \uparrow (B \cdot A), (A \cdot J) \odot B \sim A \cdot B \sim (J \cdot B) \odot A$. We refer to these conjoinability relations as the basic ones. Their validity follows from the derivability of the sequents: $(A/B) \cdot B \to A$, $B \cdot (B\backslash A) \to A$; $A \to (A \cdot B)/B$, $A \to B\backslash(B \cdot A)$; $B \odot (B \downarrow A) \to A$, $(A \uparrow B) \odot B$; $A \cdot J \to (A \cdot B) \uparrow B$, $J \cdot A \to B \uparrow (B \cdot A)$; $(A \cdot J) \odot B \to A \cdot B$, $(J \cdot B) \odot A \to A \cdot B$.

The next lemma follows from the subformula property of the displacement calculus.

Lemma 4. *Let A, B, C belong to Tp_1 and $*$ be a connective in $\{\backslash, /, \uparrow, \downarrow, \cdot, \odot\}$. Let the types A and B be conjoinable. If one of the types $A * C$ and $B * C$ belongs to Tp_1, then the other also is in Tp_1 and the types $A * C$ and $B * C$ are conjoinable.*

Below we give a series of technical lemmas concerning the properties of the conjoinability relation in $\mathbf{D}_1$.

Lemma 5.

1. *For every type $A \in \mathrm{Tp}_1$ it holds that $A \cdot I \sim A \sim I \cdot A$.*
2. *For any types $A, B \in \mathrm{Tp}_1$ it holds that $A \cdot (B/B) \sim A \cdot (B\backslash B) \sim (B/B) \cdot A \sim (B\backslash B) \cdot A$.*

Proof. 1) Follows from the fact that the sequents $A \to A \cdot I$ and $A \to I \cdot A$ are derivable. 2) Follows from 1) and the derivability of the sequents $I \to B/B$, $I \to B\backslash B$. □

Lemma 6.

1. *For any type A such that $s(A) = 0$ it holds that $A \cdot J \sim J \cdot A$.*
2. *For any types A, B such that $A \cdot B \in \mathrm{Tp}_1$ it holds that $A \cdot B \sim B \cdot A$.*
3. *For any types A, B such that $A/B \in \mathrm{Tp}_1$ it holds that $A/B \sim B\backslash A$.*
4. *For any types A, B, C such that $(A/B)/C \in \mathrm{Tp}_1$ it holds that $(A/B)/C \sim (A/C)/B$.*

Proof.
1) $A \cdot J \sim (A \cdot A) \uparrow A \sim J \cdot A$.
2) Without loss of generality we suppose that $s(A) = 0$. Then $A \cdot B \sim (A \cdot J) \odot B \sim (J \cdot A) \odot B \sim B \cdot A$.
3) Basing on 2) and the Lemma 5 we deduce from basic conjoinability relations that $A/B \sim (B \cdot (B\backslash A))/B \sim ((B\backslash A) \cdot B)/B \sim (B\backslash A) \cdot (B/B) \sim (B\backslash A)$. We used the fact that the sequent $C \cdot (D/E) \to (C \cdot D)/E$ is derivable in Lambek calculus.
4) It is not difficult to see that $(A/B)/C \in \mathrm{Tp}_1$ implies that $s(B) + s(C) \le 1$ and $s(B) + s(C) \le s(A)$, so $C \cdot B \in \mathrm{Tp}_1$ and $A/(C \cdot B) \in \mathrm{Tp}_1$. Then $(A/B)/C \sim A/(C \cdot B) \sim A/(B \cdot C) \sim (A/C)/B$. □

In fact we have proved that continuous product operation is commutative with respect to conjoinability. To finish the proof of the main result we want to reduce the conjoinability problem for the types containing discontinuous product and its residuals to the case of continuous product. Namely, for every type A from Tp_1 we find a type B without the connectives $\odot, \uparrow$ and $\downarrow$ such that the types A and B are conjoinable. Afterwards it remains to prove the criterion only for "continuous" types (probably containing J). The next lemma allows us to simplify the types with "nested" divisions.

Proposition 1. *For any types A, B, C such that $A/(B/C)$ and $(A/B) \cdot C$ belong to Tp_1, it holds that $A/(B/C) \sim (A/B) \cdot C$.*

Proof.
This chain of conjoinable types follows from Lemma 6 and basic conjoinability relations: $A/(B/C) \sim A/(C\backslash B) \sim ((A/B) \cdot B)/(C\backslash B) \sim (((A/B) \cdot C) \cdot (C\backslash B))/(C\backslash B) \sim (A/B) \cdot C$. The fact that all the types in the chain belong to Tp_1 is verified by checking all possible sorts of the types A, B, C. □

The next lemma shows the connection between continuous and discontinuous product families and plays the crucial role in the whole proof.

Lemma 7.

1. *For any types A, B such that $A \uparrow B \in \mathrm{Tp}_1$ it holds that $A \uparrow B \sim (A/B) \cdot J$.*
2. *For any types A, B such that $B \downarrow A \in \mathrm{Tp}_1$ it holds that $B \downarrow A \sim A/(B/J)$.*

Proof.
1) It is easy to see that $(A/B) \cdot J \in \mathrm{Tp}_1$. Then $A \uparrow B \sim ((A/B) \cdot B) \uparrow B \sim (((A/B) \cdot J) \odot B) \uparrow B \sim (A/B) \cdot J$.
2) Similarly to the previous case $A/(B/J) \in \mathrm{Tp}_1$. Then $B \downarrow A \sim (B/J)\backslash((B/J)\cdot (B \downarrow A)) \sim (B/J)\backslash(((B/J) \cdot J) \odot (B \downarrow A)) \sim (B/J)\backslash(B \odot (B \downarrow A)) \sim (B/J)\backslash A$. □

Now we can prove the basic result of the paper. Let $\mathrm{Pr}_0 = \{p \in \mathrm{Pr} \mid s(p) = 0\}$, $\mathrm{Pr}_1 = \{q \in \mathrm{Pr} \mid s(q) = 1\}$. We denote the elements of Pr_0 by $p_1, p_2, \ldots$ and the elements of Pr_1 by $q_1, q_2, \ldots$. We suppose that the element p_0 belongs to Pr_0 but is not used in constructing the types. The notation $Var(A_1, \ldots, A_k)$ denotes the set of all primitive types in $A_1, \ldots, A_k$ and $\#(A), \#(A)^+, \#(A)^-$ denotes the number of different primitive types, occurring A (occurring positively, occurring negatively, respectively). Also we introduce the two measures $M_1(A_1, \ldots, A_k) = \max(i \mid p_i \in Var(A_1, \ldots, A_k))$ and $M_2 = \max(i \mid q_i \in Var(A_1, \ldots, A_k))$. If the type A is of sort 0, then we denote by A^k the type $\underbrace{A \cdot A \cdot \ldots \cdot A}_{n \text{ times}}$, $A^0 = I$. For any $A \in \mathrm{Tp}_1$ we introduce auxiliary types $\widehat{A}$ and $\widetilde{A}$ which are defined below:

$$\widehat{A} = (((p_0/p_0) \cdot \prod_{\substack{i=1, \\ |A|_{p_i}^+ \geq 1}}^{M_1(A)} p_i^{|A|_{p_i}^+} \cdot \prod_{\substack{i=1, \\ |A|_{q_i}^+ \geq 1}}^{M_2(A)} (q_i/J)^{|A|_{q_i}^+}) / ((p_0/p_0) \cdot \prod_{\substack{i=1, \\ |A|_{p_i}^- \geq 1}}^{M_1(A)} p_i^{|A|_{p_i}^-} \cdot \prod_{\substack{i=1, \\ |A|_{q_i}^- \geq 1}}^{M_2} (q_i/J)^{|A|_{q_i}^-})) \cdot J^{s(A)}$$

$$\widetilde{A} = (((p_0/p_0) \cdot \prod_{\substack{i=1, \\ [A]_{p_i} > 0}}^{M_1(A)} p_i^{[A]_{p_i}} \cdot \prod_{\substack{i=1, \\ [A]_{q_i} > 0}}^{M_2} (q_i/J)^{[A]_{q_i}}) / ((p_0/p_0) \cdot \prod_{\substack{i=1, \\ [A]_{p_i} < 0}}^{M_1(A)} p_i^{-[A]_{p_i}} \cdot \prod_{\substack{i=1, \\ [A]_{q_i} < 0}}^{M_2} (q_i/J)^{-[A]_{q_i}})) \cdot J^{s(A)}$$

Let us prove some important properties of the types introduced (in the proofs we omit the symbol of the group operation $\circ$):

Lemma 8.

1. $[\widetilde{A}] = [\widehat{A}]$.
2. $[\widehat{A}] = [A]$.
3. *If $[A] = [B]$, then $\widetilde{A} = \widetilde{B}$.*

Proof.

1) $[\widetilde{A}] = \prod_{i=1}^{M_1(A)} p_i^{[A]_{p_i}} \prod_{i=1}^{M_2(A)} q_i^{[A]_{q_i}} \alpha^{s(A) - \sum_{i=1}^{M_2(A)} [A]_{q_i}} =$

$\prod_{i=1}^{M_1(A)} p_i^{|A|_{p_i}^+ - |A|_{p_i}^-} \prod_{i=1}^{M_2(A)} (q_i\alpha^{-1})^{|A|_{q_i}^+ - |A|_{q_i}^-} \alpha^{s(A)} = [\widehat{A}].$

2) We use induction on the construction of A. The base case $A \in \mathrm{Pr} \cup \{I, J\}$ is directly verified. In the induction step we prove only the case $A = B \uparrow C$ (all other variants are similar). Indeed, $[\widetilde{A}] = \prod_{i=1}^{M_1(A)} p_i^{[A]_{p_i}} \prod_{i=1}^{M_2(A)} q_i^{[A]_{q_i}} \alpha^{s(A) - \sum_{i=1}^{M_2(A)} [A]_{q_i}} =$

$\prod_{i=1}^{M_1(A)} p_i^{[B]_{p_i} - [C]_{p_i}} \prod_{i=1}^{M_2(A)} q_i^{[B]_{q_i} - [C]_{q_i}} \alpha^{(s(B) - s(C) + 1 - \sum_{i=1}^{M_2(A)} [B]_{q_i} + \sum_{i=1}^{M_2(A)} [C]_{q_i})} = [\widetilde{B}][\widetilde{C}]^{-1}\alpha =$

$[B]\alpha[C] = [B \uparrow C]$.

3) The equality $[A] = [B]$ implies that for every $p \in \mathrm{Pr}$ it holds that $[A]_p = [B]_p$
$[A]_J + [A]_\uparrow + [A]_\downarrow - [A]_\odot = [B]_J + [B]_\uparrow + [B]_\downarrow - [B]_\odot$. It is easy to show by induction that for every type C its sort equals $\sum_{q \in \mathrm{Pr}_1} [C]_q + [C]_J + [C]_\uparrow + [C]_\downarrow - [C]_\odot$.

So $s(A) = s(B)$ and it remains to use the definition of the type $\widetilde{A}$. $\qquad\square$

Proposition 2.

1. *For any types A, B such that $A \uparrow B \in \mathrm{Tp}_1$, it holds that $[A \uparrow B] = [(A/B) \cdot J]$.*
2. *For any types A, B such that $B \downarrow A \in \mathrm{Tp}_1$ it holds that $[B \downarrow A] = [A/(B/J)]$.*

Proof. It is easy to see that $\forall p \in \mathrm{Pr}\ \forall \varepsilon \in \{-, +\}\ |(A \uparrow B)|_p^\varepsilon = |(A/B) \cdot J|_p^\varepsilon$. Also $s(A \uparrow B) = s(A) - s(B) + 1 = s((A/B) \cdot J)$. Then by the definition of the interpretation $[A \uparrow B] = [(A/B) \cdot J]$. The second statement is analogous. $\qquad\square$

Lemma 9. *1) For any two types A, B such that $A \cdot B \in \mathrm{Tp}_1$ it holds that $\widehat{A \cdot B} \sim \widehat{A} \cdot \widehat{B}$.*

2) For any two types A, B such that $A/B \in \mathrm{Tp}_1$ it holds that $\widehat{A/B} \sim \widehat{A} \cdot \widehat{B}$.

3) For any two types A, B such that $B \backslash A \in \mathrm{Tp}_1$ it holds that $\widehat{B \backslash A} \sim \widehat{B} \cdot \widehat{A}$.

Proof. 1) Induction on the number of primitive types (except I and J) in C. In the basic case this number equals 0, so $\widehat{C}$ equals $((p_0/p_0)/(p_0/p_0)) \cdot J^{s(C)}$ and $\widehat{B \cdot C} = \widehat{B} \cdot J^{s(B \cdot C) = s(B)} = \widehat{B} \cdot J^{s(C)} \sim \widehat{B} \cdot ((p_0/p_0)/(p_0/p_0)) \cdot J^{s(C)} = \widehat{B} \cdot \widehat{C}$. The base is proved.

Now let C contain some zero-sorted primitive type p. Then we can write $\widehat{B} = ((E_1 \cdot p^{d_1} \cdot E_2)/(F_1 \cdot p^{d_2} \cdot F_2)) \cdot J^{s(B)}$, $\widehat{C} = ((G_1 \cdot p^{e_1} \cdot G_2)/(H_1 \cdot p^{e_2} \cdot H_2)) \cdot J^{s(C)}$. Here the types $E_1, E_2, F_1, F_2, G_1, G_2, H_1, H_2$ do not contain p, d_1 equals $|B|_p^+$, d_2 equals $|B|_p^-$ and so on. For example, E_2 contains all zero-ranked primitive types with the index greater than the index of p and all the primitive types of rank 1 which occur in B (if there are no such primitive types, then we set $B_2 = I$). Using the properties of conjoinability (i.e. commutativity of product with respect to this relation) we can prove that $\widehat{B} \cdot \widehat{C} = ((E_1 \cdot p^{d_1} \cdot E_2)/(F_1 \cdot p^{d_2} \cdot F_2)) \cdot J^{s(B)} \cdot ((G_1 \cdot p^{e_1} \cdot G_2)/(H_1 \cdot p^{e_2} \cdot H_2)) \cdot J^{s(C)} \sim ((E_1 \cdot p^{d_1 + e_1} \cdot E_2)/(F_1 \cdot p^{d_2 + e_2} \cdot F_2)) \cdot J^{s(B)} \cdot ((G_1 \cdot$

$G_2)/(H_1 \cdot H_2)) \cdot J^{s(C)}$. We denote $B' = ((E_1 \cdot p^{d_1+e_1} \cdot E_2)/(F_1 \cdot p^{d_2+e_2} \cdot F_2)) \cdot J^{s(B)}$ and $C' = ((G_1 \cdot G_2)/(H_1 \cdot H_2)) \cdot J^{s(C)}$, it is easy to see that $\widehat{C'} = C'$ and $\widehat{B'} = \widehat{B'}$. Note that C' contains fewer primitive types than C. So $\widehat{B \cdot C} = \widehat{B' \cdot C'} \sim \widehat{B'} \cdot \widehat{C'} = B' \cdot C'$. Using the properties of conjoinability it is easy to prove that $B' \sim \widehat{B} \cdot (p^{d_2}/p^{e_2})$. So we can deduce that $\widehat{B \cdot C} \sim \widehat{B} \cdot ((p^{d_2}/p^{e_2}) \cdot C') = \widehat{B} \cdot ((p^{d_2}/p^{e_2}) \cdot ((G_1 \cdot G_2)/(H_1 \cdot H_2)) \cdot J^{s(C)} \sim \widehat{B} \cdot ((G_1 \cdot p^{d_2} \cdot G_2)/(H_1 \cdot p^{e_2} \cdot H_2)) \cdot J^{s(C)} = \widehat{B} \cdot \widehat{C}$, which was required.

2, 3) The proof is analogous to case 1. $\qquad \square$

Lemma 10. *For any type* $A \in \mathrm{Tp}_1$ *the types* A *and* $\widehat{A}$ *are conjoinable.*

Proof. Induction on the construction of the type A. The basic case $A \in \mathrm{Pr} \cup \{I, J\}$ is easy to verify. In the induction step we should examine all the possible basic connectives of the type A. Due to Proposition 2 it suffices to prove the lemma for the types $A = B/C, A = C \backslash B$ and $A = B \cdot C$.

In Lemma 9 we have proved that if $A * B \in \mathrm{Tp}_1$, then $\widehat{A * B} \sim \widehat{A} * \widehat{B}$, where $*$ is an arbitrary connective from the set $\{\cdot, \backslash, /\}$. Then for an arbitrary type A where $A = B * C$ we have that $\widehat{A} = \widehat{B * C} \sim \widehat{B} * \widehat{C} \sim B * C = A$. The lemma is proved. $\qquad \square$

Lemma 11. *For any type* $A \in \mathrm{Tp}_1$ *the types* $\widehat{A}$ *and* $\widetilde{A}$ *are conjoinable.*

Proof. It suffices to show that for every type B which has the form $B = (B_1 \cdot A^k \cdot B_2)/(B_3 \cdot A^l \cdot B_4)$ for some types B_1, B_2, B_3, B_4, A such that $s(A) = 0$ it is conjoinable with the type C which equals $(B_1 \cdot A^{k-l} \cdot B_2)/(B_3 \cdot B_4)$ in the case $k > l, (B_1 \cdot B_2)/(B_3 \cdot B_4)$ — in the case $k = l$ and equals $(B_1 \cdot B_2)/(B_3 \cdot A^{l-k} B_4)$ in the case $l > k$.

Let us consider the first case. Then $B = (B_1 \cdot A^k \cdot B_2)/(B_3 \cdot A^l \cdot B_4) \sim (A^k \cdot B_1 \cdot B_2)/(A^l \cdot B_3 \cdot B_4) \sim (A^k \cdot ((B_1 \cdot B_2)/(B_3 \cdot B_4)))/A^l \sim (A^k/A^l) \cdot ((B_1 \cdot B_2)/(B_3 \cdot B_4)) \sim A^{k-l} \cdot ((B_1 \cdot B_2)/(B_3 \cdot B_4)) \sim (A^{k-l} \cdot B_1 \cdot B_2)/(B_3 \cdot B_4) \sim (B_1 \cdot A^{k-l} \cdot B_2)/(B_3 \cdot B_4)$. The second case is analogous. In the third case we have $B \sim (((B_1 \cdot B_2)/(B_3 \cdot B_4)) \cdot A^k)/A^l \sim (A^k/A^l) \cdot ((B_1 \cdot B_2)/(B_3 \cdot B_4))/A^{l-k} \sim ((B_1 \cdot B_2)/(B_3 \cdot A^{l-k} \cdot B_4))$. The lemma is proved. $\qquad \square$

Now we can prove the main result of the paper.

Theorem 1. *For any types* $A, B \in \mathrm{Tp}_1$ *the conditions* $[A] = [B]$ *and* $A \sim B$ *are equivalent.*

Proof. The necessity of the condition of equal interpretations was proved in Lemma 2. Let us prove the sufficiency. If $[A] = [B]$, then by Lemma 8 it holds that $\widetilde{A} = \widetilde{B}$, hence by Lemma 11 $\widehat{A} \sim \widehat{B}$, so by Lemma 10 we obtain $A \sim B$. The theorem is proved. $\qquad \square$

In fact we can reformulate the criterion in a slightly different form. Note that by the construction the number of α-s in the interpretation of a type equals its sort. Since types of different sorts cannot be conjoinable, we can "contract" the additional α-s without changing the conjoinability property. Let FAG_{Pr} denote the free abelian group generated by the primitive types. Let $[\cdot]_{Pr}$ be the interpretation in this group which is defined inductively (we omit the sign of the free group operation): $[p_i]_{Pr} = p_i$, $[I]_{Pr} = [J]_{Pr} = \varepsilon$, $[A/B]_{Pr} = [B \downarrow A]_{Pr} = [A \uparrow B]_{Pr} = [B \backslash A]_{Pr} = [A]_{Pr}[B]_{Pr}^{-1}$, $[A \cdot B]_{Pr} = [A \odot B]_{Pr} = [A]_{Pr}[B]_{Pr}$. Then the following statement holds (it is a direct consequence of the Theorem 1 and the arguments above).

Corollary 1. *For any types $A, B \in \mathrm{Tp}_1$ the condition $A \sim B$ holds if and only if the conditions $s(A) = s(B)$ and $[A]_{Pr} = [B]_{Pr}$ hold simultaneously.*

4 Conclusion

We have proved the conjoinability criterion in 1-discontinuous Lambek calculus. The criterion requires the types to be conjoined to have equal interpretation in the free abelian group generated by the primitive types. Practically the same criterion was already known for the Lambek-Grishin calculus. It would be interesting to formulate the conjoinability criterion in the full discontinuous Lambek calculus without any bounds on the sort of types. The criterion from the present work seems to hold in this case as well but the author does not know the complete proof of this fact. It would be also interesting to study from the algebraic point of view which properties should be possessed by a calculus with several families of operations to have a particular characterization of conjoinability.

The author is grateful to Mati Pentus for his commentaries, which helped to improve the paper. This research was partially supported by the Russian Foundation for Basic Research (grant 11-01-00958-a).

References

1. Foret, A.: Conjoinability and unification in Lambek categorial grammars. In: New Perspectives in Logic and Formal Linguistics, Proceedings Vth ROMA Workshop, Bulzoni, Roma (2001)
2. Moortgat, M., Pentus, M.: Type similarity for the Lambek-Grishin calculus. In: Proceedings of the 12th Conference on Formal Grammar, Dublin (2007)
3. Morrill, G., Valentín, O.: On calculus of displacement. In: Proceedings of the 10th International Workshop on Tree Adjoining Grammars and Related Formalisms, pp. 45–52 (2010)
4. Morrill, G., Valentín, O., Fadda, M.: The displacement calculus. Journal of Logic, Language and Information 20(1), 1–48 (2011)
5. Pentus, M.: The conjoinability relation in Lambek calculus and linear logic. ILLC Prepublication Series ML-93-03. Institute for Logic, Language and Computation, University of Amsterdam (1993)
6. Pentus, M.: Lambek grammars are context-free. In: Logic in Computer Science, Proceedings of the LICS 1993, pp. 429–433 (1993)
7. Valentín, O.: 1-discontinuous Lambek calculus: Type logical grammar and discontinuity in natural language. Master's thesis. Universitat Autònoma de Barcelona (2006)

The Hidden Structural Rules of the Discontinuous Lambek Calculus

Oriol Valentín

Universitat Politècnica de Catalunya

Abstract. The sequent calculus **sL** for the Lambek calculus **L** ([2]) has no structural rules. Interestingly, **sL** is equivalent to a multimodal calculus **mL**, which consists of the nonassociative Lambek calculus with the structural rule of associativity. This paper proves that the sequent calculus or *hypersequent* calculus **hD** of the discontinuous Lambek calculus[1] ([7], [4] and [8]), which like **sL** has no structural rules, is also equivalent to an ω-sorted multimodal calculus **mD**. More concretely, we present a faithful embedding translation $(\cdot)^\sharp$ between **mD** and **hD** in such a way that it can be said that **hD** absorbs the structural rules of **mD**.

1 The Discontinuous Lambek Calculus D and Its Hypersequent Syntax

D is model-theoretically motivated, and the key to its conception is the class **FreeDisp** of displacement algebras. We need some definitions:

(1) **Definition** (*Syntactical Algebra*)

A *syntactical algebra* is a free algebra $(L, +, 0, 1)$ of arity $(2, 0, 0)$ such that $(L, +, 0)$ is a monoid and 1 is a prime. I.e. L is a set, $0 \in L$ and $+$ is a binary operation on L such that for all $s_1, s_2, s_3, s \in L$,

$$s_1 + (s_2 + s_3) = (s_1 + s_2) + s_3 \quad \text{associativity}$$
$$0 + s = s = s + 0 \qquad \text{identity}$$

The distinguished constant 1 is called a *separator*.

(2) **Definition** (*Sorts*)

The *sorts* of discontinuous Lambek calculus are the naturals $0, 1, \dots$. The sort $S(s)$ of an element s of a syntactical algebra $(L, +, 0, 1)$ is defined by the morphism of monoids S to the additive monoid of naturals defined thus:

$$S(1) = 1$$
$$S(a) = 0 \qquad \text{for a prime } a \neq 1$$
$$S(s_1 + s_2) = S(s_1) + S(s_2)$$

---

[1] In [5] and [8], the term *displacement calculus* is used instead of Discontinuous Lambek Calculus as in [7] and [9].

C. Casadio et al. (Eds.): Lambek Festschrift, LNCS 8222, pp. 402–420, 2014.

I.e. the sort of a syntactical element is simply the number of separators it contains; we require the separator 1 to be a prime and the syntactical algebra to be free in order to ensure that this induction is well-defined.

(3) **Definition** (*Sort Domains*)

Where $(L, +, 0, 1)$ is a syntactical algebra, the *sort domains* L_i of sort i of generalized discontinuous Lambek calculus are defined as follows:

$$L_i = \{s | S(s) = i\}, i \geq 0$$

(4) **Definition** (*Displacement Algebra*)

The *displacement algebra* defined by a syntactical algebra $(L, +, 0, 1)$ is the ω-sorted algebra with the ω-sorted signature $\Sigma_D = (\oplus, \{\otimes_{i+1}\}_{i\in\omega}, 0, 1)$ with sort functionality $((i, j \to i + j)_{i,j\in\omega}, (i + 1, j \to i + j)_{i,j\in\omega}, 0, 1)$:

$$(\{L_i\}_{i\in\omega}, +, \{\times_{i+1}\}_{i\in\omega}, 0, 1)$$

where:

operation	is such that
$+ : L_i \times L_j \to L_{i+j}$	as in the syntactical algebra
$\times_k : L_{i+1} \times L_j \to L_{i+j}$	$\times_k(s, t)$ is the result of replacing the k-th separator in s by t

The sorted types of the discontinuous Lambek Calculus, **D**, which we will define residuating with respect to the sorted operations in (4), are defined by mutual recursion in Figure 1. **D** types are to be interpreted as subsets of L and satisfy what we call the *principle of well-sorted inhabitation*:

$\mathcal{F}_i ::= \mathcal{A}_i$ where $\mathcal{A}_i$ is the set of atomic types of sort i

$\mathcal{F}_0 ::= I$ Continuous unit
$\mathcal{F}_1 ::= J$ Discontinuous unit

$\mathcal{F}_{i+j} ::= \mathcal{F}_i \bullet \mathcal{F}_j$ continuous product
$\mathcal{F}_j ::= \mathcal{F}_i \backslash \mathcal{F}_{i+j}$ continuous under
$\mathcal{F}_i ::= \mathcal{F}_{i+j} / \mathcal{F}_j$ continuous over

$\mathcal{F}_{i+j} ::= \mathcal{F}_{i+1} \odot_k \mathcal{F}_j$ discontinuous product
$\mathcal{F}_j ::= \mathcal{F}_{i+1} \downarrow_k \mathcal{F}_{i+j}$ discontinuous extract
$\mathcal{F}_{i+1} ::= \mathcal{F}_{i+j} \uparrow_k \mathcal{F}_j$ discontinuous infix

Fig. 1. The sorted types of **D**

(5) | **Principle of well-sorted inhabitation:**
If A is a type of sort i, $[\![A]\!] \subseteq L_i$

Where $[\![\cdot]\!]$ is the syntactical interpretation in a given displacement algebra w.r.t. a valuation v. I.e. every syntactical inhabitant of $[\![A]\!]$ has the same sort. The connectives and their syntactical interpretations are shown in Figures 1 and 2. This syntactical interpretation is called the *standard syntactical interpretation*. Given the functionalities of the operations with respect to which the connectives are defined, the grammar defining by mutual recursion the sets $\mathcal{F}_i$ of types of sort i on the basis of sets $\mathcal{A}_i$ of atomic types, and the homomorphic *syntactical sort map* S sending types to their sorts, are as shown in Figure 3. When A is an arbitrary type, we will frequently write in latin lower-case the type in order to refer to its sort $S(A)$, i.e.:

$$a \stackrel{def}{=} S(A)$$

The syntactical sort map is to syntax what the semantic type map is to semantics: both homomorphisms mapping syntactic types to the datatypes of the respective components of their inhabiting signs in the dimensions of language in extension: form/signifier and meaning/signified.

$$
\begin{aligned}
[\![I]\!] &= \{0\} & \text{continuous unit} \\
[\![J]\!] &= \{1\} & \text{discontinuous unit} \\
[\![A]\!] &\subseteq L_i \text{ for some } i \in \omega & A \in \mathcal{A}_i
\end{aligned}
$$

$$
\begin{aligned}
[\![A{\bullet}B]\!] &= \{s_1{+}s_2 |\ s_1 \in [\![A]\!]\ \&\ s_2 \in [\![B]\!]\} & \text{(continuous) product} \\
[\![A{\backslash}C]\!] &= \{s_2 |\ \forall s_1 \in [\![A]\!], s_1{+}s_2 \in [\![C]\!]\} & \text{under} \\
[\![C/B]\!] &= \{s_1 |\ \forall s_2 \in [\![B]\!], s_1{+}s_2 \in [\![C]\!]\} & \text{over}
\end{aligned}
$$

$$
\begin{aligned}
[\![A{\odot}_k B]\!] &= \{\times_k(s_1, s_2) |\ s_1 \in [\![A]\!]\ \&\ s_2 \in [\![B]\!]\} & k > 0 \text{ deterministic discontinuous product} \\
[\![A{\downarrow}_k C]\!] &= \{s_2 |\ \forall s_1 \in [\![A]\!], \times_k(s_1, s_2) \in [\![C]\!]\} & k > 0 \text{ deterministic discontinuous infix} \\
[\![C{\uparrow}_k B]\!] &= \{s_1 |\ \forall s_2 \in [\![B]\!], \times_k(s_1, s_2) \in [\![C]\!]\} & k > 0 \text{ deterministic discontinuous extract}
\end{aligned}
$$

Fig. 2. Standard syntactical interpretation of **D** types

Observe also that (modulo sorting) $(\backslash, \bullet, /; \subseteq)$ and $(\downarrow_k, \odot_k, \uparrow_k; \subseteq)$ are residuated triples:

(6) $[\![B]\!] \subseteq [\![A{\backslash}C]\!]$ iff $[\![A{\bullet}B]\!] \subseteq [\![C]\!]$ iff $[\![A]\!] \subseteq [\![C/B]\!]$
$[\![B]\!] \subseteq [\![A{\downarrow}_k C]\!]$ iff $[\![A{\odot}_k B]\!] \subseteq [\![C]\!]$ iff $[\![A]\!] \subseteq [\![C{\uparrow}_k B]\!]$

The types of **D** are sorted into types $\mathcal{F}_i$ of sort i interpreted as sets of strings of sort i as shown in Figure 4 where $k \in \omega^+$.

If one wants to absorb the structural rules of a Gentzen sequent system in a substructural logic, one has to discover a convenient data structure for the antecedent and the succedent of sequents. We will now consider the *Hypersequent syntax*[2] from [7]. The reason for using the prefix *hyper* in the term *sequent* is that the data-structure proposed is quite nonstandard.

---

[2] Term which must not be confused with Avron's hypersequents ([1]).

$$\mathcal{F}_i ::= \mathcal{A}_i \qquad\qquad S(A) = i \qquad\qquad\qquad \text{for } A \in \mathcal{A}_i$$

$$\mathcal{F}_0 ::= I \qquad\qquad S(I) = 0$$
$$\mathcal{F}_1 ::= J \qquad\qquad S(J) = 1$$

$$\mathcal{F}_{i+j} ::= \mathcal{F}_i \bullet \mathcal{F}_j \qquad S(A \bullet B) = S(A) + S(B)$$
$$\mathcal{F}_j ::= \mathcal{F}_i \backslash \mathcal{F}_{i+j} \qquad S(A \backslash C) = S(C) - S(A)$$
$$\mathcal{F}_i ::= \mathcal{F}_{i+j}/\mathcal{F}_j \qquad S(C/B) = S(C) - S(B)$$

$$\mathcal{F}_{i+j} ::= \mathcal{F}_{i+1}\odot_k\mathcal{F}_j \quad S(A\odot_k B) = S(A) + S(B) - 1 \;\; 1 \le k \le i+1$$
$$\mathcal{F}_j ::= \mathcal{F}_{i+1}\downarrow_k\mathcal{F}_{i+j} \quad S(A\downarrow_k C) = S(C) + 1 - S(A) \;\; 1 \le k \le i+1$$
$$\mathcal{F}_{i+1} ::= \mathcal{F}_{i+j}\uparrow_k\mathcal{F}_j \quad S(C\uparrow_k B) = S(C) + 1 - S(B) \;\; 1 \le k \le i+1$$

Fig. 3. Sorted **D** types, and syntactical sort map for **D**

$$\mathcal{F}_j := \mathcal{F}_i \backslash \mathcal{F}_{i+j} \qquad [A\backslash C] = \{s_2|\ \forall s_1 \in [A], s_1+s_2 \in [C]\} \quad \text{under}$$
$$\mathcal{F}_i := \mathcal{F}_{i+j}/\mathcal{F}_j \qquad [C/B] = \{s_1|\ \forall s_2 \in [B], s_1+s_2 \in [C]\} \quad \text{over}$$
$$\mathcal{F}_{i+j} := \mathcal{F}_i \bullet \mathcal{F}_j \qquad [A\bullet B] = \{s_1+s_2|\ s_1 \in [A]\ \&\ s_2 \in [B]\} \quad \text{product}$$
$$\mathcal{F}_0 := I \qquad\qquad [I] = \{0\} \qquad\qquad\qquad\qquad\qquad \text{product unit}$$
$$\mathcal{F}_j := \mathcal{F}_{i+1}\downarrow_k\mathcal{F}_{i+j} \quad [A\downarrow_k C] = \{s_2|\ \forall s_1 \in [A], s_1\times_k s_2 \in [C]\} \quad \text{infix}$$
$$\mathcal{F}_{i+1} := \mathcal{F}_{i+j}\uparrow_k\mathcal{F}_j \quad [C\uparrow_k B] = \{s_1|\ \forall s_2 \in [B], s_1\times_k s_2 \in [C]\} \quad \text{extract}$$
$$\mathcal{F}_{i+j} := \mathcal{F}_{i+1}\odot_k\mathcal{F}_j \quad [A\odot_k B] = \{s_1\times_k s_2|\ s_1 \in [A]\ \&\ s_2 \in [B]\} \quad \text{disc. product}$$
$$\mathcal{F}_1 := J \qquad\qquad [J] = \{1\} \qquad\qquad\qquad\qquad\qquad \text{disc. prod. unit}$$

Fig. 4. Types of the Discontinuous Lambek Calculus **D** and their interpretation

We define now what we call the set of types segments:

(7) **Definition** (*Type Segments*)

In hypersequent calculus we define the *types segments* $\mathcal{SF}_k$ of sort k:

$$\mathcal{SF}_0 ::= A \quad \text{for } A \in \mathcal{F}_0$$
$$\mathcal{SF}_a ::= \sqrt[i]{A} \quad \text{for } A \in \mathcal{F}_a \text{ and } 0 \le i \le a = S(A)$$

Types segments of sort 0 are types. But, types segments of sort greater than 0 are no longer types. Strings of types segments can form meaningful logical material like the set of hyperconfigurations, which we now define. The *hyperconfigurations* $\mathcal{O}$ are defined unambiguously by mutual recursion as follows, where Λ is the empty string and $[]$ is the metalinguistic separator::

$$\mathcal{O} ::= \Lambda$$
$$\mathcal{O} ::= A, \mathcal{O} \text{ for } S(A) = 0$$
$$\mathcal{O} ::= [], \mathcal{O}$$
$$\mathcal{O} ::= \sqrt[0]{A}, \mathcal{O}, \sqrt[1]{A}, \ldots, {}^{a-1}\!\sqrt{A}, \mathcal{O}, \sqrt[a]{A}, \mathcal{O}$$
$$\text{for } a = S(A) > 0$$

The syntactical interpretation of $\sqrt[0]{A}, \mathcal{O}, \sqrt[1]{A}, \mathcal{O}, \ldots, \sqrt[a-1]{A}, \mathcal{O}, \sqrt[a]{A}$ consists of syntactical elements $\alpha_0 + \beta_1 + \alpha_1 + \cdots + \alpha_{n-1} + \beta_n + \alpha_n$ where

$$\alpha_0 + 1 + \alpha_1 + \cdots + \alpha_{n-1} + 1 + \alpha_n \in [\![A]\!]$$

and $\beta_1 \in [\![\Delta_1]\!], \ldots, \beta_n \in [\![\Delta_n]\!]$. The syntax in which set $\mathcal{O}$ has been defined, is called *string-based hypersequent syntax*. An equivalent syntax for $\mathcal{O}$ is called *tree-based hypersequent syntax* which was defined in [4], [8].

In string-based notation the *figure* $\overrightarrow{A}$ of a type A is defined as follows:

$$(8) \quad \overrightarrow{A} = \begin{cases} A & \text{if } s(A) = 0 \\ \sqrt[0]{A}, [], \sqrt[1]{A}, [], \ldots, \sqrt[a-1]{A}, [], \sqrt[a]{A} & \text{if } s(A) > 0 \end{cases}$$

The sort of a hyperconfiguration is the number of metalinguistic separators it contains. Where Γ and Φ are hyperconfigurations and the sort of Γ is at least 1, $\Gamma|_k \Phi$ ($k \in \omega^+$) signifies the hyperconfiguration which is the result of replacing the k-th separator in Γ by Φ. Where Γ is a hyperconfiguration of sort i and $\Phi_1, \ldots, \Phi_i$ are hyperconfigurations, the *generalized wrap* $\Gamma \otimes \langle \Phi_1, \ldots, \Phi_i \rangle$ is the result of simultaneously replacing the successive separators in Γ by $\Phi_1, \ldots, \Phi_i$ respectively. $\Delta\langle \Gamma \rangle$ abbreviates $\Delta(\Gamma \otimes \langle \Delta_1, \ldots, \Delta_i \rangle)$.

A *hypersequent* $\Gamma \Rightarrow A$ comprises an antecedent hyperconfiguration in string-based notation of sort i and a succedent type A of sort i. The hypersequent calculus for $\mathbf{D}$ is as shown in Figure 5 where $k \in \omega^+$. Like $\mathbf{L}$, $\mathbf{hD}$ has no structural rules.

Morrill and Valentín (2010)[4] proves Cut-elimination for the k-ary discontinuous Lambek calculus, $k > 0$. As a consequence $\mathbf{D}$, like $\mathbf{L}$, enjoys in addition the subformula property, decidability, and the finite reading property.

2 hD: Absorbing the Structural Rules of a Sorted Multimodal Calculus

We consider now a sorted multimodal calculus $\mathbf{mD}$ with a set of structural rules $\mathbf{Eq_D}$ we present in the following lines. Figure 6 shows the logical rules of $\mathbf{mD}$ and Figure 7 shows the structural rules $\mathbf{Eq_D}$ integrated in $\mathbf{mD}$. This sequent calculus is non standard in two senses. Types and structural terms are sorted. Moreover, there are two structural term constants which stand respectively for the continuous unit and discontinuous unit. Structural term constructors are of two kinds: $\circ$ (which stands for term concatenation) and $\circ_i$ (which stands for term wrapping at the i-th position, $i \in \omega^+$). Again, as in the case of sorted types, structural terms are defined by mutual recursion and the *sort map* is computed in a similar way (see (10)).

$X[Y]$ denotes a structural term with a distinguished position occupied by the structural term Y. If A, X are respectively a type and a structural term, then a and x denote their sorts. We are interested in the cardinality of the set $\mathcal{F}$ of types of $\mathbf{D}$ and their structure.

$$\frac{}{\overrightarrow{A} \Rightarrow A}\, id \qquad \frac{\Gamma \Rightarrow A \qquad \Delta\langle \overrightarrow{A}\rangle \Rightarrow B}{\Delta\langle\Gamma\rangle \Rightarrow B}\, Cut$$

$$\frac{\Gamma \Rightarrow A \qquad \Delta\langle \overrightarrow{C}\rangle \Rightarrow D}{\Delta\langle\Gamma, \overrightarrow{A\backslash C}\rangle \Rightarrow D}\, \backslash L \qquad \frac{\overrightarrow{A}, \Gamma \Rightarrow C}{\Gamma \Rightarrow A\backslash C}\, \backslash R$$

$$\frac{\Gamma \Rightarrow B \qquad \Delta\langle \overrightarrow{C}\rangle \Rightarrow D}{\Delta\langle \overrightarrow{C/B}, \Gamma\rangle \Rightarrow D}\, /L \qquad \frac{\Gamma, \overrightarrow{B} \Rightarrow C}{\Gamma \Rightarrow C/B}\, /R$$

$$\frac{\Delta\langle \overrightarrow{A}, \overrightarrow{B}\rangle \Rightarrow D}{\Delta\langle \overrightarrow{A\bullet B}\rangle \Rightarrow D}\, \bullet L \qquad \frac{\Gamma_1 \Rightarrow A \qquad \Gamma_2 \Rightarrow B}{\Gamma_1, \Gamma_2 \Rightarrow A\bullet B}\, \bullet R$$

$$\frac{\Delta\langle \Lambda\rangle \Rightarrow A}{\Delta\langle \overrightarrow{I}\rangle \Rightarrow A}\, IL \qquad \frac{}{\Lambda \Rightarrow I}\, IR$$

$$\frac{\Gamma \Rightarrow A \qquad \Delta\langle \overrightarrow{C}\rangle \Rightarrow D}{\Delta\langle\Gamma|_k \overrightarrow{A\downarrow_k C}\rangle \Rightarrow D}\, \downarrow_k L \qquad \frac{\overrightarrow{A}|_k \Gamma \Rightarrow C}{\Gamma \Rightarrow A\downarrow_k C}\, \downarrow_k R$$

$$\frac{\Gamma \Rightarrow B \qquad \Delta\langle \overrightarrow{C}\rangle \Rightarrow D}{\Delta\langle \overrightarrow{C\uparrow_k B}|_k \Gamma\rangle \Rightarrow D}\, \uparrow_k L \qquad \frac{\Gamma|_k \overrightarrow{B} \Rightarrow C}{\Gamma \Rightarrow C\uparrow_k B}\, \uparrow_k R$$

$$\frac{\Delta\langle \overrightarrow{A}|_k \overrightarrow{B}\rangle \Rightarrow D}{\Delta\langle \overrightarrow{A\odot_k B}\rangle \Rightarrow D}\, \odot_k L \qquad \frac{\Gamma_1 \Rightarrow A \qquad \Gamma_2 \Rightarrow B}{\Gamma_1|_k \Gamma_2 \Rightarrow A\odot_k B}\, \odot_k R$$

$$\frac{\Delta\langle []\rangle \Rightarrow A}{\Delta\langle \overrightarrow{J}\rangle \Rightarrow A}\, JL \qquad \frac{}{[] \Rightarrow J}\, JR$$

Fig. 5. Hypersequent calculus **hD**

Consider the following lemma:

(9) **Lemma**
The set of types $\mathcal{F}$ is countably infinite iff the set of atomic types is countable. Moreover we have that:
$$\mathcal{F} = \bigcup_{i\in\omega} \mathcal{F}_i$$
$$\mathcal{F}_i = (A_{ij})_{j\in\omega}$$

Proof. The proof can be carried out by coding in a finite alphabet the set of types $\mathcal{F}$. Of course, it is crucial that the set of sorted atomic types forms a denumerable set. $\qquad\square$

Let $\mathbf{StructTerm}_D[\mathcal{F}]$ be the ω-sorted algebra over the signature $\Sigma_D = (\{\circ\} \cup (\circ_{i+1})_{i\in\omega}, \mathbb{I}, \mathbb{J})$. The sort functionality of Σ_D is:

$$((i,j \to i+j)_{i,j\in\omega}, (i+1,j \to i+j)_{i,j\in\omega}, 0, 1)$$

Observe that the operations $\circ$ and $\circ_i$'s (with $i > 0$) are sort polymorphic. In the following, we will abbreviate $\mathbf{StructTerm}_D[\mathcal{F}]$ by $\mathbf{StructTerm}$. The set of *structural terms* can be defined in BNF notation as follows:

$$
\begin{aligned}
&\mathbf{StructTerm}_0 ::= \mathbb{I}\\
&\mathbf{StructTerm}_1 ::= \mathbb{J}\\
(10)\quad &\mathbf{StructTerm}_i ::= \mathcal{F}_i\\
&\mathbf{StructTerm}_{i+j} ::= \mathbf{StructTerm}_i \circ \mathbf{StructTerm}_j\\
&\mathbf{StructTerm}_{i+j} ::= \mathbf{StructTerm}_{i+1} \circ_k \mathbf{StructTerm}_j
\end{aligned}
$$

It is clear that the sort of $\mathbf{StructTerm}_i$ and the collections of set $(A_{ij})_{j\in\omega}$ $(i \in \omega)$ are such that:

$$
\begin{aligned}
S(\mathbf{StructTerm}_i) &= i\\
S(A_{ij}) &= i
\end{aligned}
$$

We realize that $\mathbf{StructTerm}$ looks like an ω-sorted term algebra. This intuition is correct for the ω-graduated set $\mathcal{F}$ with the collections $(A_{ij})_{j\in\omega}$ plays the role of an ω-graduated set of a variables of an ω-sorted term algebra $T_{\Sigma_D}[X]$ with signature Σ_D.

We need to define some important relations between structural terms.

(11) **Definition** (*Wrapping and Permutable Terms*)

Given the term $(T_1 \circ_i T_2) \circ_j T_3$, we say that:
(P1) $T_2 \prec_{T_1} T_3$ iff $i + t_2 - 1 < j$.
(P2) $T_3 \prec_{T_1} T_2$ iff $j < i$.
(O) $T_2 \wr_{T_1} T_3$ iff $i \le j \le i + t_2 - 1$.

Observe that in a term like $(T_1 \circ_i T_2) \circ_j T_3$, if $(P1)$ or $(P2)$ hold, (O) does not apply. Conversely, if (O) is applicable, neither $(P1)$ nor $(P2)$ hold. If $T_2 \prec_{T_1} T_3$ (respectively $T_3 \prec_{T_1} T_2$), we say that T_2 and T_3 (respectively T_3 and T_2) *permute* in T_1. Otherwise, if (O) holds, we say that T_2 *wraps* T_3 in T_1.

(12) **Example**
Suppose that $T_1 = A$ where A is an arbitrary type of sort 3, and T_2, T_3 are arbitrary structural terms. Let $a_0 + 1 + a_1 + 1 + a_2 + 1 + a_3$ be an element of $[\![A]\!]$ in a displacement model $\mathcal{M}$. Suppose $S(T_2) = 3$. Consider now:

$$(A \circ_2 T_2) \circ_5 T_3$$

According to definition (11), $T_2 \prec_A T_3$, for $2 + S(T_2) - 1 = 4 < 5$. The intuition of this relation is the following. Interpreting in M we have that:
(13) $[\![(A \circ_2 T_2) \circ_5 T_3]\!] = a_0 + 1 + a_1 + [\![T_2]\!] + a_2 + [\![T_3]\!] + a_3$

We clearly see that the string $[\![T_2]\!]$ precedes the occurrence of $[\![T_3]\!]$. Similarly, if we have $T_3 \prec_A T_2$ in $(A \circ_i T_2) \circ_j T_3$, the occurrence of $[\![T_3]\!]$ precedes $[\![T_2]\!]$. Finally, if $T_2 \between_A T_3$ then $[\![T_2]\!]$ wraps $[\![T_3]\!]$, i.e. $[\![T_3]\!]$ is intercalated in $[\![T_2]\!]$.

We define the following relation between structural terms $\sim$:

(14) $T \sim S$ iff S is the result of applying one structural rule to a subterm of T

$\sim^*$ is defined to be the reflexive, symmetric and transitive closure of $\sim$.

2.1   The Faithful Embedding Translation $(\cdot)^\sharp$ between mD and hD

We consider the following embedding translation from **mD** to **hD**:

$$(\cdot)^\sharp : \mathbf{mD} = (\mathcal{F}, \mathbf{StructTerm}, \to) \longrightarrow \mathbf{hD} = (\mathcal{F}, \mathcal{O}, \Rightarrow)$$
$$T \to A \qquad\qquad \mapsto \qquad (T)^\sharp \Rightarrow (A)^\sharp$$

$(\cdot)^\sharp$ is such that:

$$A^\sharp = \overrightarrow{A} \text{ if } A \text{ is of sort strictly greater than } 0$$
$$A^\sharp = A \text{ if A is of sort } 0$$
$$(T_1 \circ T_2)^\sharp = T_1^\sharp, T_2^\sharp$$
$$(T_1 \circ_i T_2)^\sharp = T_1^\sharp |_i T_2^\sharp$$
$$\mathbb{I}^\sharp = \Lambda$$
$$\mathbb{J}^\sharp = []$$

Collapsing the Structural Rules

Let us see how the structural rules are absorbed in **hD**. We show here that structural postulates of **mD** collapse into the same textual form when they are mapped through $(\cdot)^\sharp$. Later we will see that:

$$\text{If } T \sim^* S \text{ then } T^\sharp = S^\sharp$$

Moreover will see that for every $A, B, C \in \mathcal{F}$ the following hypersequents are provable in **hD**:

(15)   – **Continuous associativity**

$$\overrightarrow{A \bullet (B \bullet C)} \Rightarrow (A \bullet B) \bullet C \text{ and } \overrightarrow{(A \bullet B) \bullet C} \Rightarrow A \bullet (B \bullet C)$$

– **Mixed associativity** If we have that $B \between_A C$:

$$\overrightarrow{A \odot_i (B \odot_j C)} \Rightarrow (A \odot_i B) \odot_{i+j-1} C \text{ and } \overrightarrow{(A \odot_i B) \odot_{i+j-1} C} \Rightarrow A \odot_i (B \odot_j C)$$

– **Mixed permutation** If we have that $B \prec_A C$:

$$\overrightarrow{(A \odot_i B) \odot_j C} \Rightarrow (A \odot_{j-b+1} C) \odot_i C \text{ and } \overrightarrow{(A \odot_{j-b+1} C) \odot_i C} \Rightarrow (A \odot_i B) \odot_j C$$

If we have that $C \prec_A B$:

$$\overrightarrow{(A \odot_i B) \odot_j C} \Rightarrow (A \odot_j C) \odot_{i+c-1} C \overrightarrow{(A \odot_j C) \odot_{i+c-1} C} \Rightarrow (A \odot_i B) \odot_j C$$

$$A \to A \text{ Id} \qquad \frac{S \to A \qquad T[A] \to B}{T[S] \to B} \, Cut$$

$$\frac{T[\mathbb{I}] \to A}{T[I] \to A} \, IL \qquad \frac{}{\mathbb{I} \Rightarrow I} \, IR$$

$$\frac{T[\mathbb{J}] \to A}{T[J] \to A} \, JL \qquad \frac{}{\mathbb{J} \Rightarrow J} \, JR$$

$$\frac{X \to A \qquad Y[B] \to C}{Y[X \circ A \backslash B] \to C} \, \backslash L \qquad \frac{A \circ X \to B}{X \to A \backslash B} \, \backslash R$$

$$\frac{X \to A \qquad Y[B] \to C}{Y[B/A \circ X] \to C} \, /L \qquad \frac{X \circ A \to B}{X \to B/A} \, /R$$

$$\frac{X \to A \qquad Y[B] \to C}{Y[B \uparrow_i A \circ_i X] \to C} \, \uparrow_i L \qquad \frac{X \circ_i A \to B}{X \to B \uparrow_i A} \, \uparrow_i R$$

$$\frac{X \to A \qquad Y[B] \to C}{Y[X \circ_i A \downarrow_i B] \to C} \, \downarrow_i L \qquad \frac{A \circ_i X \to B}{X \to A \downarrow_i B} \, \downarrow_i R$$

$$\frac{X[A \circ B] \to C}{X[A \bullet B] \to C} \, \bullet L \qquad \frac{X \to A \qquad Y \to B}{X \circ Y \to A \bullet B} \, \bullet R$$

$$\frac{X[A \circ_i B] \to C}{X[A \odot_i B] \to C} \, \odot_i L \qquad \frac{X \to A \qquad Y \to B}{X \circ_i Y \to A \odot_i B} \, \odot_i R$$

Fig. 6. The Logical rules of **mD**

- **Split wrap:**

$$\overrightarrow{A \bullet B} \Rightarrow (A \bullet J) \odot_{a+1} B \text{ and } \overrightarrow{(A \bullet J) \odot_{a+1} B} \Rightarrow A \bullet B$$

and:

$$\overrightarrow{(J \bullet B) \odot_1 A} \Rightarrow A \bullet B \text{ and } \overrightarrow{A \bullet B} \Rightarrow (J \bullet B) \odot_1 A$$

- **Continuous unit and discontinuous unit:**

$$\overrightarrow{A \bullet I} \Rightarrow A \text{ and } \overrightarrow{A} \Rightarrow A \bullet I \text{ and } \overrightarrow{I \bullet A} \Rightarrow A \text{ and } \overrightarrow{A} \Rightarrow I \bullet A$$

and:

$$\overrightarrow{A \odot_i J} \Rightarrow A \text{ and } \overrightarrow{A} \Rightarrow A \odot_i J \text{ and } \overrightarrow{J \odot_1 A} \Rightarrow A \text{ and } \overrightarrow{A} \Rightarrow J \odot_1 A$$

That **hD** absorbs the rules is proved in the following theorem:

(16) **Theorem (hD** *Absorption of* **Eq$_D$** *Structural Rules*)

For any $T, S \in$ **StructTerm**, if $T \sim^* S$ then $(T)^\sharp = (S)^\sharp$.

Proof. We define a useful notation for vectorial types which will help us to prove the theorem. Where A is an arbitrary type of sort greater than 0:

$$(17) \quad \vec{A}_i^j = \begin{cases} \sqrt[i]{A}, \text{ if } i = j \\ \vec{A}_i^{j-1}, [], \sqrt[j]{A}, \text{ if } j - i > 0 \end{cases}$$

Note that $\vec{A} = \vec{A}_0^a$. Now, consider arbitrary types A, B and C. As usual we denote their sorts respectively by a, b and c. We have then:

- Continuous associativity:

$$\begin{cases} ((A \circ B) \circ C)^\sharp = (\vec{A}, \vec{B}), \vec{C} = \vec{A}, \vec{B}, \vec{C} \\ (A \circ (B) \circ C))^\sharp = \vec{A}, (\vec{B}, \vec{C}) = \vec{A}, \vec{B}, \vec{C} \end{cases}$$

- Discontinuous associativity: Suppose that $B \between_A C$

We have that:

$$\vec{B} |_j \vec{C} = \vec{B}_0^{i-1}, \vec{C}, \vec{B}_i^b$$
$$\vec{A} |_i (\vec{B} |_j \vec{C}) = \vec{A}_0^{i-1}, \vec{B}_0^{j-1}, \vec{C}, \vec{B}_j^b, \vec{A}_i^a$$

On the other hand, we have that:

$$\vec{A} |_i \vec{B} = \vec{A}_0^{i-1}, \vec{B}, \vec{A}_i^a = \vec{A}_0^{i-1}, \vec{B}_0^{j-1}, \underbrace{[]}_{(i+j-1)\text{-th } []}, \vec{B}_j^b, \vec{A}_i^a$$

It follows that:

$$(\vec{A} |_i \vec{B})|_{i+j-1} \vec{C} = \vec{A}_0^{i-1}, \vec{B}_0^{j-1}, \vec{C}, \vec{B}_j^b, \vec{A}_i^a$$

Summarizing:

$$\begin{cases} (A \circ_i (B \circ_j C))^\sharp = \vec{A}_0^{i-1}, \vec{B}_0^{j-1}, \vec{C}, \vec{B}_j^b, \vec{A}_i^a \\ ((A \circ_i B) \circ_{i+j-1} C)^\sharp = \vec{A}_0^{i-1}, \vec{B}_0^{j-1}, \vec{C}, \vec{B}_j^b, \vec{A}_i^a \end{cases}$$

Hence:

$$(A \circ_i (B \circ_j C))^\sharp = ((A \circ_i B) \circ_{i+j-1} C)^\sharp$$

For the case $(A \circ_i B) \circ_k C$, if one puts $k = i + j - 1$ one gets $j = k - i + 1$. Therefore, changing indices: we have that:

$$((A \circ_i B) \circ_j C)^\sharp = (A \circ_i (B \circ_{j-i+1} C))^\sharp$$

This ends the case of discontinuous associativity.

Structural rules for units

- Continuous unit:

$$\frac{T[X] \to A}{T[\mathbb{I}\circ X] \to A} \qquad \frac{T[\mathbb{I}\circ X] \to A}{T[X] \to A} \qquad \frac{T[X] \to A}{T[X\circ\mathbb{I}] \to A} \qquad \frac{T[X\circ\mathbb{I}] \to A}{T[X] \to A}$$

- Discontinuous unit:

$$\frac{T[X] \to A}{T[\mathbb{J}\circ_1 X] \to A} \qquad \frac{T[\mathbb{J}\circ_1 X] \to A}{T[X] \to A} \qquad \frac{T[X] \to A}{T[X\circ_i\mathbb{J}] \to A} \qquad \frac{T[X\circ_i\mathbb{J}] \to A}{T[X] \to A}$$

Continuous associativity

$$\frac{X[(T_1\circ T_2)\circ T_3] \to D}{X[T_1\circ(T_2\circ T_3)] \to D} \ Assc_c \qquad\qquad \frac{X[T_1\circ(T_2\circ T_3)] \to D}{X[(T_1\circ T_2)\circ T_3] \to D} \ Assc_c$$

Split-wrap

$$\frac{T_1[T_2\circ T_3] \to D}{T_1[(\mathbb{J}\circ T_3)\circ_1 T_2] \to D} \ SW \qquad\qquad \frac{T_1[(\mathbb{J}\circ T_3)\circ_1 T_2] \to D}{T_1[T_2\circ T_3] \to D} \ SW$$

$$\frac{T_1[T_2\circ T_3] \to D}{T_1[(T_2\circ\mathbb{J})\circ_{t_2+1} T_3] \to D} \ SW \qquad\qquad \frac{T_1[(T_2\circ\mathbb{J})\circ_{t_2+1} T_3] \to D}{T_1[T_2\circ T_3] \to D} \ SW$$

Discontinuous associativity $T_2 \, \lozenge_{T_1} \, T_3$

$$\frac{S[T_1\circ_i(T_2\circ_j T_3)] \to C}{S[(T_1\circ_i T_2)\circ_{i+j-1} T_3)] \to C} \ Assc_d 1 \qquad\qquad \frac{S[(T_1\circ_i T_2)\circ_j T_3] \to C}{S[T_1\circ_i(T_2\circ_{j-i+1} T_3)] \to C} \ Assc_d 2$$

Mixed permutation 1 case $T_2 \prec_{T_1} T_3$

$$\frac{S[(T_1\circ_i T_2)\circ_j T_3] \to C}{S[(T_1\circ_{j-S(T_2)+1} T_3)\circ_i T_2] \to C} \ MixPerm1 \qquad\qquad \frac{S[(T_1\circ_i T_3)\circ_j T_2] \to C}{S[(T_1\circ_j T_2)\circ_{i+S(T_2)-1} T_3] \to C} \ MixPerm1$$

Mixed permutation 2 case $T_3 \prec_{T_1} T_2$

$$\frac{S[(T_1\circ_i T_2)\circ_j T_3] \to C}{S[(T_1\circ_j T_3)\circ_{i+S(T_3)-1} T_2] \to C} \ MixPerm2 \qquad\qquad \frac{S[(T_1\circ_i T_3)\circ_j T_2] \to C}{S[(T_1\circ_{j-S(T_3)+1} T_2)\circ_i T_3] \to C} \ MixPerm2$$

Fig. 7. Structural Rules of **mD**

– Mixed permutation:

There are two cases: $B \prec_A C$ or $C \prec_A B$. We consider only the first case, i.e. $B \prec_A C$. The other case is analogous. Let us see $((A \circ_i B) \circ_j C)^\sharp$:

$$\vec{A}|_i \vec{B} = \vec{A}_0^{i-1}, \vec{B}, \vec{A}_i^{k-1}, \underbrace{[]}_{j\text{-th }[]}, \vec{A}_k^a$$

We have therefore:

$$j = k - 1 + b \text{ iff } k = j - b + 1$$

$$((A \circ_i B) \circ_j C)^\sharp = \vec{A}_0^{i-1}, \vec{B}, \vec{A}_i^{k-1}, \vec{C}, \vec{A}_k^a$$

Hence:

$$(\vec{A} \circ_{j-b+1} \vec{C})^\sharp = \vec{A}_0^{i-1}, [], \vec{A}_i^{k-1}, \vec{C}, \vec{A}_k^a$$

It follows that:

$$((A \circ_{j-b+1} C) \circ_i B)^\sharp = \vec{A}_0^{i-1}, \vec{B}, \vec{A}_i^{k-1}, \vec{C}, \vec{A}_k^a$$

Summarizing:

$$\begin{cases} ((A \circ_i B) \circ_j C)^\sharp = \vec{A}_0^{i-1}, \vec{B}, \vec{A}_i^{k-1}, \vec{C}, \vec{A}_k^a \\ ((A \circ_{j-b+1} C) \circ_i B)^\sharp = \vec{A}_0^{i-1}, \vec{B}, \vec{A}_i^{k-1}, \vec{C}, \vec{A}_k^a \end{cases}$$

Hence

$$((A \circ_i B) \circ_j C)^\sharp = ((A \circ_{j-b+1} C) \circ_i B)^\sharp$$

Putting $i = j - b + 1$ we have that $j = i + b - 1$. Hence:

$$((A \circ_i C) \circ_j B)^\sharp = ((A \circ_j C) \circ_{i+b-1} B)^\sharp$$

This ends the case of mixed permutation.
– Split-wrap:
We have:

$$((A \circ J) \circ_{a+1} B)^\sharp = (\vec{A}, [])|_{a+1} \vec{B} = \vec{A}, \vec{B}$$
$$((J \circ B) \circ_1 A)^\sharp = ([], \vec{B})|_1 \vec{A} = \vec{A}, \vec{B}$$

Hence:

$$((A \circ J) \circ_{a+1} B)^\sharp = (A \circ B)^\sharp$$
$$\text{and}$$
$$((J \circ B) \circ_1 A)^\sharp = (A \circ B)^\sharp$$

This ends the case of split-wrap.

– Units:

$$(\mathbb{I} \circ A)^{\sharp} = \vec{A} \qquad = (A \circ \mathbb{I})^{\sharp}$$
$$(\mathbb{J} \circ_1 A)^{\sharp} = ([]|_1 \vec{A}) = \vec{A} = \vec{A}|_i[] = (A \circ_i \mathbb{J})^{\sharp}$$

We recall that types play the role of variables of structural terms. Now, we have seen that structural rules for arbitrary type variables collapse into the same textual form. This result generalizes to arbitrary structural terms by simply using type substitution.

More concretely, we have proved that: if $T \sim S$ (i.e. S is the result of applying a single structural rule to T) then $T^{\sharp} = S^{\sharp}$. Suppose we have $T \sim^* S$ (we omit the trivial case $T \sim^* T$). We have then a chain:

$$T := T_1 \sim T_2 \sim \cdots \sim T_{i-1} \sim T_i =: S \text{ for } i \geq 2$$

Applying $(\cdot)^{\sharp}$ to each $T_k \sim T_{k+1}$ $(1 \leq k \leq i-1)$ we have proved that:

$$(T_k)^{\sharp} = (T_{k+1})^{\sharp}$$

We have therefore a chain of identities:

$$(T)^{\sharp} = (T_1)^{\sharp} = (T_2)^{\sharp} = \ldots = (T_i)^{\sharp} = (S)^{\sharp}$$

This completes the proof.

□

We will now prove the associativity theorems of **hD** displayed in (15). Other theorems corresponding to the structural postulates of **mD** have similar proofs.

– Continuous associativity is obvious as in the Lambek calculus. The only difference is that types are sorted and in our notation the antecedent of hypersequents have the vectorial notation.
– Discontinuous associativity: we suppose that $B \, \mathbb{Q}_A \, C$. The following hypersequents are provable:

$$\overrightarrow{(A \odot_i B) \odot_{i+j-1} C} \Rightarrow A \odot_i (B \odot_j C)$$

And:

$$\overrightarrow{A \odot_i (B \odot_j C)} \Rightarrow (A \odot_i B) \odot_{i+j-1} C$$

By the previous lemma the identity $\vec{A}|_i(\vec{B}|_j\vec{C}) = (\vec{A}|_i\vec{B})|_{i+j-1}\vec{C}$ holds. We have the two following hypersequent derivations:

$$\cfrac{\vec{A} \Rightarrow A \qquad \cfrac{\cfrac{\vec{B} \Rightarrow B \qquad \vec{C} \Rightarrow C}{\vec{B}|_j\vec{C} \Rightarrow B \odot_j C} \odot_j R}{\cfrac{\cfrac{\cfrac{\vec{A}|_i(\vec{B}|_j\vec{C}) = (\vec{A}|_i\vec{B})|_{i+j-1}\vec{C} \Rightarrow A \odot_i (B \odot_j C)}{(\vec{A}|_i\vec{B})|_{i+j-1}\vec{C} \Rightarrow A \odot_i (B \odot_j C)}}{\cfrac{A \odot_i \vec{B}|_{i+j-1}\vec{C} \Rightarrow A \odot_i (B \odot_j C)}{(A \odot_i B) \odot_{i+j-1} \vec{C} \Rightarrow A \odot_i (B \odot_j C)} \odot_{i+j-1}L} \odot_i L}} \odot_i R}$$

and

$$\cfrac{\cfrac{\cfrac{\cfrac{\vec{A} \Rightarrow A \quad \vec{B} \Rightarrow B}{\vec{A}|_i\vec{B} \Rightarrow (A \odot_i B)} \odot_i R \quad \vec{C} \Rightarrow C}{(\vec{A}|_i\vec{B})|_{i+j-1}\vec{C} = \vec{A}|_i(\vec{B}|_j\vec{C}) \Rightarrow (A \odot_i B) \odot_{i+j-1} C} \odot_{i+j-1} R}{\vec{A}|_i(\overrightarrow{B \odot_j C}) \Rightarrow (A \odot_i B) \odot_{i+j-1} C} \odot_j L}{\overrightarrow{A \odot_i (B \odot_j C)} \Rightarrow (A \odot_i B) \odot_{i+j-1} C} \odot_i L$$

(18) **Theorem** (*Equivalence Theorem for* **StructTerm**)

Let R and S be arbitrary structural terms. The following holds:

$$R \sim^* S \text{ iff } (R)^\sharp = (S)^\sharp$$

Proof. We have already seen in (16) the *only if* case, which is the fact that **hD** absorbs the **Eq$_D$** structural rules. The *if* case is more difficult and needs some technical machinery from sorted universal algebra. For details, see [9].   □

(19) **Lemma** $((\cdot)^\sharp$ *is an Epimorphism*)

For every $\Delta \in \mathcal{O}$ there exists a structural term[3] T_Δ such that:

$$(T_\Delta)^\sharp = \Delta$$

Proof. This can be proved by induction on the structure of hyperconfigurations. We define recursively T_Δ such that $(T_\Delta)^\sharp = \Delta$:

- Case $\Delta = \Lambda$ (the empty tree): $T_\Delta = \mathbb{I}$.
- Case where $\Delta = A, \Gamma$: $T_\Delta = A \circ T_\Gamma$, where by induction hypothesis (i.h.) $(T_\Gamma)^\sharp = \Gamma$.
- Case where $\Delta = [], \Gamma$: $T_\Delta = \mathbb{J} \circ T_\Gamma$, where by i.h. $(T_\Gamma)^\sharp = \Gamma$.
- Case $\Delta = \vec{A} \otimes \langle \Delta_1, \cdots, \Delta_a \rangle, \Delta_{a+1}$. By i.h. we have:

$$(T_{\Delta_i})^\sharp = \Delta_i \text{ for } 1 \leq i \leq a+1$$

$T_\Delta = (A \circ_1 T_{\Delta_1}) \circ T_{\Delta_2}$ if $a = 1$
$T_\Delta = ((\cdots ((A \circ_1 T_{\Delta_1}) \circ_{1+d_1} T_{\Delta_2}) \cdots) \circ_{1+d_1+\cdots+d_{a-1}} T_{\Delta_a}) \circ T_{\Delta_{a+1}}$ if $a > 1$

□

By induction on the structure of **StructTerm**, we have the following intuitive result on the relationship between structural contexts and hypercontexts:

(20) $(T[S])^\sharp = T^\sharp \langle S^\sharp \rangle$

---

[3] In fact there exists an infinite set of such structural terms.

These two technical results we have seen above are necessary for the proof of the faithful embedding translation $(\cdot)^\sharp$ of theorem (24). We prove now an important theorem which is crucial for the mentioned theorem (24).

(21) **Theorem** (*Visibility for Extraction in* **StructTerm**)

Let $T[A]$ be a structural term with a distinguished occurrence of type A. Suppose that:

$$(T[A])^\sharp = \Delta|_i \overrightarrow{A}$$

where $\Delta \in \mathcal{O}$ and $A \in \mathcal{F}$. Then A is visible for extraction in $T[A]$, i.e. there exist a structural term T' and an index i such that:

$$T[A] \sim^* T' \circ_i A$$

Proof. Let T_Δ be a structural term such that $(T_\Delta)^\sharp = \Delta$. This is possible by lemma (19). We have $(T_\Delta \circ_i A)^\sharp = \Delta|_i \overrightarrow{A}$. We have then $(T_\Delta \circ_i A)^\sharp = (T[A])^\sharp$. By the equivalence theorem (18) it follows that $T[A] \sim^* T_\Delta \circ_i A$. Put $T' := T_\Delta$. We are done. $\qquad\square$

(22) **Theorem** (*Uniqueness of Extractability*)

Suppose that $T[A] \sim S \circ_i A$ and $T[A] \sim S' \circ_j A$, where A. Then:

$$S \sim^* S'$$
$$i = j$$

Proof. We have that $(S \circ_i A)^\sharp = \Delta|_i \overrightarrow{A} = \Delta|_j \overrightarrow{A} = (S' \circ_j A)^\sharp$. Hence $i = j$ and $(S)^\sharp = (S')^\sharp$. By theorem (18), $S \sim^* S'$. We are done. $\qquad\square$
Theorems (21) and (22) will be crucial for the proof of the $(\cdot)^\sharp$ embedding theorem (24).

Before proving theorem (24), it is worth seeing what is the intuition behind the structural rules of **Eq$_D$**. This intuition is exemplified by a constructive proof of theorem (21):

Proof. Constructive proof of theorem (21): By induction on the structural complexity of $T[A]$: The cases are as follows:

i) $T[A] = A$.

We put $T' = \mathbb{J}$ and hence :

$$T[A] \sim^* \mathbb{J} \circ_1 A$$

ii) $T[A] = S[A] \circ R$.

By induction hypothesis (i.h.), $S[A] \sim^* S' \circ_k A$ for some $k > 0$. We have the following equational derivation:

$$T[A] \sim^* (S' \circ_k A) \circ R$$
$$\sim^* (\mathbb{J} \circ R) \circ_1 (S' \circ_k A) \quad \text{by } \mathbf{SW}$$
$$\sim^* ((\mathbb{J} \circ R) \circ_1 S') \circ_k A) \quad \text{by } \mathbf{Assc_d}$$
$$\sim^* (S' \circ R) \circ_k A \quad\quad \text{by } \mathbf{SW}$$

iii) $T[A] = S \circ R[A]$

By i.h. $R[A] \sim^* R' \circ_k A$ for some term R' and $k > 0$. It follows that:

$$T[A] \sim^* S \circ (R' \circ_k A)$$
$$\sim^* (S \circ \mathbb{J}) \circ_{S(S)+1} (R' \circ_k A) \quad\quad \text{by } \mathbf{SW}$$
$$\sim^* ((S \circ \mathbb{J}) \circ_{S(S)+1} R') \circ_{S(S)+k} A \quad \text{by } \mathbf{Assc_d}$$
$$\sim^* (S \circ R') \circ_{S(S)+k} A \quad\quad\quad\quad \text{by } \mathbf{SW}$$

iv) $T[A] = S[A] \circ_i R$ for some term $S[A]$, R and $i > 0$.

By i.h. $S[A] \sim^* S' \circ_k A$ for some S' and $i > 0$. We derive the following equation:

$$T[A] \sim^* (S' \circ_k A) \circ_i R$$

If $R = \mathbb{J}$ we are done. Suppose that $R \neq \mathbb{J}$. In this case A must permute with R in S', i.e. $A \prec_{S'} R$ or $R \prec_{S'} A$, for otherwise $(T[A])^\sharp = \Delta|_i \overrightarrow{A}$ would not hold. Without loss of generality, let us suppose that $A \prec_{S'} R$. In that case we have:

$$T[A] \sim^* (S' \circ_{i-S(A)+1} R) \circ_k A \text{ by } \mathbf{MixPerm1}$$

Hence A is permutated to right periphery in $T[A]$.

v) $T[A] = S \circ_i R[A]$ for some terms S and $R[A]$ and $i > 0$. By i.h. $R[A] \sim^* R' \circ_k A$. Then:

$$T[A] \sim^* S \circ_i (R' \circ_k A)$$
$$\sim^* (S \circ_i R') \circ_{i+k-1} A \text{ by } \mathbf{Assc_d}$$

$\square$

(23) **Remark**
Interestingly, the constructive proof for extractability does not use continuous associativity. Therefore, a priori a non-associative discontinuous Lambek calculus could be considered. This remark needs further study.

(24) **Theorem** (*Faithfulness of* $(\cdot)^\sharp$ *Embedding Translation*)

Let A, X and Δ be respectively a type, a structural term and a hyperconfiguration. The following statements hold:
   i) If $\vdash_{\mathbf{mD}} X \to A$ then $\vdash_{\mathbf{hD}} (X)^\sharp \Rightarrow A$
   ii) For any X such that $(X)^\sharp = \Delta$, if $\vdash_{\mathbf{hD}} \Delta \Rightarrow A$ then $\vdash_{\mathbf{mD}} X \to A$

Proof.

i) Logical rules in **mD** translate without any problem to **hD**. We need recall only that if X and Y are structural terms then $(X \circ Y)^\sharp = (X)^\sharp, (Y)^\sharp$ and $(X \circ_i Y)^\sharp = (X)^\sharp |_i (Y)^\sharp$. Structural rules in **mD** collapse in the same textual form as theorem (16) proves. Finally, the Cut rule has no surprise. This proves i).

ii) This part of the theorem becomes easy if we use the following four facts:

- Lemma (19) which states that for any hyperconfiguration Δ there is a structural term T_Δ such that $(T_\Delta)^\sharp = \Delta$.
- The fact (20) we stated before which gives the relationship between structural terms and hypercontexts $(T[A])^\sharp = T^\sharp \langle \vec{A} \rangle$.
- Theorem (18).
- Theorem (21).

The proof is by induction on the length of **hD** derivations. The three first facts prove the induction of all the rules but the right rule of the connectives $\uparrow_i$. Suppose the last rule of a **hD** derivation is $\uparrow_i R$:

$$\frac{\Delta|_i \vec{A} \Rightarrow B}{\Delta \Rightarrow B\uparrow_i A} \uparrow_i R$$

Let $T[A]$ be such that $(T[A])^\sharp = \Delta|_i \vec{A}$. We know by induction hypothesis that $\vdash_{\mathbf{mD}} T[A] \Rightarrow B$. By the last fact of above, i.e. theorem (21) of visibility of extraction, since $(T[A])^\sharp = \Delta|_i \vec{A}$, we know there exist T' and i such that $T[A] \sim^* T' \circ_i A$. It follows that in **mD**:

$$\frac{T[A] \to B}{\vdots \quad \textbf{Sequence of structural rules}}$$
$$\frac{T' \circ_i A \to B}{T' \to B\uparrow_i A} \uparrow_i R$$

Hence, $\vdash_{\mathbf{mD}} T' \to B\uparrow_i A$. And T' is in fact T_Δ, and therefore $(T')^\sharp = \Delta$. Moreover, for any S such that $(S)^\sharp \sim^* T'$, we have that applying a finite number of structural rules we obtain the **mD** provable sequent $S \to B\uparrow_i A$, and of course $(S)^\sharp = \Delta$. This completes the proof of ii).

□

(25) Example

Let B, D, E, C, A five arbitrary atomic types of sort respectively 2, 2, 1, 0 and 0. The following two derivations have the following end-sequent and end-hypersequent:

$$\vdash_{\mathbf{mD}} (((B\uparrow_2 A \circ_1 D) \circ_4 E) \circ_3 (J \circ C \backslash A)) \to ((B \odot_1 D) \odot_3 E)\uparrow_3 C$$
$$\vdash_{\mathbf{hD}} \sqrt[0]{B\uparrow_2 A}, \vec{D}, \sqrt[1]{B\uparrow_2 A}, [], C\backslash A, \sqrt[2]{B\uparrow_2 A}, \vec{E}, \sqrt[3]{B\uparrow_2 A} \Rightarrow ((B \odot_1 D) \odot_3 E)\uparrow_3 C$$

The above multimodal sequents are in correspondence through the mapping $(\cdot)^{\sharp}$. Derivations (26) and (27)/(28) illustrate theorem (24). Notice the sequence of structural rules in derivation (26) in order to extract type C.

(26)
$$
\cfrac{
\cfrac{
\cfrac{
\cfrac{
\cfrac{
\cfrac{
\cfrac{
\cfrac{
\cfrac{C \circ C \backslash A \to A \quad B\uparrow_2 A \circ_2 A \to B}{B\uparrow_2 A \circ_2 (C \circ C \backslash A) \to B}\uparrow_2 \quad D \to D
}{(B\uparrow_2 A \circ_2 (C \circ C \backslash A)) \circ_1 D \to B \odot_1 D}\odot_1 \quad E \to E
}{((B\uparrow_2 A \circ_2 (C \circ C \backslash A)) \circ_1 D) \circ_3 E \to (B \odot_1 D) \odot_3 E}\odot_3
}{((B\uparrow_2 A \circ_1 D) \circ_3 (C \circ C \backslash A)) \circ_3 E \to (B \odot_1 D) \odot_3 E}\text{MixPerm}
}{((B\uparrow_2 A \circ_1 D) \circ_4 E) \circ_3 (C \circ C \backslash A) \to (B \odot_1 D) \odot_3 E}\text{MixPerm}
}{((B\uparrow_2 A \circ_1 D) \circ_4 E) \circ_3 ((J \circ C \backslash A) \circ_1 C) \to (B \odot_1 D) \odot_3 E}\text{SW}
}{(((B\uparrow_2 A \circ_1 D) \circ_4 E) \circ_3 (J \circ C \backslash A)) \circ_3 C \to (B \odot_1 D) \odot_3 E}\text{Assc}_d
}{(((B\uparrow_2 A \circ_1 D) \circ_4 E) \circ_3 (J \circ C \backslash A)) \to ((B \odot_1 D) \odot_3 E)\uparrow_3 C}\uparrow_3
$$

(27)
$$
\cfrac{
\cfrac{C, C\backslash A \Rightarrow A \quad \sqrt[0]{B\uparrow_2 A}, [], \sqrt[1]{B\uparrow_2 A}, A, \sqrt[2]{B\uparrow_2 A}, [], \sqrt[3]{B\uparrow_2 A} \Rightarrow B}{\sqrt[0]{B\uparrow_2 A}, [], \sqrt[1]{B\uparrow_2 A}, C, C\backslash A, \sqrt[2]{B\uparrow_2 A}, [], \sqrt[3]{B\uparrow_2 A} \Rightarrow B}\uparrow_2 \quad \vec{D} \Rightarrow D
}{\sqrt[0]{B\uparrow_2 A}, \vec{D}, \sqrt[1]{B\uparrow_2 A}, C, C\backslash A, \sqrt[2]{B\uparrow_2 A}, [], \sqrt[3]{B\uparrow_2 A} \Rightarrow B \odot_1 D}\odot_1
$$

(28)
$$
\cfrac{
\cfrac{\sqrt[0]{B\uparrow_2 A}, \vec{D}, \sqrt[1]{B\uparrow_2 A}, C, C\backslash A, \sqrt[2]{B\uparrow_2 A}, [], \sqrt[3]{B\uparrow_2 A} \Rightarrow B \odot_1 D \quad E \Rightarrow E}{\sqrt[0]{B\uparrow_2 A}, \vec{D}, \sqrt[1]{B\uparrow_2 A}, C, C\backslash A, \sqrt[2]{B\uparrow_2 A}, \vec{E}, \sqrt[3]{B\uparrow_2 A} \Rightarrow (B \odot_1 D) \odot_3 E}\odot_3
}{\sqrt[0]{B\uparrow_2 A}, \vec{D}, \sqrt[1]{B\uparrow_2 A}, [], C\backslash A, \sqrt[2]{B\uparrow_2 A}, \vec{E}, \sqrt[3]{B\uparrow_2 A} \Rightarrow ((B \odot_1 D) \odot_3 E)\uparrow_3 C}\uparrow_3
$$

3 Conclusions

It is not a priori a trivial task to find out a set of structural rules $\mathcal{E}$ that makes the hypersequent calculus $\mathbf{hD}$ equivalent to a multimodal calculus with the structural rules of $\mathcal{E}$. The faithful embedding translation $(\cdot)^{\sharp}$ between $\mathbf{mD}$ and $\mathbf{hD}$ is then, we think, a remarkable discovery. The equivalent multimodal calculus $\mathbf{mD}$ gives $\mathbf{D}$ the Moot's powerful proof net machinery almost for free (see [3]). It must be noticed that this proof net theory approach for $\mathbf{D}$ is completely different from the one in [6]. Finally, the discovery of $\mathbf{mD}$ can be very useful to investigate new soundness and completeness results for $\mathbf{D}$ (see [9]).

References

1. Avron, A.: Hypersequents, Logical Consequence and Intermediate Logic form Concurrency. Annals of Mathematics and Artificial Intelligence 4, 225–248 (1991)

2. Lambek, J.: The mathematics of sentence structure. American Mathematical Monthly 65, 154–170 (1958); Reprinted in Buszkowski, W., Marciszewski, W., van Benthem, J. (eds.): Categorial Grammar. Linguistic & Literary Studies in Eastern Europe, vol. 25, pp. 153–172. John Benjamins, Amsterdam (1988)
3. Moot, R.: Proof nets for display logic. CoRR, abs/0711.2444 (2007)
4. Morrill, G., Valentín, O.: Displacement Calculus. Linguistic Analysis 36(1-4), 167–192 (2010)
5. Morrill, G., Valentín, O., Fadda, M.: Dutch grammar and processing: A case study in tlg. In: Bosch, P., Gabelaia, D., Lang, J. (eds.) TbiLLC 2007. LNCS, vol. 5422, pp. 272–286. Springer, Heidelberg (2009)
6. Morrill, G., Fadda, M.: Proof Nets for Basic Discontinuous Lambek Calculus. Logic and Computation, 239–256 (2008)
7. Morrill, G., Fadda, M., Valentín, O.: Nondeterministic Discontinuous Lambek Calculus. In: Geertzen, J., Thijsse, E., Bunt, H., Schiffrin, A. (eds.) Proceedings of the Seventh International Workshop on Computational Semantics, IWCS 2007, pp. 129–141. Tilburg University (2007)
8. Morrill, G., Valentín, O., Fadda, M.: The Displacement Calculus. Journal of Logic, Language and Information 20(1), 1–48 (2011), doi:10.1007/s10849-010-9129-2.
9. Valentín, O.: Theory of Discontinuous Lambek Calculus. PhD thesis, Universitat Autònoma de Barcelona, Barcelona (2012)

Author Index

Abramsky, Samson 1
Abrusci, V. Michele 14

Bastenhof, Arno 28
Béchet, Denis 51
Bernardi, Raffaella 63
Blute, Richard F. 90
Bonato, Roberto 108
Buszkowski, Wojciech 136

Casadio, Claudia 156

Foret, Annie 172

Guglielmi, Alessio 90

Hines, Peter 188

Ivanov, Ivan T. 90

Jacobs, Bart 211

Kiślak-Malinowska, Aleksandra 156
Kissinger, Aleks 235
Kołowska-Gawiejnowicz, Mirosława 253
Kuznetsov, Stepan 268

Moortgat, Michael 279
Moot, Richard 297
Morrill, Glyn 331

Panangaden, Prakash 90
Pavlovic, Dusko 353
Pentus, Mati 368

Retoré, Christian 108
Reyes, Gonzalo E. 381

Sadrzadeh, Mehrnoosh 1
Silva, Alexandra 211
Sorokin, Alexey 393
Straßburger, Lutz 90

Valentín, Oriol 402